STUDIES IN INTERFACE SCIENCE

Novel Methods to Study Interfacial Layers

STUDIES IN INTERFACE SCIENCE

SERIES EDITORS

D. Möbius and R. Miller

Vol. 1
Dynamics of Adsorption at Liquid Interfaces
Theory, Experiment, Application
by S.S. Dukhin, G. Kretzschmar and R. Miller

Vol. 2
An Introduction to Dynamics of Colloids
by J.K.G. Dhont

Vol. 3
Interfacial Tensiometry
by A.I. Rusanov and V.A. Prokhorov

Vol. 4
New Developments in Construction and Functions of Organic Thin Films
edited by T. Kajiyama and M. Aizawa

Vol. 5
Foam and Foam Films
by D. Exerowa and P.M. Kruglyakov

Vol. 6
Drops and Bubbles in Interfacial Research
edited by D. Möbius and R. Miller

Vol. 7
Proteins at Liquid Interfaces
edited by D. Möbius and R. Miller

Vol. 8
Dynamic Surface Tensiometry in Medicine
by V.N. Kazakov, O.V. Sinyachenko, V.B. Fainerman, U. Pison and R. Miller

Vol. 9
Hydrophile-Lilophile Balance of Surfactants and Solid Particles
Physicochemical Aspects and Applications
by P. M. Kruglyakov

Vol. 10
Particles at Fluid Interfaces and Membranes
Attachment of Colloid Particles and Proteins to Interfaces and Formation of Two-Dimensional Arrays
by P.A. Kralchevsky and K. Nagayama

Vol. 11
Novel Methods to Study Interfacial Layers
by D. Möbius and R. Miller

Novel Methods to Study Interfacial Layers

Edited by

D. Möbius

Max-Planck Institut für Biophysikalische Chemie
P.O. Box 2841, 37077 Göttingen, Germany

R. Miller

Max-Planck Institut für Kolloid- und Grenzflächenforschung
Am Mühlenberg 1, 14476 Potsdam/Golm, Germany

2001
ELSEVIER
Amsterdam – London – New York – Oxford – Paris – Shannon – Tokyo

ELSEVIER SCIENCE B.V.
Sara Burgerhartstraat 25
P.O. Box 211, 1000 AE Amsterdam, The Netherlands

First edition 2001

Library of Congress Cataloging in Publication Data
A catalog record from the Library of Congress has been applied for.

ISBN: 0 444 50948 8
ISSN: 1384 7309

∞ The paper used in this publication meets the requirements of ANSI/NISO Z39.48-1992 (Permanence of Paper).
Printed in The Netherlands.

FOREWORD

In August 2000 the international conference „LB9 - International Conference on Organized Molecular Films" was held in Potsdam, Germany. The meeting was jointly organised by the Max Planck Institute of Colloids and Interfaces in Golm and the University of Potsdam. The topics comprised of course much more than only Langmuir Blodgett layers, which only gave the traditional name of this series of conferences. Actually the scientific fields spanned over many aspects of monolayers and adsorption layers at various liquid and solid interfaces. More than 450 scientists from 28 countries participated in the LB9. Many new groups, especially from Eastern Europe and Asia jointed the meeting, and imported new ideas and enthusiasm. The next conference in this series, LB 10 will be organised in 2003 in Beijing, China, chaired by Junbai Li and Zongfang Liu and organised under the auspices of the Chinese Academy of Sciences.

This book presents a number of selected papers given at the LB9 conference and dedicated to new techniques and methodologies for studying interfacial layers. One group of manuscripts deals with the application of surface plasmons at solid interfaces, used for example in resonance spectroscopy and light scattering. New applications of various types of Atomic Force Microscopy were reported making use of various modifications of tips. A number of chapters is dedicated to light emitting diodes built with the help of LB layers. The target of these studies is the improvement of efficiency. Electrochemical methods were described as tools for developing sensors, in particular miniaturised pH or gas sensors.

The application of synchrotron X-ray and NMR techniques have been described in detail in two extended chapters. It is demonstrated how molecular information can be detected by these methods for various types of interfacial layers.

To fabricate nano-engineered films on colloidal particles a layer-by-layer adsorption of oppositely charged macromolecules is used. Different templates can be coated with multilayer films and decomposed to form hollow capsules with defined size, shape and shell thickness. As example, polyelectrolyte capsules can be used as carrier for biological species, for a controlled release and targeting of drugs and as micro-containers to perform chemical reactions in restricted volumes.

Drop and bubble methods have been developed significantly during the last years. These techniques provide access to dynamic properties of liquid interfaces. The drop and bubble shape technique as well as the fast oscillating drops and bubbles are described essentially as tools for dilational surface rheology.

In addition to the chapters of this book about 130 papers have been submitted for publication as original contributions in the journals *Colloids and Surfaces A* and *Colloids and Surfaces B*, depending on whether the topic is of physicochemical and engineering or biochemical relevance. These papers will be published in special issues of the journals. All together, the two special issues of *Colloids and Surfaces A* and *Colloids and Surfaces B* and this monograph represent the proceedings of the LB9.

The editors of this volume want to use the opportunity to thank the authors of the chapters for their cooperation when assembling the book, and to express their gratitude to Ludmila Makievski for her invaluable support in the overall preparation of the proceedings consisting of an immense number of manuscripts.

Contents

Novel Methods to Study Interfacial Layers
D. Möbius and R. Miller (Editors)

Ellipsometry in Interface Science

H. Motschmann and R. Teppner

Max-Planck-Institut für Kolloid- und Grenzflächenforschung,
Am Mühlenberg 1, 14426 Golm, Germany

Contents

1 What is it all about?

The following chapter presents the basics of ellipsometry and discusses some recent advances. The article covers the formalism and theory used for data analysis as well as instrumentation. The treatment is also designed to familiarize newcomers to this field. The experimental focus is on adsorption layers at the air-water and oil-water interface. Selected examples are discussed to illustrate the potential as well as the limits of this technique. The authors hope, that this article contributes to a wider use of this technique in the colloidal physics and chemistry community. Many problems in our field of science can be tackled with this technique.

Ellipsometry refers to a class of optical experiments which measure changes in the state of polarization upon reflection or transmission on the sample of interest. It is a powerful technique for the characterization of thin films and surfaces. In favorable cases thicknesses of thin films can be measured to within Å accuracy, furthermore it is possible to quantify submonolayer surface coverages with a resolution down to 1/100 of a monolayer or to measure the orientation adopted by the molecules on mesoscopic length scales. The high sensitivity is remarkable if one considers that the wavelength of the probing light is on the order of 500 nm. The data accumulation is fast and allows to monitor the kinetics of adsorption processes. The technique can also be extended to a microscopy. Imaging ellipsometry allows under certain conditions a direct visualization of surface inhomogeneities as well as quantification of the images. Many samples are suitable for ellipsometry and the only requirement is that they must reflect laser light. Its simplicity and power makes ellipsometry an ideal surface analytical tool for many objects in interface science.

2 Polarized light

Light is an electromagnetic wave and all its features relevant for ellipsometry can be described within the framework of Maxwell's theory [1]. The relevant material properties are described by the complex dielectric function ϵ or alternatively by the corresponding refractive index n.

An electromagnetic wave consists of an electric field $\vec{E}$ and a magnetic field $\vec{B}$. The field vectors are mutually perpendicular and also perpendicular to the propagation direction as given by the wave vector $\vec{k}$. All states of polarization are classified according to the trace of the electrical field vector during one period. Linearly polarized light means the electrical field vector oscillates within a plane, elliptically polarized light means that the trace of the electric field vector during one period is an ellipse. A convenient mathematical representation of a given state of polarization is based on a superposition of two linearly polarized light waves within an arbitrarily chosen orthogonal coordinate system.

$$\vec{E}(\vec{r},t) = \begin{pmatrix} |E_\mathrm{p}|\cos(2\pi\nu t - \vec{k}\cdot\vec{r} + \delta_\mathrm{p}) \\ |E_\mathrm{s}|\cos(2\pi\nu t - \vec{k}\cdot\vec{r} + \delta_\mathrm{s}) \end{pmatrix} \tag{1}$$

$|E_\mathrm{p}|$ and $|E_\mathrm{s}|$ are the amplitudes, δ_p and δ_s the phases, $|\vec{k}| = 2\pi/\lambda$ is the magnitude of the wave vector and ν the frequency. Only the phases $\delta_\mathrm{p}, \delta_\mathrm{s}$ and the amplitudes are required for a representation of the state of polarization; the time dependence is not of importance and can be neglected. The so called Jones vector reads

$$\vec{E} = \begin{pmatrix} |E_\mathrm{p}|e^{i\delta_\mathrm{p}} \\ |E_\mathrm{s}|e^{i\delta_\mathrm{s}} \end{pmatrix} = \begin{pmatrix} E_\mathrm{p} \\ E_\mathrm{s} \end{pmatrix} \tag{2}$$

Different states of polarization are depicted in Fig. 1. The state of polarization is

a) linear, if $\delta_\mathrm{p} - \delta_\mathrm{s} = 0$ or $\delta_\mathrm{p} - \delta_\mathrm{s} = \pi$,

b) elliptical, if $\delta_\mathrm{p} \neq \delta_\mathrm{s}$ and $|E_\mathrm{p}| \neq |E_\mathrm{s}|$,

c) circular for the special case $\delta_\mathrm{p} - \delta_\mathrm{s} = \pi/2$ and $|E_\mathrm{p}| = |E_\mathrm{s}|$.

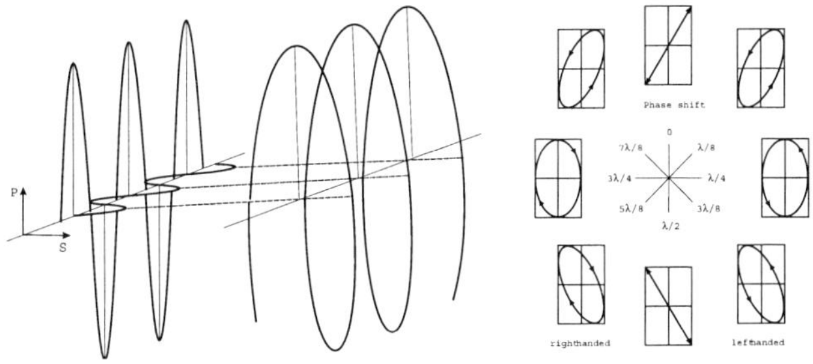

Figure 1: The Jones representation of polarized light represents any state of polarization as a linear combination of two orthogonal linearly polarized light waves.

An alternative representation of a given state of polarization uses the quantities ellipticity ω and azimuth α. The ellipticity is defined as the ratio of the length of the semi-minor axis to that of the semi-major axis as shown in Fig. 2. The azimuth angle α is measured counterclockwise from the $\hat{x}$-axis. Light is assumed to propagate in positive $\hat{z}$ direction so that $\hat{x}$, $\hat{y}$ and $\hat{z}$ define a right handed coordinate system. Some authors prefer this representation and for this reason the conversion formulas between both notations are listed below. The derivations require a vast number of tedious algebraic manipulations and can be found in [2].

$$\tan 2\alpha = \frac{2E_x E_y cos(\delta_y - \delta_x)}{E_x^2 - E_y^2} \qquad -90° \le \alpha \le 90° \tag{3}$$

$$\sin(2\omega) = \frac{2E_x E_y \sin(\delta_y - \delta_x)}{E_x^2 + E_y^2} \qquad -45° \le \omega \le 45° \tag{4}$$

$$\tan(\delta_y - \delta_x) = \frac{\tan(2\omega)}{\sin(2\alpha)} \qquad -180° \le \delta_y - \delta_x \le 180° \tag{5}$$

$$\cos\left(2\arctan\left(\frac{E_y}{E_x}\right)\right) = \cos(2\omega)\cos(2\alpha) \qquad 0° \le \arctan(E_x/E_y) \le 90° \tag{6}$$

Ambiguities arising from the inversion of trigonometric functions are settled if the upper and lower limits of all quantities are considered.

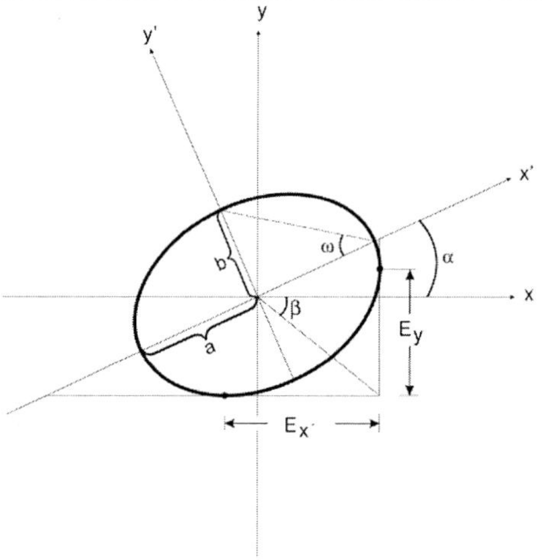

Figure 2: Elliptically polarized light defined by the azimuth α and the ellipticity ω

3 Basic equation of ellipsometry

A typical ellipsometric experiment is depicted in Fig. 3. Light with a well defined state of polarization is incident on a sample. The reflected light usually differs in its state of polarization and these changes are measured and quantified in an ellipsometric experiment.

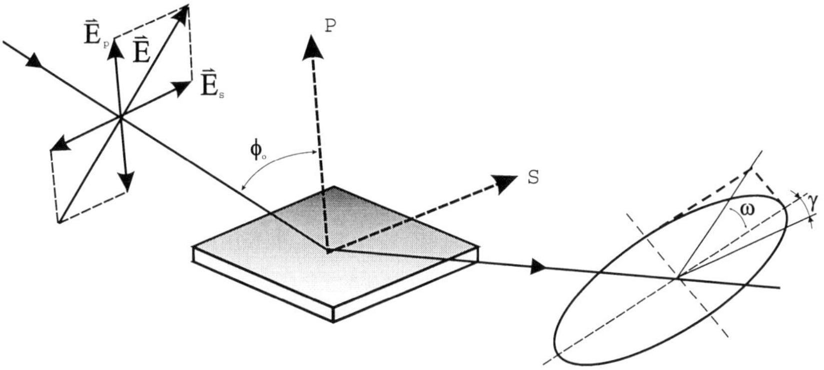

Figure 3: Ellipsometric experiment in reflection mode

The mathematical description is best done within the laboratory frame of reference defined by the plane of incidence. The propagation direction of the beam and the normal of the reflecting surface define the plane of incidence. Light with an electric field vector oscillating within the plane of incidence ($\hat{p}$-light) remains linearly polarized upon reflection and the same holds for $\hat{s}$-light with $\vec{E}$ perpendicular to the plane of incidence. For this reason $\hat{p}$- and $\hat{s}$-light are also called Eigen-polarizations of isotropic media or uniaxial perpendicular media. This consideration makes it obvious that this frame of reference is distinct. Incident and reflected beam can be described by their corresponding Jones vector :

$$\vec{E}_{\text{inc}} = \begin{pmatrix} |E_{\text{p}}^{\text{i}}|e^{i\delta_{\text{p}}^{\text{i}}} \\ |E_{\text{s}}^{\text{i}}|e^{i\delta_{\text{s}}^{\text{i}}} \end{pmatrix} \qquad \vec{E}_{\text{refl}} = \begin{pmatrix} |E_{\text{p}}^{\text{r}}|e^{i\delta_{\text{p}}^{\text{r}}} \\ |E_{\text{s}}^{\text{r}}|e^{i\delta_{\text{s}}^{\text{r}}} \end{pmatrix} \tag{7}$$

Two quantities Ψ and Δ are introduced in order to describe the changes in the state of polarization.

$$\Delta = (\delta_{\text{p}}^{\text{r}} - \delta_{\text{s}}^{\text{r}}) - (\delta_{\text{p}}^{\text{i}} - \delta_{\text{s}}^{\text{i}}) \tag{8}$$

$$\tan \Psi = \frac{|E^{\mathrm{r}}_{\mathrm{p}}|/|E^{\mathrm{i}}_{\mathrm{p}}|}{|E^{\mathrm{r}}_{\mathrm{s}}|/|E^{\mathrm{i}}_{\mathrm{s}}|} \tag{9}$$

Changes in the ratio of the amplitudes are described as the tangent of the angle Ψ . It will turn out later that Ψ can be directly measured.

The reflectivity properties of a sample within a given experiment are given by the corresponding reflection coefficients r_p and r_s. The reflection coefficient is a complex quantity that accounts for changes in phase and amplitude of the reflected electric field E^{r} with respect to the incident one E^{i}.

$$r_{\mathrm{p}} = \frac{|E^{\mathrm{r}}_{\mathrm{p}}|}{|E^{i}_{\mathrm{p}}|} e^{i(\delta^{\mathrm{r}}_{\mathrm{p}} - \delta^{i}_{\mathrm{p}})} \qquad r_{\mathrm{s}} = \frac{|E^{\mathrm{r}}_{\mathrm{s}}|}{|E^{i}_{\mathrm{s}}|} e^{i(\delta^{\mathrm{r}}_{\mathrm{s}} - \delta^{i}_{\mathrm{s}})} \tag{10}$$

Interference cannot be observed between orthogonal beams and hence $\hat{p}$- and $\hat{s}$-light do not influence each other and can be separately treated. With these definitions the basic equation of ellipsometry is obtained

$$\tan \Psi \cdot e^{i\Delta} = \frac{r_{\mathrm{p}}}{r_{\mathrm{s}}} = \rho = \Re(\rho) + i\Im(\rho) \tag{11}$$

Eqn. (11) relates the quantities Ψ and Δ with the reflectivity properties of the sample. The following two section discuss various ways to measure Δ and Ψ and the theory and algorithm used for a calculation of the complex reflectivity coefficient. Some ellipsometers are designed in a way that they measure directly the real and imaginary part of the complex quantity ρ instead of Ψ and Δ. Eqn. (11) allows a conversion between both notations.

4 Design of an ellipsometer

Many different designs of ellipsometers have been suggested and a good overview is presented in Azzam and Bashara [3]. Here we discuss common roots of all arrangements and the underlying theory.

The layout of a typical ellipsometer is depicted in Fig. 4. The main components are a polarizer P which produces linearly polarized light, a compensator C which introduces a defined phase retardation of one field component with respect to the orthogonal one, the sample S, the analyzer A and a detector.

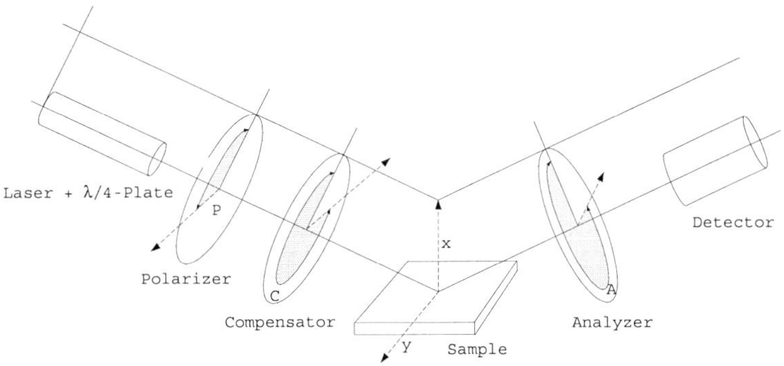

Figure 4: Ellipsometer in a PCSA-configuration

This setup allows the determination of the unknown ellipsometric angles and can be operated in various modes. Each optical component modifies the state of polarization. Since any state of polarization can be represented by a complex Jones vector consisting of two columns, the effect of each optical components is described by a complex 2×2 matrix. The Jones formalism provides an elegant means for a quantitative description [4].

$$\begin{pmatrix} E_x \\ E_y \end{pmatrix}^{Ja} = \begin{pmatrix} T_{11} & T_{12} \\ T_{21} & T_{22} \end{pmatrix}^{J} \begin{pmatrix} E_x \\ E_y \end{pmatrix}^{Je} = \mathbf{T}^J \vec{E}^{Je} \tag{12}$$

The superscript J refers to the optical component $J \in$[PCSA]; the subscripts e and a refer to the $\vec{E}$-vector before and after the component. Each of the optical components including the sample possesses a distinct coordinate system in which the corresponding matrix is diagonal. For example a compensator consist of a birefringent material cut to a thin plate of a defined thickness t. Optically it possesses a fast and a slow axis. Linearly polarized light oscillating parallel to either axis remains linearly polarized, however, since the corresponding refractive indices of fast n_f and slow axis n_s differ, they travel with different speed which leads to a phase shift $\phi = (n_\mathrm{f} - n_\mathrm{s}) \cdot t \cdot 2\pi/\lambda$ between the two components. Similar expressions can be given for each component:

component	**coordinate system**	**Jones-Matrix**
Polarizer	t_P = transmission axis e_P = extinction axis	$\mathbf{T}^P = \begin{pmatrix} 1 & 0 \\ 0 & 0 \end{pmatrix}$
Compensator	s_c = slow axis l_c = fast axis	$\mathbf{T}^C = \begin{pmatrix} 1 & 0 \\ 0 & \rho_c \end{pmatrix}$

$$\text{with} \qquad \rho_c = t_c e^{i\delta_c} = \frac{|E_s^{Ca}|}{|E_l^{Ca}|} e^{i(\delta_s - \delta_l)}$$

Sample	eigenpolarization p parallel to plane of incidence s perpendicular to plane of incidence	$\mathbf{T}^S = \begin{pmatrix} r_\mathrm{p} & 0 \\ 0 & r_\mathrm{s} \end{pmatrix}$
Analyzer	t_A = transmission axis e_A = extinction axis	$\mathbf{T}^A = \begin{pmatrix} 1 & 0 \\ 0 & 0 \end{pmatrix}$

The simple diagonal matrix is only valid in the coordinate system of the component. A matrix **R** is required to transform the vector between the coordinate systems of adjacent components.

$$\vec{E}_{xy}^{J+1,e} = \mathbf{R}(\alpha)\vec{E}_{xy}^{J,a} \qquad \text{with} \qquad \mathbf{R}(\alpha) = \begin{pmatrix} \cos\alpha & \sin\alpha \\ -\sin\alpha & \cos\alpha \end{pmatrix} \tag{13}$$

The setting of the optical components is defined by the angles P, A and C of its distinct axis, with respect to the plane of incidence. An angle $C = -45°$ means the fast axis of the compensator is set to an angle of $-45°$ with respect to the plane of incidence.

With these tools we can describe the E-vector at the detector as a function of the setting of all components including the unknown reflectivity properties of the sample.

$$\vec{E}^{A,\,a}_{e_A\,t_A} = \mathbf{T}^A\mathbf{R}(A)\mathbf{T}^S\mathbf{R}(-C)\mathbf{T}^C\mathbf{R}(C-P)\vec{E}^{P,a}_{e_P\,t_P} \tag{14}$$

To understand Equation (14), one must first realize that the multiplication always goes from right to left. Hence, the above mathematical formula can be described as follows: linear polarized light exits the polarizer in the polarizer's frame of reference, then is rotated to the coordinate system of the compensator by the matrix operator $R(C-P)$, then the compensator acts on the state of polarization as given by T_C, then the exiting light from the compensator is rotated to the coordinate system of the sample by the matrix operator $R(-C)$ an so on. The multiplication given by Equation (14) yields:

$$\vec{E}^{A,\,a}_{e_A\,t_A} = \begin{pmatrix} E^{A,\,a}_{t_A} \\ 0 \end{pmatrix} = \begin{pmatrix} 1 \\ 0 \end{pmatrix} \gamma E^{P,\,a}_{t_P}\{\Omega_1 + \Omega_2\} \tag{15}$$

$$\begin{aligned} \Omega_1 &= R_\mathrm{p}\cos A[\cos C\cos(C-P) - \rho_c \sin C \sin[(C-P)] \\ \Omega_2 &= R_\mathrm{s}\sin A[\sin C\cos(C-P) + \rho_c \cos C \sin[(C-P)] \end{aligned}$$

γ accounts for the attenuation of the light intensity. Additional components in the optical path, for example the cell windows, should not change the state of polarization and can therefore be neglected. The intensity at the detector is proportional to

$$I \propto |\vec{E}^{A,\,a}_{e_A\,t_A}|^2 \tag{16}$$

The Jones matrix algorithm leads to the desired relation between the intensity at the detector and the setting of all optical components. The unknown reflectivity coefficients can be retrieved in various manners and the applied measurement scheme names the method. *Rotating analyzer* means recording the intensity as a function of the setting of the analyzer and work out the unknown ellipsometric angles by a Fourier analysis. *Polarization modulation* ellipsometry uses a variable phase retardation δ_c for a calculation of the ellipsometric angles. Polarization modulation uses an electro-optic or acusto-optic modulator driven at a high frequency. The measurement is fast, however, there are also some inherent problems due to an undesired interferometric contribution of the modulator to the signal which cannot be separated from the contribution of the sample. The technique is very well suited to follow relative changes.

A particular successful implementation is *Nullellipsometry* which eliminates many intrinsic errors due to slight misalignments of the sample. Within Nullellipsometry the setting of the optical components is chosen such that the light at the detector vanishes. This requires that the angle dependent term $\{\Omega_1 + \Omega_2\}$ of eqn. (15) vanishes. A given elliptical state of polarization of the incident light leads to linear polarized light after reflection and can be completely extinguished with an analyzer.

$$\frac{r_\mathrm{p}}{r_\mathrm{s}} = -\tan A \frac{\tan C + \rho_c \tan(C-P)}{1 - \rho_c \tan C \tan(C-P)} \qquad \text{for} \quad I = 0 \tag{17}$$

This equation can be further simplified by using a high precision quarter waveplate as a compensator ($t_C = 1$, $\delta_C = \pi/2 \Longrightarrow \rho_c = -i$) fixed to $C = \pm 45°$. With eqn. (11) a further simplification of the eqn. (17) can be achieved.

$$\begin{aligned} \tan \Psi e^{i\Delta} &= \tfrac{r_\mathrm{p}}{r_\mathrm{s}} = \tan A_0 \exp[i(2P_0 + \tfrac{\pi}{2})] \qquad &\text{if} \quad C = -45° \\ \tan \Psi e^{i\Delta} &= \tfrac{r_\mathrm{p}}{r_\mathrm{s}} = -\tan A_0 \exp[i(\tfrac{\pi}{2} - 2P_0)] \qquad &\text{if} \quad C = 45° \end{aligned} \tag{18}$$

Eqn. (18) links the quantities Δ an Ψ to the null settings of the polarizeP_0 and analyzer A_0. Once a setting (P_0, A_0) has been determined which provides a complete cancellation of the light, then the same holds for the pair ($\tilde{P}_0$, $\tilde{A}_0$).

$$(\tilde{P}_0,\ \tilde{A}_0) = (P_0 + 90°,\ 180° - A_0) \qquad \text{if} \quad I = 0 \quad \text{for} \quad (P_0,\ A_0) \tag{19}$$

These nontrivial pairs of nullsettings are refered as ellipsometric zones. Measurements in various zones lead to a high accuracy in the determination of absolute values. Many intrinsic small errors due to misalignment are cancelled by this scheme.

So far we introduced two quantitites which account for changes in the state of reflection upon reflection. We also discussed how these quantities can be measured. The next chapter deals with the theory of reflection and illustrates means for a calculation of the ellipsometric angles of a given optical layer system.

5 Theory of reflection

In the following section a procedure for a calculation of the reflectivity coefficients r_p and r_s (and consequently the ellipsometric angles) first published by Lekner [6] will be described.

This method is certainly not the easiest one for the calculation of the reflectivity properties of a single homogenous layer at the interface between two bulk phases, but is advantageous, if the layer has an inner structure, e.g. a continuously varying refractive index normal to the interfaces. For just a few refractive index profiles it is possible to calculate r_p and r_s exactly, but in most cases approximations must be used: A continuously varying refractive index profile is subdivided into many thin layers. This method uses matrices to relate the electrical field and its derivative in between adjacent layer and matrices to account for changes of the phase in each layer. In every layer the refractive index is assumed to be constant. Obviously this approximation gets the better the finer the subdivision is.

For the derivation we use a rectangular coordinate system with the $\hat{z}$-axis normal to the interfaces pointing from the incident medium to the substrate. This means that the interfaces of the layer structure are planes of constant z_n. The $\hat{x}, \hat{z}$-plane is the plane of incidence.

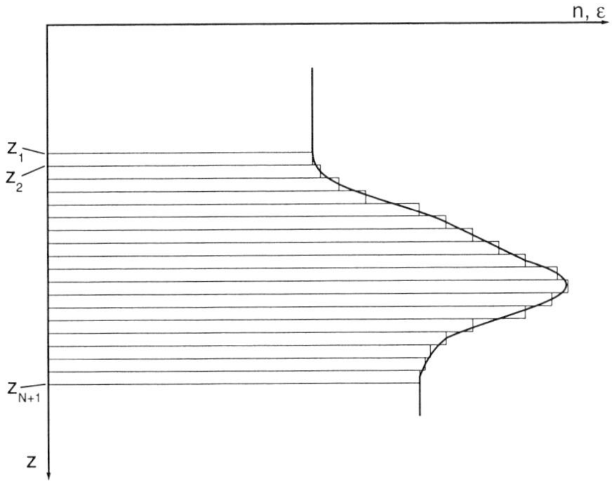

Figure 5: Subdivision of a refractive index profile into step profiles

We start with Maxwell's equation. In the case of no magnetization, no free charges, no current in and linear polarizability of the sample they read:

$$\nabla \times \vec{B} = \frac{\epsilon}{c^2} \cdot \frac{\partial \vec{E}}{\partial t} \tag{20}$$

$$\nabla \times \vec{E} = -\frac{\partial \vec{B}}{\partial t} \tag{21}$$

$$\nabla \cdot \vec{B} = 0 \tag{22}$$

$$\nabla \cdot \vec{E} = 0 \tag{23}$$

with $\vec{B}$ representing the magnetic field, $\vec{E}$ the electric field, $\epsilon(z) = n(z)^2$ the dielectric function and c the velocity of light in vacuo.

For the $\vec{B}$- and the $\vec{E}$-field a plane wave ansatz is used:

$$\vec{B} = \vec{B}_0 \cdot e^{i(\vec{k}\cdot\vec{r}-\omega t)}, \qquad \vec{E} = \vec{E}_0 \cdot e^{i(\vec{k}\cdot\vec{r}-\omega t)} \tag{24}$$

This ansatz fulfills Maxwell's equations, if the following relationship between the wave vector $\vec{k}$ and the angular frequency ω of the propagating wave is obeyed:

$$\vec{k}^2 + \epsilon\frac{\omega^2}{c^2} = 0, \tag{25}$$

and if additionally $\vec{B}_0$, $\vec{E}_0$ and $\vec{k}$ are normal to each other. Thus eqn. (20) and eqn. (21) can be further simplified:

$$\nabla \times \vec{B} = -i\frac{\epsilon\omega}{c^2}\vec{E} \qquad \nabla \times \vec{E} = i\omega\vec{B} \tag{26}$$

Let's consider $\hat{s}$-polarized light, that can be represented in the following way in the coordinate system:

$$\vec{E}_s = \begin{pmatrix} 0 \\ E_y \\ 0 \end{pmatrix} \tag{27}$$

Inserting $\vec{E}_s$ in eqn. (26) results in three equations for E_y:

$$-\frac{\partial E_y}{\partial z} = i\frac{\omega}{c}B_x \qquad \frac{\partial E_y}{\partial x} = i\frac{\omega}{c}B_z \qquad \frac{\partial B_x}{\partial z} - \frac{\partial B_z}{\partial x} = -i\frac{\omega}{c}E_y \tag{28}$$

After elimination of B_x and B_z one obtains a differential equation in E_y which can be separated:

$$\frac{\partial^2 E_y}{\partial x^2} + \frac{\partial^2 E_y}{\partial z^2} + \epsilon\frac{\omega^2}{c^2}E_y = 0 \tag{29}$$

The ansatz $E_y(x, z, t) = e^{i(k_x x - \omega t)} \cdot E(z)$ leads to a differential equation in $E(z)$:

$$\frac{d^2 E(z)}{dz^2} + q^2 E(z) = 0, \tag{30}$$

with q being the $\hat{z}$-component of the wave vector:

$$q^2 = \epsilon\frac{\omega^2}{c^2} - k_x^2 = k^2 - k_x^2 \quad \text{or} \quad q = \sqrt{\epsilon}\frac{\omega}{c} \cdot \cos\varphi \tag{31}$$

and φ representing the angle of incidence (defined as the angle between the surface normal and the direction of propagation of the light beam). Since $E(z)$ is continuous at every interface, it is obvious from eqn. (30), that also $dE(z)/dz$ is continuous, if $d\epsilon \neq \infty$.

Eqn. (30) can thus be split into two coupled differential equations of first order:

$$\frac{dE(z)}{dz} = D(z) \quad \text{und} \quad \frac{dD(z)}{dz} = -q^2 E \tag{32}$$

If q takes the value q_n within a layer located between z_n and z_{n+1}, and E_n and D_n are the values of the electric field and its derivative at z_n, the solution of eqn. (32) in $z_n \leq z \leq z_{n+1}$ is:

$$E(z) = E_n \cos q_n(z - z_n) + \frac{D_n}{q_n}\sin q_n(z - z_n), \tag{33}$$

$$D(z) = D_n \cos q_n(z - z_n) - E_n q_n \sin q_n(z - z_n). \tag{34}$$

Since E and D are continuous at the interfaces, it follows that

$$\begin{aligned} E_{n+1} &= E_n \cos\delta_n + \tfrac{D_n}{q_n}\sin\delta_n \\ D_{n+1} &= D_n \cos\delta_n - E_n q_n \sin\delta_n, \end{aligned} \tag{35}$$

with

$$\delta_n = q_n(z_{n+1} - z_n) \tag{36}$$

representing the phase shift encountered while propagating through the layer. The eqn. (35) can be expressed in a matrix form:

$$\begin{pmatrix} E_{n+1} \\ D_{n+1} \end{pmatrix} = \begin{pmatrix} \cos\delta_n & \frac{\sin\delta_n}{q_n} \\ -q_n \sin\delta_n & \cos\delta_n \end{pmatrix} \begin{pmatrix} E_n \\ D_n \end{pmatrix} = \mathbf{M}_{ns} \cdot \begin{pmatrix} E_n \\ D_n \end{pmatrix} \tag{37}$$

where $\mathbf{M}_{ns}$ describes the influence of the n-th layer on the $\hat{s}$-polarized wave.

The matrices for $\hat{p}$-polarized light can be calculated analogously. Since $\hat{p}$-light has non-vanishing $\hat{x}$- and $\hat{z}$-components of the electrical field in the chosen coordinate system, it is more convenient to use the linearly coupled magnetic field $\vec{B}$, which has just one nonvanishing component, instead of $\vec{E}$.

$$\vec{B} = \begin{pmatrix} 0 \\ B_y \\ 0 \end{pmatrix} = \begin{pmatrix} 0 \\ e^{i(k_x x - \omega t)} B(z) \\ 0 \end{pmatrix} \tag{38}$$

Insertion of $\vec{B}$ into Maxwell's equations and simplification lead to a differential equation in $B(z)$:

$$\frac{d}{dz}\left(\frac{1}{\epsilon}\frac{dB(z)}{dz}\right) + \frac{q^2}{\epsilon}B(z) = 0, \tag{39}$$

that can be split into two equations of first order again:

$$\frac{1}{\epsilon}\frac{dB(z)}{dz} = C \qquad \text{and} \frac{dC(z)}{dz} = -\frac{q^2}{\epsilon}B(z) \tag{40}$$

Their solutions within a layer n resemble those of the $\hat{s}$-polarized light (eqn. (33) and eqn. (34)):

$$B(z) = B_n \cos q_n(z - z_n) + \frac{\epsilon_n}{q_n} C_n \sin q_n(z - z_n) \tag{41}$$

$$C(z) = C_n \cos q_n(z - z_n) - \frac{q_n}{\epsilon_n} B_n \sin q_n(z - z_n) \tag{42}$$

It follows from eqn. (20) and eqn. (39) that $B(z)$ and $C(z)$ are continuous at the layers' interfaces, if $d\epsilon \neq \infty$. This can be used again to relate the values of B and C at neighboring interfaces.

$$\begin{pmatrix} B_{n+1} \\ C_{n+1} \end{pmatrix} = \begin{pmatrix} \cos \delta_n & \frac{\epsilon_n}{q_n} \sin \delta_n \\ -\frac{q_n}{\epsilon_n} \sin \delta_n & \cos \delta_n \end{pmatrix} \begin{pmatrix} B_n \\ C_n \end{pmatrix} = \mathbf{M}_{np} \begin{pmatrix} B_n \\ C_n \end{pmatrix} \tag{43}$$

Now it is clear how to calculate the reflectivity coefficients of such a layer structure: The layers' matrices $\mathbf{M}_n$ have to be multiplied consecutively, separately for $\hat{p}$- and $\hat{s}$-light:

$$\mathbf{M} = \begin{pmatrix} m_{11} & m_{12} \\ m_{21} & m_{22} \end{pmatrix} = \mathbf{M}_N \cdot \mathbf{M}_{N-1} \cdots \mathbf{M}_2 \cdot \mathbf{M}_1 \tag{44}$$

The resulting matrices $\mathbf{M}_s$ and $\mathbf{M}_p$ relate the fields before and after the layer structure.

Before it (f) there are incident and reflected wave, behind it (h) just the transmitted one:

$$\begin{aligned}
E_1 &= e^{iq_f z_1} + r_s e^{-iq_f z_1} \implies & D_1 &= iq_f(e^{iq_f z_1} - r_s e^{-iq_f z_1}) \\
E_{N+1} &= t_s e^{iq_h z_{N+1}} \implies & D_{N+1} &= iq_h t_s e^{iq_h z_{N+1}} \\
B_1 &= e^{iq_f z_1} - r_p e^{-iq_f z_1} \implies & C_1 &= iq_f(e^{iq_f z_1} + r_p e^{-iq_f z_1}) \\
B_{N+1} &= \sqrt{\tfrac{\epsilon_h}{\epsilon_f}} t_p e^{iq_h z_{N+1}} \implies & C_{N+1} &= iq_h \sqrt{\tfrac{\epsilon_h}{\epsilon_f}} t_p e^{iq_h z_{N+1}}
\end{aligned} \tag{45}$$

Thus the following relations evolve:

$$\begin{pmatrix} t_s e^{iq_h z_{N+1}} \\ iq_h t_s e^{iq_h z_{N+1}} \end{pmatrix} = \begin{pmatrix} m_{11s} & m_{12s} \\ m_{21s} & m_{22s} \end{pmatrix} \begin{pmatrix} e^{iq_f z_1} + r_s e^{-iq_f z_1} \\ iq_f(e^{iq_f z_1} - r_s e^{-iq_f z_1}) \end{pmatrix} \tag{46}$$

$$\begin{pmatrix} \sqrt{\frac{\epsilon_h}{\epsilon_f}} t_p e^{iq_h z_{N+1}} \\ iq_h \sqrt{\frac{\epsilon_h}{\epsilon_f}} t_p e^{iq_h z_{N+1}} \end{pmatrix} = \begin{pmatrix} m_{11p} & m_{12p} \\ m_{21p} & m_{22p} \end{pmatrix} \begin{pmatrix} e^{iq_f z_1} - r_p e^{-iq_f z_1} \\ iq_f(e^{iq_f z_1} + r_p e^{-iq_f z_1}) \end{pmatrix} \tag{47}$$

These equations can be solved for r_s and r_p, resulting in two formally identical expressions for the two coefficients:

$$r_s = e^{2iq_f z_1} \frac{q_f q_h m_{12s} + m_{21s} - iq_h m_{11s} + iq_f m_{22s}}{q_f q_h m_{12s} - m_{21s} + iq_h m_{11s} + iq_f m_{22s}} \tag{48}$$

$$-r_p = e^{2iq_f z_1} \frac{\frac{q_f q_h}{\epsilon_f \epsilon_h} m_{12p} + m_{21p} - i\frac{q_h}{\epsilon_h} m_{11p} + i\frac{q_f}{\epsilon_f} m_{22p}}{\frac{q_f q_h}{\epsilon_f \epsilon_h} m_{12p} - m_{21p} + i\frac{q_h}{\epsilon_h} m_{11p} + i\frac{q_f}{\epsilon_f} m_{22p}} \tag{49}$$

Up to now the only simplification is the discretization of the continuous refractive index profile. The next step is a reduction of the computational effort by using a Taylor-approximation up to the second order in the phaseshift δ_n. This approximation is valid if the discrete layer thicknesses are much smaller than the wavelength of the light: in other words a sufficiently high number of layers is required.

$$\mathbf{M}_n = \begin{pmatrix} \cos\delta_n & \frac{\sin\delta_n}{q_n} \\ -q_n \sin\delta_n & \cos\delta_n \end{pmatrix} \quad \text{and} \quad \begin{pmatrix} \cos\delta_n & \frac{\epsilon_n}{q_n}\sin\delta_n \\ -\frac{q_n}{\epsilon_n}\sin\delta_n & \cos\delta_n \end{pmatrix} \tag{50}$$

for $\hat{s}$- and $\hat{p}$-polarized light simplify to

$$\mathbf{M}_n \approx \begin{pmatrix} 1 - \frac{\delta_n^2}{2} & \frac{\delta_n}{q_n} \\ -q_n\delta_n & 1 - \frac{\delta_n^2}{2} \end{pmatrix} \quad \text{and} \quad \begin{pmatrix} 1 - \frac{\delta_n^2}{2} & \frac{\epsilon_n \delta_n}{q_n} \\ -\frac{q_n \delta_n}{\epsilon_n} & 1 - \frac{\delta_n^2}{2} \end{pmatrix} \tag{51}$$

With just a slight increase of computational effort the approximation can be greatly improved by using layers with linearly varying dielectric functions ϵ as depicted in fig.6 instead of those with constant ϵ.

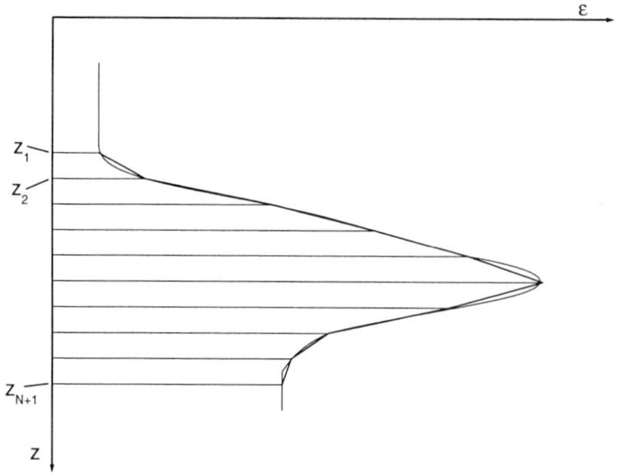

Figure 6: Approximation of the real dielectric function by a system of layers with linearly varying dielectric function ϵ

Using this ansatz the components of the individual layer matrices adopt the following form:

$$
\begin{aligned}
m_{11s} &= 1 + (z_{n+1} - z_n)^2 \left(\tfrac{k_x^2}{2} - \tfrac{\omega^2}{c^2} \tfrac{2\epsilon_n + \epsilon_{n+1}}{6} \right) \\
m_{12s} &= z_{n+1} - z_n \\
m_{21s} &= (z_{n+1} - z_n) \left(k_x^2 - \tfrac{\omega^2}{c^2} \tfrac{\epsilon_n + \epsilon_{n+1}}{2} \right) \\
m_{22s} &= 1 + (z_{n+1} - z_n)^2 \left(\tfrac{k_x^2}{2} - \tfrac{\omega^2}{c^2} \tfrac{\epsilon_n + 2\epsilon_{n+1}}{6} \right) \\
m_{11p} &= 1 + (z_{n+1} - z_n)^2 \left(k_x^2 \tfrac{2\epsilon_n + \epsilon_{n+1}}{6\epsilon_n} - \tfrac{\omega^2}{c^2} \tfrac{\epsilon_n + 2\epsilon_{n+1}}{6} \right) \\
m_{12p} &= (z_{n+1} - z_n) \tfrac{\epsilon_n + \epsilon_{n+1}}{2} \\
m_{21p} &= (z_{n+1} - z_n) \left(\tfrac{k_x^2}{2} (\tfrac{1}{\epsilon_n} + \tfrac{1}{\epsilon_{n+1}}) - \tfrac{\omega^2}{c^2} \right) \\
m_{22p} &= 1 + (z_{n+1} - z_n)^2 \left(k_x^2 \tfrac{\epsilon_n + 2\epsilon_{n+1}}{6\epsilon_{n+1}} - \tfrac{\omega^2}{c^2} \tfrac{2\epsilon_n + \epsilon_{n+1}}{6} \right)
\end{aligned}
\tag{52}
$$

Setting $\epsilon_n = \epsilon_{n+1}$ within a layer in eqn. (52)yields the step profile approximation eqn. (51).

>From the equations for r_p and r_s, eqn. (49) and eqn. (48), the ellipsometric angles Δ and Ψ of arbitrary layer structures can be calculated using the basic equation of ellipsometry eqn. 11.

6 Ellipsometry applied to ultrathin films

The previous section described an algorithm for the calculation on the reflectivity coefficients of given refractive index profiles. Profiles are only of importance if their characteristic length scale is comparable to that of the probing beam. In many cases, i.e. adsorption layer of nonionic surfactants at the air-water interface, there is a striking mismatch between interfacial height h and the wavelength of light λ. As a result certain peculiarities exist which are discussed in this section [8].

The most striking limitation is a reduction of the measurable quantities. The presence of an organic monolayer (refractive index 1.3-1.6) with a thickness below 2.5 nm does not change the reflectivity $|r_i|^2$ and as a consequence there are no detectable changes in Ψ . In the thin film limit$h \ll \lambda$ the data analysis relies only on a single parameter, namely changes in the phase Δ. Unfortunately the number of independent data cannot be increased. Neither spectroscopic ellipsometry nor a variation of the angle of incidence yield new independent data, instead all quantities remain strongly coupled. A sound treatment is given in [9]. However, the sensitivity of an ellipsometric measurement can be significantly increased by the choice of the angle of incidence. A sensitivity analysis is given in [10].

The exact formula relating the reflectivity coefficients of a single homogeneous layer with refractive index $n_1 = \sqrt{\epsilon_1}$ in between two infinite media ($n_0 = \sqrt{\epsilon_0}$ and $n_2 = \sqrt{\epsilon_2}$) at an angle of incidence φ is given by :

$$\Delta = \arctan \frac{Im\left(\frac{r_p}{r_s}\right)}{Re\left(\frac{r_p}{r_s}\right)} \quad \text{with} \quad \begin{array}{l} r_p = |r_p| \cdot e^{i\delta_{r,p}} = \frac{r_{0,1,p}+r_{1,2,p}e^{-i2\beta}}{1+r_{0,1,p}r_{1,2,p}e^{-i2\beta}} \\ r_s = |r_s| \cdot e^{i\delta_{r,s}} = \frac{r_{0,1,s}+r_{1,2,s}e^{-i2\beta}}{1+r_{0,1,s}r_{1,2,s}e^{-i2\beta}} \end{array} \tag{53}$$

where the reflectivity coefficients $r_{0,1,p}$, $r_{1,2,p}$, $r_{0,1,s}$ and $r_{1,2,s}$ describing the reflection at refractive index jumps $n_0 \rightarrow n_1$ and $n_1 \rightarrow n_2$ for $\hat{p}$- and $\hat{s}$-light are given by Fresnel's laws. $\beta = 2\pi\frac{h}{\lambda}\sqrt{n_1^2 - n_0^2 \sin^2\varphi}$ accounts for the phase shift occuring in a single pass within the adsorption layer.

If the layer thickness h is much smaller than the wavelength λ of light it is justified to expand the complex reflectivity coefficients in a power series in terms of h/λ. The first term in this expansion describes reflection at a monolayer.

$$\Delta \approx \frac{4\sqrt{\epsilon_0}\epsilon_2\pi\cos\varphi\sin^2\varphi}{(\epsilon_0-\epsilon_2)((\epsilon_0+\epsilon_2)\cos^2\varphi-\epsilon_0)} \cdot \frac{(\epsilon_1-\epsilon_0)(\epsilon_2-\epsilon_1)}{\epsilon_1} \cdot \frac{h}{\lambda} \tag{54}$$

If the refractive index is varying over the height of the layer, the term

$$\frac{(\epsilon_1-\epsilon_0)(\epsilon_2-\epsilon_1)}{\epsilon_1} \cdot h$$

within eqn. (63) has to be replaced by an integral η across the interface:

$$\eta = \int \frac{(\epsilon-\epsilon_0)(\epsilon_2-\epsilon)}{\epsilon}\, dz \tag{55}$$

An ellipsometric experiment on a monolayer yields a quantity proportional to η. A simplification of eqn. (55) reveals its physical meaning. Quite often, as for example in case of adsorption of organic compounds onto solid supports [7], ϵ_2 exceeds ϵ of the monolayer while $\epsilon \approx \epsilon_0$.

Under these circumstances eqn. (55) can be further simplified:

$$\eta = \frac{\epsilon_2-\epsilon_0}{\epsilon_0} \int (\epsilon-\epsilon_0)\, dz \tag{56}$$

A linear relationship between ϵ and the prevailing concentration c of amphiphile within the adsorption layer is well established [11]

$$\epsilon = \epsilon_0 + c\frac{\mathrm{d}\epsilon}{\mathrm{d}c} \tag{57}$$

This relation yields a direct proportionality between the quantity η and the adsorbed amount Γ.

$$\eta = \frac{\epsilon_2-\epsilon_0}{\epsilon_0}\frac{\mathrm{d}\epsilon}{\mathrm{d}c}\int c\, dz = \frac{\epsilon_2-\epsilon_0}{\epsilon_0}\frac{\mathrm{d}\epsilon}{\mathrm{d}c}\Gamma \tag{58}$$

None of the assumptions which lead to eqn. (58) apply for adsorption layers at the liquid-air interface and hence eqn. (55) cannot be further simplified. The relationship between monolayer data and recorded changes remains obscure with no further simplifications possible on the basis of Maxwell's equations. A proportionality between Δ and Γ may hold but it cannot be established within this theoretical framework.

7 Microscopic model for reflection

Quantities such as refractive index or thickness are macroscopic quantities. Their meaning at sub-monolayer coverage is not an obvious one. To bridge these inherent difficulties several approaches have been developed aiming for a calculation of the optical properties of a monolayer based on microscopic quantitities [12]. We follow here mainly a model originally derived by Dignam et.al. [13, 14]. Explicit formulas are derived relating the changes in Δ to the polarizibility tensor of the adsorbed molecule and their number density at the interface.

The adsorption layer is treated as a two dimensional sheet of dipoles driven by the external laser field. The oscillating dipoles act as sources of radiation and the summation of all their contributions yields the reflected beam in the far field.

The adsorption layer is considered as homogeneous, fluctuations in the orientation or density occur on a length scale much smaller than the wavelength of light. These conditions are usually fulfilled for adsorption layers of soluble surfactants.

The decisive new quantitity introduced by Dignam et. al. is the surface susceptibility tensor $\vec{\gamma}$

$$\vec{\gamma} E_0 = 4\pi P t \tag{59}$$

which links the external electric field E_0 to the resulting polarization per volume P, t represents the thickness of the adsorbed layer. The polarization is given by the vector sum of all dipole moments μ_j over all molecules j:

$$\vec{\gamma} E_0 = 4\pi \sum_j \mu_j \tag{60}$$

The local field at each dipole has two origins, the external laser field E_0 and the dipolar contribution which is proportional to E_0.

$$E_L = E_0 + \vec{\beta} E_0 \tag{61}$$

Each adsorbed molecule is characterized by its own tensor β. The induced dipole moment of a molecule with a polarizability α_j is given by

$$\mu_j = \vec{\alpha}_j E_{jL} = \vec{\alpha}_j (1 + \vec{\beta}_j) E_0 \tag{62}$$

with this notation the susceptibilty tensor $\vec{\gamma}$ reads:

$$\vec{\gamma} = 4\pi \sum_j \vec{\alpha}_j (1 + \vec{\beta}_j) \tag{63}$$

In order to calculate β it is desirable to introduce the field E_{ij}, which accounts for the electric field at the position r_i generated by the dipole of molecule j.

$$E_{ij} = \vec{f}_{ij} \mu_j \tag{64}$$

$$\vec{f}_{ij} = (3\vec{\rho}_{ij} - 1)/r_{ij}^3 \quad \text{for } i \neq j \tag{65}$$

$$\vec{f}_{ij} = 0 \quad \text{for } i = j \tag{66}$$

The projection operator $\vec{\rho}$ projects the induced dipole moment on the connection of the molecules at site i and j.

$$E_{ij} = \vec{f}_{ij} \vec{\alpha}_{ij} (1 + \vec{\beta}_j) E_0 \tag{67}$$

With the relation:

$$\sum_j E_{ij} = \vec{\beta}_i E_o \tag{68}$$

$\vec{\beta}_i$ is given by

$$\vec{\beta}_i = \sum_j \vec{f}_{ij} \vec{\alpha}_{ij} \left(1 + \vec{\beta}_j\right) \tag{69}$$

Each molecule is characterized by a tensor β_j. For a further calculation certain assumptions about the molecular distribution must be introduced. For adsorption layers of soluble surfactants at the air-water interface it is a good approximation to consider β for all molecules equal. The underlying molecular pircture is is that the adsorption layer consists of one species of amphiphiles with a narrow angular distribution.

Under these conditions:

$$\beta = \sum_j \vec{f}_{ij}\vec{\alpha}_j\,(1+\beta) \tag{70}$$

meaning that

$$1+\beta = \left(1 - \sum_j \vec{f}_{ij}\vec{\alpha}\right)^{-1} \tag{71}$$

With equation 63 the surface susceptibility tensor reads:

$$\vec{\gamma} = \frac{4\pi N\vec{\alpha}}{(1-\sum_j \vec{f}_{ij})\vec{\alpha}} \tag{72}$$

The number of molecules per unit area is denoted by N. For an uniaxial film with its optical axis parallel to the surface normal the tangential t and normal n component of the susceptibility tensor read:

$$\gamma_t = \frac{4\pi N\alpha_t}{(1-4\pi/3(N/t_e)\alpha_t} \tag{73}$$

$$\gamma_n = \frac{4\pi N\alpha_n}{(1-8\pi/3(N/t_e)\alpha_n} \tag{74}$$

The quantity t_e can be regarded as an effective thickness as defined as

$$t_e = \frac{8/3\pi N}{\sum_{i\neq j} 1/r_{ij}^3} \tag{75}$$

If the adsorption layer can be mathematically represented as adsorbed molecules on a rectangular grid with an average next neighbour distance of a

$$t_e = \frac{8/3\pi a^3 N}{(i^2+j^2)^{-3/2}} = 0.935a^3 N \tag{76}$$

At monolayer coverage $N = 1/a^2$ the effective thickness is $t_e = 0.935a$. The adsorption of molecules at the interface changes the optical properties of the system. The changes in the ellipsometric angles are given by

$$\ln\left(\frac{\tan\psi}{\tan\psi_0}\right) + i\,(\Delta - \Delta_0) = \delta L \tag{77}$$

In the thin film limit only changes in Δ occur:

$$\delta L_v = \frac{i4\pi}{\lambda}\cos\varphi\frac{\delta\gamma_t}{1-\epsilon_2}\left(1 + \delta_v, p\frac{1-\epsilon_s\delta\gamma_n/\delta\gamma_t}{\cot^2\varphi - 1/\epsilon_2}\right) \tag{78}$$

$\delta\gamma_{n,t}$ denotes the difference of the susceptibility tensor between the film covered and the bare surface, $i = \sqrt{-1}$, φ is the angle of incidence and λ is the wavelength of light.

Changes in the ellipsometric angle Δ are given by the difference in L parallel and perpendicular to the plane of incidence.

$$\delta\Delta = -i(\delta L_p - \delta L_s) \tag{79}$$

The parallel component reads

$$-i\delta L_p = \frac{4\pi}{\lambda}\cos\varphi\frac{\delta\gamma_t}{1-\epsilon_2}\left(1 + \frac{1-\epsilon_2\delta\gamma_n/\delta\gamma_t}{\cot^2\varphi - 1/\epsilon_2}\right) \tag{80}$$

whereas the perpendicular component is given by

$$-i\delta L_s = \frac{4\pi}{\lambda}\cos\varphi\frac{\delta\gamma_t}{1-\epsilon_2} \tag{81}$$

This model reflects the optical response at the molecular level, but the model suffers from its complexity and high number of parameters, the values of which can only be assumed. Since the macroscopic ansatz resulting in eqn. (54) comes to similar results with a properly chosen dependence of the refractive index of the layer on the adsorbed amount, the use of the microscopic model is not common in practical applications. In the following section we will discuss some representative examples.

8 Adsorption layers of soluble surfactants

8.1 Importance of purification

Due to the peculiarities of surfactant synthesis, many surfactants contain trace impurities of higher surface activity than the main component. These trace impurities do not influence most bulk properties. However, at the surface they are enriched and impurities may even dominate the properties of the interface. This behaviour was first recognized by Mysels [15] and a purification scheme using foam fractionation was proposed [16]. A detailed discussion on artifacts caused by impurities can be found in [17].

In studies performed in our lab, we use a fully automated purification device developed by Lunkenheimer et. al. which ensures a complete removal of these unwanted trace impurities [18]. The aqueous stock solution undergoes numerous of purification cycles consisting of a) compression of the surface layer, b) its removal with the aid of a capillary, c) dilation to an increased surface and d) formation of a new adsorption layer. At the end of each cycle the surface tension σ_e is measured. The solution is referred to as surface chemically pure grade if σ_e remains constant in between subsequent cycles. Quite frequently more than 300 cycles and a total time of several days are required to achieve the desired state. The sample preparation is time consuming and tedious but mandatory for the investigation of equilibrium properties of adsorption layers of soluble surfactants at the air-liquid interface.

8.2 Physicochemical properties of our model systems

In the following we discuss the interfacial properties of two related amphiphiles, the cationic amphiphile 1-dodecyl-4-dimethylaminopyridinium bromide, C12-DMP, and the closely related nonionic 2-(4'dimethylaminopyridinio)-dodecanoate, C12-DMP betaine. The corresponding chemical structures are depicted in fig. 7 together with their equilibrium surface tension isotherms. The synthesis is described in [19].

Both components are classical amphiphiles and resemble all common features such as the existence of a critical micelle concentration *cmc*. The members of the homologous series of the alkyl-dimethylaminopyridinium bromide are strong electrolytes and follow the predictions of Debye-Hückel theory as experimentally verified by conductivity measurements. For solubility reasons we used 1-butyl-4-dimethylaminopyridinium bromide instead of the C_{12} representative of the homologous series which gave us experimental access to a wider concentration range. Debye-Hückel predicts a proportionality of the activity coefficient to the square root of the ionic strength for aqueous solutions of 1:1 electrolytes which is indeed observed in our experiment.

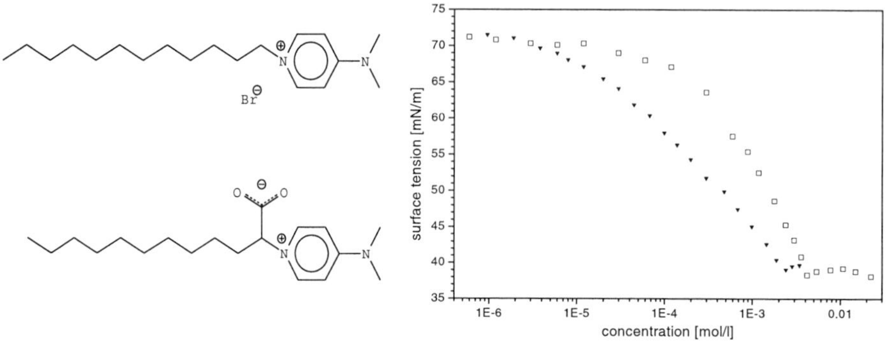

Figure 7: Chemical structures of the cationic amphiphile 1-dodecyl-4-dimethylaminopyridinium bromide, C12-DMP bromide, and the closely related nonionic 2-(4'dimethylaminopyridinio)-dodecanoate, C12-DMP betaine. The equilibrium surface tension σ_e of a purified aqueous solution of the ionic C12-DMP bromide(squares) and the nonionic C12-DMP betaine(triangles) as a function of the bulk concentration c_o. (redrawn from [20])

8.3 Adsorption layer of a nonionic surfactant

The ellipsometric isotherm $\Delta - \Delta_0$ of the nonionic amhiphile C12-DMP betaine is shown in Fig. 8. It decreases in a monotoneous fashion and reaches a limiting value at higher concentration. In order to assess its physical meaning we compare $\Delta - \Delta_0$ to the surface excess retrieved from the surface tension isotherm Fig. 7. According to Gibbs' fundamental law:

$$\Gamma = -\frac{1}{mRT} \cdot \frac{d\sigma_e}{d\ln a} \approx -\frac{1}{mRT} \cdot \frac{d\sigma_e}{d\ln c} \tag{82}$$

the total surface excess Γ is given by the derivative of the surface tension isotherm $\sigma_e(c)$.

Γ is then compared to the changes of the ellipsometric quantity $\Delta - \Delta_0$. The inset of Fig. 8 presents the result. The surface excess as given by the slope of the surface tension isotherm is compared to the ellipsometric response at each concentration. Obviously the relation between both quantities can be described by a straight line. Hence, experimental evidence has been provided that ellipsometry measures indeed the surface excess of the adsorption layer of the soluble nonionic amphiphile. Since optics measures refractive indices which are not very sensitive to molecular details we anticipate that this holds for a wider class of materials.

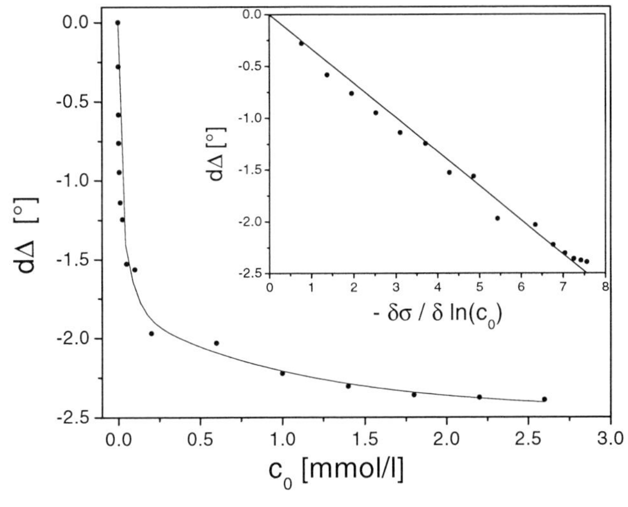

Figure 8: Ellipsometric isotherm of the nonionic C12-DMP betain. The ellipsometric isotherm decreases in a monotonic fashion and is proportional to the surface excess as can be seen in the inset, which compares the surface excess according to Gibbs and the ellipsometric response. (redrawn from [20])

8.4 Ionic surfactant at the air-water interface

In the previous section it was shown that in spite of the uncertainties from the mathematical point of view concerning the interpretation of eqn. (55) a proportionality between Δ and Γ can be established by a comparison of thermodynamic and ellipsometric data for a nonionic surfactant at the air-water interface. In the following we will discuss a measurement carried out on adsorption layers of an ionic surfactant. It will be demonstrated that there are major differences between the two model systems.

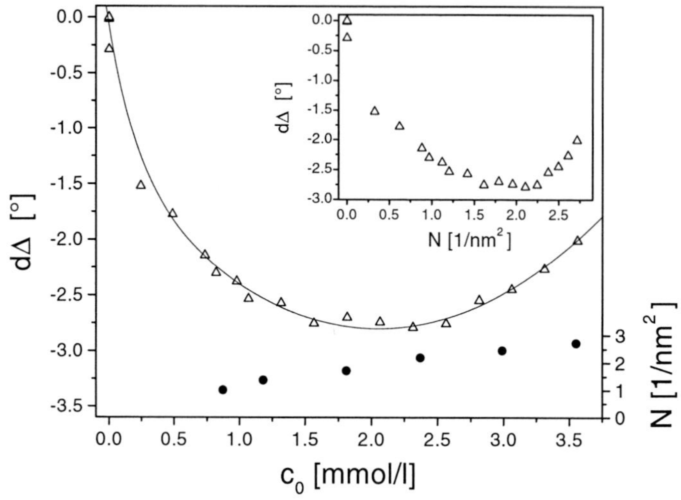

Figure 9: Characterization of the equilibrium properties of the adsorption layer by ellipsometry and Surface second harmonic generation, SHG. The SHG-signal $\sqrt{I^{2\omega}(P=45, A=90)}$ (circles) is proportional to the surface coverage and increases monotonously with the bulk concentration. The ellipsometric quantity $d\Delta = \Delta - \Delta_0$ (triangles) shows an extremum at an intermediate concentration far below the *cmc*. The inset clearly shows the nonmonotonic dependency of $d\Delta$ on the adsorbed amount. (redrawn from [20])

Fig. 9 shows the ellipsometric isotherm $\Delta - \Delta_0$(triangles) of the cationic surfactant C12-DMP bromide. A pronounced non-monotonic behaviour is shown with an extremum at an intermidiate concentration far below the *cmc*. Also shown is the number density of amphiphiles adsorbed to the interface (circles) as determined by Surface second harmonic generation (SHG). At these bulk concentrations the measured number density equals the surface excess Γ. SHG reveals a monotoneous increase in the surface excess in qualitative agreement to a thermodynamic analysis within the Gibb's framework. The data also clearly prove that the ellipsometric quantity need not be proportional to the adsorbed amount for a soluble ionic surfactant. What causes the nonmonotonous behaviour and how can it be understood?

In the following we discuss some possible scenarios:

Scenario 1: Filling up the adsorption layer

The adsorption layer is described as an isotropic optical layer of constant thickness with a refractive index n_{layer} which depends on the surface coverage [21]. Within this model there are two distinct surface concentrations which lead to a vanishing $d\Delta = \Delta - \Delta_0 = 0°$. At very low coverage the refractive index of the layer matches the one of air $n_{layer} = n_{air} = 1$ and at an intermediate surface coverage the surface layer adopts the very same refractive index as the water bulk phase $n_{layer} = n_{water} = 1.332$. Consequently there has to be an extremum in between. The resulting $d\Delta(n_{layer})$-curve is depicted in Fig. 10. The geometrical dimension of the molecules has been used as thickness of the layer.

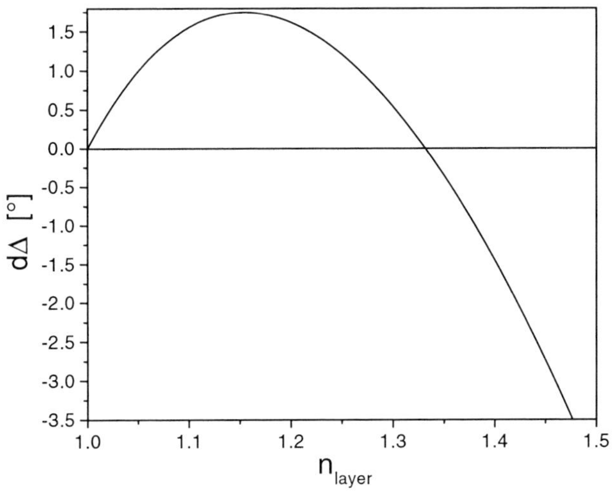

Figure 10: Simulation of the effect of a changing refractive index of a layer of constant thickness $h = 2.1nm$ on the ellipsometric signal $d\Delta$.

Obviously this scenario is not suitable to explain the measurement, the model predicts even a wrong sign of $d\Delta_{max} \approx 1.75°$! The experimental data require that $n_{layer} > n_{water}$ at all concentration. This means, water instead of air is the effective environment of this particular amphiphile within the adsorption layer.

Scenario 2: Effect of anisotropy

The following provides an estimation if changes in the molecular orientation may be responsible for the surprising features in the ellipsometric isotherm. The following assumptions were made in order to estimate an upper limit for this effect:

- The optical model is that of an uniaxial layer with the optical axis normal to the interface. The molecular arrangement is $C_{\infty V}$ which has been experimentally verified for the headgroups by polarization dependent SHG measurements.
- The thickness of the adsorbed layer and the mean tilt angle of the molecules within the layer change with their number density N at the interface. Within the investigated number density range the thickness increases from 1nm to 1.9nm proportional to the cosine of the tilt angle which is assumed to change from around 70° to 40°.
- The whole molecule, including the head group, is assumed to change its tilt angle. The molecules are assumed to be all-trans and perfectly aligned which would yield the maximum possible change in anisotropy.
- The refractive index for an E-vector along the length axis of the molecule is $n_{axis} = 1.56$, while the refractive index for an E-vector perpendicular to the long axis of the molecule is $n_{perp} = 1.48$. Both refractive indices have been taken from Riegler et.al. [22] and rely on a combination of X-ray reflection data with ellipsometric measurements of monolayers of behenic acid at the air-water interface for a densely packed and perfectly oriented apmphiphiles. In our case the actual refractive indices will depend on the prevailing volume concentration and the data are therefore an upper limit.

The calculated $d\Delta(N)$ versus density-curve is plotted in Fig. 11. Obviously a change of the molecular tilt leads to changes in $d\Delta$, but cannot account for the measured pronounced extremum. In addition it is impossible to reach $d\Delta = -2.77°$, which is the minimum of the measured curve, with reasonable parameters for the anisotropic layer. An increase in anisotropy has the same impact on the ellipsometric measurements as a reduction of

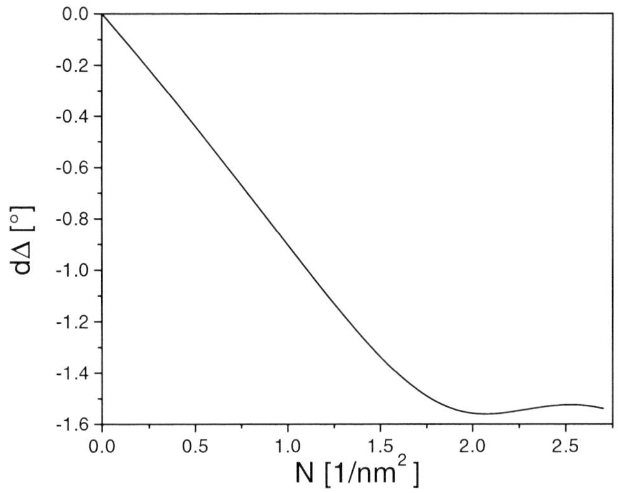

Figure 11: Investigation of the impact of anisotropy on the ellipsometric signal. The ellipsometric signal $d\Delta$ calculated in dependence of the number density of adsorbed molecules N. A molecular length of 2.1 nm and a refractive index of $n_{axis} = 1.56$ for an E-vector along and of $n_{perp} = 1.48$ for an E-vector perpendicular to the molecular axis were used. The layer thickness (1.0nm to 1.9nm), the tilt angle (70° down to 40°), n_s and n_p were all assumed to be dependent on the surface coverage.

the layer thickness; hence, the only way to get nearer to $d\Delta = -2.77°$ is to diminish the anisotropy!

At this point we would like to point out that the assumptions made for this simulation are highly exaggerating the effect of the anisotropy. The SHG measurements reveal that the heads do not change their tilt angle at all. In other words that only the tails may change their tilt, which results in an even thinner anisotropic layer and therefore a smaller effect. Additionally the tails are certainly neither all-trans nor perfectly oriented even at the highest coverage measured $0.37 nm^2$/molecule. For these reasons Fig. 11 represents an upper limit of the impact of anisotropy. The real effect is by far smaller for the surface coverages encountered here, which means that anisotropy cannot account for the surprising feature of the ellipsometric isotherm.

Scenario 3: Changing the counterion distribution

Ellipsometry probes the complete interfacial architecture. The reflected light is generated within the transition region between the two adjacent bulk phases, air and aqueous surfactant solution. The electric double layer has to be explicetely considered. Hence, changes in the ion distribution at higher concentrations may cause the observed feature in the ellipsometric isotherm.

The classical model of a charged double layer has been developed by Gouy, Chapman and Stern. The interface is described by two distinct regions: a compact layer consisting of the positively charged adsorbed amphiphiles with some directly adsorbed counterions and a diffuse layer of counterions.

The ion distribution within the diffuse layer is given by the solution of the Poisson equation which relates the divergence of the gradient of the electric potential Φ to the charge density ρ at that point (see for instance [23, 24]):

$$\text{div grad}\Phi = \Delta\Phi = -\frac{\rho}{\epsilon_o\epsilon_r} \tag{83}$$

The compact layer is positively charged and in contact with an electrolyte solution which forms a diffuse layer of charges. The ion concentration distribution within the electrical potential Φ obeys Boltzmann:

$$c^- = c_o^- \, e^{\frac{z^- e\Phi}{k_B T}} \qquad c^+ = c_o^+ \, e^{-\frac{z^+ e\Phi}{k_B T}} \tag{84}$$

where z^- and z^+ are the valencies of the anions and cations respectively. For a symmetric electrolyte solution ($-z^- = z^+ = z$ and $c_o^- = c_o^+ = c_o$) eqn.(84) leads to a net charge of the ion cloud of

$$\rho = ze(c^+ - c^-) = -2c_o ze \sinh\frac{ze\Phi}{k_B T} \tag{85}$$

The combination of eqn.(85) with the Poisson eqn.(83) yields a differential equation in the electric potential Φ. In our case the potential is only a function of the normal coordinate to the surface x. It is convenient to define a reduced potential

$$y = \frac{ze\Phi}{k_B T} \qquad y_o = \frac{ze\Phi_o}{k_B T} \tag{86}$$

which further simplifies the equations and results in

$$\frac{d^2y}{dx^2} = \frac{2c_o z^2 e^2}{k_B T \epsilon_o \epsilon_r} \sinh y = \kappa^2 \sinh y \tag{87}$$

The integration of equation (87) with the boundary conditions ($y = 0$ and $dy/dx = 0$) for $x = \infty$ and $y = y_o$ at $x = 0$ lead to the Gouy-Chapmann solution of the reduced potential within the diffuse layer:

$$e^{y/2} = \frac{e^{y_o/2} + 1 + \left(e^{y_o/2} - 1\right) e^{-\kappa x}}{e^{y_o/2} + 1 - \left(e^{y_o/2} - 1\right) e^{-\kappa x}} \tag{88}$$

Hence, knowing the charge of the compact layer the ion distribution can be modelled. The crux of Stern's treatment is the estimation to which extent ions enter the compact layer and reduce the surface potential. The ion distribution between compact and diffuse layer cannot be easily experimentally established, for example surface potential measurements cannot measure this distribution.

The optical analysis requires the translation of the prevailing distribution of molecules and ions within the interphase into a corresponding refractive index profile. The reflectivity coefficients can then be calculated on the basis the previously discussed numerical algorithms for stratified media. Two elements dominate the refractive index profile and hence the reflectivity properties, the topmost monolayer of the amphiphile and the distribution of ions within the diffuse layer. The excess of ions within the diffuse layer leads to a slightly elevated refractive index as compared to the bulk of the aqueous surfactant solution. The typical dimensions of the diffuse layer are on the order of 10 nm and this fairly wide extension leads to a profound impact on Δ, since the topmost monolayer has a normal extension of only 2 nm. The solution of the potential Φ plugged into equation (85) yields the distribution of anions and cations within the diffuse layer. The prevailing charge distribution is determined by the surface charge of the topmost monolayer. The refractive index profile is then determined by the ion distribution by a multiplication with the refractive index increment dn/dc which has been independently measured with an Abbe refractometer.

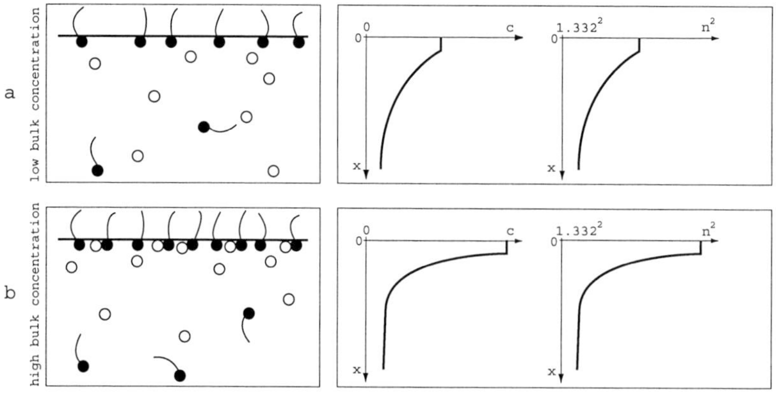

Figure 12: The interphase of a soluble cationic surfactant at the air-water interface at low (a) and high (b) bulk concentration. It consists of a charged topmost cationic monolayer, a diffuse layer of counterions and at higher concentrations a compact layer of directly adsorbed counterions. The charge density of the topmost monolayer reduced by the charge of the inner Stern layer determines the ion distribution within the diffuse layer. The prevailing ion distribution is given by solution of the nonlinear Poisson-Boltzmann equation. The excess of ions can be readily translated in a corresponding refractive index profile. The profile determines the reflectivity properties. Ellipsometric data modeled within this framework allow an estimation of the extent to which ions enter the compact layer.

Fig. 12 sketches the model for the interface. At lower surface coverage most of the ions are spread out within the diffuse layer leading to a pronounced refractive index profile. The situation is sketched in Fig. 12a. At higher concentration some ions enter the adsorption layer and form ion pairs with the headgroups accounted with a lower dn/dc-value as compared to the bulk phase. The formation of this Stern layer effectively reduces the surface charge and as a consequence the extension and the magnitude of the refractive index profile of the diffuse layer decrease. This situation is sketched in Fig. 12b.

This model has been used for monitoring the formation of the Stern layer. The ion distribution within the diffuse layer is determined by the total effective charge density at the interface. The SHG measurements yield the number density and charge density

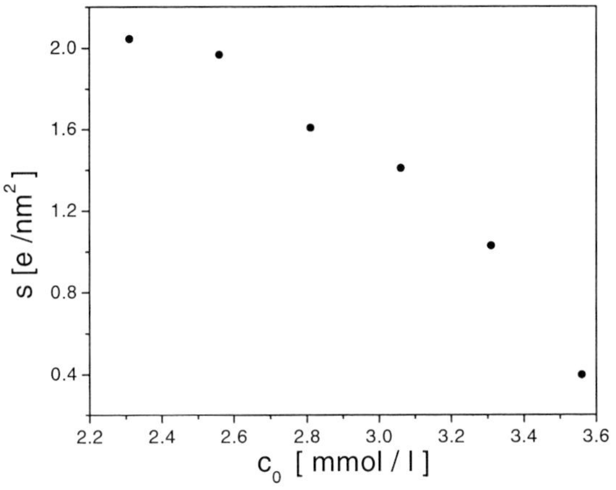

Figure 13: The charge density of the compact Stern layer has been retrieved by optical means. The surface charge refers to the number density of the cationic amphiphile reduced by the number of directly bound ions within the Stern layer.

produced by the cationic amphiphilic monolayer. The effective charge density at the interface is given by this value reduced by the number density of counterions within the Stern layer. The corresponding refractive index of Stern and topmost monolayer is given by an effective medium approach. Hence the optical properties of the monolayer are known by independent means. Furthermore the refractive index profile of the diffuse layer is given once the effective charge density at the interface is known. We used the experimental data in order to retrieve the corresponding effective surface charge density. The results are depicted in Fig. 13 starting in the vicinity of the extremum of the ellipsometric isotherm. Hence, with this model we are able to monitor by purely optical means the extent to which ions enter the compact Stern layer.

In short, ellipsometry applied to adsorption layers of ionic soluble surfactants does not measure the surface excess. The ellipsometric signal may show a non-monotonic behaviour which is caused by a redistribution of the ions between compact and diffuse layer. The data analysis within the classical model of a charged double layer yields an estimate of the prevailing ion distribution.

Ellipsometry applied to adsorption layers of nonionic surfactant directly yields the surface excess. It is therefore a valuable alternative to surface tension measurements, especially if one considers rheological properties which require higher derivatives of the surface tension isotherm.

8.5 Kinetics of absorption

Rapidly expanding liquid surfaces occur in many technical processes such as foam formation, spraying or painting, they are also omnipresent in nature, i.e. the oxygen exchange in the lung. Surfactants play a crucial role in these processes. They stabilize the new surfaces by a reduction of the surface tension. Inhomogeneities in surface coverage lead to gradients in surface tension which in turn have a strong impact on bulk hydrodynamic flow, a phenomenon also known as Marangoni flow. The overall dynamic behaviour is very complex and despite many efforts far from understood [25]. The investigation requires time resolution within the ms regime putting severe limitations on the choice of the experimental technique. The maximum bubble pressure method [25] is most often used to perform these types of studies but ellipsometry offers an interesting alternative to supplement these investigations. This method relies on surface tension measurements and the surface coverage is obtained using the equilibrium surface isotherm. The underlying assumption is that the surface is locally at equilibrium, however, in the time regime of interest this may not be the case. As outlined in the theoretical section ellipsometry measures directly the surface coverage for nonionic surfactants and provides in addition the requested time resolution. The challenging part is the design of an experiment with precisely defined boundary conditions required for modeling the underlying physics leading to the observed kinetic behavior. The most promising approaches produce an interface with a non-equilibrium surface coverage, and maintain that surface coverage through the flow of fluid in and out of the control volume. To satisfy the requirement of well-defined boundary conditions, at the beginning of the flow a freshly formed surface must be created. Several arrangements have been suggested such as the inclined plate [25], the overflowing cylinder [26] or the use of a jet [27]. In all these approaches the surface has a defined age at a given spot, hence the desired dynamic picture of the surfactant adsorption is obtained by moving the sample relative to the light beam.

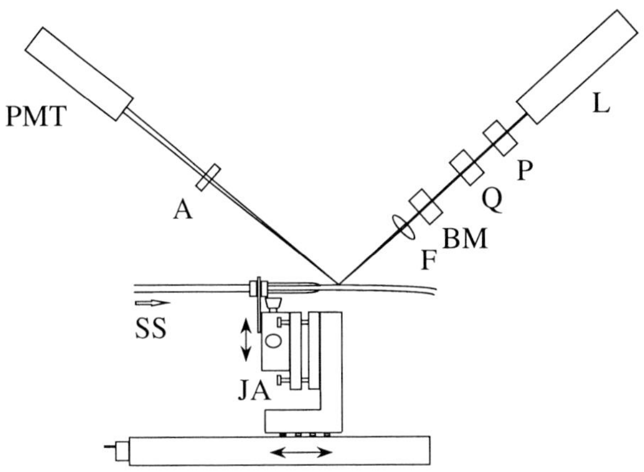

Figure 14: Setup for an ellipsometric measurement on a jet. The distance between laser spot and nozzle determines the age of the surface. (redrawn from *Langmuir* **15**, 7530-7533 (1999))

The overflowing cylinder with its flat surface is very well suited for optical reflection technique. Water is pumped vertically trough a cylinder and allowed to flow over the rim in a radial flow pattern. This experiment covers the time regime of $0.1 - 1s$ and has been successfully used by Penfold et.al. [28] in neutron reflectometry and ellipsometric experiments. A precision in the determination of the dynamic surface excess of $2 \cdot 10^{-8}$mol/m^2 has been demonstrated. Neutron reflectivity enabled the calibration of the ellipsometric data which is important to avoid artifacts caused by changes in the ion distribution of ionic surfactants. The more interesting time range of $1 - 20ms$ is covered by jet techniques. The surfactant solution is pumped with a high speed through a nozzle. An ellipsometric experiment performed on a jet is quite demanding and many subtle problems are caused by the curved surface acting as a cylindrical lens. A recent feasibility study of Hutchison et.al. demonstrates that these problems can be overcome [29]. The adsorption of a cationic surfactant was monitored and the time dependence was found to obeye the Ward-Tordai equation [30] providing evidence for a diffusion controlled adsorption kinetics of the surfactant in this time regime. These studies suffer from a simultaneous measurement of the flow field which is required to assess Maragoni currents induced by surface tension gradients.

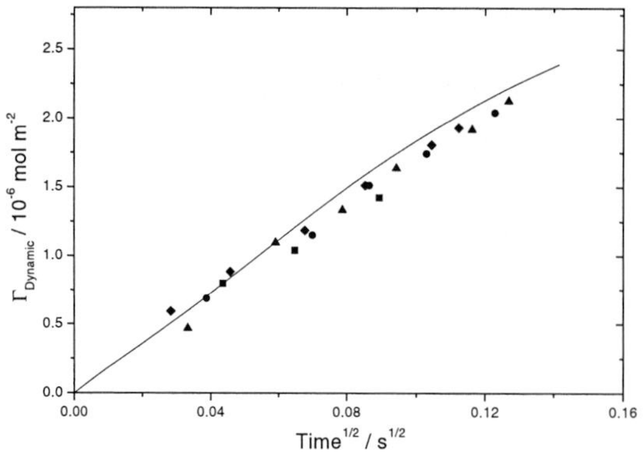

Figure 15: Surface coverage measured with the setup sketched in Fig. 14. (redrawn from *Langmuir* **15**, 7530-7533 (1999))

9 Principle of imaging ellipsometry

Some surfaces under investigation are laterally inhomogeneous on a micrometer scale due to variations in the thickness or surface composition or due to changes in the orientational order of the molecules at the interface. In this case the lateral inhomogeneity is imparted to the properties of the reflected light. The most well-known imaging technique for a visualization of these features is Brewster Angle Microscopy BAM [31, 34] which has been successfully employed for characterizing the phase diagrams and the morphology of Langmuir films. BAM is based on the principle that the reflectivity of the $\hat{p}$-polarized light is zero at the Brewster angle. Any modification of the Brewster conditions, as for instance the presence of a single monolayer, modifies the reflectivity. A decisive advantage of this method vs. fluorescence microscopy [35] is that fluorescent labelling is not required. Also the internal structures of domains can be assessed.

Ellipsometry can also be extended to an imaging technique which offers a wider field of applications [9]. In contrast to BAM, imaging is not bound to the existence of a Brewster angle and can even be employed for the investigation of monolayers on highly reflecting supports.

Fig. (16a) provides a Nullellipsometric image of a self assembled monolayer of (Dimethyl-

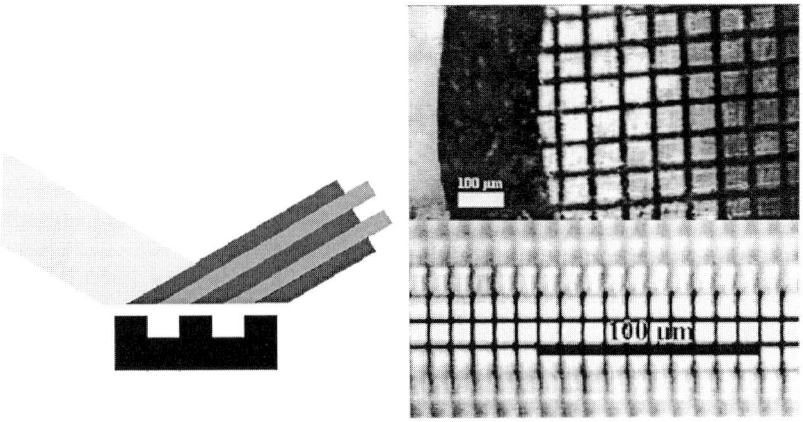

Figure 16: The lateral inhomogeneitity is transformed in the state of polarization. a) Nullellipsometric image of a self assembled monolayer on a silicon wafer. The monolayer was patterned with the aid of a mask and UV-light. The difference in the thickness between the dark and bright regions is $0.8nm$ b) At high magnification problems arise due to the limited depth of field. The width of the bars are $1.9\mu m$, the mesh size is $10\mu m$ and imaging is performed at an angle of 53°

chlorosilyl)-2-(p,m-chloromethylphenyl)ethan on silicon. The monolayer was photochemically pattererned using UV-irradiation and an electron microscopy grid as a mask. The difference in the thickness between dark and bright regions is 0.8 nm demonstrating the high vertical resolution of this technique. At a higher magnification there are certain problems arising from the limitation of the depth-of-field.

9.1 Depth of field problem

All the above mentioned imaging techniques work under an oblique angle of incidence and as a result some peculiarities exist. Imaging under an oblique angle of incidence imposes certain restrictions on the diameter and working distance of the objective. As a result the resolution is limited to about $1 - 3\mu m$. A discussion of this issue can be found in [32].

The depth of field, t, of an objective is given by the numerical aperture A, the refractive index n of the ambient medium, and the wavelength λ according to [33]:

$$t = \frac{n\lambda}{A^2} \tag{89}$$

Depending on the angle of incidence, φ, and the numerical aperture, A, only a region of the illuminated part of the sample is in focus. This region is of the order of $1 - 50\mu m$. This problem is illustrated in Figure (16b). The size of a bar is $1.9\mu m$, the mesh size is $10\mu m$ and imaging was performed at $\varphi = 53°$.

The depth-of-field problem is a severe limitation at higher magnification. A straightforward solution consists of a scan of the objective [34]. Only the area in focus is recorded and the complete image is constructed by many stripes by imaging processing software. This procedure yields sharp images of static objects. At the oil-water or air-water interface it is only of limited use due to the prevailing convection.

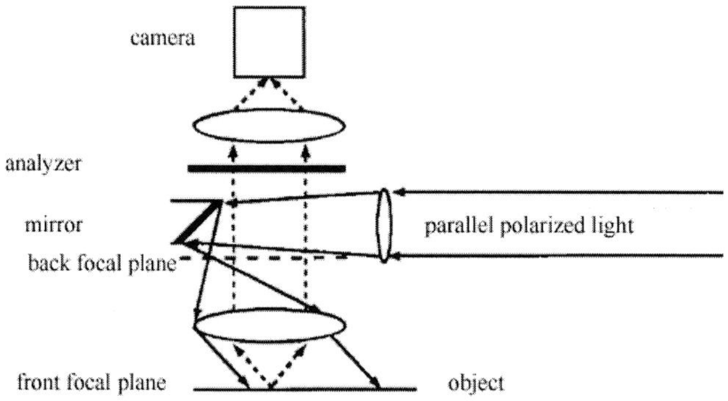

Figure 17: Principle of ellipsometric microscopy. Full arrows symbolise the light path of illumination, broken arrows stand for the observation, respectively. A parallel beam of polarised light is focussed into an off-axis spot in the back focal plane of the lens. In the front focal plane of the lens, where the object is located, this results in a parallel polarised beam of light hitting the object under a shallow angle of incidence. Light reflected by the object is collected by the lens, passes a motorised polarization analyzer and is focused onto a digital CCD camera. For clarity, several optical elements are omitted. (redrawn from [36])

An elegant approach which overcomes these difficulties was suggested in the group of Sackmann et al.. A parallel polarized laser beam was focused to an off axis spot in the back focal plane of a microscope objective of high numerical aperture. The sample which is located in the front plane is then illuminated by a parallel polarized beam of light incident on the sample under an oblique angle. The exact angle of incidence is controlled by the exact position of the laser focus in the back focal plane. The image is formed by the same objective used for illumination and the complete image is in focus [36].

The performance of this microscope is shown in Fig. 18. The sample consists of a photopatterned LB film with dimensions indicated in the figure. The image on the left hand side sketches the inner structure of the sample.

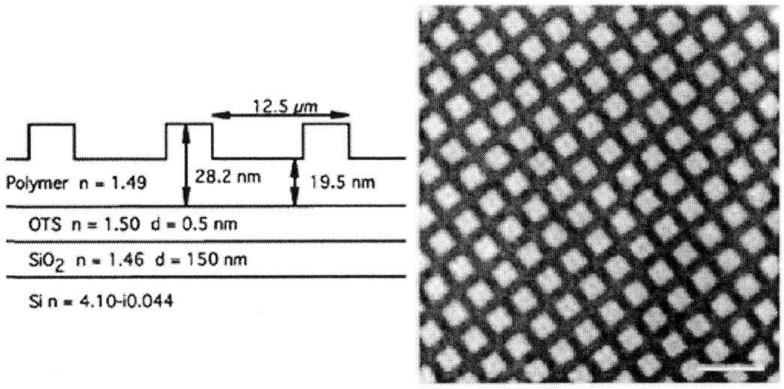

Figure 18: Schematic view of the micropatterned Langmuir-Blodgett film with all relevant dimensions; Ellipsometric microscopy images of the sample. The angle of incidence was $\varphi = 55.8°$. The gray scales from black to white represent a change in Δ of $120 - 130°$. The scale bar has a length of $25\mu m$. (redrawn from [36])

Another elegant approach was suggested by Meunier *et. al.* who presented a custom made objective based on a modified Schwarzschild objective. Its optic axis is perpendicular to the studied layer and consequently the complete area is in focus. The design is ideally suited for dynamic investigations and the alignment is easier than for the previously discussed solution.

9.2 Beyond the diffraction limit

The high vertical resolution of ellipsometry is not matched by the achieved lateral resolution. The upper limit is given by diffraction and on the order of the wavelength of light. Recently an experiment has been described which overcomes the diffraction limit and allows a direct visualization of refractive index patterns with a much finer resolution

[37]. The arrangement is based on a combination of Atomic Force Microscopy (AFM) and Ellipsometry and the AFM tip dimensions determines the resolution.

The sample of interest is mounted on the base of a prism. The ellipsometer is operated in the null mode and the total reflected light at the prism base is completely cancelled by the setting of the polarization optics. This null setting is kept constant during the course of the experiment. The metal AFM tip couples to the evanescent field and depending on the local optical properties of the sample the null setting is out of tune. The intensity at the detector is monitored while carrying out a topographic scan with a conventional Atomic Force Microscope. Thus two images are simultaneously generated: a topographic image by the AFM tip and an ellipsometric image which relates the intensity reading at the detector with the x,y position of the tip. The latter contains information on refractive index inhomogenity. This technique allowed for instance a visualization of 10 nm CDTe particles within a polymer matrix. The topographic scan was not able to identify the particles within the spincoated polymer film. However, due to the differences in the refractive index it could be visualized by the ellipsometric image.

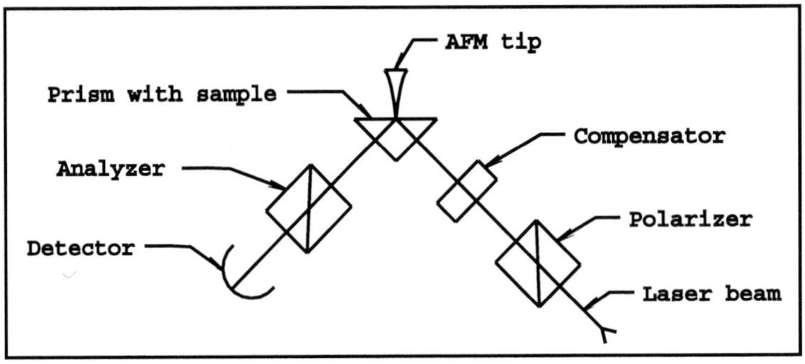

Figure 19: Sketch of the used apparatus

References

[1] M. Born, *Optik*, Springer Verlag, New York, Heidelberg (1998)

[2] D. S. Kliger, J. W. Lewis, C. Randall, *Polarized Light in optics and spectroscopy*, Academic Press, Harcout Brace Javanovich Publishers, Boston (1990)

[3] R. M. Azzam, N.M. Bashara, *Ellipsometry and Polarized Light*, North Holland Publication, Amsterdam (1979)

[4] R.C. Jones, *J. Opt. Soc. Am* **16**, 488 (1941)

[5] S.N. Jasperson, S.E. Schnatterly, *Rev. Sci. Instr.* **40** , 761 (1969)

[6] J. Lekner, *Theory of Reflection*, Martinus Nijhoff Publishers, Boston, (1987)

[7] T. Kull, T. Nylander, F. Tiberg, N. Wahlgren, *Langmuir* **13**, 5141 (1997)

[8] R. Teppner, S. Bae, K. Haage, H. Motschmann, *Langmuir* **15**, 7002 (1999)

[9] R. Reiter, H. Motschmann, H. Orendi, A. Nemetz, W. Knoll, *Langmuir* **8**, 1784 (1992)

[10] C. Flueraru, S. Schrader, V. Zauls, H. Motschmann, *Thin Solid Films* **379**, 15 (2000)

[11] H. Motschmann, M. Stamm, C. Toprakcioglu, *Macromolecules* **24**, 3681 (1991)

[12] J. Lekner, P.J. Castle, *Physica* **101A**, 89 (1980)

[13] M.J. Dignam, M. Moskovits, *J. Chem. Soc. Faraday II* **69**, 56 (1973)

[14] M.J. Dignam, J. Fedyk, *J. Phys.* (Paris) **38**, C5-57 (1977).

[15] P.H. Elworthy, K.J. Mysels, *J. Coll. Int. Sci.* , 331 (1966)

[16] K.J. Mysels, A. Florence, *J. Coll. Int. Sci.* **43**, 577 (1973)

[17] K. Lunkenheimer, *J. Coll. Int. Sci.* **131**, 580 (1989)

[18] K. Lunkenheimer, H.J. Pergande, H. Krüger, *Rev. Sci. Instr.* **58**, 2313 (1987)

[19] K. Haage, H. Motschmann, S. Bae, E. Gründemann, *Colloids and Surfaces* (in press)

[20] R. Teppner, K. Haage, D. Wantke , H. Motschmann,
J.Phys.Chem.B **104**, 11489 (2000)

[21] T. Pfohl, H. Möhwald, H. Riegler, *Langmuir* **14**, 5285 (1998)

[22] M. Paudler, J. Ruths, H. Riegler, *Langmuir* **8**, 184 (1992)

[23] D.F. Evans, H. Wennerström, *The Colloidal Domain*,
VCH Publishers, New York (1994)

[24] A.W. Adamson, *Physical Chemistry of Surfaces*, Wiley & Sons, New York (1993)

[25] S.S. Dukhin, G. Kretzschmar, R. Miller,
Dynamics of Adsorption at Liquid Interfaces, Elsevier, Amsterdam (1995)

[26] D.J.M. Bergink-Martens, H.J. Bos, A. Prins, B.C. Schulte, *J. Coll. Int. Sci.* 138 (1990); D.J.M. Bergink-Martens, H.J. Bos, A. Prins, *J. Coll. Int. Sci.* **165**, 221 (1994)

[27] B.A. Noskov, *Adv. Colloid Interface Sci.* **69**, 63 (1996)

[28] S. Manning-Benson, S. R. W. Parker, C. D. Bain, J. Penfold, *Langmuir* **14**, 990 (1998)

[29] J. Hutchison, D. Klenerman, S. Manning-Benson, C. Bain, *Langmuir* **15**, 7530 (1999)

[30] A.F.H. Ward, L. Tordai, *J. Chem. Phys.* **14**, 453 (1946)

[31] D. Hönig, D. Möbius, *J. Phys. Chem* **95**, 4590 (1991)

[32] M. Harke, R. Teppner, O. Schulz, H. Orendi, H. Motschmann,
Rev. Sci. Instrum. **68**, 8, 3130 (1997)

[33] H. Riesenberg *Handbook of Microscopy*, VEB Verlag Technik, Berlin (1988)

[34] S. Henon, J. Meunier, *Rev. Sci. Instr.* **62**, 936 (1991)

[35] M. Lösche, E. Sackmann, H. Möhwald, *Ber. Bunsenges. Phys. Chem.* **10**, 848 (1983)

[36] K.R. Neumaier, G. Elender, E. Sackmann, R. Merkel,
Europhysics Letters **49 (1)**, 14 (2000)

[37] P. Karageorgiev, H. Orendi, B. Stiller, L. Brehmer, *Appl. Phys. Lett.*, (2001) in press

Novel Methods to Study Interfacial Layers
D. Möbius and R. Miller (Editors)

DETECTION OF EVANESCENT FIELDS ON ARACHIDIC ACID LB FILMS ON AL FILMS CAUSED BY RESONANTLY EXCITED SURFACE PLASMONS

Takayuki Nakano[a], Hazime Kobayashi[a], Futao Kaneko[a,*], Kazunari Shinbo[a], Keizo Kato[a], Takahiro Kawakami[a], Takashi Wakamatsu[b]

[a] Faculty of Engineering, Niigata University, Ikarashi 2-8050, Niigata 950-2181, Japan

[b] Ibaraki National College of Technology, Nakane866, Hitachinaka 312-8508, Japan

Surface plasmons (SPs) were resonantly excited for arachidic acid (C20) LB ultrathin films on Aluminum (Al) thin films in a Kretschmann configuration of the attenuated total reflection (ATR) method. ATR properties were measured at 488 nm for the Al thin films and the Al/C20 LB ultrathin films. Evanescent fields generating on the Al thin films and the Al/C20 LB thin films due to SP excitations at the resonant incident angles were detected as luminescent light from a metallic needle probe covered with CdS. The luminescent intensities exponentially decayed from the sample surfaces, and the exponential decay lengths were about 230nm. Furthermore, the experimental results were compared with the calculated ones.

*Corresponding author: Tel/Fax:+81-25-262-6741, E-mail:fkaneko@eng.niigata-u.ac.jp

Key words: evanescent field, surface plasmon, attenuated total reflection, aluminium thin film, arachidic acid LB film.

1. INTRODUCTION

Surface plasmons (SPs or surface plasmon polaritons: SPPs) are known as non-radiative optical modes due to plasma oscillations of free electrons resonantly excited at a metal surface and propagate along the surface with evanescent electromagnetic fields [1]. Since the SPs localized at the metal surface are strongly influenced by conditions of the surface, the attenuated total reflection (ATR) method resonantly exciting the SPs is used to evaluate structure and optical properties of ultrathin films on metal thin films. We have already investigated structure, optical properties and surface roughness of various LB thin films [2]-[5] and have also estimated orientations of liquid crystal molecules on rubbing-free polyimide LB aligning layers [6] using the ATR method due to excitation of SPs.

Evanescent fields decay rapidly in the distance of sub-microns from some localized surface and cannot be usually observed, but they are enhanced by excitation of SPs and strongly depend upon nanometric structure of ultrathin films. Recently, evanescent fields are of great interest to evaluate nanometric spatial characteristics of ultrathin films and to fabricate ultrahigh-density optical memory, nano-devices, and so on [7]-[9]. And many types of scanning near-field optics microscope (SNOM) utilizing evanescent fields have been developed [10]-[12].

In this study, ATR properties of Al thin films and C20 LB thin films on the Al films were investigated. Evanescent fields generating on the Al film and the C20 LB thin films on the Al films were detected using a system constructed in this research. The experimental results were compared with the calculated ones.

2. EXPERIMENTAL DETAILS

2.1 Evanescent Fields and the ATR method

Fig. 1 shows the Kretschmann configuration of the attenuated total reflection (ATR) method. The ATR method was reported in previous papers [1]-[6]. An Ar^{+} laser beam at 488.0 nm was used to excite SPs on metal thin films in this measurement and the p-polarized beam was directed onto the samples on the substrates through the half-cylindrical prism. The prism was mounted on a rotating stage. The incident angles of the laser beam were scanned using a computer-controlled pulse motor. The reflected intensities of the laser light, that is, the ATR signals were detected at intervals of 0.1 degrees as a function of the incident angle. When the SP is excited at a resonant incident angle (θ_{SP}) of the p-polarized light in the total reflection region, the reflected light is attenuated since considerable energy of the incident light is converted into the SP excitation on metal thin films, and the electromagnetic fields, that is, evanescent fields are enhanced on the surfaces. The evanescent fields strongly depend upon structure of metal thin films and/or dielectric films on them.

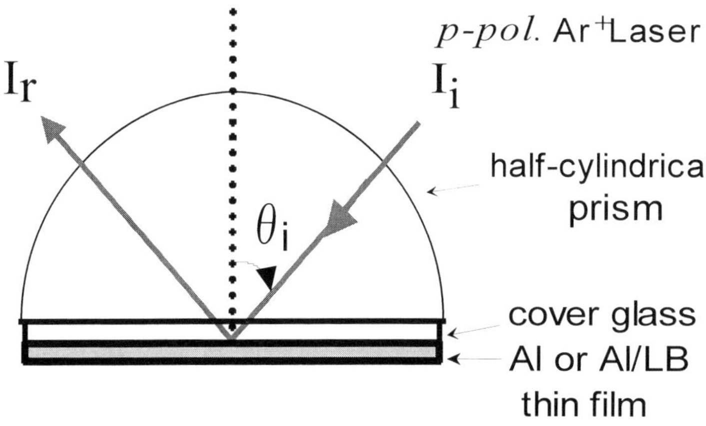

Fig. 1 The ATR configuration.

A typical ATR curve is shown in Fig. 2. This curve was calculated for an Al thin film at the wavelength of 488nm using Fresnel's formula [1,13], and the following values were used; the dielectric constants of the glass prism and the Al film:ε_p =2.316 andε_{Al} =□26.3+i6.90, respectively, and the thickness of the Al: d=20nm. In the configuration, the critical angle of the total reflection (θ_C) is 41.3°. The SP at the Al/air interface is resonantly excited at the incident angle of 42.3°, where a deep dip in the ATR curve is observed.

We calculated the intensities of the evanescent fields as a function of the distance z from the Al surface at various incident angles marked in Fig. 2, using a transfer-matrix method [1,13]. The squared fields correspond to the energy, and the result is shown in Fig. 3. At the incident angles marked with (a) lower than θ_C and with (f) considerably higher than θ_{SP} the fields are very weak, but the fields marked with (b), (c), (d) and (e) are larger and decay exponentially. Here note that the field (c) at the slightly lower angle is higher than one with (d) at θ_{SP} as the imaginary part of the dielectric constant is large for the Al film [14,15]. Figure 3 also exhibits that the decay lengths of the evanescent fields are about one or two hundred nanometers from the sample surface.

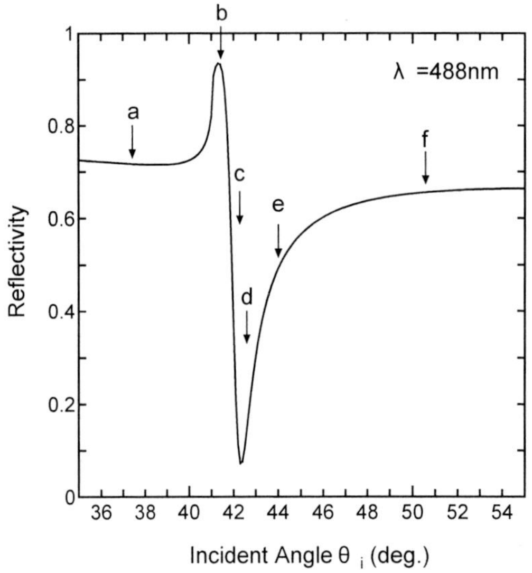

Fig. 2 A theoretical ATR curve at 488nm for the Al thin film.

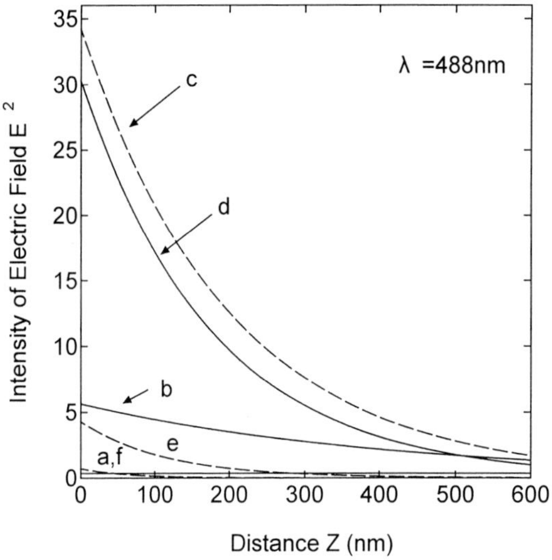

Fig. 3 Intensities of the squared evanescent fields as a function of the distance from the metal surface.

2.2 A sample configuration

Microscopic cover glasses covered with thin film samples were placed on a half-cylindrical prism (BK-7, n=1.522 at 488nm) using refractive matching oil (OLYMPUS immersion oil) as shown in Fig.1. Silver (Ag) thin films of about 50nm thick have been generally used as SP active layer because the ATR property exhibits a sharp dip at the resonant angle in the reflectivity. However, Ag thin films are very easy to form island structure and scattered light from surface roughness of Ag thin films, that is, noise light has been observed [5]. In contrast with Ag thin films, the ATR property using Al thin films does not exhibit a sharp dip, but surfaces of Al thin films are relatively flat. And there is another advantage that various LB thin films can be easily deposited on Al thin films. From the reasons, Al thin films were used to excite SPs in this study. Al thin films of about 20 nm thick were vacuum-evaporated on substrates of microscopic cover glasses, and cadmium arachidate (C20: $CH_3(CH_2)_{18}$ COO 1/2Cd) Langmuir-Blodgett (LB) films were deposited on the Al thin films by means of a vertical dipping LB method [16]. In this study, monolayer, three and five layers of the C20 LB films were deposited on the Al thin films. The LB films on the opposite side to the evaporated Al films were swept off. The C20 LB films are transparent and have no absorption bands in visible light.

2.3 A system to detect evanescent fields

Fig. 4 shows the sample configuration in order to detect evanescent fields [14, 15]. A 45°-right angle prism of BK-7 glass was used in this measurement. The samples, that is, Al thin films and C20 LB films on the Al films were prepared as the same ones in the ATR method. The experimental system to detect the evanescent fields that was constructed in our research group is shown in Fig. 5 [14,15]. The prism was set on a rotating stage because SPs can be excited as a function of the incident angle of the laser beam similar to the ATR method. And the reflected laser beam was measured with the photodiode (PD).

In this study, a platinum (Pt) needle probe covered with cadmium sulfide (CdS) was used to detect the evanescent fields and the needle tip size was about 10μm. And the luminescent light at 652 nm was observed from CdS excited by the evanescent fields due to SP excitations in the ATR configuration using the incident Ar^+ laser beam at 488 nm. The luminescent light at 652 nm was collected by a lens of a microscope, and guided into an optical fiber. Intensities of the luminescent light were detected using a photon counting unit.

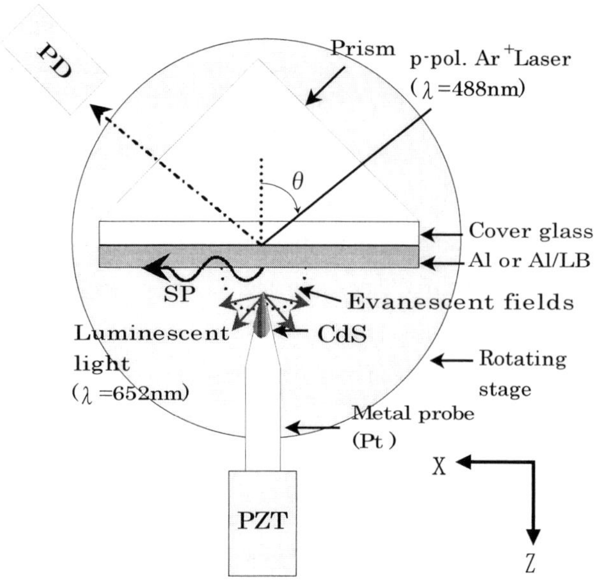

Fig. 4 The needle probe tip and the prism.

The needle probe tip was attached on a top of a piezo tube with a position sensor (Physik Instrumente, Germany, P-841), which was fixed with a holder on the z stage mounted on the x one. The rough approach of the probe to within piezo range was performed with two micrometers in the z direction. One micrometer was used for coarse positioning up to 10 μm and another was for more approach to within the distance less than 1 μm from the sample surface. The maximum displacement of the piezo translator is 15 μm and the position can be controlled with a linearity of about 0.2% by applying voltages in the range of 0 to 100V to the piezo translator, using a feedback control. As the voltage source a piezo driver (PI E-610) was used and the output voltage was controlled by an analogue signal through a 12-bit D/A converter from a computer. The position of the probe can be moved with accuracy of about 7nm in this measurement.

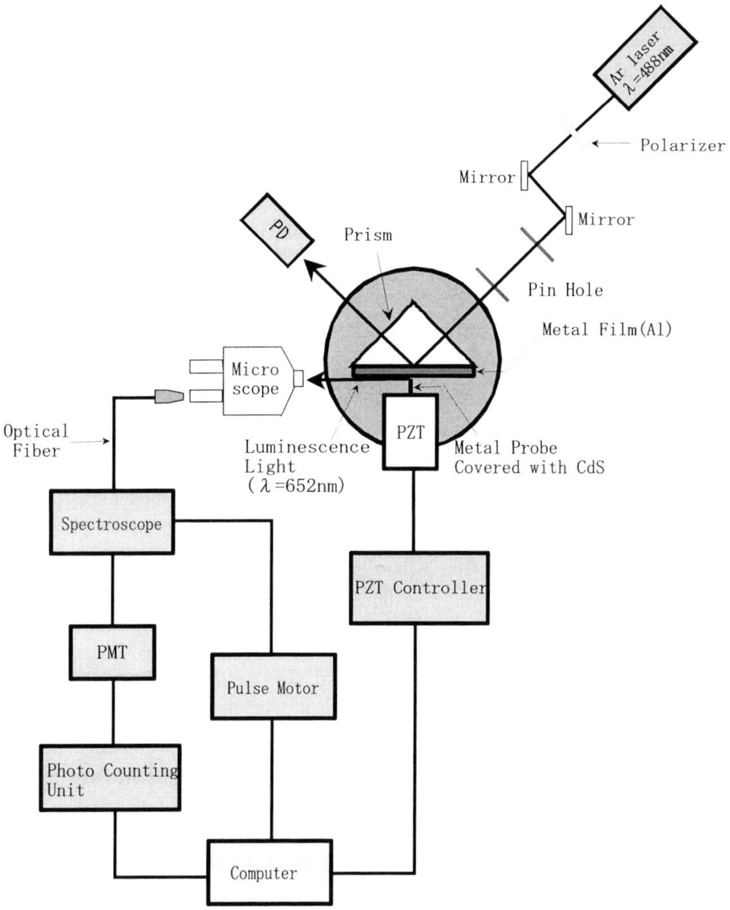

Fig. 5 A system to detect evanescent fields.

3. RESULTS AND DISCUSSION

3.1 The ATR properties

ATR properties for the Al thin film and the Al/C20 LB thin films were shown as closed circles in Fig.6. The minimum reflectivities were observed at each resonant angle, where SPs were resonantly excited at the surfaces of the Al thin film and the Al/C20 LB thin films by the laser beam at 488 nm. Solid curves in Fig. 6 show the theoretical reflectivities that were calculated for the multilayers using a transfer-matrix method [1,13].

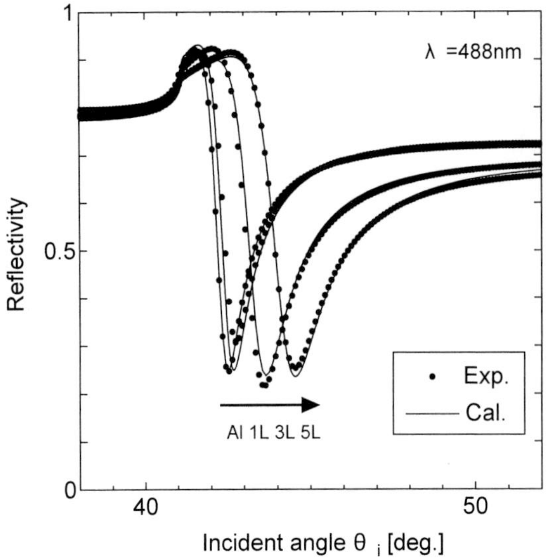

Fig. 6 ATR properties for the Al thin film and the Al/C20 LB thin films with monolayer, three and five layers.

The complex dielectric constants and the thicknesses of the Al thin film and the C20 LB thin films were decided from the curve fittings. The constants and the thicknesses of the Al thin film were $\varepsilon_{Al} = -25.22 + i6.62$ and 25.5 nm, respectively. The Al thin film had a native oxide, Al_2O_3, and the obtained dielectric constant and the thickness for the oxide layer were $\varepsilon_{Al\ oxide}$ = 3.1+ i0 and 2 nm, respectively. Thicknesses of the C20 LB films were evaluated assuming the dielectric constant of C20 LB films to be 2.34 [17]. The thicknesses of the C20 LB films of monolayer, three and five layers were 1.6, 8.9 and 14.2nm, respectively.

3.2 Evanescent fields on the Al thin film and on the Al/C20 LB thin films

Fig. 7 shows the luminescent intensities at 652nm as a function of the distance, z, of the needle probe from the sample surfaces at each dip angle. The luminescent intensities decayed exponentially near the sample surfaces, and the decay lengths were about 230nm.

Fig. 8 shows the squared electric field intensities calculated as a function of the distance from the sample surfaces at each dip angle. Both the intensities of the fields at the sample surfaces and the decay lengths decrease as the number of the C20 LB monolayers increases.

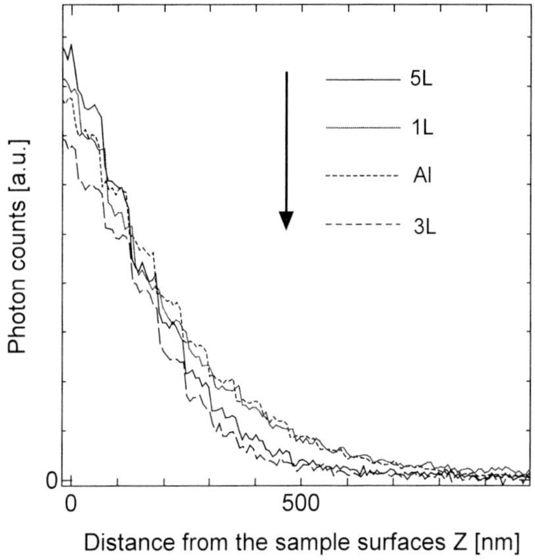

Fig. 7 Luminescent intensities as a function of the distance from the sample surfaces at each dip angle.

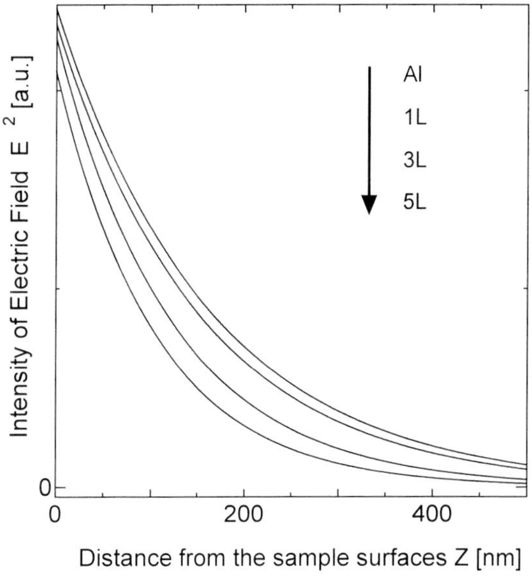

Fig. 8 Squared evanescent field intensities calculated as a function of the distance from the sample surfaces at each dip angle.

The calculated exponential decay lengths were 160, 150, 120 and 105nm for the Al thin film and the Al/C20 LB films with monolayer, three and five layers, respectively. The experimental results were slightly larger than the calculated ones. It was thought that the difference was caused by the large size of the probe tip. The results showed that evanescent fields on the Al thin films and the Al/C20 thin films could be detected by means of our measuring system.

4. CONCLUSION

Evanescent fields generating on Al thin films and Al/C20 LB films have been investigated using a system constructed in our research group. The evanescent fields due to SPs resonantly excited in the ATR method were detected as luminescent light from a Pt needle probe covered with CdS. The experimental decay lengths were about 230nm that were slightly lager than the calculated ones. It was concluded that evanescent fields due to SP excitations in the Kretschmann configuration could be detected using this system.

5. ACKNOWLEDGEMENTS

This work was partly supported by Grand-in-Aid for Scientific Research from the Ministry of Education, Science and Culture of Japan.

6. REFERENCES

1. V.M. Agranovich and D.L. Mills(eds.), Surface Polaritons, North-Holland, Amsterdam,1982.
2. F. Kaneko, S. Honda, T. Fukami, K. Kato, T. Wakamatsu, K. Shinbo and F. Kaneko, Thin Solid Films, 284 -285 (1996) 417.
3. K. Kato, Y. Aoki, K. Ohashi, K. Shinbo and F. Kaneko, Jpn. J. Appl. Phys., 35 (1996) 5466.
4. T. Wakamatsu, K. Saito, Y. Sakakibara and H. Yokoyama, Jpn. J. Appl. Phys., 36 (1997) 155.
5. Y. Aoki, K. Kato, K. Shinbo, F. Kaneko and T. Wakamatsu, Thin Solid Films, 327 - 329 (1998) 360.

6. A. Baba, F. Kaneko, K. Shinbo, K. Kato, S. Kobayashi and T. Wakamatsu, Jpn. J. Appl. Phys., 37 (1998) 2581.

7. J. Hwang, L. K. Tamm, C. Böhm, T.S. Ramalingam, E. Betzig and M. Edidin, Science 270 (1995) 610.

8. E. Betzig, J. K. Trautmann, R. Wolfe, E.M. Gyorgy, P. L. Finn, M. H. Kryder and C. H. Chang, Appl. Phys. Lett., 61 (1992) 142.

9. S. Jiang, J. Ichihashi, H. Monobe, M. Fujihara and M. Ohtsu, Opt. Commun, 106 (1994) 173.

10. Y. Inoue and S. Kawata, Opt. Lett. 19, (1994) 159.

11. K. Nakajima, R. Miccheletto, K. Mitsui, T. Isoshima, M. Hara, T. Wada, H. Sasabe and W. Knoll , Jpn. J. Appl. Phys., 38 (1999) 3949.

12. T. Okamoto and I. Yamaguchi, Jpn. J. Appl. Phys., 37 (1998) 2581.

13. M. Born and E. Wolf, Principles of Optics, Pergamon, Oxford, 1974.

14. T. Wakamatsu, T. Nakano, K. Shinbo, K. Kato and F. Kaneko, Rev. Sci. Ins., 70 (1999) 3962.

15. T. Nakano, T. Wakamatsu, H. Kobayashi, F. Kaneko, K. Shinbo, K. Kato and T. Kawakami, Mol. Cryst. Liq. Crsst., 349 (2000) 235.

16. M.C. Petty, Langmuir-Blodgett Films, Cambridge University Press, 1996.

17. I. Pockrand, J.D. Swalen, J.G. Gordon and M.R. Philpott, Surface Science, 74 (1978) 237.

Novel Methods to Study Interfacial Layers
D. Möbius and R. Miller (Editors)

SIMULTANEOUS OBSERVATION OF THE ELECTROPOLYMERISATION PROCESS OF CONDUCTING POLYMERS BY SURFACE PLASMON RESONANCE SPECTROSCOPY, SURFACE PLASMON ENHANCED LIGHT SCATTERING AND CYCLIC VOLTAMMETRY

Akira Baba[1], Rigoberto C. Advincula[2] and Wolfgang Knoll[1]

[1]Max-Planck-Institut für Polymerforschung, Ackermannweg 10, D-55128, Mainz, Germany

[2]Department of Chemistry, University of Alabama at Birmingham, Birmingham, AL 35294-1240, USA

In this study, we report the combined use of surface plasmon resonance spectroscopy (SPR) and cyclic voltammetry to investigate the electropolymerization process of aniline onto a gold electrode *in situ*. Electropolymerization of aniline was achieved by applying a cycling potential known from cyclic voltammetry monitored by SPR simultaneously. The potential cycling resulted in an oscillation of the SPR kinetic reflectivity curve. The time differential SPR kinetic reflectivity curve was correlated with the cyclic voltammogram at higher potential. At lower potential, a large difference between the SPR kinetic reflectivity curve and the cyclic voltammogram was observed. It is shown that this difference is caused by scattered light enhanced by the SPR and is due to the change of the optical refractive index at the polyaniline/electrolyte interface. The technique promises to be an important tool in determining the mechanism for the electropolymerization processes of conducting polymers.

1. INTRODUCTION

Conducting polymers such as polyaniline have been attracting much interest owing to their wide variety of possible application, e.g., as battery electrodes [1], as electrochromic displays [2], light-emitting devices [3,4], etc. The film growth process of polyaniline by potential cycling involves oxidation of the monomers and a redox process of the deposited polymer. For all of the aforementioned applications, understanding the electropolymerization mechanism of conducting polymers is of considerable importance.

Surface plasmon resonance (SPR) spectroscopy has been shown to be a technique of high sensitivity for characterizing ultrathin films at the nanometer thickness scale [5]. Recently, the combination of the SPR with electrochemical techniques for the simultaneous characterization and manipulation at an electrode/electrolyte interfaces has been investigated [6, 7]. In addition, roughness of the metal surfaces could be evaluated by light scattering from metal surfaces, which is enhanced by SPR [8].

In this study, we have attempted to demonstrate by *in situ* investigations using the combination of SPR spectroscopy and cyclic voltammetry the electropolymerization process of aniline onto a gold electrode. Electropolymerization of aniline was initiated by applying a cycling potential as in cyclic voltammetry and was monitored by SPR simultaneously. By *in situ* investigations of the light enhanced by SPR and scattered off the sample interface, we were able to investigate the morphology at the polyaniline/electrolyte interface and simultaneously correlate it with the SPR kinetic reflectivity curve and the cyclic voltammogram.

2. EXPERIMENTAL DETAILS

2.1 Plasmon surface polaritons at metal/electrolyte interface

Figure 1(a) shows the Kretschmann configuration [9] for the excitation of plasmon surface polaritons (surface plasmons for short) [10] in the attenuated total reflection (ATR) mode. When a p-polarized laser beam is irradiated at the (internal) incident angle θ_i from the prism of a refractive index n_p above θ_c, a strong nonradiative electromagnetic wave, i.e. a surface plasmon is excited at the resonant angle which propagates at the metal /electrolyte interface.

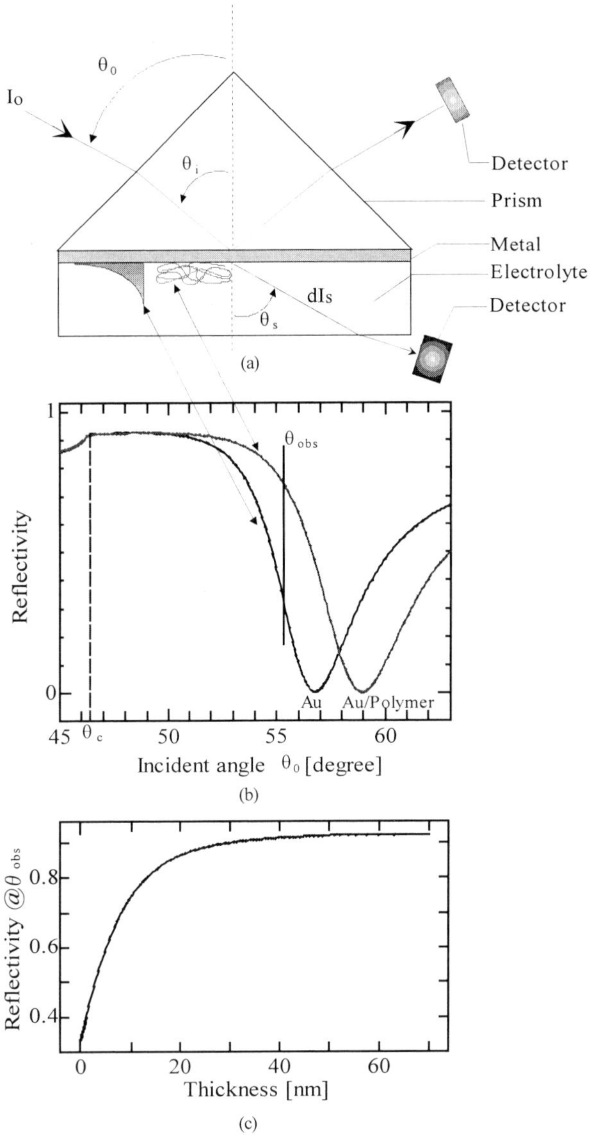

Fig. 1 (a) ATR setup for the excitation of surface plasmons in the Kretschmann geometry: a thin metal film is evaporated onto the base of a glass prism and acts as a resonator driven by the photon field.
(b) Calculated reflectivity for Au(50nm)/electrolyte(ε = 1.75 + i0) and Au(50nm)/deposited film (10nm, ε = 2.25 + i0)/electrolyte systems. Kinetic information on the layer formation can be obtained by monitoring the reflected intensity at a fixed angle, θ_{obs} as shown in (c); (c) Calculated reflectivity change as a function of the thickness of the deposited film on the Au film at the fixed angle θ_{obs} given in (b).

The coupling angle is given by the energy- and momentum-matching condition between photons and surface plasmons:

$$k_{sp} = k_{ph} = n_p k_0 \sin\theta_i, \tag{1}$$

where, $k_0 = \omega/c$, $\omega = 2\pi c/\lambda_0$, ω the angular frequency, c the light velocity in vacuum, and λ_0 the wavelength in vacuum. A detailed theory for the surface plasmons in the ATR measurements has been given previously [11]. As the energy of the incident light is transferred to the surface plasmon, a large reduction the reflectivity in the ATR scan is observed at the resonant angle as shown in Fig. 1(b). The resonance character of this excitation gives rise to an enhancement of the electric field at the interface by more than an order of magnitude, which is the origin of the remarkable sensitivity enhancements, e.g., in scattered light [12], fluorescence [13,14] or Raman spectroscopies [15,16] when working with surface plasmon light.

If the formation of such a thin coating is observed *in situ*, e.g. during an electropolymerization process of conducting polymers at the metal electrode/electrolyte interface, kinetic information about the deposition of the film, or the diffusional transport of mass adjacent to the interface by cycling potentials can be obtained by monitoring the reflected intensity at a fixed angle of observation θ_{obs} (see Fig. 1(c)) [7].

2.2 Surface plasmon enhanced light scattering at metal/electrolyte interface

The experimental setup for monitoring surface plasmon enhanced scattered light is also shown in Fig. 1 (a) [17]. The intensity ratio of the surface plasmon enhanced scattered light dI_s at θ_s per solid angle $d\Omega$ can be given by

$$\frac{dI_s}{I_0\, d\Omega} = \left(\frac{\omega}{c}\right)^4 \frac{4\sqrt{\varepsilon_p}}{\cos\theta_i} |F|^2 \, |s(\Delta k)|^2, \tag{2}$$

where

$$\Delta k = \frac{\omega}{c}\left(\sqrt{\varepsilon_p}\,\sin\theta_i - \sqrt{\varepsilon_e}\,\sin\theta_s\right), \tag{3}$$

with ε_p and ε_e the dielectric constant of the prism and the electrolyte solution, respectively. F is a function of Fresnel's reflection and transmission coefficients and also depends on θ_i and θ_s. If we assume a Gaussian distribution as the correlation function that describes the inhomogeneity at the interface, the roughness function $|s(\Delta k)|$ can be obtained from the Fourier transform of a Gaussian distribution and is given by,

$$|s(\Delta k)|^2 = \frac{\sigma^2 \delta^2}{4\pi} \exp\left[-\frac{\sigma^2 (\Delta k)^2}{4} \right], \tag{4}$$

with σ the transverse correlation length and δ the surface corrugation depth. Eq. (2) can be extended for polymer coating inhomogeneities surfaces [18]. In our experiments, $dI_s/I_0 d\Omega$ during the electropolymerization process of a conducting polymer at the metal electrode/electrolyte interface is measured as a function of incident angle θ_0 at a fixed angle $\theta_s = 0$, or at a fixed angle of observation θ_{obs} and $\theta_s = 0$ for *in situ* investigation.

Figure 2(a) and (b) show calculated examples of the surface plasmon enhanced scattered light from the polymer/electrolyte interface assuming different roughness parameters as a function of incident angle θ_0 at a fixed angle $\theta_s = 0$. For comparison the reflected intensity, i.e. the usual ATR scan, is also simulated for this sample. The assumed laser wavelength was $\lambda = 633$nm. In this spectral range the dielectric function of Au allows for a field enhancement of a factor 20. The evanescent character of this enhanced optical field gives rise to an exponential decay into the polymer/electrolyte solution with a characteristic length of ca. 200 nm. This means that the measured reflectivity and scattered light also depend on the change of the electrolyte solution, i.e. ion concentration and mass transfer etc. in this region.

Figure 2(b) shows that the peak intensity of the scattered light is at a slightly lower incident angle than the minimum in reflectivity. The reason of this is related to the resonance character of the surface plasmon excitation. Since we are observing in reflection a coherent superposition of a partial wave directly reflected from the metal/prism interface with (a fraction of) the surface mode re-radiated via the, the minimum in the total reflection then is the destructive interference of two partial waves differing in phase by 180° which is reached just above the angle of maximum surface plasmon intensity. Therefore, we measure at a slightly lower incident angle than the peak intensity of the scattered light for the on-line monitoring measurement, because it is thought that the reflectivity and the scattered light becomes high and low, respectively at observed angle θ_{obs} if the deposited film becomes thicker and the inhomogeineity at the polymer/electrolyte interface does not change with the film thickness, i.e. in a very simple case (as shown in the calculation, see Fig. 1(c)).

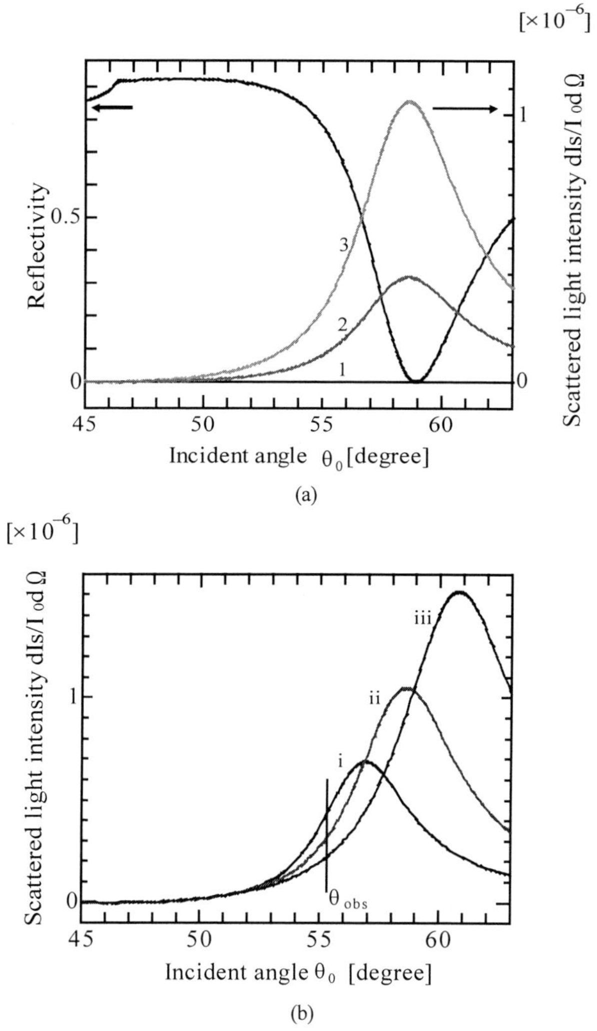

Fig. 2 (a) Calculated reflectivity and scattered light intensity $dI_s/I_0d\Omega$ from the deposited film(10nm, $\varepsilon = 2.25 + i0$)/electrolyte($\varepsilon = 1.75 + i0$) solution interface for several roughness parameters (at $\theta_s = 0°$), σ(nm)/δ(nm): 0 nm/0 nm (Curve 1), 50 nm/1 nm (Curve 2) and 100 nm/1 nm (curve 3) (b) Calculated light intensity scattered off the deposited film/electrolyte solution interface for several thicknesses of the deposited film (i:2 nm, ii:10 nm and iii:20nm). Roughness parameters [σ(nm)/δ(nm)] are 100 nm/1 nm.

2.3 Sample preparation

Figure 3 shows an attenuated total reflection (ATR) setup for the excitation of surface plasmons in the Kretschmann configuration combined with an electrochemical cell.

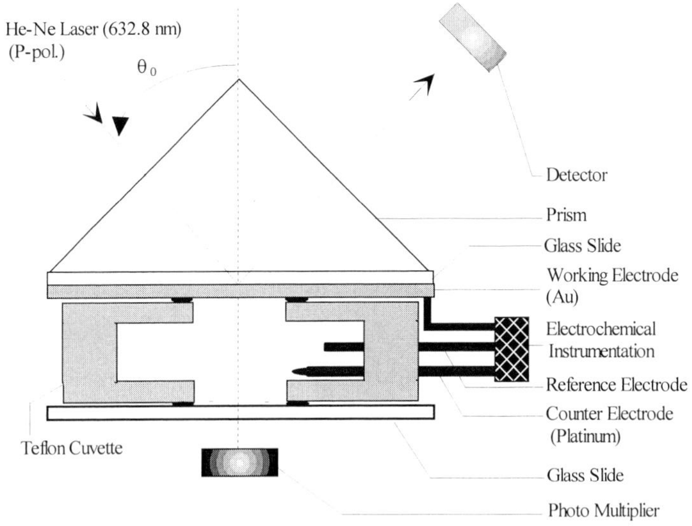

Fig. 3 Surface plasmon resonance (SPR) spectroscopy. A schematic in the Kretschmann configuration combined with an electrochemical cell.

The Au film ca. 50 nm thick evaporated onto BK7 glass slides with a 1-2 nm thick chromium adhesion layer was used as the working electrode. The Au/glass substrates were clamped against the Teflon cuvette with an o-ring providing a liquid-tight seal. The cuvette was then mounted for investigation by SPR. Details of the setup have been described previously [19]. Surface plasmons are excited at the metal-dielectric interface, upon total internal reflection of a He-Ne laser beam (λ = 632.8 nm). The electropolymerization processes on the gold were detected in situ by monitoring reflectivity changes as a function of time at a fixed incident angle θ. This angle of incidence light was selected such that excitation was at a slightly lower angle than the resonance minimum position. Electrochemical experiments were performed in a conventional 3-electrode cell with an Au/glass working electrode, platinum wire counter electrode and an Ag/AgCl (3 M NaCl) reference electrode. A potentiostat (Princeton Applied Research 263A, EG&G) was used for the cyclic voltammetric experiments.

Electropolymerization of polyaniline at a gold surface was done by applying potential cycles at 20 mV/s from –0.2 V to 1.0 V.

3. MEASUREMENT CONSIDERATIONS

3.1 Cyclic voltammograms during the electropolymerization

Figure 4 shows the cyclic voltammograms during the electropolymerization of 0.1 M aniline in 0.5 M HCl solution using repeated potential cycling between –0.2 and 1.0 V at a scan rate of 20 mVs^{-1}.

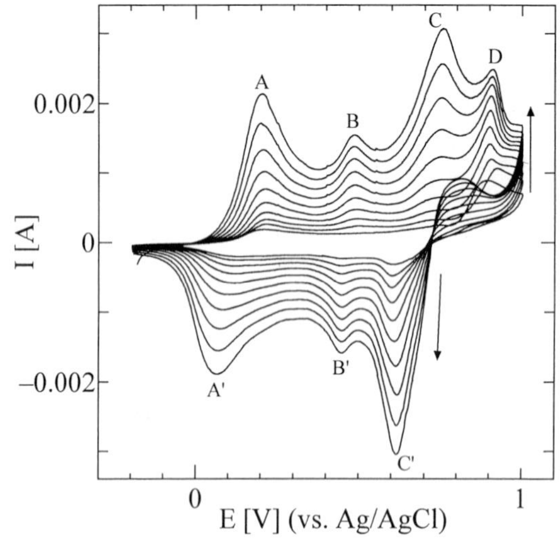

Fig. 4 Cyclic voltammograms during the electropolymerization of 0.1 M aniline in 0.5 M HCl solution using repeated potential cycling between –0.2 and 1.0 V at a scan rate of 20 mVs^{-1}

It was observed that the peak current increased with the number of potential cycles. This indicates that an electroactive polyaniline film was deposited onto the gold surface. The anodic peak at 0.91 V (peak D) has no corresponding cathodic peak on the cyclic voltammograms, which means that this anodic peak refers to the oxidation of the aniline monomer to produce polyaniline. Another three anodic peaks were observed at 0.2 V (peak A), 0.49 V (peak B) and 0.76 V (peak C) respectively, and these peaks also increased with the number of potential cycle. This indicates oxidation of the deposited polyaniline on the gold electrode [20,21]. The three cathodic dips (dips A', B' and C') are due to reduction of polyaniline as also shown in the figure.

3.2 Simultaneous observation of cyclic voltammogram and surface plasmon resonance

Figure 5 shows the simultaneous observation of a SPR kinetic reflectivity curve and a cyclic voltammogram during the electropolymerization from the 1st to 5th cycle of 0.1 M aniline in 0.5 M HCl solution at a scan rate of 20 mVs^{-1}.

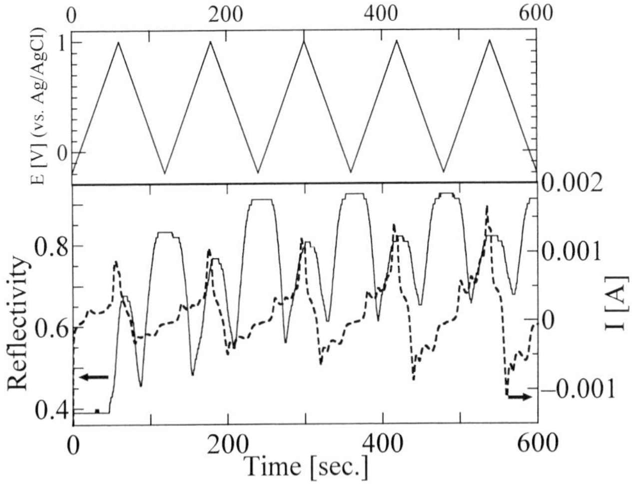

Fig. 5 Simultaneous observation of a SPR kinetic curve and a cyclic voltammmogram during the electropolymerization of 0.1 M aniline in 0.5 M HCl solution at a scan rate of 20 mVs^{-1}. Top: potential ramp vs. time; bottom: a SPR kinetic curve as a function of potential cycling (solid line) and current vs. time (dotted) Cyclic voltammograms

As shown in this figure, the potential cycling resulted in a sensitive oscillation of the SPR kinetic reflectivity curve, but also in a successive overall increase. The latter indicates the shift of R(θ) to higher angles [5] owing to the growth of the polyaniline film on the Au electrode during the continuous potential cycling. In the anodic scan of 1st cycle, the SPR kinetic reflectivity curve began to increase abruptly at around 0.72 V. This value was close to the onset of oxidation for the aniline monomer obtained from the cyclic voltammogram.

3.3 Time differential SPR kinetic curve and cyclic voltammogram

Iwasaki et al. reported that by using the simple system of time differential SPR data (Δθ/Δt), a Faradaic current behavior can be observed based on the same potential dependence behavior with the voltammetric currents [22]. Figure 6 shows the time differential SPR kinetic reflectivity curve (dR/dt) as a function of potential cycling between –0.2 and 1.0 V at a scan rate of 20 mVs^{-1} together with the cyclic voltammogram of 1st cycle, 2nd cycle and 5th cycle.

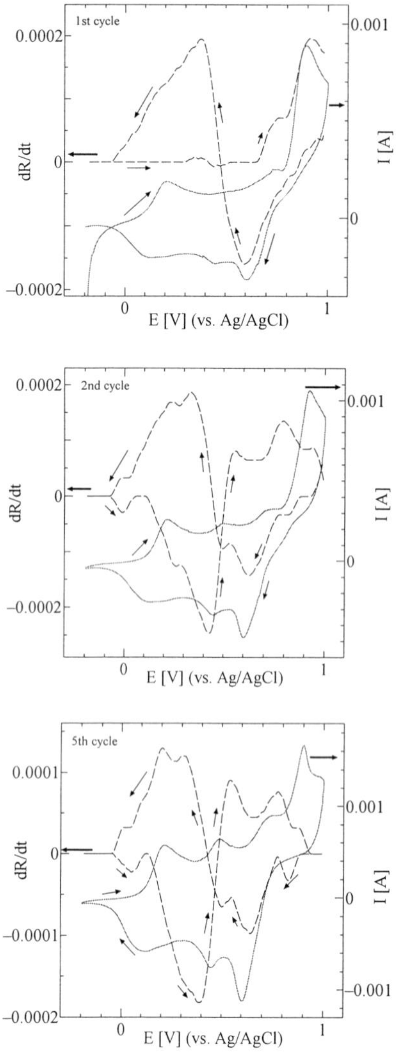

Fig. 6 The time differential SPR kinetic curve (dR/dt) as a function of potential cycling between –0.2 and 1.0 V at a scan rate of 20 mVs^{-1} and the cyclic voltammogram of 1st cycle (a), 2nd cycle (b) and 5th cycle (c).

The reason for the small peak at 0.2 V in the cyclic voltammogram is not clear yet. In the first scan cycle, the monomer oxidation peak at 0.9 V and reduction peak of polyaniline at 0.6 V in the time differential SPR kinetic reflectivity curve is almost coincident with the cyclic voltammogram. It was observed that time differential SPR kinetic reflectivity curves in the 2nd and 5th cycles have time delays as compared with oxidation peaks of the polyaniline film at 0.49 and 0.76 V. On the other hand, the reduction peak of polyaniline at 0.45 and 0.6 V have time delays as compared with time differential SPR kinetic reflectivity curves in the 2nd and 5th cycles. It is thought that the time differential SPR kinetic reflectivity curve is dependent upon the ion moving away from the electrode surface within the region of the penetrating field of surface plasmons [7].

3.4 Simultaneous observation of the SPR kinetic curve and scattered light kinetic curve

However, there is no relation below ca. 0.4 V between the cyclic voltammograms and time differential SPR kinetic reflectivity curves in all cycles. To further probe this behavior, we measured the scattered light intensity enhanced by SPR from the polyaniline/electrolyte interface during the electropolymerization. Figure 7 shows the angular dependence of the measured reflectivity and scattered light intensity from the substrate/gold/electrolyte system before the electropolymerization.

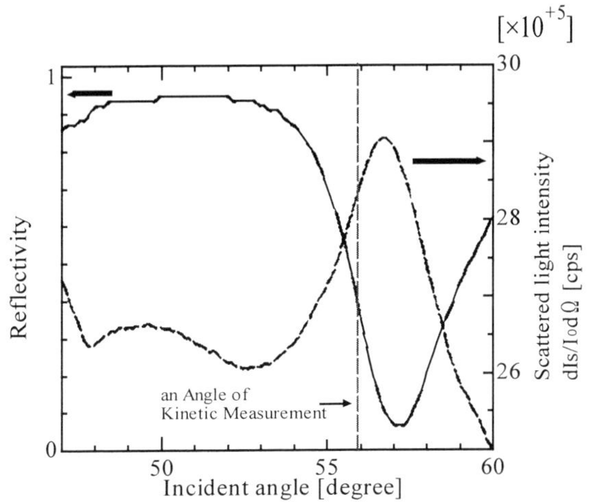

Fig. 7 Angular SPR curves and scattered light intensity of substrate/gold/solution system, i.e., before the electropolymerization.

Under these conditions, the electrolyte has no absorption at λ = 632.8 nm. Therefore, it is concluded that this light detected from the backside of the prism is light scattered at the substrate/gold/electrolyte system enhanced by SPR. The following kinetic reflectivity measurements and the kinetic scattered light intensity recordings during electropolymerization were taken at a slightly lower angle than the resonance minimum as shown in this figure. This figure also implies that the scattered light intensity in the kinetic measurement should become weaker if the film thickness becomes thicker during the electropolymerization.

Figure 8 shows the simultaneous observation of the SPR kinetic reflectivity curve and the kinetic scattered light intensity curve (Is) during the electropolymerization of 0.1 M aniline in 0.5 M HCl solution at a scan rate of 20 mVs^{-1}. We also measured simultaneously the SPR kinetic reflectivity curve and photoluminescent property using a low-cut filter (λ = 665 nm) during electropolymerization. However, the intensity of the photoluminescence was considerably weaker as compared to the scattered light intensity.

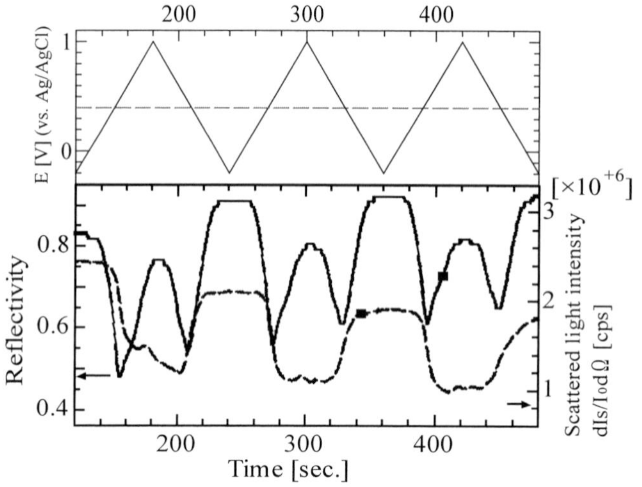

Fig. 8 Simultaneous observation of a SPR kinetic curve and a scattered light (Is) during the electropolymerization of 0.1 M aniline in 0.5 M HCl solution at a scan rate of 20 mVs^{-1}. Top: potential ramp vs. time; bottom: a SPR kinetic curve as a function of potential cycling (solid line) and scattered light (dotted)

The photoluminescence of polyaniline film in air after 10 potential cycles was very weak. Therefore, it was thought that the detected light as shown in Fig. 6 was mainly due to scattered light.

As shown in Fig. 8, an obvious increase of the scattered light intensity was observed below ca. 0.4 V at every cycle, and the SPR kinetic reflectivity curve also increased in this region at every cycle. Two possible explanations can be considered.

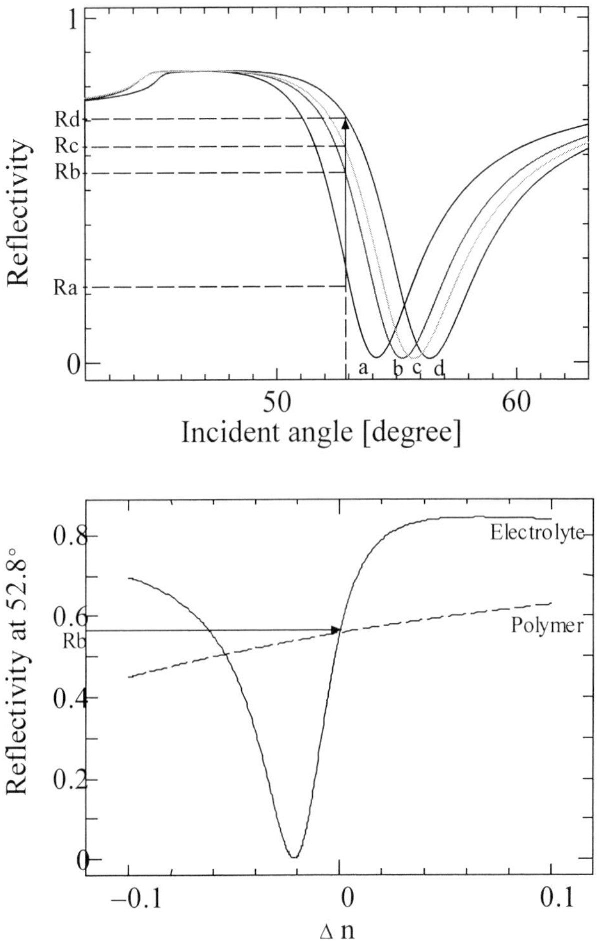

Fig. 9 (a) Theoretical surface plasmon resonance angular reflectivity curves, Curve a: Au/electrolyte 1, Curve b: Au/polymer 1/electrolyte 1, Curve c: Au/polumer 2/electrolyte 1, Curve d: Au/polymer 1 electrolyte 2 (electrolyte 1: $n + ik = 1.3 + i0$, electrolyte 2: $n + ik = 1.31 + i0$, polymer 1: $n + ik = 1.5 + i0$, $d = 5$ nm, polymer 2: $n + ik = 1.6 + i0$, $d = 5$ nm); (b) Calculated reflectivity changes as a function of the refractive index n of the electrolyte and the polymer, respectively at the fixed angle θ_{obs} and the constant thickness given in the Fig. 9(a) (curve b).

Figure 9(a) and (b) show examples of the theoretical surface plasmon resonance angular reflectivity curves for several cases of the Au/polymer/electrolyte system. One is the change of the optical refractive index of the deposited polyaniline films, which is caused by a redox process in this region. Namely, some defects in the polyaniline film increased after the ion de-doping at 0.45 V and decreased after the ion doping at 0.2 V ($R_b \rightarrow R_c$ as shown in Fig. 9(a)). The second possibility is the change of the optical refractive index of the electrolyte. Namely, the high ion concentration in the region of penetration depth of the surface plasmon, i.e., ca. 200 nm from the surface into the electrolyte as shown in Fig. 9(a) ($R_b \rightarrow R_d$). In this case, it was thought that the scattered light was enhanced by the contrast between the polyaniline/electrolyte interface due to the high ion concentration. Based on these factors, we attribute the increased SPR kinetic reflectivity curves below ca. 0.4 V in Fig. 3 and Fig. 6 were due to the increased scattered light. Thus, large difference between the cyclic voltammograms and SPR kinetic reflectivity curves in all cycles below ca. 0.4 V was observed.

4. CONCLUSIONS

Our studies have shown the possibilities of doing a combined *in situ* surface plasmon resonance and cyclic voltammetry during electropolymerization of conducting polymer. The potential cycling resulted in the oscillation of the SPR kinetic reflectivity curve. The SPR kinetic reflectivity data during electropolymerization can be divided into two regions which sensitively depend upon the electrochemical property and the morphology at the polyaniline/electrolyte interface. The time differential SPR kinetic reflectivity curve was correlated with the cyclic voltammogram. The scattered light enhanced by the surface plasmon depends upon the change of the optical refractive index at the polyaniline/electrolyte interface during electropolymerization. These results indicate that the simultaneous observation of the SPR kinetic reflectivity curve, the measurement of the scattered light, and the recording of cyclic voltammograms will allow for the elucidation of the electrochemical property and surface morphology during the electropolymerization. Further investigations will be made by our groups on various electropolymerizable monomers.

5. ACKNOWLEDGMENT

Akira Baba acknowledges the Alexander von Humboldt Stiftung for a post-doctoral fellowship.

6. REFERENCES

1. A.G. MacDiarmid, S.L. Mu, N.L.D. Somasiri and W. Mu, Mol. Cryst. Liq. Cryst., 121 (1985) 187.
2. T. Kobayashi, H. Yoneyama and H. Tamura, J. Electroanal. Chem., 161 (1984) 419.
3. H.L. Wang, A.G. Macdiarmid, Y.Z. Wang, D.D. Gebler and A.J. Esptein, Synth. Met., 78 (1996) 33.
4. S. Karg, J.C. Scott, J.R. Salem and M. Angelopoulos, Synth. Met., 80 (1996) 111.
5. W. Knoll, Annu. Rev. Phys. Chem., 49 (1998) 569.
6. T.M. Chinowsky, S.B. Saban and S.S. Yee, Sens. Actuat., B 35-36 (1996) 37.
7. A.Badia, S. Arnold, V. Scheumann, M. Zizlsperger, J. Mack, G. Jung and W. Knoll, Sens. Actuat. B, 54 (1999) 145.
8. E. Fontana and R.H. Pantel, Phys. Rev. B, 37 (1988) 3164.
9. E. Kretschmann, Opt. Commun. 6 (1972) 185.
10. E. Burstein, W.P. Chen, Y.J. Chen and A. Hartstein, J. Vac. Sci. Technol. 11 (1974) 1004.
11. H. Raether, in: M.H. Francome, R.W. Hoffmann and G. Hass II (Eds.), Physics of Thin Films, vol. 9, Wiley, New York, 1977.
12. E. Kretschmann, Opt. Commun. 5 (1972) 331.
13. J.W. Atteidge, P.B. Daniels, J.K. Deacon, G.A. Robinson and G.P. Davidson, Biosensors Bioelectron. 6 (1991) 201.
14. T, Liebermann and W. Knoll, Coll. Surf. A, 171 (2000) 115.
15. S. Ushioda and Y. Sasaki, Phys. Rev., B27 (1983) 1401.
16. H. Knobloch, C. Duschl and W. Knoll, Chem. Phys., 91 (1989) 3810.

17. H. Raether: *Surface Polaritons*, ed. V.M. Agranovich and D.L. Milles (North-Holland, Amsterdam, 1982) Chap. 9.

18. Y. Aoki, K. Kato, K. Shinbo, F. Kaneko and T. Wakamatsu, IEICE Trans, Electron, E81-C (1998) 1098.

19. E.F. Aust, S. Ito, M. Sawondny and W. Knoll, Trends, Polym. Sci., 2 (1994) 9.

20. Y. Wei, Y. Sun and X. Tang, J. Phys. Chem., 93 (1989) 4878.

21. S. Mu, C. Chen and J. Wang, Synth. Met., 88 (1997) 249.

22. Y. Iwasaki, T. Horiuchi, M. Morita and O. Niwa, Sens.Actuat. B, 50 (1998) 145.

7. LIST OF SYMBOLS AND ABBREVIATIONS

θ_0	Incident angle of He-Ne laser beam
θ_I	Internal incident angle of He-Ne laser beam
θ_s	Angle of scattered light
$dIs/I_0 d\Omega$	Scattered light intensity
dR/dt	Time differential reflectivity
SPR	Surface plasmon resonance
ATR	Attenuated total reflection

Novel Methods to Study Interfacial Layers
D. Möbius and R. Miller (Editors)

SCATTERED LIGHT AND EMISSION FROM AG THIN FILM AND MEROCYANINE LANGMUIR-BLODGETT FILM ON AG THIN FILM DUE TO SURFACE PLASMON POLARITON EXCITATION

Kazunari Shinbo[a,*], Kazushige Honma[a], Mitsuru Terakado[a], Takayuki Nakano[a], Keizo Kato[a], Futao Kaneko[a] and Takashi Wakamatsu[b]

[a] Department of Electrical and Electronic Eng., Niigata University, Ikarashi 2-8050, Niigata, 950-2181, Japan
[b] Department of Electric Eng., Ibaraki National College of Technology, Hitachinaka, 312-8508, Japan

Contents

Keywords: surface plasmon polariton, attenuated total reflection, emission, merocyanine, scattered light

* Tel: +81-25-262-7543, Fax: +81-25-262-6741, E-mail: kshinbo@eng.niigta-u.ac.jp

Attenuated total reflection (ATR) properties, scattered light and emission due to surface plasmon polariton (SPP) excitation from Ag film and merocyanine (MC) LB film on Ag film were investigated. A Kretschmann configuration of the ATR measurement, prism/Ag film/air or prism/Ag film/MC LB film/air structures were used for the SPP excitation. The thicknesses and the dielectric constants of the Ag and the MC LB films were obtained from the ATR measurement. Scattered lights due to the SPP excitation were also observed from the Ag thin film and the MC LB film in the ATR measurement. Furthermore, emission was observed through the prism when a laser beam was irradiated to the sample from the air. The angles of the emission peaks almost corresponded to the resonant angles of the ATR curves. It was considered that the emission was due to the SPP excitation and depended on the surface roughness of the films. The intensities of the emission from the Ag film strongly depended on the incident angle of the laser beam.

1. Introduction

Recently, many studies of electrical and optical devices of organic thin films have been reported [1, 2]. For development of organic devices with high efficiency, it is quite important to evaluate structure and optical functions of the films. The surface plasmon resonance spectroscopy, that is, the attenuated total reflection (ATR) measurement utilizing surface plasmon polariton (SPP) excitations is one of very useful methods to evaluate structure and dielectric properties of ultrathin films of several tens nano-meter thick [3-8]. If sample films containing surface roughnesses and/or inhomogeneities in the Kretschmann configuration of the ATR measurement, scattered light from the sample surface is emitted because resonant conditions of the SPP excitation are disturbed [3]. From analysis of the angular distribution of the scattered light, surface roughnesses can been evaluated for metal thin film [9-12] and LB ultrathin films on the metal films [13-15]. Furthermore, when a laser beam is irradiated from the air, emission through the prism is observed. Although the emissions are also considered to be due to SPP excitations [16, 17], detailed mechanism is not clarified yet. So, it is important to investigate relations between structure of the films and emission.

In this study, properties of ATR, scattered light and emission from Ag sputtered film due to SPP excitation in the Kretschmann configuration were observed and the influence of the surface roughness to the scattered light and emission were investigated. Furthermore, the properties of ATR, scattered light and emission from Merocyanine (MC) LB film on Ag film were also observed.

2. EXPERIMENTAL DETAILS

2.1 Preparations of the samples

The Ag thin films were fabricated on glass substrates using RF sputtering method. In order to investigate the contribution of the surface roughness to the scattered lights and the emissions, two kinds of Ag sputtered films with same thickness and with different surface structures were prepared by controlling the applied RF power and deposition time [18]. The applied RF power and the sputtering times for the Ag film fabrications were 20 W and 40 s for sample A, and 40 W and 20 s for sample B, respectively.

The samples of MC LB thin films on Ag-evaporated films were also fabricated to investigate the influence of the organic dye film to the scattered light and the emission. Fig. 1 shows the structure of the MC dye used in this work. MC is one of photosensitizing dyes and was purchased from Japanese Research Institute for Photosensitizing Dyes Co., Ltd., Japan. LB films containing MC and arachidic acid (C20) mixed to molar ratio of 1:2 were deposited on vacuum evaporated Ag film covered with 2 layers of cadmium arachidate (C20Cd) LB film. The C20Cd LB films were deposited for the easy deposition of the MC LB film.

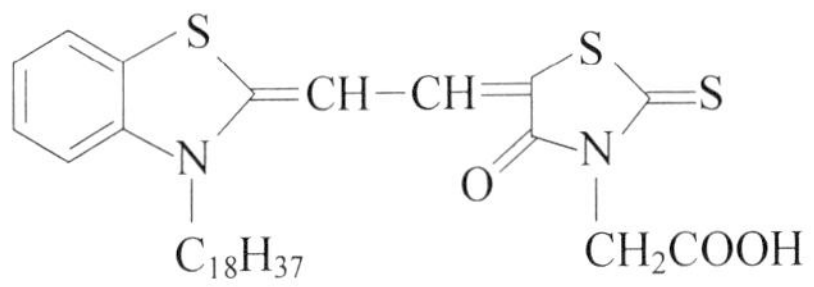

Fig. 1 Chemical structure of merocyanine (MC) dye used in this work.

2.2 Measuring system of the ATR, scattered light and emission properties

Fig. 2 (a) shows the sample configuration for the measurements of ATR and scattered light due to SPP excitation. The glass substrates with Ag thin films or the MC LB films on Ag thin films were brought into optical contact with half-cylindrical glass prisms using an index matching fluid. These samples have the Kretschmann configuration [3]. The experimental system for the measurements of ATR and scattered light is shown in Fig. 2 (b). He-Ne lasers with wavelengths of 632.8 nm and 594.1 nm and Ar^+ laser of 488 nm were used as the incident radiation. The samples shown in Fig. 2 (a) were mounted on a computer-controlled goniostage and the incident angle θ_{i1} and θ_{i2} were changed by rotating the goniostage.

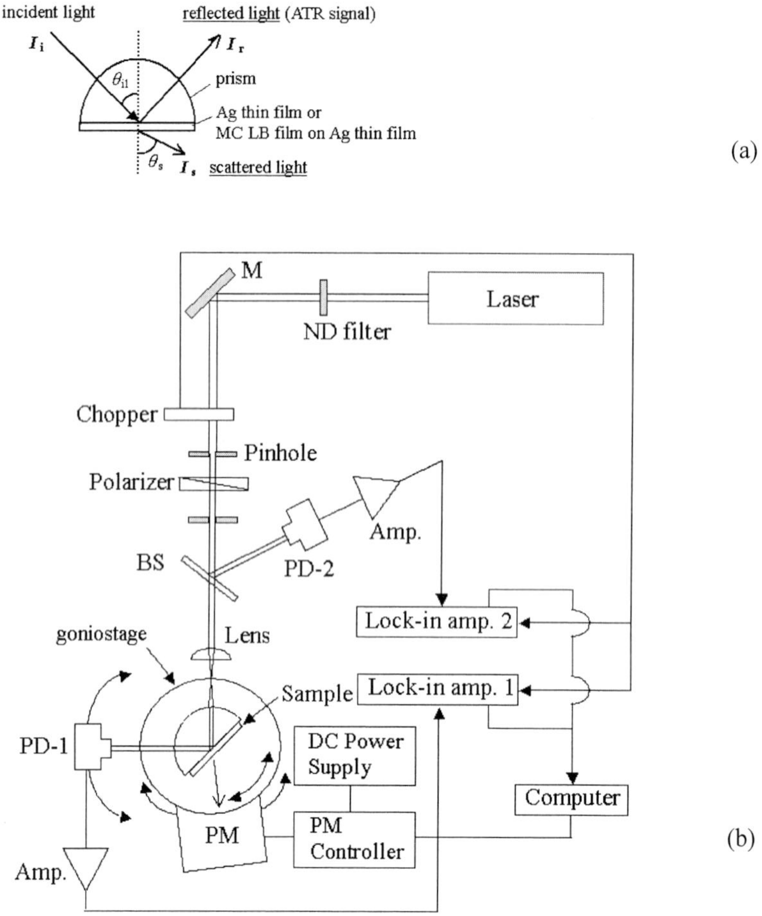

Fig. 2 Sample configurations for the measurements of the ATR and scattered light (a) and the experimental setup (b).

P-polarized laser beams were directed onto the back surface of the Ag thin film through the half-cylindrical prism for the ATR and scattered light measurements. The intensity I_r of the reflected light of the incident laser beams was observed as a function of the incident angle θ_{i1} of the laser beam using photodiode (PD1) at intervals of 0.1°. The intensity I_i of the incident laser beam was also monitored using a photodiode (PD2). The reflectivity, that is, the ATR value was obtained from the ratio of I_r to I_i.

The scattered light I_S was observed as a function of the scattering angle θ_S from -90° to 90° at intervals of 1.0° using photomulitiplier (PM). The incident angles θ_{i1} for the scattered light measurements were set as the resonant angles of the ATR curves.

Surface roughnesses of the films can be evaluated from the angular distribution of the scattered light [9-15]. Here, we explain the method of the estimation of the surface roughness using the scattered light measurements. We assume the Gaussian distribution of the transverse correlation length σ and the surface corrugation depth δ for the autocorrelation function indicating the characteristics of the roughness. For simplicity, it is also assumed that any lights are not scattered from the insides of the MC LB films or the Ag films and that the lights are scattered from only roughness at the interfaces of the Ag-LB and the LB-air or Ag-air. According to the Kröger and Kretschmann theory [9], the roughness parameters, σ and δ, can be evaluated from the angular distribution of the scattered light intensity using the surface current model [9] that currents flow within a thin vacuum layer between two media. The scattered light intensity dI_s at θ_s is expressed by [9]

$$\frac{dI_s}{I_i d\Omega} = \frac{4\varepsilon_p^{1/2}}{\cos\theta_{i1}} \frac{\pi^4}{\lambda^4} |F|^2, \tag{1}$$

where $d\Omega$ is the solid angle, I_i is the p-polarized incident light intensity at the incident angle θ_{i1}, ε_p is the dielectric constant of the prism and λ is the optical wavelength of the incident light. F is a function of Fresnel's reflection and transmission coefficients and depends on θ_{i1} and θ_S. The scattered light intensities are proportional to $|s(\Delta k)|^2$, that is, a roughness spectrum, where $s(\Delta k)$ is the Fourier component of the surface roughness and $\Delta k = (2\pi/\lambda)(\varepsilon_p^{1/2}\sin\theta_{i1} - \sin\theta_S)$. Using a Gaussian distribution as the autocorrelation function with the transverse correlation length σ and the surface corrugation depth δ, the roughness spectrum $|s(\Delta k)|^2$ is given by

$$|s(\Delta k)|^2 = \frac{\delta^2\sigma^2}{4\pi}\exp\left[-\frac{\sigma^2}{4}(\Delta k)^2\right]. \tag{2}$$

Eq. (1) can be extended for the multilayered rough surfaces. If the roughness is small enough to neglect the higher order scattering, the scattered light intensity can be expressed by a linear superposition of the lights scattered from each interface. The surface contains several roughnesses with different pairs of σ and δ [11]. Assuming that the scattered light intensity is described by a linear superposition of the several roughness spectra, several pairs of the roughness parameters can be obtained. Therefore, roughness parameters for LB films on Ag thin films or Ag thin films can be described by several pairs of roughness parameters for the interfaces of Ag-LB and LB-air or Ag-air.

The sample configuration for the measurements of emission through the prism is shown in Fig.3 (a). The experimental setup shown in Fig.2 (b) was used and the sample was rotated on the goniostage as shown in Fig.3 (b).

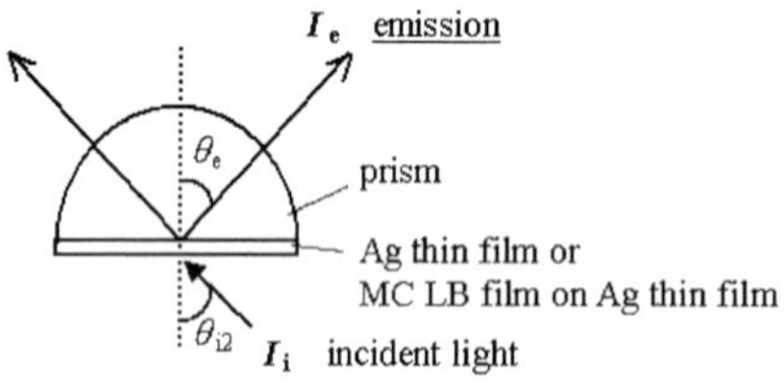

(a)

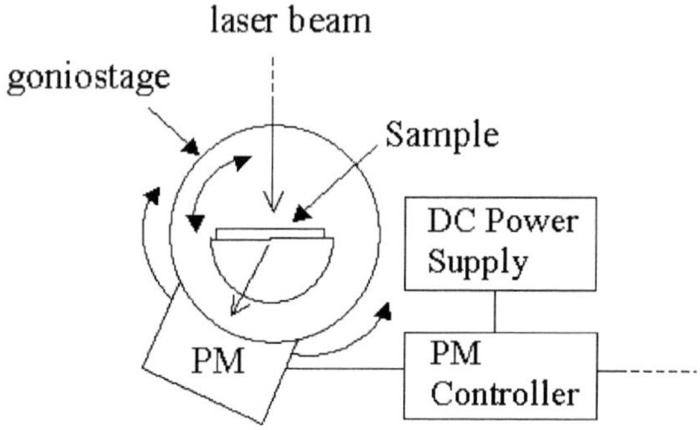

(b)

Fig. 3 Sample configurations for the measurements of the emission (a) and the experimental setup (b).

The p-polarized lights were irradiated directly from the air to the front surface of the Ag thin film or the Ag thin film. The intensity I_e of the emission was observed using the photomulitiplier as a function of the emission angle θ_e from 38° to 60° at intervals of 1.0°. The incident angle of the laser beam θ_{i2} was set as 0° or 40°.

3. Results and discussion

3.1 ATR, scattered light and emission properties of Ag sputtered film

Fig. 4 shows the ATR properties of the Ag sputtered films measured using He-Ne laser beam of the wavelength at 632.8 nm. The closed circles represent experimental data and the solid lines represent calculated ones. The theoretical calculations of the ATR curves were carried out from Fresnel's formula using transfer matrix method [20]. The dielectric constants and the film thicknesses obtained by curve fittings are 38.05 nm and -17.017+i•0.595 for sample A and 38.51 nm and -17.523+i•0.704 for sample B, respectively.

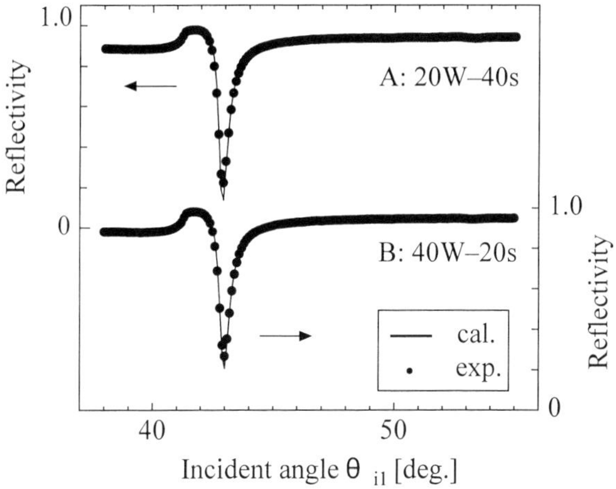

Fig. 4 ATR properties of the Ag sputtered thin films.

The emission properties of the Ag films at incident angles of θ_{i2} = 0° and 40° are shown in Fig.5. The dashed and solid curves represent the results measured at the incident angles of θ_{i2} = 0° and 40°, respectively. The peak angle of the emission almost corresponded to the resonant angle of the ATR curve. It was considered that SPPs were excited on the Ag film and part of the energy of the SPP was emitted through the prism. Although the film thicknesses of the films were almost the same, larger emission was observed for sample A.

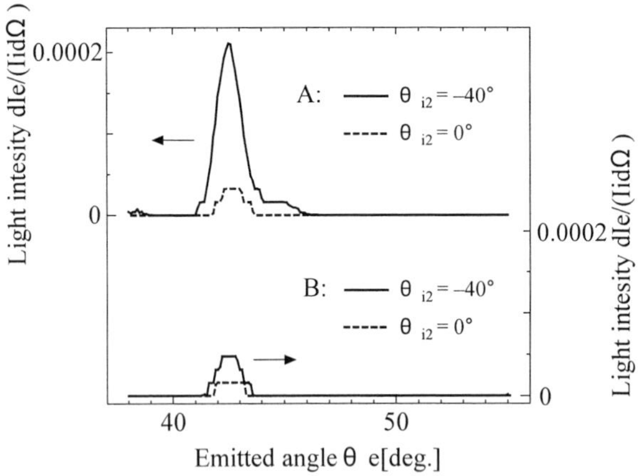

Fig. 5 Emission from the Ag sputtered thin films.

Fig. 6 shows the scattered light properties of the Ag films. From the results, roughness parameters δ[nm]/σ[nm] were estimated as a pair of 0.87/40, 0.2/750, 0.34/2000 and 1.05/7000 for sample A and a pair of 0.4/40, 0.25/2000 and 1.0/7000 for sample B.

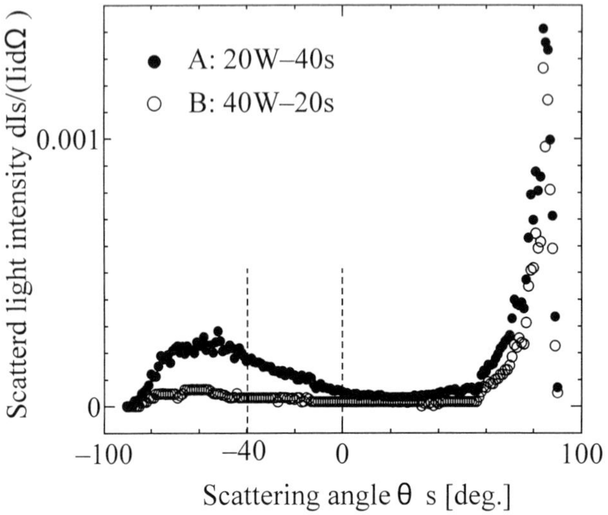

Fig. 6 Scattered light properties of the Ag sputtered thin films.

It was considered that the roughness of the film contributed to the excitation of SPP at the emission light measurement and that the strong emission light from sample A was due to the large surface roughness. Surface roughness was also observed by the atomic force microscopy (AFM), and the sample A was found to have larger roughness than the sample B. The surface roughness was also observed by Atomic Force Microscopy (AFM), and the sample A was found to have larger roughness than sample B. Moreover, larger emissions from the Ag film at incident angle θ_{12} = 40° were observed than those at θ_{12} = 0°. It was estimated that the large emissions were caused by the strong SPP excitation due to the large wavenumber component parallel to the Ag film of the incident light at θ_{12} = 40°.

3.2 ATR, scattered light and emission properties of MC LB film on Ag film

Fig. 7 shows an optical absorption spectrum of MC LB films with 16 monolayers. A sharp absorption peak at 590 nm was observed due to J aggregate of the MC dye [12]. The ATR, emission lights and scattered light were investigated for wavelengths at 488.0, 594.1 and 632.8 nm.

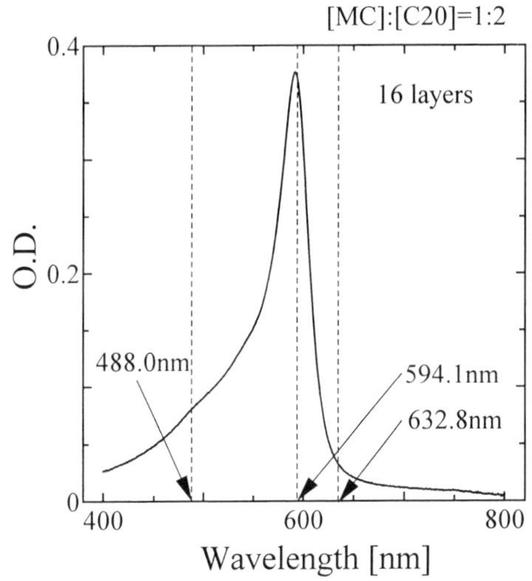

Fig. 7 Absorption spectrum of the MC LB films with 16 monolayers.

The ATR properties of the MC LB film measured at 488.0 nm are shown in Fig. 8. The MC LB films were deposited on the evaporated Ag thin films covered with 2 monolayers of C20Cd LB films. MC4L and MC8L in the figure indicate the results for the MC LB films with 4 and 8 monolayers, respectively. The ATR property of the Ag thin film is also shown in Fig. 8.

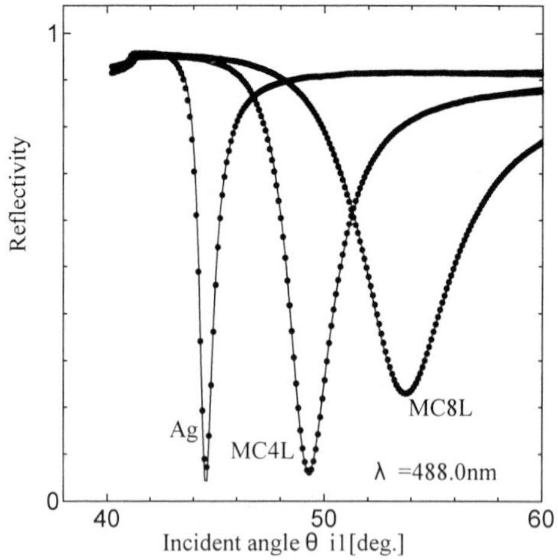

Fig. 8 ATR properties of the MC LB films on Ag thin films.

The closed circles and the solid curves represent the theoretical and calculated results, respectively. The resonant angle of the ATR curve increased with the film deposition and the dip became shallow and wide because of the light absorption of the MC LB film. The dielectric constants were also calculated on the assumption that the each monolayer had same thickness of 2.76 nm and the results are shown in Table 1.

Table 1 Dielectric constants and thicknesses evaluated for each layer of the MC LB film sample.

	complex dielectric constant			thickness
	632.8 nm	594.1 nm	488.0 nm	[nm]
Ag thin film	-17.37 + i0.631	-14.75 + i0.510	-8.85 + i0.475	47.57
C20Cd LB film (2 monolayers)	2.34 + i0.000	2.34 + i0.000	2.34 + i0.000	5.52
MC LB film (4 monolayers)	2.27 + i0.0178	2.08 + i0.656	1.78+ i0.0825	11.04

The emission properties of the MC LB film on the Ag film are shown in Fig. 9. The dashed and solid curves represent the results measured at the incident angles of θ_{i2} = 0° and 40°, respectively. The peak angles of the emission corresponded to the resonant angles of the ATR curves in Fig. 8. Same tendency was observed for the ATR and the emission light properties measured at 594.1 nm and 632.8 nm. It also indicates that the emissions are derived from SPP excitations.

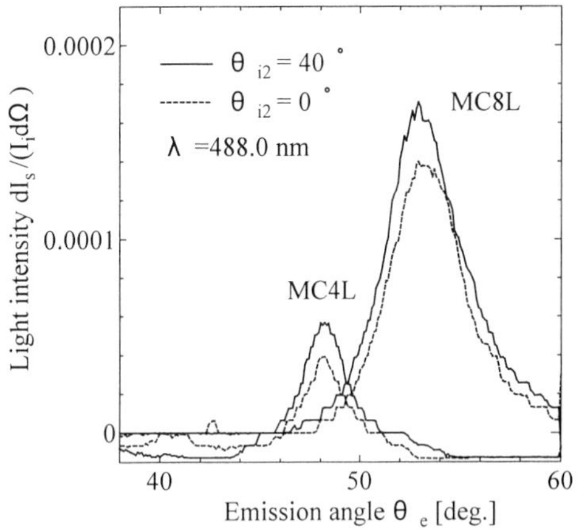

Fig. 9 Emission from the MC LB films on Ag thin films.

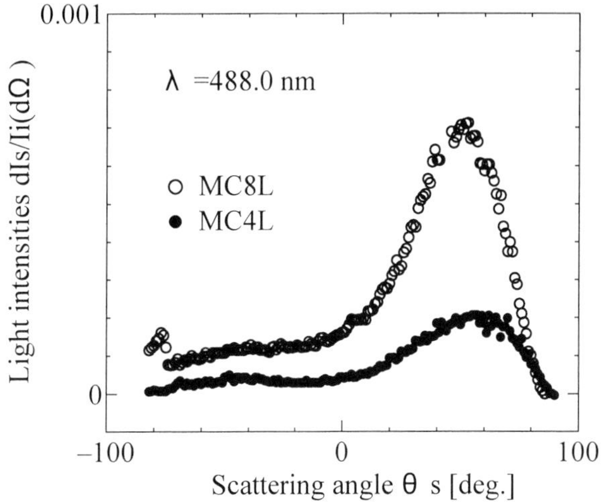

Fig. 10 Scattered light properties of the MC LB films on Ag thin films.

Scattered light distributions were observed for the MC LB films and are also shown in Fig.10. The scattered light of the MC8L sample was larger than that of the MC4L sample. It indicated the MC8L sample had larger surface roughness that induced the intense emission. Roughness parameters could not be obtained because of large inhomogeneity of the Ag film in the samples.

Larger emissions from the films at incident angle $\theta_{i2} = 40°$ were observed than those at 0°. However, the dependence on θ_{i2} was much smaller than that of the Ag sputtered films. The results suggest that mechanism of the SPP excitation is changed with deposition of the MC LB film. Although it has not been clear yet, it might be due to some energy interactions between the MC LB film and Ag film.

4. CONCLUSIONS

The properties of ATR, scattered light and emission were investigated for Ag films and MC LB films on Ag films. The film thickness and the dielectric constants of the films were obtained from the ATR measurement. The angles of the emission peaks corresponded to the resonant angles of the ATR curves and the emission indicated to be derived from SPP excitation at the Ag film surface or at the interface of the MC LB films and the Ag thin films. It was also found that the emission was closely related to the surface roughness and might be due to some energy interactions between the MC LB films and the Ag thin films. Further investigation on the emission due to SP excitation is under way.

5. REFERENCES

1. G. J. Ashwell (ed.), Molecular Electronics, Research Studies Press, Taunton, 1992.
2. T. H. Richardson (ed.), Functional Organic and Polymeric Materials, John Wiley & Sons, Chichester, 2000.
3. V. M. Agranovich and D. L. Mills (eds.), Surface Polaritons, North-Holland, Amsterdam, 1982.
4. K. Kato, Y. Aoki, K. Ohashi, K. Shinbo and F. Kaneko, Jpn. J. Appl. Phys. 35 (1996) 5466.
5. F. Kaneko, S. Honda, T. Fukami, K. Kato, T. Wakamatsu, K. Shinbo and S. Kobayashi, Thin Solid Films, 284-285 (1996) 417.

6. K. Kato, H. Saiki, H. Okuchi, F. Kaneko, T. Wakamatsu, K. Shinbo and S. Kobayashi, Thin Solid Films, 284-285 (1996) 420.
7. A. Baba, F. Kaneko, K. Shinbo, K. Kato, S. Kobayashi and T. Wakamatsu, Jpn. J. Appl. Phys. 37 (1998) 2581.
8. A. Baba, F. Kaneko, K. Shinbo, T. Wakamatsu, K. Kato and S. Kobayashi, Mater. Sci. & Eng. C 8-9 (1999) 145.
9. E. Kröger and E. Kretschmann, Z.Phys. 237 (1970) 1.
10. E. Fontana, R. H. Pantel, Phys. Rev. B 37 (1988) 3164.
11. Y. Naoi and M. Fukui, J. Phys. Soc. Jpn. 58 (1989) 4511.
12. Y. Hirao, Y. Naoi, Y. Nagano, M. Fukui, J. Phys. Soc. Jpn. 60 (1991) 4366.
13. Y. Aoki, K. Kato, K. Shinbo, F. Kaneko and T. Wakamatsu, IEICE Trans. Electron., E81-C (1998) 1098.
14. Y. Aoki, K. Kato, K. Shinbo, F. Kaneko and T. Wakamatsu: Thin Solid Films, 327-329 (1998) 360.
15. Y. Aoki, K. Kato, K. Shinbo, F. Kaneko and T. Wakamatsu, Mol. Cryst. Liq. Cryst., 327 (1999) 127.
16. S. Hayashi, T. Kume, T. Amano and K. Yamamoto, Jpn. J. Appl. Phys., 35 (1996) L331.
17. H. Raether, "Surface Plasmons on Smooth and Rough Surfaces and on Gratings", Springer-Verlag, Berlin, 1988.
18. J. L. Vossen and W. Kern (eds.), Thin Film Processes, Academic Press, New York, 1978.
19. M. Born and E. Wolf (eds.), Principles of Optics: Electromagnetic Theory of Propagation, Interference and Diffraction of Light, Pergamon Press, Oxford, 1980.
20. H. Nakahara, H. Uchimi, K. Fukuda, N. Tamai and I. Yamazaki, Thin Solid Films, 178 (1989) 549.

Novel Methods to Study Interfacial Layers
D. Möbius and R. Miller (Editors)

Enhancement of Photocurrents in Merocyanine LB Film Cell Utilizing Surface Plasmon Polariton Excitations

Futao Kaneko[a*], Kazunari Shinbo[a], Keizo Kato[b], Takaaki Ebe[b], Hironori Tsuruta[b], Satoshi Kobayashi[c] and Takashi Wakamatsu[d]

[a] Department of Electrical and Electronic Eng., Niigata University, Niigata, 950-2181, Japan

[b] Graduate School of Science and Technology, Niigata University, Niigata, 950-2181, Japan

[c] Department of Material Science and Technology, Niigata University, Niigata, 950-2181, Japan

[d] Department of Electrical Eng., Ibaraki National College of Technology, Hitachinaka,312-0011 Japan

Contents

Key words: surface plasmon polariton, attenuated total reflection method, photocurrent, merocyanine LB film, optical absorption.

* Corresponding author. Tel: +81-25-262-6741, Fax: +81-25-262-6741,
E-mail: fkaneko@eng.niigata-u.ac.jp

Short-circuit photocurrents (I_{sc}) utilizing surface plasmon polariton (SPP) excitations were investigated for the merocyanine (MC) LB film photoelectric cell. The cell has a prism/MgF_2/Al/MC LB film/Ag structure. SPPs were resonantly excited at the interfaces between MgF_2 and Al (MgF_2/Al) and between Ag and air (Ag/air) in the attenuated total reflection (ATR) configurations. Short-circuit photocurrents, I_{SC}s, were observed simultaneously during the ATR measurements as a function of the incident angle of the laser beam. In the I_{SC} curves, large and small peaks were observed, and the peak angles of the I_{SC} almost corresponded to the dip angles of the ATR curves due to the SPP excitations. Reflectivities and optical absorptions in the cell were calculated. The calculations exhibited that the incident light could be efficiently absorbed into the cell due to the SPP excitations. Moreover, the calculated absorptions in the MC layer corresponded to the I_{SC} curves. It was estimated that the photocurrents were enhanced by strong optical absorption due to the SPP excitations.

1. INTRODUCTION

Recently, organic materials have been attracting a lot of attention for their variety of functions, such as non-liner optics, luminescence, photoelectric effect and so on [1]. Several researches of organic photoelectric cells have been reported [2-4]. However, high efficiency in such conventional cells has not been obtained because of high transmission of light through organic thin layers, high reflection at the electrodes in the cells and so on.

In the attenuated total reflection (ATR) method, surface plasmon polaritons (SPPs) are resonantly excited on surface of metal thin films by the incident light, and remarkably strong absorption of the incident light occurs due to the resonantly excited SPPs[5,6]. Therefore, photoelectric effects in the cells are expected to be improved by utilizing the ATR method exciting the SPPs[7,8].

In this study, short-circuit photocurrents (I_{SC}) have been investigated for merocyanine (MC) LB film photoelectric cells having the ATR configurations.

2. EXPERIMENTAL DETAILS

2.1 Attenuated total reflection measurement

Figs. 1 (a) and (b) show the Kretschmann and the Otto configurations of the ATR measurement [5]. Evanescent waves generating at the prism surface under the condition of the total reflection of the p-polarized incident light resonantly excite the SPP on the metal surface at a resonant incident angle. The totally reflected light is attenuated at the resonant angles of the SPP because energy of the light is spent on the excitation of the SPP. The angle and the half-width of the dip in the reflectivity curve sensitively change with the thickness and the optical constants of the metal and/or the thin film on the metal. The film thickness and the optical constants of the films can be evaluated by fitting the experimental curves to the theoretical ones.

2.2 Fabrication of the photoelectric cell

Merocyanine (MC) LB films of the Y-type were deposited using the conventional vertical dipping method [9]. Since MC dyes exhibit p-type conduction, the Schottky and Ohmic contacts are obtained at the interfaces between the MC LB film and Aluminum (MC/Al) and between the LB film and Silver (MC/Ag), respectively. In this study, Schottky photoelectric cell was constructed as the ATR sample.

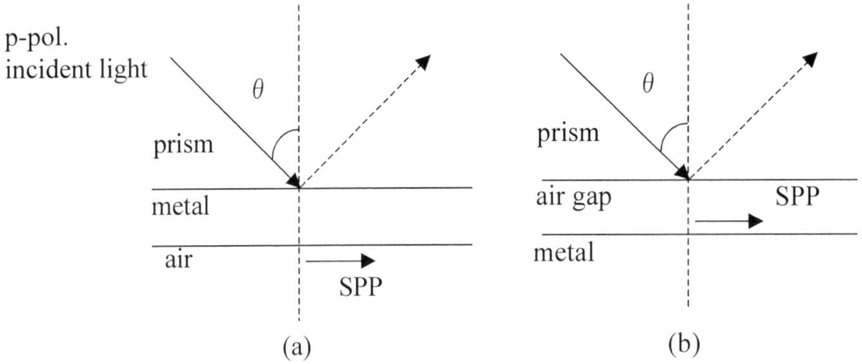

Fig.1 The Kretschmann (a) and the Otto configurations (b) for SPP excitations.

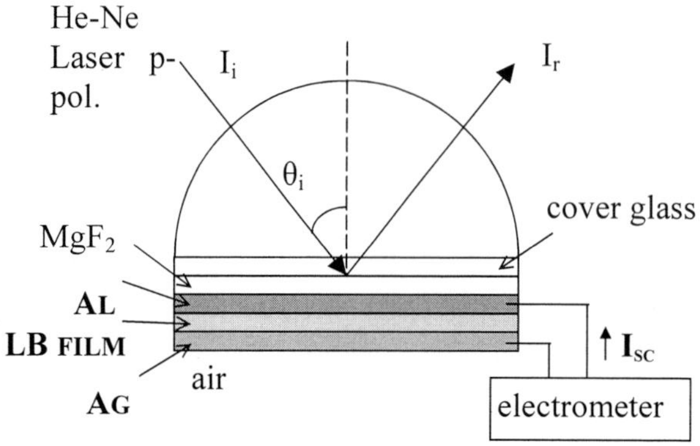

Fig.2 The structure of the photoelectric cell and the ATR configuration.

The structure of the photoelectric cell fabricated in this study is shown in Fig. 2. The MgF_2 and the Al lower electrode were evaporated onto a slide glass at first. Then, the MC LB films were deposited and the Ag upper electrodes were evaporated. The effective junction area of the cell was about 0.13 cm^2. The glass was set on a half-cylindrical glass prism using an index matching fluid. The slide glass and the prism were made of BK-7 glass that has a refractive index of 1.5150 and 1.5167 at wavelengths of 632.8 and 594.1nm, respectively. The layer structure of the cell enables the SPP excitations on the Al surface at the MgF_2/Al interface due to the Otto configuration and on the Ag surface at the Ag/air interface due to the Kretschmann configuration, respectively.

2.3 Measurements of ATR curves and the short circuit photocurrents

The MC LB film photoelectric cell was set on the ATR measurement system shown in Fig. 3. The ATR properties and the I_{SC} in the device were simultaneously measured. Two He-Ne lasers were used at 594.1 and 632.8 nm, where the MC LB films show a strong and sharp absorption peak due to J-aggregation of MC and weak absorption out of the peak, respectively. The p-polarized laser beams were directed onto the back surface of the cell through the prism as shown in Fig. 2. The cell on the prism was mounted on a rotating stage and the incident angle $\theta$of the laser beams was scanned at an interval of 0.1° using a pulse motor. The reflected intensities in the ATR curves and the I_{SC} were detected as a function of the incident angle θusing a photodiode (PD2) and an electrometer (Keithley K6517), respectively. The currents were indicated in the direction shown in Fig. 2.

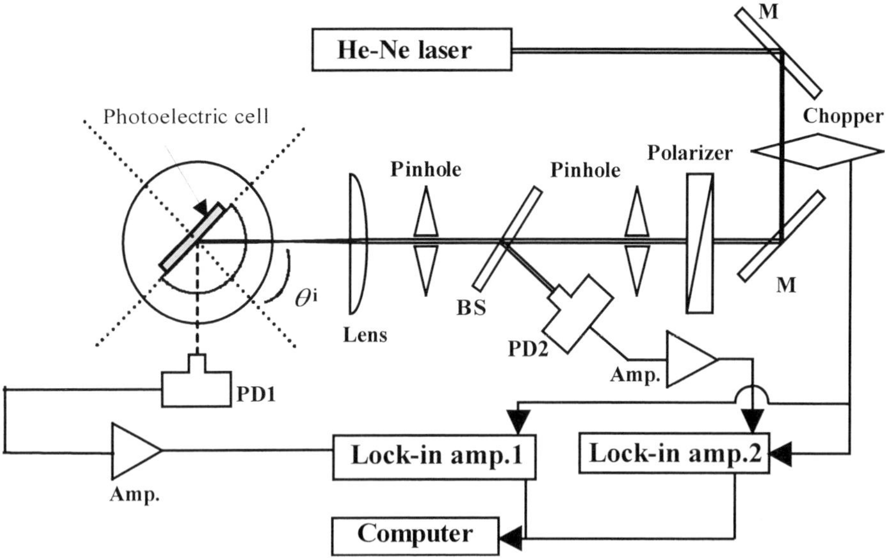

Fig.3 The experimental system used for the ATR and the Isc measurement.

3. RESULTS AND DISCUSSIONS

3.1 The ATR curve of photoelectric cell

Figs. 4 (a) and (b) show ATR curves of the photoelectric cell measured at 594.1nm and at 632.8nm, respectively. There were no much differences in the ATR curves due to the wavelengths of 594.1nm and 632.8nm [9]. A large dip at around 74 degr. and small one at around 43 degr. were caused by the SPP excitations due to strong evanescent fields from the prism at the MgF_2/Al and the Ag/air interfaces, respectively [7-9]. Theoretical ATR curves were calculated at 594.1nm and at 632.8nm for the structure of the prism/MgF_2/Al/Al_2O_3/MC LB film/Ag using a transfer matrix method [10, 12], and very good fittings were obtained.

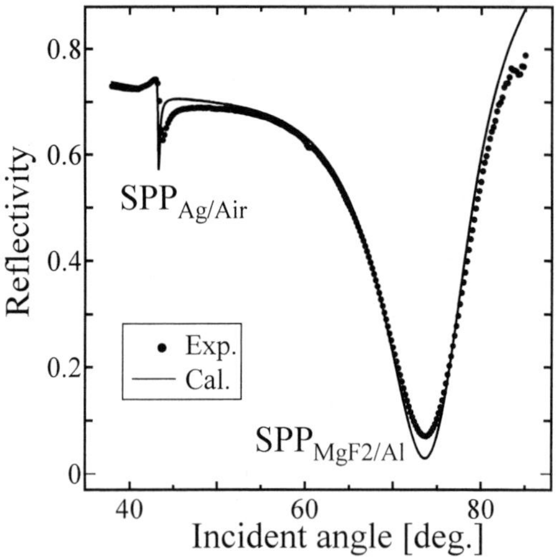

a)

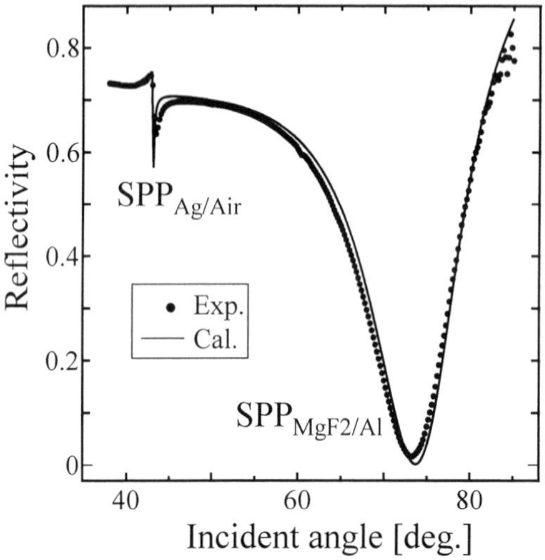

b)

Fig.4 ATR curves of the cell at 594.1nm (a) and at 632.8nm (b).

3.2 The short-circuit photocurrent (I_{SC}) curves of the photoelectric cell

Fig. 5 shows the I_{SC} curves as a function of the incident angle of the two laser beams at 594.1 nm and at 632.8 nm. The angles of the I_{SC} peaks almost corresponded to the dip angles of the ATR curves due to the SPP excitations in Fig. 4. The large peak at around 74 degr. and the small one at around 43 degr. were considered to be due to the strong and weak SPP excitations at the MgF_2/Al interface and the Ag/air interface, respectively. From the results, it was estimated that the SPP excitations caused the I_{SC} in the cell.

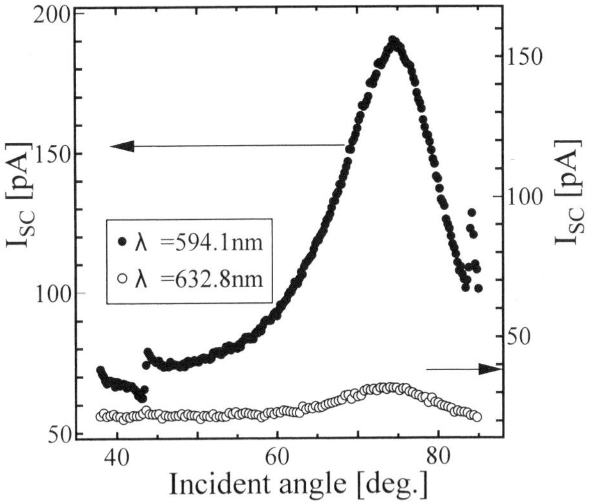

Fig.5 I_{SC} properties as a function of the incident angle of the laser beams at 594.1nm and 632.8nm.

3.3 Optical absorption in the cell

Electric fields and optical absorption in the cell were calculated using a transfer matrix method [10-12]. Much difference was not observed in the calculated electric fields in the MC LB layer at 594.1 nm and at 632.8 nm[9]. Fig. 6 shows the calculated reflectivities for the cell used in this experiment and a supposed cell without the prism and MgF_2 layer that can not excite the SPPs. As the cell without the SPP excitations exhibits large reflectivities, almost all of the incident light is not absorbed in the cell. The calculations also show that there is no dependence of the wavelengths.

Fig. 7 shows the calculated total absorptions in the cell as a function of the incident angle at 594.1 nm and at 632.8 nm. The calculated profiles almost corresponded to the I_{SC} shown in Fig. 5. From the results, it was estimated that the absorption due to the SPP excitations in the MC LB layer dominated the I_{SC}. 594.1nm and at 632.8 nm.

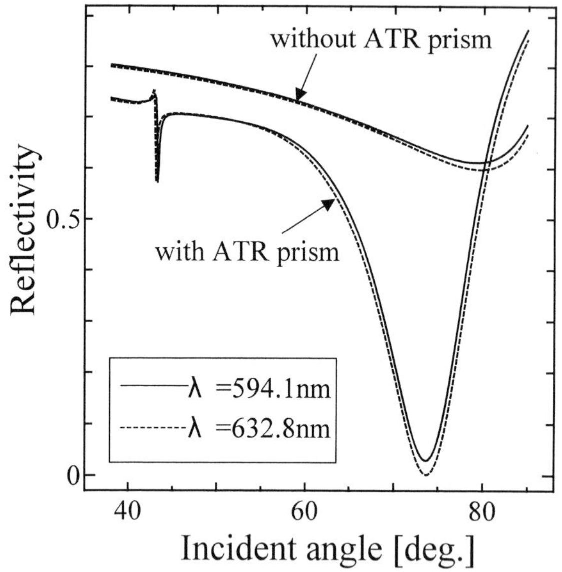

Fig.6 The calculated reflectivities with and without the prism and the MgF_2 .

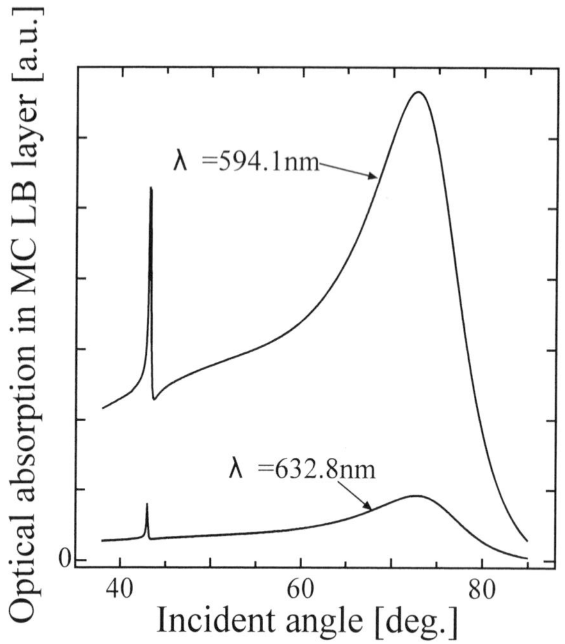

Fig.7 The calculated absorption spectra in the cell as a function of the incident angle of the laser beams at 594.1nm and at 632.8 nm.

4. Conclusion

The short-circuit photocurrents (I_{SC}) were investigated for the MC LB photoelectric cell in the ATR configuration. The I_{SC} profiles as a function of the incident angle corresponded to the ATR properties in the cell. From the calculations, it was estimated that the I_{SC} was caused by strong optical absorption in the cell related to the SPP excitations.

5. Acknowledgement

This study is partly supported by Grant-in-Aid for Scientific Research from the Ministry of Education, Science, Sports and Culture of Japan.

6. References

1. G.J. Ashwell (Ed.), "Molecular Electronics", Research Studies Press, Taunton, 1992.
2. A. Ulman ed., "An Introduction to Ultrathin Organic Films from Langmuir-Blodgett to Self-Assembly", Academic Press, New York, 1991.
3. R.O. Loutfy and J.H.Sharp, "Photovoltaic Properties of Metal-Free Phthalocyanine. I. Al/H_2Pc schottky barrier solar cells", J.Chem.Phys., 71 (1979) 1211.
4. A.P. Piechowski, G.R. Bird, D.L. Morel and E.L. Stogryn, "Desirable Properties of Photovoltaic Dyes", J. Phys. Chem., 88 (1984) 934.
5. See for example, V.M. Agranovich and D.L. Mills eds., "Surface Polaritons", North Holland, Amsterdam, 1982.
6. A. Baba, F. Kaneko, K. Shinbo, K. Kato and S. Kobayashi, "Evaluation of Liquid Crystal Molecules on Polyimide LB Films Using Attenuated Total Reflection Measurement", Thin Solid Films, 327-329 (1998.) 353.
7. T. Wakamatsu, K. Saito, Y. Sakakibara and H. Yokoyama, "Surface Plasmon-Enhanced Photocurrent in Organic Photoelectric Cells", Jpn.J.Appl.Phys., 36 (1997) 155.
8. T. Wakamatsu, K. Saito, Y. Sakakibara and H. Yokoyama, "Enhanced Photocurrent in Organic Photoelectric Cells Based on Surface Plasmon Excitations", Jpn. J. Appl. Phys., 34 (1995) 1467.

9. K. Shinbo, T. Ebe, F. Kaneko, K. Kato and T. Wakamatsu, "A Photoelectric Property of Merocyanine LB Film Cell Utilizing Surface Plasmon Excitation", IEICE Trans. Electron., E83-C (2000) 1081.

10. P. Yen, "Optical Waves in Layered Media", Wiley, New York, 1988.

11. J. Derov Y.Y. Teng and A.S. Karakashian, “Angular Scan Spectrum of a Surface Plasma Excitation on a Schottky Diode”, Phys. Lett. A, 95 (1983) 197.

12. M. Born and E. Wolf (Eds.), Principle of Optics, Pergamon, 1974.

Novel Methods to Study Interfacial Layers
D. Möbius and R. Miller (Editors)

LOCAL PIEZOELECTRIC RESPONSE AND SURFACE POTENTIAL OF DIELECTRIC AND FERROELECTRIC LANGMUIR-BLODGETT FILMS STUDIED BY ELECTROSTATIC FORCE MICROSCOPY

L.M. Blinov [1,2]*, R. Barberi[2], S.P. Palto[1], Th. Rasing[3], M.P. De Santo[2] and S.G. Yudin[1]

[1] Institute of Crystallography, Russian Academy of Sciences, Moscow 117333, Russia

[2] INFM, Universita' della Calabria, I-87036, Rende (Cs), Italy

[3] RIM, University of Nijmegen, Box 9010, 6500GL, Nijmegen, The Netherlands

Contents

Keywords: electrostatic force microscopy, LB films, piezo-effects, ferroelectricity.

*Corresponding author: fax: +7-095-1351011; e-mail: palto@online.ru

A comparative study of ultrathin dielectric (an azo-compound) and ferroelectric (copolymer P(VDF-TrFE)) Langmuir-Blodgett (LB) films has been carried out by Electrostatic Force Microscopy (EFM). Films were poled locally by a strong d.c. field applied between a conductive tip of an Atomic Force Microscope (AFM) and the bottom Al electrode. The electrically poled domain was studied by EFM using a weak a.c. electric field and a lock-in amplifier technique. Two modes, a contact and non-contact ones, allowed for the measurement of field E_a in the air gap between the film and the tip and the piezoelectric distortion of the film due to the d.c. field aligned spontaneous polarization. Simultaneously the topographic relief of the same area was imaged. The results confirm unequivocally a possibility to switch ferroelectric LB film locally by an AFM tip.

1. Introduction

Recently an indication of two-dimensional ferroelectricity in ultrathin Langmuir-Blodgett (LB) films as thin as two molecular monolayers (about 1nm) was discovered [1, 2] for a copolymer vinylidene fluoride with trifluoroethylene P(VDF-TrFE). The observations of the spontaneous polarization[2] of the order of $P_s \approx 0.1 C/m^2$, the first order ferroelectric phase transition [2], a critical point [3], a surface phase transition[4] and electric switching [1, 2, 5] were reported.

The structural unit of the copolymer $-(-CH_2-CF_2)_n-(-CF_2-CHF-)_m-$ contains *n* and *m* corresponding monomer links. The ferroelectric properties are attributed to transverse dipole moments, formed by positive hydrogen and negative fluorine atoms. Below the temperature of the ferroelectric phase transition (about 80-100°C), the main chain of the polymer is in all-trans form and the dipole moments are parallel, at least, within ferroelectric domains separated from each other by domain walls. The ferroelectric switching is due to an electric field induced, collective flip-flop of the dipoles around the backbone of the polymer. Several recent studies were devoted to a local ferroelectric switching of the domains in *cast* P(VDF-TrFE) films [6-8]. To this effect, a powerful technique, called Electrostatic Force Microscopy (EFM) [9] was used which was developed for studies of domains in thin ferroelectric films, see papers [10, 11] and references therein.

The LB films of P(VDF-TrFE) were also studied by EFM technique [12]. A film was switched locally by a voltage applied between a tip and the Al bottom electrode (writing process). Then the electric relief of the switched area was read out by EFM with simultaneous control of the topographic relief. The recorded lines show a contrast depending on poling conditions as is typical of the cast films of the same material [8]. For comparison, the same technique was applied to pure dielectric films, not showing ferroelectric properties[13]. The images were obtained even in this case. In order to explain the image contrast, a simple model for a thin dielectric and ferroelectric films was developed which took into account the remanent polarization, the screening charge and interface work functions.

In [12, 13] there was only used a non-contact regime of imaging [9] which probed a surface potential influenced by many factors. In the present paper we make a comparative study of ultrathin dielectric and ferroelectric LB films poled locally by a strong d.c. field using two modes, a contact and non-contact ones. The latter allows for the measurement of the piezoelectric distortion of the film due to the d.c. field aligned spontaneous polarization [6, 14]. The piezoelectric response has, indeed, been observed on ferroelectric LB films. Therefore, their local ferroelectric switching by an AFM tip has been confirmed.

2. Experiment

Samples. In this work we used three different films, a dielectric LB film made of an amphiphilic azo-benzene compound (ABC), a ferroelectric polymer LB film (FLB) and a PZT ceramic film (for comparison).

The ABC film was prepared from amphiphilic dye p-nonyl-amino-p'-carboxyazobenzene whose molecule contains an azobenzene chromophore, a polar head (-COOH) with electron acceptor properties and hydrophobic tail (-NHC9H19) with electron donor properties. In this work, as earlier [12], we used an X-type LB film consisting of 30 monolayers transferred onto an ITO covered glass substrate by the Schäfer technique at surface pressure 3-5 mN/m. The thickness of the film estimated from the optical absorbance at λ=460nm and by AFM measurements is 40nm.

The FLB film was made of copolymer P(VDF-TrFE)(70:30 mol %) deposited onto an n-type silicon substrate supplied with 2mm wide aluminum stripes evaporated in vacuum. The solution of the copolymer in dimethyl-sulfoxide of 0.01% weight concentration was spread over the pure water in a home-made LB trough. The film consisted of 6 Y-type bilayers was made by their subsequent transfer from the water surface at surface pressure 2.3 dyn/cm and temperature 23.5°C.

The PZT film on a Pt covered Si substrate was 200nm thick and made of $Pb(ZrO_{0.53}TiO_{0.47})O_3$ ceramics by a sol-gel process with subsequent annealing at 650°C.

Measurements. To write and read lines we used a standard AFM device (Autoprobe CP, Park Scientific Instrument) and commercially available silicon tips (PSI Ultralevers) covered with a home made layer of silver (50nm) evaporated in vacuum. First, we recorded a line along the y-direction in a contact regime, with a d.c. electric voltage applied between a cantilever and a bottom electrode of a film.

To read out the image we used two modes, a contact and non-contact ones, allowed for the measurement of field E_a in the air gap between the film and the tip (surface potential mode) and the piezoelectric distortion of the film due to the d.c. field aligned spontaneous polarization. To measure E_a we applied between the tip and the sample an electric voltage (1.5 V rms) at frequency f_e=2 kHz and made an x,y-scan of the corresponding area using the non-contact regime of AFM. To measure a piezoelectric response we applied the same voltage and made an x,y-scan of the corresponding area using the contact regime of AFM. A laser beam reflected from the cantilever was modulated at two frequencies and detected by a photodiode. The cantilever oscillation signal with frequency about 80kHz was used for a surface topography study and the low frequency one, after a low-pass filter, was delivered to a lock-in amplifier (EG&G Instrument, model 7260) to study an electric relief. The digital output of the amplifier (time constant 5ms) was connected to an input channel of the AFM. Therefore, two images (the topographic and electric relief) were displayed simultaneously.

3. A MODEL

For a quantitative analysis of the electrical images due to a field distribution between an air gap d and an electrically poled ferroelectric layer of thickness l, consider a simple model [12, 13] for a field distribution between an air gap d and an electrically poled ferroelectric layer of thickness l, Fig. 1.

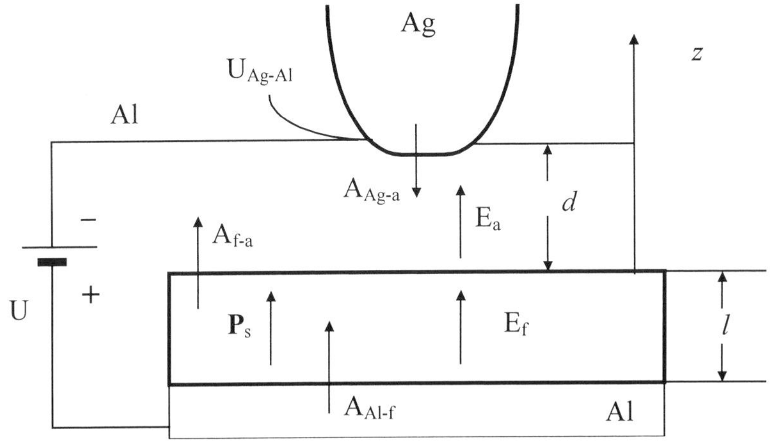

Fig. 1 A model for a ferroelectric film places between a rear electrode and AFM tip.

A voltage U is applied between the silver tip and the bottom Al electrode of the ferroelectric layer The ferroelectric film may be poled along the z-axis and, consequently, possesses polarization P_s. For simplicity, imagine that the tip has a flat bottom of apparent area S* (depending on real geometry of the experiment) and the electric field $E=E_z$. The condition of the zero potential difference along the close electric contour reads :

$$A_{Al-f} + E_f l + A_{f-a} + E_a d - A_{Ag-a} + U_{Ag-Al} = U \tag{1}$$

where $A_{Al\text{-}f}$, $A_{f\text{-}a}$, $A_{Ag\text{-}a}$ are work functions of the Al-ferroelectric, ferroelectric-air and Ag-air interfaces (expressed in Volts), $U_{Ag\text{-}Al}$ is contact potential difference for a Ag-Al junction. E_f and E_a the field strengths in ferroelectric and air, assumed to be permanent due to the absence of free charges in the bulk of ferroelectric and the air gap.

In the absence of free charges the continuity of the electric displacement vector (Maxwell equation) reads:

$$D = \varepsilon_0 E_f + P \cong \varepsilon_f \varepsilon_0 E_f + P_s = \varepsilon_0 E_a \tag{2a}$$

where ε_f is relative dielectric permittivity of the ferroelectric, $\varepsilon_0 = 8.85 \cdot 10^{-12}$ F/m is dielectric permittivity of vacuum (or air). In (2) we have left only the linear-in-field contribution to the induced polarization. When a surface charge of density σ screening the polarization is present on the free surface of a ferroelectric, the displacement vector is no more continuous and Eq. (2a) should be modified :

$$\Delta D \cong \varepsilon_f \varepsilon_0 E_f + P_s - \varepsilon_0 E_a = \sigma \tag{2b}$$

From (1) and (2b) follows that the field strength in the air gap and in a ferroelectric is:

$$E_a = \frac{U - \Delta A + (P_s - \sigma) l / \varepsilon_0 \varepsilon_f}{d + l / \varepsilon_f} \tag{3}$$

$$E_f = \frac{U - \Delta A - (P_s - \sigma) d / \varepsilon_0}{d \varepsilon_f + l} \tag{4}$$

where

$$\Delta A = A_{Al-f} + A_{f-a} - A_{Ag-a} + U_{Ag-Al} \tag{5}$$

It is of great importance to note that due to the presence of a field in a ferroelectric, ΔA does not vanish, as it might seem from the first glance.

From Eqn.(3), we come to a conclusion that the field in the gap depends not only on two geometrical parameters d and l, and dielectric constant ε_f but also on work functions A_{Al-f} and A_{f-a} dependent on the polarization state of a ferroelectric and screening charges accumulated at the surface. A careful consideration of ΔA is especially important for ultrathin ferroelectric films with small value of spontaneous (or remanent) polarization , when

$$l \leq \frac{\varepsilon_0 \varepsilon_f \Delta A}{(P_s - \sigma)} \tag{6}$$

The force of the electrostatic interaction between the tip and the surface (additional to the short-range forces) is obtained by variation of electrostatic energy W_e consisting of two terms, the energy in the air gap W_a and the energy in ferroelectric W_f.

$$f_e = -\frac{\delta}{\delta z}(W_a + W_f) = -\varepsilon_0 \int_{V_a} E_a \frac{\delta E_a}{\delta z} dV - \int_{V_f} (\varepsilon_f \varepsilon_0 E_f + P_s - \sigma) \frac{\delta E_f}{\delta z} dV \tag{7}$$

Substituting (3) and (4) into (7) we obtain the force

$$f_e = -\varepsilon_f S^* \frac{\delta E_f}{\delta z} (\varepsilon_0 \varepsilon_f E_f + P_s - \sigma)(d + l/\varepsilon_f) =$$
$$= \frac{\varepsilon_0 S^*}{(d + l/\varepsilon_f)^2} \left[(U - \Delta A) + \frac{(P_s - \sigma) l}{\varepsilon_f \varepsilon_0} \right]^2 \tag{8}$$

Now, consider a typical situation, when a ferroelectric is already poled locally by a d.c. field and an electric relief is read out with a small a.c. voltage $U_m \sin\omega t$ applied between a conductive tip and a rear electrode. Then, with a lock-in amplifier one can measure the amplitude of the low frequency cantilever oscillation on the first and second harmonics of the applied voltage which are proportional to the corresponding components of force (8):

$$f_e(\omega) = \frac{2\varepsilon_0 S^*}{(d + l/\varepsilon_f)^2} \left(\frac{(P_s - \sigma) l}{\varepsilon_f \varepsilon_0} - \Delta A \right) U_0 \sin\omega t \tag{9}$$

$$f_e(2\omega) = -\frac{\varepsilon_0 S^*}{2(d + l/\varepsilon_f)^2} U_0^2 \cos 2\omega t \tag{10}$$

The observed signal at 2ω is independent of work functions and polarization and allows for the determination of effective area S^* as soon as the other parameters (d, l and ε_f) are measured [13]. The signal at frequency ω contains the most important term (in parentheses) dependent on parameters of a material studied, P_s, σ and ΔA. We see that the sign of the signal may be either positive or negative (in phase with voltage or shifted by π). As a rule, the switching of polarization is incomplete and we may substitute remanent polarization value P_r for spontaneous polarization P_s in (2)-(10).

In case of a non-ferroelectric (dielectric) film, P_s (or P_r) vanishes, however the signal at the first harmonic may still be observed due to a surface charge σ deposited by a tip at dielectric-air interface, which may also modify the corresponding work function at that interface maintaining ΔA finite. Therefore, it is not easy to distinguish a switched ferroelectric film from a poled dielectric film using solely measurements of the field in the air gap. To do this unambiguously a piezoelectric technique has to be used. In this case, a cantilever is brought in contact with the film and the a.c. voltage applied directly to the film causes a piezo-electric compression and dilatation of the film $\Delta l/l=d_{33}E$ (or $\Delta l=d_{33}U$) , which is to be measured through a deviation of a cantilever. With d_{33}=-33pC/N (for cast PVDF films) and U=1.5V we have $\Delta l \approx 0.5$Å, the value several times exceeding the noise level of our EFM set-up. The piezo-effect is a bulk property and always observed in the materials with polar symmetry (and some others belonging to piezoelectric crystallographic classes).

4. Results and Discussion.

Eqs. (3) and (9) allow one to explain the observed variations in the contrast of images obtained in a non-contact (surface potential) mode, in particular, the inverse contrast (small P_s, large ΔA and σ) which, otherwise, would require an ill-defined concept of "overscreening" the polarization by free charges [8, 14]. We can illustrate this with two EFM images shown in Figs. 2 and 3.

The first image (Fig. 2) is "normal" but quite rare; it corresponds to an observation of a black line written by negative voltage (-10V) on the bottom film electrode with respect to the grounded tip. It corresponds to the switched ferroelectric polarization (which is negative, that is oriented downward, opposite to that shown in the sketch, Fig. 1). We met such a situation only few times. The second image is quite typical, the line recorded with the same polarity (- 15V) is white; this would correspond to the so-called "overscreened" negative polarization. In fact, there is no "overscreening", just ΔA term in (9) dominates because the difference (P_s - σ) in the first term of (9) is small.

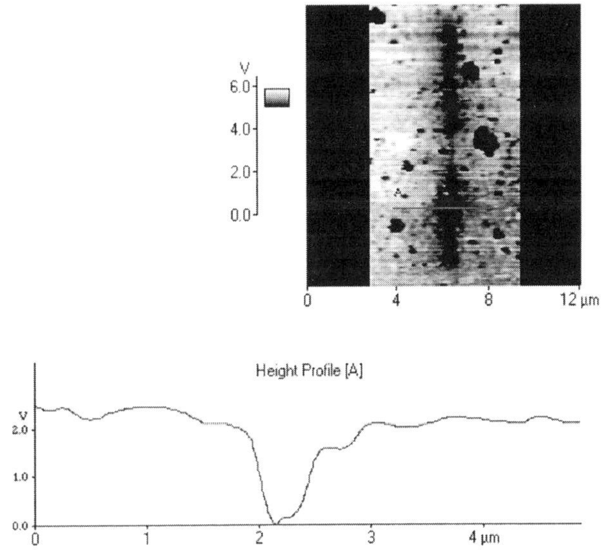

Fig. 2. Surface potential image and horizontal relief of a vertical line written on the ferroelectric LB film in a contact mode with U_{dc}= -15V applied to the bottom electrode with respect to the grounded cantilever. Reading in a non-contact mode.

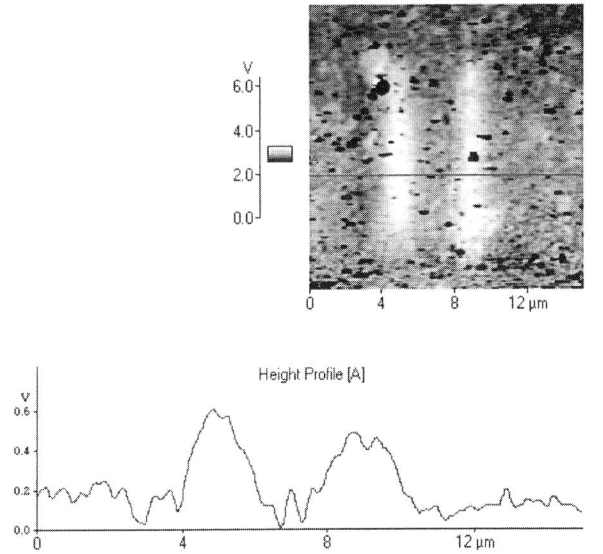

Fig. 3. Surface potential image and horizontal relief of two lines written in another place of the same film as in Fig. 2 in a contact mode with the same polarity U_{dc}= -10V. Reading in a non-contact mode.

In case of the inverse contrast, the field oriented ferroelectric domains are looking as areas charged by electric current from the tip and only a comparison with piezoelectric measurements may finally show whether a ferroelectric has been switched or has not. Below we present the results of such a comparison (note, for all images below the tip is grounded, i.e. has a zero potential).

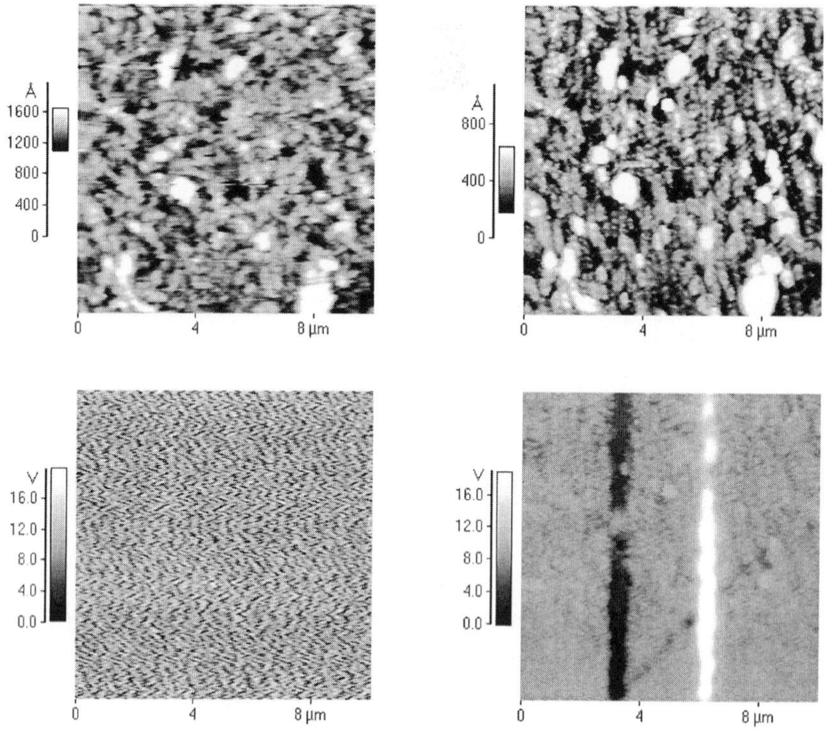

Fig. 4. Topographic (top) and EFM (bottom) images of two lines written with ±15V on a dielectric ABC Langmuir-Blodgett film. The left pair of images is taken in the contact (piezo-electric) mode, the right pair in non-contact (surface potential) mode.

Fig. 4 shows topographic (top) and EFM (bottom) images of two lines written on a non-ferroelectric ABC Langmuir-Blodgett film. The left pair of images has been read out in the contact (piezo-electric) mode, the right pair in the non-contact (surface potential) mode. The lines were written with +15V (left black line in the lower-right image) and -15V (white right line) on the bottom electrode with respect to the grounded tip. Topographic images are quite similar and a very thin vertical scratch may only be seen in the upper-right pattern. As expected, the surface potential clearly shows the presence of the charge deposited from the tip, the negative one (black line) coming from the negative tip and positive one (white line) deposited by the positive tip. The high brightness of the lines is indicative of a good dielectric properties of the ABC film, the charge relaxes very slowly due to low conductivity of the film. Note, that no signal is seen in the lower-left image, because the film, even poled, is not piezoelectric. Fig. 5 shows the images of the lines written in the same way as those in Fig. 4 but on a film of ferroelectric PZT ceramics.

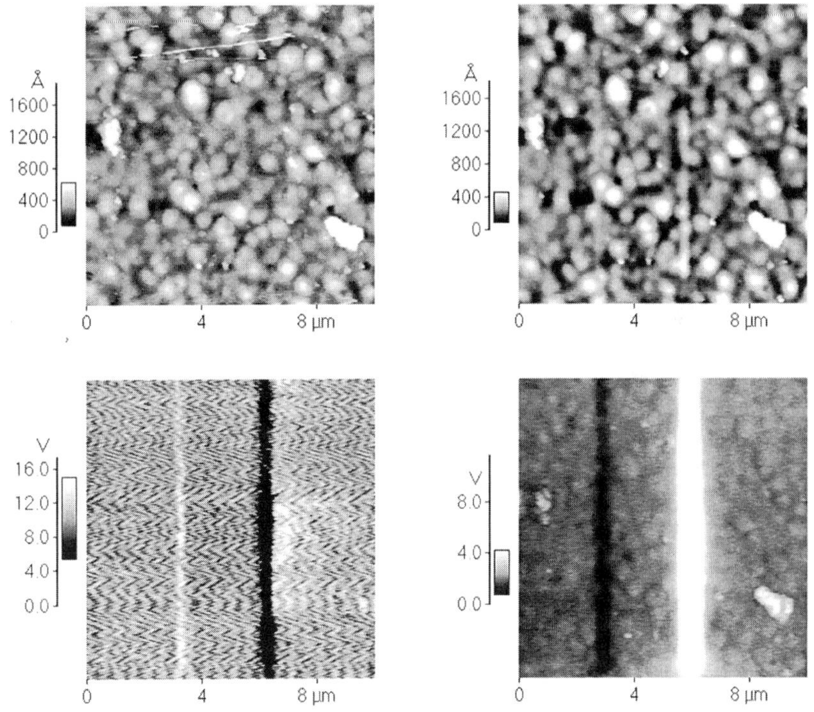

Fig. 5. Topographic (top) and EFM (bottom) images of two lines written with ±15V on a ferroelectric PZT ceramic film. The left pair of images is taken in the contact (piezo-electric) mode, the right pair in non-contact (surface potential) mode.

Again, only weak traces of the scratches are seen on the upper right (non-contact, topographical) image. Surface potential (lower-right) signal shows high contrast lines of the same sign as in the case of a dielectric film. In terms of references [8, 14] it would correspond to "overscreening" the polarization, the effect which can alternatively be explained with Eqn.(3) for E_a (a case when $\Delta A>(P_s-\sigma)l/\varepsilon_0\varepsilon_f$). At the same time, from lower-left figure is evident that the PZT film is poled by an external voltage, the areas poled with +15V (left) and -15V (right) show piezoelectric signals of opposite phase, pointing to the opposite film polarity. This is a picture usually observed on PZT films in the piezoelectric mode.

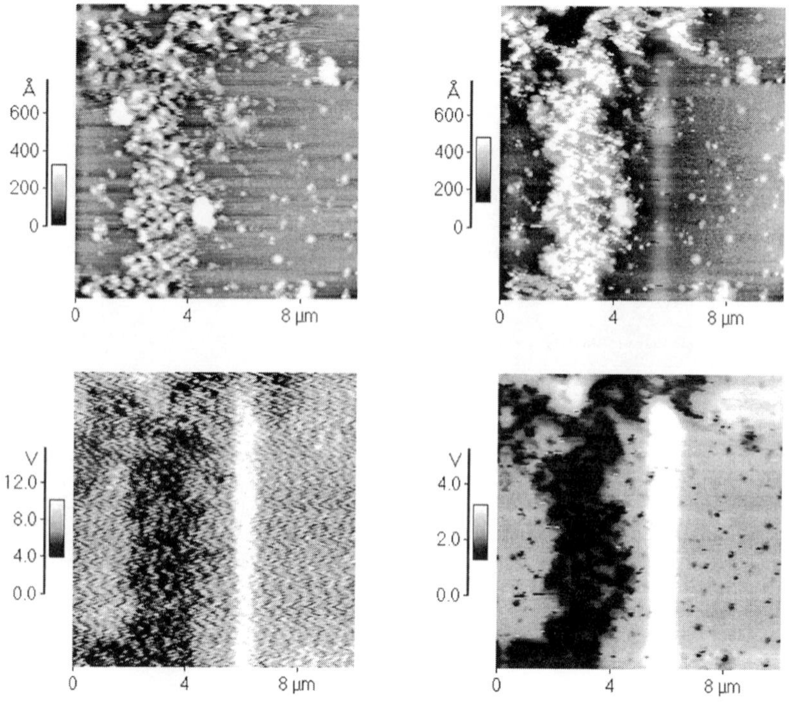

Fig. 6. Topographic (top) and EFM (bottom) images of two lines written with ±12V on a ferroelectric PVDF-TrFE Langmuir-Blodgett film. The left pair of images is taken in the contact (piezo-electric) mode, the right pair in non-contact (surface potential) mode.

Finally, Fig. 6 shows the images of the lines written in the same way but with a voltage of ±12V on a film of ferroelectric PVDF-TrFE LB film. The film is not uniform as seen in topographic (top) images. For this reason, the EFM images are blurred. However, the contrast on the surface potential image (lower-right), has the same features as in the previous figures (the left line is again black and the right one is white). The piezoelectric image, Fig. 6 (lower-left), shows that poled areas are piezoelectric as expected for a ferroelectric film, however, now the "piezoelectric lines" has opposite contrast with respect to the PZT image shown in Fig. 5. The reason is simple: the two materials have piezoelectric coefficient d_{33} of the opposite sign [15]. Therefore, the piezoelectric measurements clearly show that the polymer LB film is locally switched by the field from the AFM tip.

5. CONCLUSION

In conclusion, a comparative study of locally poled ultrathin dielectric and ferroelectric Langmuir-Blodgett films has been carried out by Electrostatic Force Microscopy (EFM). The electrically poled domains were observed using a lock-in amplifier technique. Two modes for image reading out, a contact and non-contact ones, allowed for the measurement of both the field E_a in the air gap between a film and a tip and the piezoelectric distortion of a film due to the d.c. field aligned spontaneous polarization. Simultaneously the topographic relief of the same area was imaged. The results confirm unequivocally a possibility for switching ferroelectric LB films locally by an AFM tip. The films appear to be very interesting from the point of view of applications. First, due to an extremely small thickness and high dielectric constant their electric capacity is huge and the films are very good for low voltage thin film capacitors. Second, they can be used in nonvolatile memory devices controlled by low voltages as well. They can also be used for information recording at the mesoscopic scale.

ACKNOWLEDGEMENTS.

L.B., S.P. and S.G. are grateful to Prof. V. Fridkin (Institute of Crystallography, Moscow) and Dr. S.Ducharme (Nebraska University) for many interesting discussions on ferroelectric LB films. L.B. thanks Prof. R. Bartolino for hospitality in Calabria University. The work has been carried out in framework of the INFM "Progetto Sud" and RFBR grant 99-02-16484.

6. References

1. S. Palto, L. Blinov, E. Dubovik, V. Fridkin, N. Petukhova, A.Sorokin, K. Verkhovskaya, S. Yudin and A. Zlatkin, *Europhys. Lett.* **34** (1996) 465
2. A. Bune, V.M. Fridkin, S. Ducharme, L.M. Blinov, S.P.Palto, A.V.Sorokin, S.G. Yudin and A. Zlatkin, *Nature* **391** (1998) 874.
3. S. Ducharme, A. Bune, L.M. Blinov, V.M. Fridkin, S.P. Palto, A.V. Sorokin and S.G. Yudin. Phys. Rev. B **57** (1998) 25
4. J. Choi, P. A. Dowben, S. Pebley, A.V. Bune, S. Ducharme, V.M. Fridkin, S.P. Palto and N. Petukhova, Phys. Rev. Lett. **80** (1998) 1328
5. S. Ducharme, V.M. Fridkin, A.V. Bune, S.P. Palto, L.M. Blinov, N. Petukhova and S.G. Yudin, Phys. Rev. Lett. **84** (2000) 175
6. P. Güthner and K. Dransfeld, Appl. Phys. Lett. **61** (1992) 1137
7. T. Kajiyama, N. Khuwattanasil and A. Takahara, J. Vac. Sci. Technol. **B16** (1998) 121
8. X. Chen, H. Yamada, T. Horiuchi, and K. Matsushige, Jpn. J. Appl. Phys. (Pt.1) **37**(1998) 3834; **38** (1999) 3932
9. F. Saurenbach and B.D. Terris, Appl. Phys. Lett. **56** (1990) 1703
10. R. Lüthi, H. Häfke, K.-P. Meyer, E. Meyer, L. Howald, and H.-J. Güntherodt, J. Appl. Phys. **74** (1993) 7461
11. C.H. Ahn, T. Tybell, L. Antongnazza, K. Char, R.H. Hammond, M.R. Beasly, O. Fisher and J.-M. Triskone, Science **276** (1997) 1100
12. L.M. Blinov, R. Barberi, S. P. Palto, M.P. De Santo and S.G. Yudin, J. Appl. Phys., 89 (2001) 3960
13. M.P. De Santo, R.Barberi, L. M. Blinov, S.P. Palto and S.G. Yudin, Mol. Materials, **12** (2000) 359
14. T. Tybell, C.H. Ahn and J.-M. Triscone, Appl. Phys. Lett. **75** (1999) 856
15. H.L. W.Chan, P.K.L. Ng and C.L. Choy, Appl. Phys. Lett. **74** (1999) 3029

Novel Methods to Study Interfacial Layers
D. Möbius and R. Miller (Editors)

Development of EPR Method for Examination of Paramagnetic Complex Ordering in Films

E.G. Boguslavsky, S.A. Prokhorova and V.A. Nadolinny

Institute of Inorganic Chemistry, Novosibirsk, Russia

Contents

A procedure is proposed for quantitative description of partially ranked paramagnetic samples. The characterization is based on the assumption that selected orientations for the major axes of magnetic resonance tensors exist. Another assumption is the random scattering of major axes with respect to this direction according to the Gauss law. The orientation and disordering parameters are optimised by simulation of theoretical ESR spectra and comparison with experimental spectra. The mathematical algorithm developed for the simulation program is also discussed. Results of this approach were applied to experimental data for films of copper phthalocyanine and dipivaloil methanate, obtained by sedimentation on quartz plates.

Key words: EPR; ESR; ranked system; molecule ordering; ordered films; copper phthalocyanine; copper dipivaloilmethanate; ESR spectra simulation;

1. Introduction

The location and shape of EPR spectrum lines of paramagnetic centres in paramagnetic single crystals often depends strongly on sample orientation in the spectrometer magnetic field. In powder EPR spectra we can see the total signal as a sum of all possible orientations with equal probability. It makes the resulting spectrum independent of the magnetic field direction relative to the sample frame.

Considerable changes in EPR spectra of flat films during the angular variation between the normal line to the sample plane and magnetic field direction arise from orientation ordering of paramagnetic molecules. In other words, paramagnetic fragments are predominantly aligned under particular angle to the sample plane. Such matters are named textured, i.e. partially ranked, the phase is characterized by a more or less strong orientation order in the absence of complete ordering as it takes place in a single crystal. The ordering arises during the preparation or modification of the sample and is determined by specific details of these processes. Thus, the analysis of EPR spectra can help us to study formation and modification processes in films in detail. This paper describes a new designed procedure, which allows one to get information about the construction of paramagnetic films.

The problem in interpretation of EPR spectra of textured paramagnetic complexes in liquid crystals, or radicals in prolated polymeric compounds was already discussed in literature. We can find some experimental and theoretical papers on this problem [1-3]. From the point of view of an experimental technique the EPR spectra acquisition and mathematical simulation algorithm for paramagnetic centers in a flat surface film do not differ from centres in oriented liquid crystal solutions or inserted in prolated polymeric compounds. Solving the problem in a general form requires triple integration and is possible only by numerical methods. Attempts to solve this problem analytically were successful only in very particular cases. The techniques proposed in these papers based on theoretical tables and plots is not obvious enough.

Our approach is based on numerical simulations and became possible only with the development of high performance computer technology. We have carried out analytical calculations necessary to design a simulation algorithm.

2. MATHEMATICAL PART

The interpretation of magnetic resonance experimental data is usually based on a spin Hamiltonian (SH) method. The concrete SH selection is determined by features of an explored paramagnetic centre. In our program we were restricted to objects with one unpaired electron and one magnetic nucleus

$$H = \beta H\{g\}S + S\{A\}I, \quad (1)$$

where {g} and {A} are tensors, and β is the Bohr magneton.

The value of hyperfine interaction is assumed to be small enough as compared with the Zeeman interaction energy of an unpaired electron with a stationary magnetic field.

In the given example, only the first part of the SH was used. For an anisotropic g-factor, the position of the resonant transition line for a single paramagnetic centre can be found by the following equation

$$H_{res} = (h\nu/\beta)/(g_{zz}^2 k^2 + g_{yy}^2 l^2 + g_{xx}^2 m^2)^{1/2}, \quad (2)$$

where k, l and m define the direction cosines of Euler's angles in a natural frame ***h ν*** - quantum energy of a microwave frequency. This means that, during a magnetic field variation, the position of a magnetic resonance line will vary from a minimum value corresponding to orientation with the maximal g-factor up to a value due to the minimum g-factor.

To find an EPR spectrum of a sample we need to sum up the contributions of all orientations, with allowance for their probabilities and anisotropy of the individual line width. First of all, for this purpose it is necessary to find expressions for transformation of dependencies, given in different frames. The probability of a paramagnetic centre presence and EPR signal strength is linked with the sample frame. The magnetic field of the spectrometer is given in a laboratory frame. The resonance position on the magnetic field scale is linked with the main frame of the paramagnetic centre.

Such a transformation is illustrated in Fig. 1. The sample plane is preset by the axes ***v*** and ***h***, which, together with the normal line ***n***, make the frame of the sample. The sample is located in a magnetic field which can be rotated in a plane ***nh*** from the direction along the normal line ***n*** up to a position parallel to the axis ***h*** in the sample plane. Generally, the field vector makes an angle γ with the normal line. The paramagnetic centre can be oriented arbitrarily to the axes of the sample system.

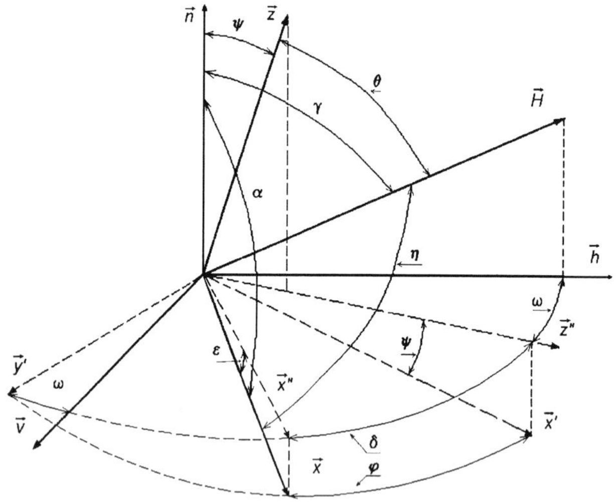

Fig. 1. Correlation of a sample (vhn) frame with laboratory (H) and molecular (xyz) frames.

The current position of its z axis is deflected by an angle ψ relative to the normal line $\boldsymbol{n}$, and the projection of the axis z onto the plane $\boldsymbol{vh}$ makes an angle ω with the axis $\boldsymbol{h}$. The angles ψ and ω are integration variables, they range from 0 up to π for ψ, and from 0 up to 2π for ω. Taking into account symmetry reasons of the sample relative to the magnetic field, the range can be decreased from 0 up to $\pi/2$ for ψ and from 0 up to π for ω. After the z axis is fixed by the angles ψ and ω, the paramagnetic centre still has rotation freedom of the $\boldsymbol{x}$ and $\boldsymbol{y}$ axes around the z axis. Thus, the angle between the axes $\boldsymbol{x}$ and $\boldsymbol{n}$ varies from a maximum value $\pi/2 + \psi$ up to minimum $\pi/2 - \psi$, the same is true for the $\boldsymbol{y}$ axis. For a unique assignment of orientation of a paramagnetic centre frame, the angle φ is introduced, as an angle of rotation of the $\boldsymbol{x}$ axis around the z axis from the position of a maximal angle with the $\boldsymbol{n}$ axis, at which the projections of axes the $\boldsymbol{x}$ and z onto the plane of the sample $\boldsymbol{vh}$ coincide. The range of angular variation of φ is 2π. Simulation of the texture spectrum of a texture requires integration over the entire range of φ.

To find the resonance line positions, one has to define direction cosine squares of the vector H in the coordinates ***xyz***. It is sufficient to define only two values out of three and to use the normalization condition: $k^2 + l^2 + m^2 = 1$. In Fig. 1, these are the angles θ and η

$$k = \text{Cos}(\theta) = \text{Cos}(\psi)\,\text{Cos}(\gamma) + \text{Sin}(\psi)\,\text{Sin}(\gamma)\,\text{Cos}(\omega) \tag{3}$$

$$m = \text{Cos}(\eta) = \text{Cos}(\alpha)\,\text{Cos}(\gamma) + \text{Sin}(\alpha)\,\text{Sin}(\gamma)\,\text{Cos}(\omega + \delta) \tag{4}$$

where

$$\text{Sin}(\delta) = \text{Sin}(\varphi)\,/(\text{Sin}^2(\varphi) + \text{Cos}^2(\varphi)\text{Cos}^2(\psi))^{1/2}, \tag{5}$$

$$\text{Cos}(\delta) = \text{Cos}(\varphi)\text{Cos}(\psi)/(\text{Sin}^2(\varphi) + \text{Cos}^2(\varphi)\text{Cos}^2(\psi))^{1/2}, \tag{6}$$

$$\text{Sin}(\varepsilon) = \text{Sin}(\psi)\text{Cos}(\varphi), \tag{7}$$

$$\text{Cos}(\varepsilon) = (\text{Sin}^2(\varphi) + \text{Cos}^2(\varphi)\text{Cos}^2(\psi))^{1/2}, \tag{8}$$

$$\delta = \pi/2 + \varepsilon. \tag{9}$$

As a result we obtain

$$\text{Cos}(\eta) = \text{Cos}(\omega)\text{Cos}(\varphi)\text{Cos}(\psi)\text{Sin}(\gamma) - \text{Cos}(\varphi)\text{Sin}(\psi)\text{Cos}(\gamma) - \text{Sin}(\omega)\text{Sin}(\varphi)\,\text{Sin}(\gamma). \tag{10}$$

The probability function is determined in the sample frame depending on the angles between the axes z and x, and the normal $\boldsymbol{n}$. The dimension of the distribution function depends on the symmetry of paramagnetic centres described by it. In the axial case of symmetry of the sample and axial symmetry of paramagnetic centres we have $W(\psi)$ depending on one parameter. If the symmetry of paramagnetic centres is lower than axial, the probability function depends on two parameters – $W(\psi,\varphi)$. In textured samples, the probability of finding a paramagnetic centre with some orientation of the magnetic field relative to its main frame may be essentially different. It is determined by the paramagnetic centre construction and the way of sample preparation. In the our program, we have used a two-dimensional function of the form:

$$W(\psi,\varphi) = W_{\text{rand}} + W_0\exp\{-\ln(2)(\psi-\psi_0)^2/\sigma^2_{\psi 0}\}\,\exp\{-\ln(2)(\varphi-\varphi_0)^2/\sigma^2_{\varphi 0}\} + W_1 + \ldots \tag{11}$$

The transformation to an axial symmetry of a paramagnetic centre means that the dispersion σ_φ tends to infinity and all φ orientations become equally probable. The total EPR spectrum is calculated by summation of discrete orientation spectra, after passing all the grid of angles ψ, ω, φ. The individual line shape function can be Lorentzian or Gaussian.

The mathematical model offered is rather flexible and does not impose significant restrictions on the sample construction. Under the restrictions described, the model is applicable for description of ordering for the majority of copper(II) complexes (in the absence of additional splitting on the ligand nuclei), long living organic nitroxylic radicals with one nitrogen nucleus split, and films of crystalline paramagnetic matters.

Hereinafter the mathematical model was transformed to a computer program, which allows one to simulate spectra, to compare them with experimental data and to store spectra in the program WinEpr format (Bruker corporation).

The program runs in a dialog mode and allows one to change orientation and comparison spectrum without leaving it. Thus, during a session one can find all spin Hamiltonian parameters and orientation performances of a texture. The final spectrum satisfying the researcher can be saved in the memory for further usage.

The principal values of g- and A- tensors or even part of them can often be found from an EPR spectrum of a powdered sample. In the present case we proceeded as follows. The paramagnetic film from the examined or test sample was mechanically removed with the help of an inert material powder. After that we recorded the powder spectrum of the analysable matter together with an inert powder and simulated the powder EPR spectrum. For simulation we use both our own program and the program Simfonia from Bruker corporation (Shareware version), the results are identical. If the main values of magnetic resonance parameters are known, the fitting operation is reduced to ordering function fitting.

We deliberately restrict the opportunities of our program to an electronic spin 1/2 and only one magnetic nucleus without second order corrections, to speed up calculations of particular systems. With occurrence of more complex experimental systems, further development of the program opportunities is implied.

3. Experimental part

The purpose of the experimental part of the project was a functional test of the theoretical model on real films. We chose films of divalent copper phthalocyanine and dipivaloyl methanate obtained by sedimentation of volatile complexes on the surface of quartz plates. We tried to obtain the most ordered films and for this purpose changed the temperature of the evaporator, time of sedimentation, distance between the evaporator and sample, and the film thickness. Typical experimental values were: evaporator temperature 400 – 450^0 for copper phthalocyanine and 100 – 110^0 for copper dipivaloil methanate, quartz plate was at ambient temperature 20 – 70 mm apart from the evaporator. Ordered films with 0.5 up to 1 μm width were obtained in vacuum ($5 \cdot 10^{-5}$ torr) during 1 – 5 minutes. The orderliness control was provided by EPR spectra simulation. Only the most ordered samples are shown below.

EPR spectra were obtained by the EPR-spectrometer Varian E-109. To increase splitting, we have used the Q-range of the EPR. The serial spectrometer was equipped with an AD-converter and original software for PC, permitting to accumulate weak signals and to save experimental spectra in WinEpr format. The device construction allows to change with good precision the magnetic field direction in relation to the sample plane by turning the magnet.

The films were polycrystalline by nature with well expressed angular dependence of the EPR spectra. Hyperfine splitting of spectra was not observed, which corresponds to a magnetically concentrated phase with considerable exchange.

3.1. Copper phthalocyanine film on a quartz slice surface

Copper(П) phthalocyanine is a very stable volatile complex, which has a number of remarkable properties. Axially symmetric phthalocyanine molecules in the solid phase are packed in stacks from which the crystalline lattice is created. The angle between the straight line bridging copper atoms in a stack and the direction of the z axis of an individual ion frame differs for α and β crystalline modifications (cf. Fig. 2). It is known from literature [4] that phthalocyanine complex stacks on the substrate surface may be ordered in different ways depending on the substrate nature. They may lie along the surface in one case and stay orthogonal to it on another surface.

The magnetic exchange is considerable only inside stacks, but is not extended between them. As the individual z axes of Cu ions in stacks are parallel to one another, such exchange results in averaging of the hyperfine splitting of EPR spectra, but the anisotropy of Zeeman interaction of an individual ion is completely retained.

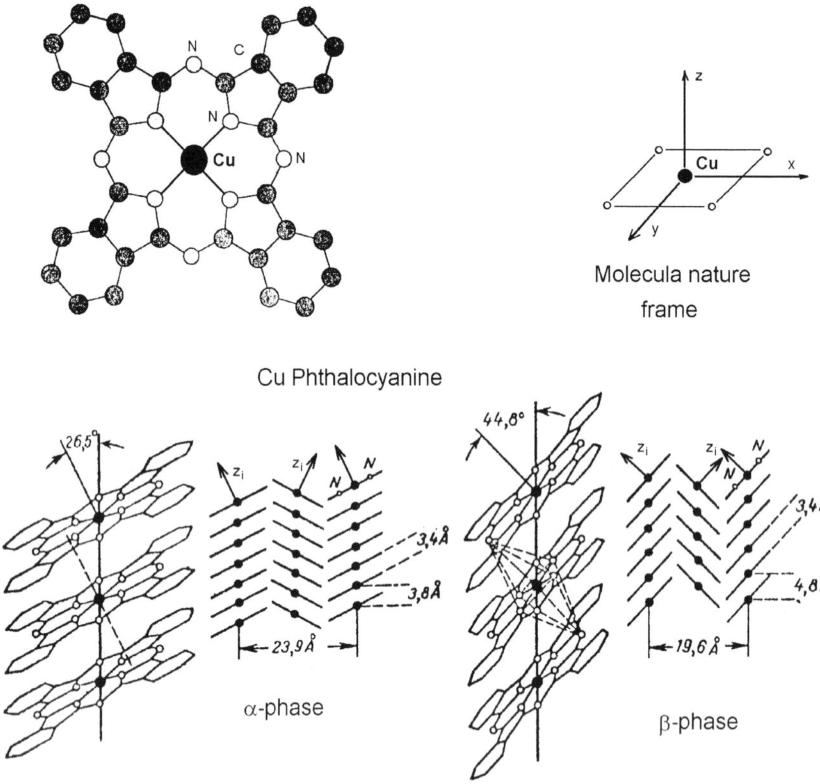

Fig. 2. The chemical formula of a copper phthalocyanine and layout of molecules in a crystalline lattice.

The angular dependence of EPR spectra of copper phthalocyanine films and the results of simulation with the help of our program are shown in Fig. 3. This remarkable picture was obtained using the following angular distribution function (cf. Fig. 4).

The ordered phase contains 95% of the total amount of copper phthalocyanine in the sample. The obtained angle of 26^0 and very small dispersions allow us to conclude that this sample consists of a practically pure ordered α-phase. Its crystalline axis b is oriented along the normal line to the plane of the slice. The degree of ordering depends on the film thickness - thin films had no such a good ordering. It was also necessary to move the target away from the evaporator as far as possible and to decrease the deposition rate of matter.

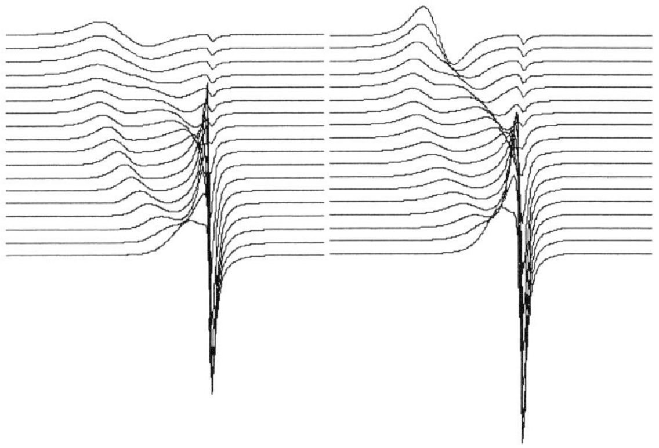

Fig. 3 EPR spectra angular dependence (left) and the result of EPR spectra simulation (right) for Copper phthalocyanine film

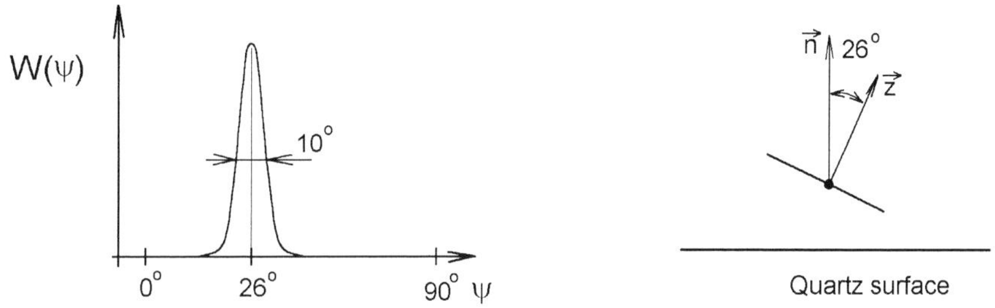

Fig. 4. Orientation probability distribution (left) and ordering of molecule on the surface (right).

3.2. Copper dipivaloil methanate film on a quartz slice surface.

Copper(П) dipivaloil methanate film is a planar electrically neutral chelate complex. The central paramagnetic copper(II) ion is surrounded by four nearest oxygen atoms lying in one plane. The nearest environment of the Cu ion compensates for its charge. The individual magnetic *z* axes of the ion are located normally to the plane of oxygen atoms. In a crystalline state, the matter consists of molecular layers, where the *z* axes of the copper ions are parallel to one another, but not normal to the layer. The molecular frames in two neighboring layers are not parallel (cf. Fig. 5).

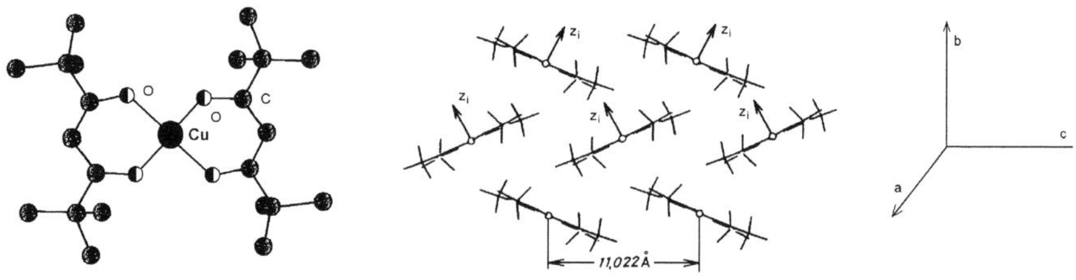

Fig. 5. The chemical formula of copper dipivaloil methanate (left), layout of molecules in the crystalline lattice (centre), crystalline and magnetic frame (right).

In contrast to the copper phthalocyanine complex, magnetic exchange in the copper dipivaloil methanate complex is significant in all directions and averages not only the hyperfine but also the Zeeman part of interactions of all ions in a given direction. The resulting g-factor has a three-axis anisotropy and its main frame coincides with the axes of a crystalline lattice frame.

The angular dependence of EPR spectra and the results of simulations are shown in Fig. 6. These results were obtained using the two-dimensional distribution function shown in Fig. 7. It corresponds to orientation of microcrystals as shown in the figure. The crystalline *c* axis in such a film lies along the sample surface and the normal line ***n*** cross over the *ab* plane of the crystalline frame with the angles 19^0 and 71^0.

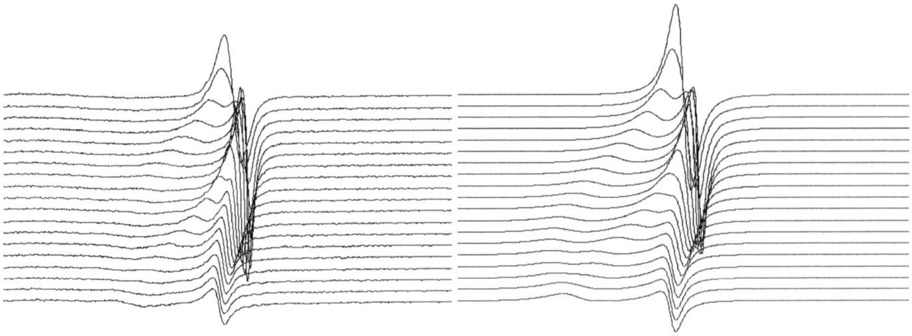

Fig. 6. Angular dependence of EPR spectra of a copper dipivaloilmethanate (left) and results of simulation (right).

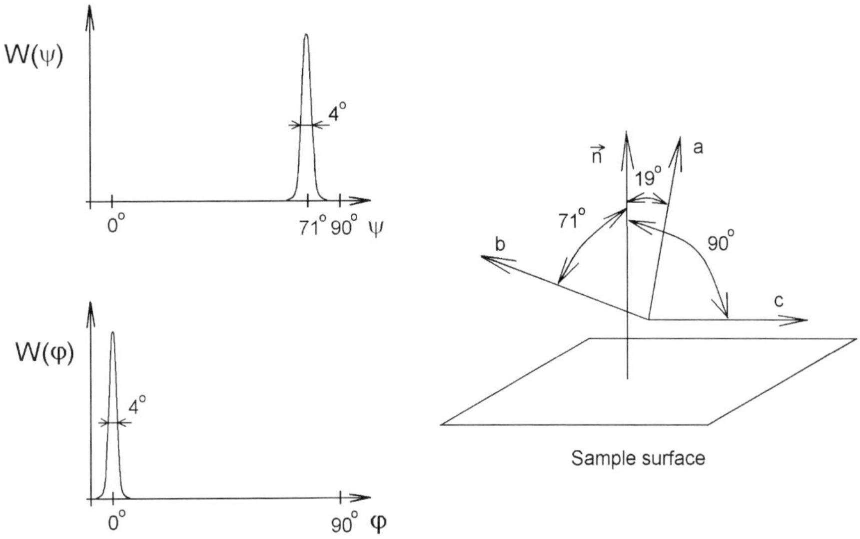

Fig. 7. Two-dimensional probability distribution of the copper dipivaloil methanate in the film (left) and main orientation of crystallites on the surface (right).

In this case, the film was even more ordered. When the deposition rate and the film thickness were increased, a non-oriented phase appeared in the EPR spectra. The contents of the non-oriented phase in the investigated samples varied from 0 up to 80%. The first actually obtained film contained a significant amount of the disordered phase, but later we found a mode of deposition to completely remove it. Thin films had the best ordering, and in this aspect the copper dipivaloil methanate differ from the copper phthalocyanine.

4. Conclusion

The designed procedure allows us to obtain information on the construction of paramagnetic films. The obtained information has been used for the optimisation of volatile compound deposition processes from the gas phase to make the highly ordered films of copper phthalocyanine and dipivaloil methanate. Optimal conditions for the acquisition of films for the studied compounds were found to be different. In the copper phthalocyanine case the films grow up along the crystallographic axis b, and thick films are the most ordered. The allocation of the crystallographic axis is quite different for copper dipivaloil methanate, and thin films are maximally ordered.

5. Acknowledgments

Financial support for this study was provided by the Russian Foundation of Basic Research, grant N00-03-32553, and the Netherlands organization for scientific research (NOW) under grant "*Development of Electron Spin Probes for in Vivo Oxymetry, pH Measurements, ESR Imaging and Localized Spectroscopy*".

6. References

1. G.M. Zhidomirov, Ja.S. Lebedev and S.N. Dobrjakov. Interpretation of complex EPR spectra, Moscow, "Nauka", 1975, 216 pp (russ).

2. I.V. Ovchinnicov. EPR of coordination compounds in molecular mediums with the partial orientation order, Kazan,. Kazan State University, 1980 (russ).

3. V.N. Konstantinov, I.V. Ovchinnicov and N.E. Domracheva, Russian J. Structural Chemistry, 25 (1984) 19.

4. M. Nakamura and H Tokumoto, Surface Science 398 (1998) 143

Novel Methods to Study Interfacial Layers
D. Möbius and R. Miller (Editors)

ELECTRONIC STRUCTURE OF ORDERED LANGMUIR-BLODGETT FILMS OF AN AMPHIPHILIC DERIVATIVE OF 2,5-DIPHENYL-1,3,4-OXADIAZOLE

M.B. Casu[1]*, P. Imperia[1], S. Schrader[1], B. Schulz[1], F. Fangmeyer[2], H. Schürmann[2]

[1]Universität Potsdam, Institut für Physik, Lehrstuhl Physik kondensierter Materie, Potsdam, Germany

[2]Universität Osnabrück, Fachbereich Physik, Osnabrück, Germany

Contents

Keywords: Conjugated molecules, Semi-empirical models and model calculations, Synchrotron radiation photoelectron spectroscopy, Near edge X-ray absorption fine structure

*Corresponding author: tel: +49 331 977 1046; fax: +44 331 977 1083;
e-mail: casu@rz.uni-potsdam.de

Electronic structure of ordered films obtained from an amphiphilic substituted 2,5-diphenyl-1,3,4-oxadiazole (NADPO-11) by Langmuir-Blodgett (LB) technique were studied, for the first time, by means of ultraviolet photoelectron spectroscopy (UPS). The complete one-dimensional valence band structure has been determined from the UPS spectra. The spectral features that correspond to the top of the valence band and that dominate the electronic properties of this material were assigned to the different molecular eigenstates by comparison with simulated spectra from quantum chemical calculations. Good agreement between simulated and experimental curves allowed to assign UPS peaks to certain groups of molecular orbitals (MOs) and to understand the electronic bonding state (σ- or π- type) of the involved MOs. Furthermore, we performed UPS measurements on in-situ evaporated amorphous films of the same substance, comparing their valence band features to the previous ones. The correlation between ionisation potential and electron affinity, determined by combined ultraviolet photoelectron spectroscopy/optical spectroscopy, and carrier-transport and light emitting properties of these material are also discussed.

1. Introduction

A promising way to improve the performance of organic light emitting devices (OLEDs) is to match suitable electron transporting (ET) and hole transporting (HT) layers in a heterolayer device in order to optimise the recombination in the emissive material [1].

This approach requires a detailed characterisation of the materials candidates to act as ET or HT layers in OLEDs. Angle Resolved Ultraviolet Photoelectron Spectroscopy (ARUPS) on ordered films provides not only their experimental electronic band structure but also parameters like ionisation potential (I_P) and electron affinity (E_A) that are of crucial importance in organising better OLED configurations [2-4]. In this work we investigated an amphiphilic derivative of 2,5-diphenyl-1,3,4-oxadiazole by means of Ultraviolet Photoelectron Spectroscopy (UPS) and ARUPS. This structure is based on a very stable moiety [5, 6] and the family of substituted 2,5-diphenyl-1,3,4-oxadiazole can be used as emitting as well as hole blocking/electron transporting layer in OLEDs [7].

2. Principles of Ultraviolet Photoelectron Spectroscopy

The photoemission process is the emission of electrons from a material when it interacts with a proper incident electromagnetic radiation, $E_{inc} = h\nu$; this is a process that, in principle, should be described with a quantum mechanical approach. On the other hand a semi-classical model, the three step model [8], gives the opportunity to interpret in a simpler but effective way the photoemission events.

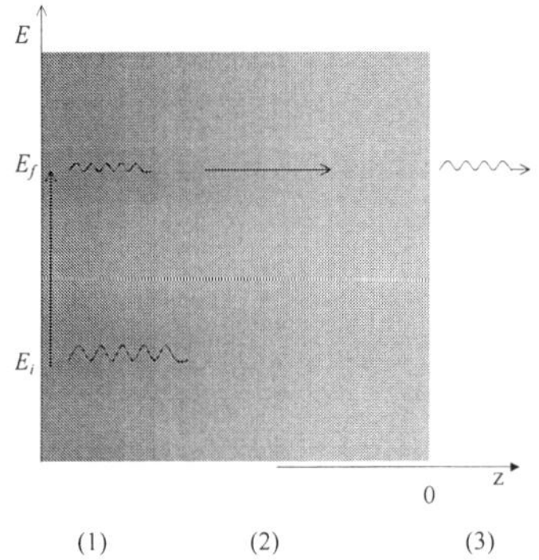

Fig. 1 Scheme of the three steps model: (1) optical excitation of a wave packet; (2) diffusion; (3) transmission through the surface.

In the three steps model (Fig. 1) the photoemission process is broken up into three sequential steps: the optical excitation of the electron inside the bulk, the diffusion of the electron to the surface and its transmission through the surface into the vacuum. Only those electrons that have enough kinetic energy along the direction normal to the surface are able to overcame the surface potential barrier and to escape, the others are totally reflected.

In this work we will analyse photoemission process occurred in organic films in the UPS-regime: the aim of this article is not to give a deep picture of these technique that is already done in several books and articles [2, 9-12], but we will briefly describe the basic principle and what is possible to figure out from UPS investigations building a proper background in order to explain our results.

In UPS-regime $h\nu$ is below 100 eV, the photon momentum is negligible and the photoexcited electrons have the same momentum in the initial and in the final state, it means that direct transitions occur; UPS spectra show energy and intensity variations depending on the incident energy and this gives the opportunity, by choosing proper methods that we will show later, to map out the band structure of well ordered materials.

The kinetic energy of the emitted electrons is given by the Einstein relation:

$$E_{kin} = h\nu - E_b = E_{inc} - E_b \tag{1}$$

where E_b is the binding energy of the electrons before excitation relative to the vacuum level *VL*. In a metal the maximum kinetic energy, $E_{kin}^{\max}$, that one electron escaping from the emitter can have is:

$$E_{kin}^{max} = h\nu - \Phi_m, \tag{2}$$

where Φ_m is the work function of the metal.

If a thin organic layer is deposited on the metal surface, the photoelectrons come from the organic layer and the photoelectrons with highest kinetic energy, $E_{kin\ org}^{\max}$, come from the highest occupied molecular orbital (HOMO), consequently it is possible to obtain the ionisation potential, I_P, of the organic layer as:

$$I_P = h\nu - E_{kin\ org}^{\max} \tag{3}$$

Several other physical parameters can be obtained from the UPS spectra like the alignment of the HOMO of the organic layer with respect to the Fermi level, E_F, of the metal; the *VL* energy shift of the organic layer at the interface metal/organic with respect to the *VL* of the metal and also the energy of the *VL* of the organic layer relative to the E_F [2].

This means that UPS is a very powerful technique to be used to study the interfacial electronic structure and the energy alignment at the organic/metal and organic/organic interfaces.

But mainly, UPS spectra are strongly dependent on valence and conduction bands and they give direct information on the valence band structure. In particular when a well ordered thin organic film is investigated with this technique, the spectra are very sensitive to the variation of the incident energy and to the variation of the angles of incidence, α, the take off angle, θ, and the azimuth angle, φ, (cf. Fig. 2), therefore ARUPS allows to determine the energy-band dispersion relation $E(k)$ between the energy E and the wave vector k of the photoelectrons.

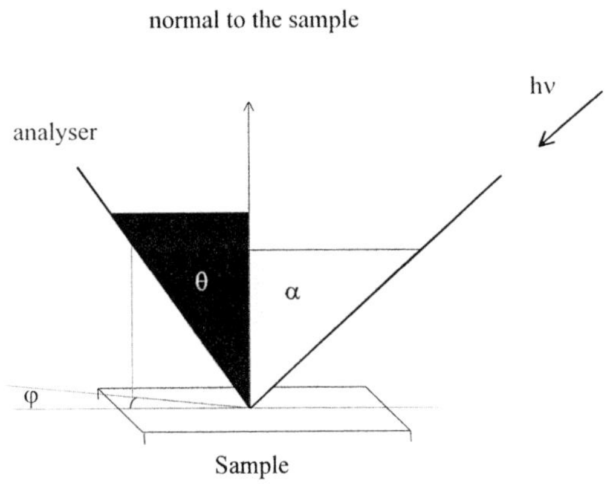

Fig. 2 Geometry of the UPS experiment.

To obtain the *k*-resolved band structure, the peaks from the recorded spectra are to be related to *k*-conserving transitions from occupied (E_i) to unoccupied bands (E_f) of the material, satisfying the equation

$$E_f = E_i \pm h\nu \tag{4}$$

in which direct transitions are assumed.

The measured quantity is the kinetic energy of the photoelectrons, E_{kin}, detected for a given energy and fixed angles α, θ and φ, relatively to the normal to the sample:

$$E_f = E_{kin} + \Phi \tag{5}$$

where Φ is the work function of the organic layer.

The kinetic energy and the momentum of the electron in the vacuum are related according to

$$E_{kin} = \frac{\hbar^2}{2m_e}\left(k_\perp^2 + k_\parallel^2\right) \tag{6}$$

where $k_\perp$ and $k_\parallel$ are the components of the momentum of the electron, perpendicular and parallel, respectively, to the surface.

Since in a photoemission process only the component of the wave vector parallel to the surface is conserved, because the translation symmetry is broken by the surface, this component will be given by the following equation:

$$k_{\|} = \left(\frac{2m_e}{\hbar^2} E_{kin} \right)^{1/2} \sin\theta \tag{7}$$

To determine $k_{\perp}$ the free-electron model has been widely used [13, 14]. In this model the dispersion relation of the final state is assumed to be:

$$E_f = \frac{\hbar^2}{2m_e} (\mathbf{k} + \mathbf{G})^2 + E_0 \tag{8}$$

where $\mathbf{G}$ is a reciprocal lattice vector, $E_v - E_0$ is the potential well depth and the energies are measured with respect to the Fermi level. The only parameter to be determined is the inner potential $V_0 = E_v - E_0 = |E_0| + \Phi$.

In normal emission $k_{\|} = 0$, therefore the momentum normal to the surface is given by

$$k_{\perp}^2 = \frac{2m_e}{\hbar^2} (E_{inc} - E_b + V_0) \tag{9}.$$

We estimated the inner potential, by combining theoretical calculations on one molecule with optical spectroscopy and UPS data on solid state, as an adjustable parameter that takes into account several factors like the potential step at the surface and the polarisation energies due to the multi-electronic effect in the condensed state.

By means of equations 7 and 9, it is possible to map out the complete band structure of the materials, using different crystal faces.

3. Experimental

We investigated an amphiphilic substituted 2,5-diphenyl-1,3,4-oxadiazole (NADPO-11), (Fig. 3) as ordered film (LB NADPO-11), obtained by Langmuir-Blodgett (LB) technique, and as well as amorphous film (AM NADPO-11).

The UPS measurements were performed at the beam line TGM-2 at BESSY in Berlin. This line was characterised by a three gratings monochromator which covered the photon energy range from 8 to 180 eV. The photoelectrons were collected with a Vacuum Generators ADES 400 angle resolving spectrometer system at room temperature. The base pressure during the experiment was 2 x 10^{-10} mbar. The resolution of the system monochromator-analyser (80 meV) was determined from the measured Fermi edge of a freshly evaporated gold film. The photoelectron spectra were measured in normal emission for various incident energies and with a fixed incident energy for various angles of emission (Fig. 2).

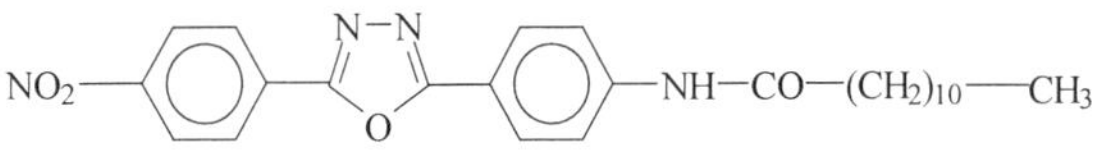

Fig. 3 Chemical structure of investigated substance.

The LB film deposition was performed on a NIMA 622 alternating trough (NIMA Technology, Coventry, UK) equipped with a Wilhelmy plate surface pressure sensor. The spreading solutions were obtained by dissolving 0.9 mg/ml of substance in chloroform. The monolayers were formed by spreading 200 μl of the solution on the water subphase provided by a Milli-Q system (Millipore). The subphase temperature was held constant at 21 °C. The films were compressed with a constant barrier speed of 50 cm^2/min. Multilayers of 10 monolayers were deposited on cleaned silicon wafers that were hydrophobized by treatment with hexamethyldisilazane (HMDS) before the deposition. A target pressure of 28 mN/m and a dipper speed of 5 and 10 mm/min yielded an observed transfer ratio between 0.8 and 1 (Y-type transfer).

The amorphous material was evaporated in-situ, in a preparation chamber attached to the main measuring unit, from pinhole-sources onto fresh deposited Au films on flat silicon wafers. Film thickness was monitored with a quartz micro-balance (120 nm). During the evaporations the pressure increased up to 10^{-6} mbar. The direct transfer from the preparation chamber to the measuring unit was possible, without breaking the vacuum, using a magnetic transfer line.

We also directly determined the ionisation potential from the HOMO-onset by a linear extrapolation and the secondary-electron cutoff and the electron affinity defined as the ionisation potential minus the optical band gap.

4. RESULTS AND DISCUSSION

In Fig. 4 an energy series performed on LB NADPO-11 is shown, the incident energy was varied between 27.9 and 72.1 eV, the spectra were recorded in normal emission.

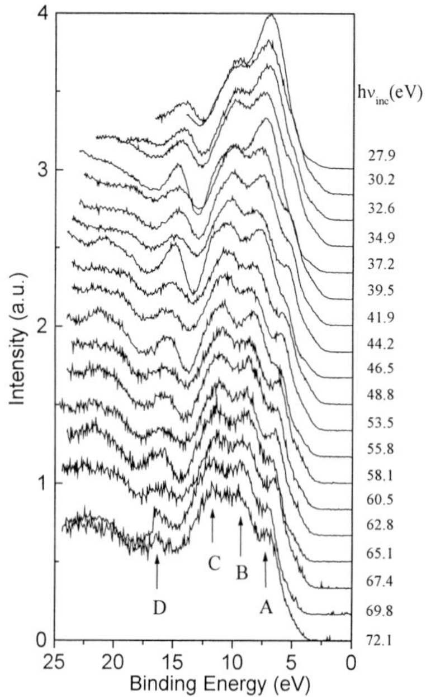

Fig. 4 UPS spectra of LB NADPO-11. They were measured with an angle of incidence of 45° and in normal emission. The incident photon energy is also indicated. The valence band is given with respect to the vacuum level set at zero. All spectra are normalised by the intensity of peak D. Vertical offsets are used for clarity.

The binding energy is referred to the vacuum level set at zero. Figure 5 shows the ARUPS spectra of LB NADPO-11. The incident energy was 46.5 eV and the take off angle θ was varied between 45° and 115°.

The spectra show four well defined peaks, labelled A, B, C and D, respectively, and a broader one at higher energy. As expected, because of the high ordered structure of the LB film, the peaks disperse with the increasing of photon energy (Fig. 4) or with increase of the take off angle (Fig. 5). In particular, depending on the incident energy, the peak labelled A starts to be evident when the incident energy is higher then 34.9 eV; the ratio of the intensity of peaks B and C is also energy dependent: 39.5 eV is a threshold value at which the intensity ratio between peaks B and C changes in favour of the latter.

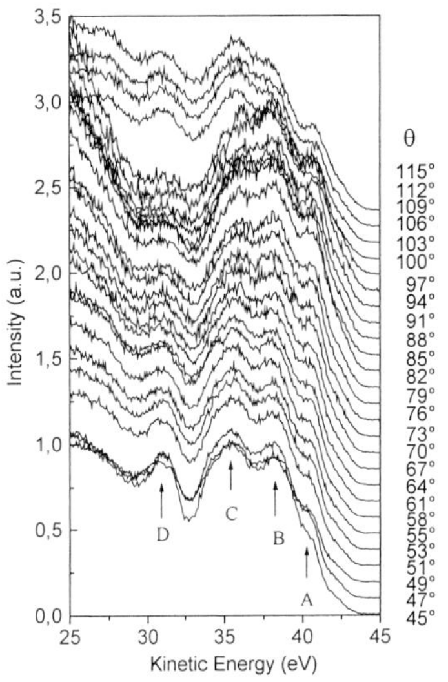

Fig. 5 ARUPS spectra of LB NADPO-11. The incident energy was 46.5 eV and the take off angle θ was varied between 45° and 115°. All spectra are normalised by the intensity of peak D. Vertical offsets are used for clarity.

As discussed before, by using ARUPS spectra and equations 7 and 9, it is possible to obtain the band structure of LB NADPO-11.

We estimated the inner potential of NADPO-11 to be $V_0 = (5.0 \pm 0.5)\,\text{eV}$ and we used this value in the equation 9.

In Fig. 6 the experimentally determined band structure of LB NADPO-11, along the direction perpendicular to the surface, is shown. The result in off-normal emission is shown in Fig. 7.

The comparison of the bands along the two directions gives the evidence that the molecules are oriented in the direction perpendicular to the surface since the peaks are less sensitive to the change of the take off angle θ in off-normal emission.

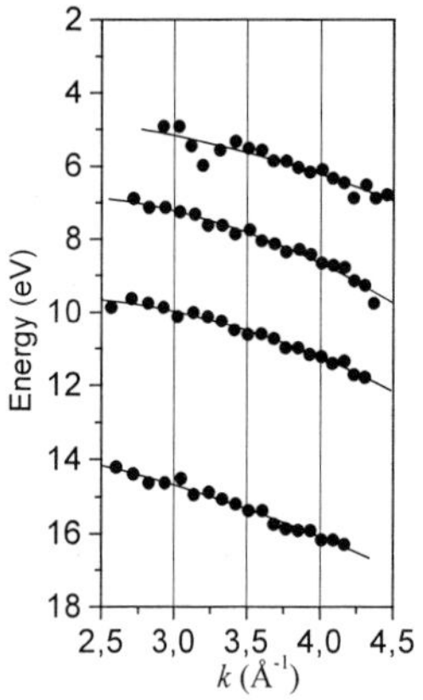

Fig. 6 Experimental band structure of LB NADPO-11 along the direction perpendicular to the surface. The straight line is a guide for the eyes.

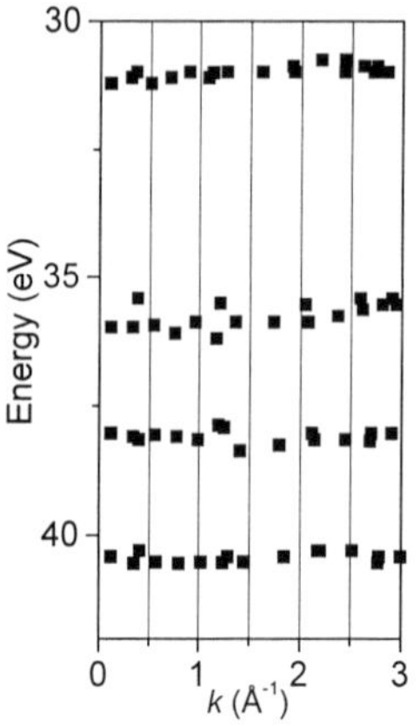

Fig. 7 Experimental band structure of LB NADPO-11 along the direction parallel to the surface.

To have a more deep feature of the band structure of LB NADPO-11, the further step will be also to calculate the theoretical bands and to compare them with the experimental ones.

Figure 8 shows the valence band spectra of AM NADPO-11, depending on the incident energies. They are also characterised by the four peaks A, B, C, D. As expected, investigating an amorphous sample, the energy series do not show any dispersion of the peaks with incident photon energy. This also indicates that no observable charging effects occurred. We directly determined the ionisation potential, 7.9 eV, from the UPS spectra and, by using optical spectroscopy, we also obtained the values for the electron affinity, 4.8 eV. These values are quite high in comparison with that ones of typical hole transporting materials like polyparaphenylenevinylene [15, 16].

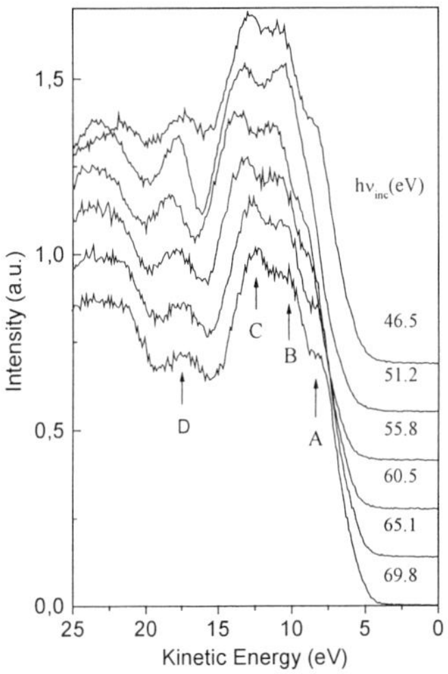

Fig. 8 UPS spectra of AM NADPO-11. They were measured with an angle of incidence of 45° and in normal emission. The incident photon energy is also indicated. Vertical offsets are used for clarity

Furthermore we estimated the energy level alignment [2] at the organic-Au interface, determining the shift in the UPS spectrum of the organic material in comparison with the spectrum that comes from gold. The results are summarised in Fig. 9 that presents the obtained energy diagram.

In order to interpret the experimental features that characterise the spectra of LB NADPO-11 and AM NADPO-11, we also performed semi-empirical quantum chemical calculations using the AM1 method. The calculations were performed taking into account only a single molecule. They gave the complete set of molecular eigenstates. From a superposition of multiple Gaussians, FWHM=0.5 eV, one for each molecular orbital, we obtained the theoretical valence band spectrum. The effect due to Auger electrons was not considered in the simulations. The theoretical curve, compared with the experimental one, is shown in Fig. 10. The good agreement between experimental findings and theoretical results allows to relate each UPS peak with the contributions resulting from distinct groups of occupied molecular orbitals. In particular the lowest binding energy peak is essentially due to the HOMO. This emission is characterised, therefore, mainly by π–orbitals. These states influence the optical properties of the molecule and, hence, the properties of the material when it is used in a device.

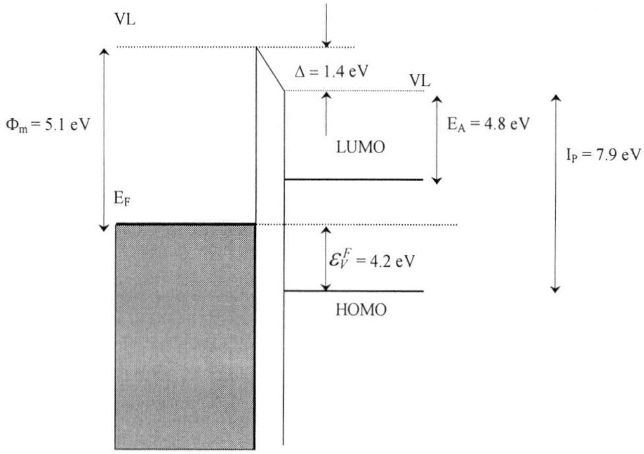

Fig. 9 Energy diagram of the organic-metal interface. VL: vacuum level; E_F: Fermi level of the metal; Φ_m: work function of the metal; I_P: ionisation potential; E_A: electron affinity; Δ: VL shift at the interface; HOMO: highest occupied molecular orbital; LUMO: lowest unoccupied molecular orbital; ε_V^F: energy of the HOMO relative to the E_F.

The main contribution to the peak B is also related to π–orbitals of the two phenyl rings. The neighbouring peak C is related with the emission from multi-centre σ-states that involve all the molecule. The peak at higher energy has a double nature contribution: on one hand it is due to emission from π–orbitals involving all the molecule, on the other hand to multi-centre σ-states again both from the oxadiazole moiety as well as from the substitute.

Taking into account the overall electronic feature of NADPO-11, obtained from UPS studies together with quantum chemical calculations, we can discuss its correlation with the charge transport properties of the material in a light emitting device. For this discussion we can use as a parameter the ionisation potential as it can give a hint about the behaviour of the materials in a OLED depending on its value [17]. The quite high values of ionisation potential and, consequently, of electron affinity, are expected because of the presence of the oxadiazole ring

[17]. This makes NADPO-11 a promising material which can be used as a hole blocking/electron transporting layer in emitting devices, presumably with very good electron injecting properties.

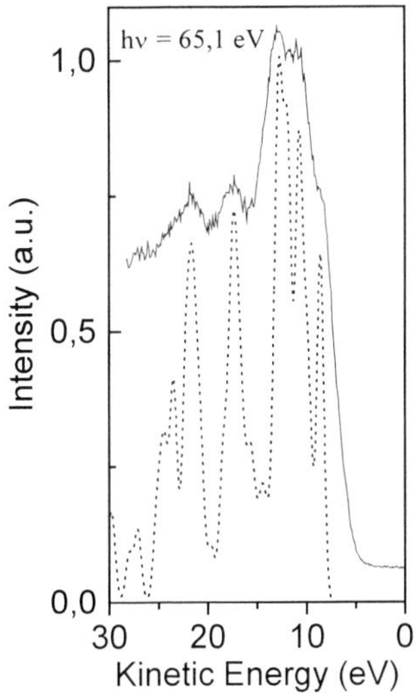

Fig. 10 Valence electronic spectrum (full lines) compared with the simulated spectrum (AM1, dot line) of NADPO-11. The incident photon energy is 65.1 eV. Vertical offsets are used for clarity.

5. Conclusions

We carried out UPS measurements on a Langmuir-Blodgett film of an amphiphilic substituted 2,5-diphenyl-1,3,4-oxadiazole, determining its complete *k*-resolved band structure. We also performed UPS measurements on an amorphous film of the same material, comparing them with the previous ones and determining the ionisation potential, the electron affinity and the energy alignment at the organic-metal interface. Furthermore, we were able to relate the main features of UPS spectrum to their molecular orbital contributions. The synergy between the use of experimental findings and theoretical calculation not only gives a complete picture of the electronic band structure of NADPO-11, but it also allows to understand its behaviour in OLEDs. The quite high value of I_P characterises NADPO-11 as a very interesting "hole blocking/electron transporting" material for OLEDs.

6. Acknowledgements

The authors would like to thank B. Dietzel, IDM Teltow-Berlin, Germany, for LB NADPO-11 sample preparation; BESSY staff, in particular M. Mast and O. Schwarzkopf for their help concerning experimental details. We would also like to thank Prof. M. Neumann, University of Osnabrück, Germany, Dr. C. Flueraru, Steacie Institute for Molecular Science, Ottawa, Canada, Prof. L. Brehmer, University of Potsdam, Germany, for helpful discussions. Dr. W. Regenstein, University of Potsdam, for his experimental support with regards to optical spectroscopy. Financial support of the European Commission under contract number FMRX-CT97-0106 (TMR-*EUROLED*) is gratefully acknowledged.

7. References

1. M.S. Weaver, D.G. Lidzey, T.A. Fisher, M.A. Pate, D.O. Brien, A. Bleyer, A. Tajbakhsh, D.D.C. Bradley, M.S. Skolnick and G. Hill, Thin Solid Films, 273 (1996) 39.
2. H. Ishii, K. Sugiyama, E. Ito and K. Seki, Adv. Mater. 11 (1999) 605.
3. A. Rajagopal and A. Kahn, Adv. Mater. 10 (1998) 140.
4. R. Schlaf, P.G. Schroeder, M.W. Nelson, B.A. Parkinson, P.A. Lee, K.W. Nebesky and N.R. Armstrong, J. Appl. Phys. 86 (1999) 1499.
5. Y. Hamada, C. Adachi, T. Tsutsui and S. Saito, Jpn. J. Appl. Phys. 31 (1992) 1812.
6. B. Schulz, M. Bruma and L. Brehmer, Adv. Mater. 9 (1997) 601.
7. Y. Kaminorz, B. Schulz and L. Brehmer, Synth. Met. 111 (2000) 75.
8. C.N. Berglund and W.E. Spicer, Phys. Rev. A 136 (1964) 1030.
9. R. Courths and S. Hüfner, Physics Reports 112 (1984) 53.
10. G. Ertl and J. Küppers, *Low energy electrons and surface chemistry*, 2nd edition, VCH Verl. Ges., Weinheim (1985).
11. W.R. Salaneck, S. Stafstrom and J.-L. Bredas, *Conjugated Polymer Surfaces and Interfaces: Electronic and Chemical Structure of Interfaces for Polymer Light Emitting Devices* , Cambridge Univ. Pr. (1996).

12. M. Pope and C.E. Swenberg, Monographs on the Physics and Chemistry of Materials 56, *Electronic Process in Organic Crystals and Polymers*, 2nd edition, Oxford University Press, New York (1999).

13. H. Dröge, M. Nagelstraßer, J. Nürnberger, W. Faschinger, A. Fleszar and H.P. Steinrück, Surface Science 454-456 (2000) 477

14. N. Ueno, K. Seki, N. Sato, H. Fujimoto, T. Kuramochi, K. Sugita and H. Inokuchi, Phys. Rev. B 41 (1990) 1176

15. D.D.C Bradley, Synth. Met. 54 (1993) 401.

16. J.L. Bredas, R.R. Chance, H. Baughman and R. Silbey, J. Chem. Phys. 76 (1982) 3673.

17. K. Sugiyama, D. Yoshimura, T. Miyamae, T. Miyazaki, H.H. Ishii, Y. Ouchi and K. Seki, J. Appl. Phys. 83 (1998) 4928.

Novel Methods to Study Interfacial Layers
D. Möbius and R. Miller (Editors)

STUDY OF ADHESION PROPERTIES OF POLYPROPYLENE SURFACES BY ATOMIC FORCE MICROSCOPY USING CHEMICALLY MODIFIED TIPS : IMAGING OF FUNCTIONAL GROUP DISTRIBUTION

Anne-Sophie Duwez* and Bernard Nysten

Unité de chimie et de physique des hauts polymères, Université catholique de Louvain,
Place Croix du Sud, 1, B-1348 Louvain-la-Neuve, Belgium.

Contents

Polypropylene surfaces were investigated by atomic force microscopy (AFM) using chemically modified probes. We show in this contribution that it is possible to locally detect additives on the surface of polypropylene, using tips modified with methyl and hydroxyl terminated self-assembled alkanethiol monolayers. The response of pure additives towards the chemical probes, and the medium wherein the force measurements are realized (water or nitrogen atmosphere), were determined in order to be able to further recognize and localize these compounds on polymer surfaces. The pull-off force measurements have shown that it is possible to discriminate between antioxidants, which show a hydrophobic behaviour, and UV-light stabilizers, which show a more hydrophilic behaviour. Similarly we have measured pull-off forces on a melt-pressed polypropylene sample stabilized with two antioxidants and one UV-light stabilizer. These adhesion forces measurements show that the extreme surface of the polymer presents a hydrophilic behaviour and is mainly composed of an UV-light stabilizer layer. We were able to obtain adhesion maps showing the distribution of these additives on the surface.

1. INTRODUCTION

Thoroughly controlled fabrication of highly defined ultra thin polymer films is of increasing importance in some technical fields such as microelectronics and sensors. The study of the reactivity and the chemical heterogeneity of polymer surfaces requires the development of surface chemical imaging tools allowing the analysis of complex and multifunctional systems. Owing to their capacity to give original information on surfaces at the nanoscopic scale, scanning probe microscopies have an important role to play in this context. A promising approach seems to be the combination of the unique lateral resolution of atomic force microscopies with designed chemical interactions between tip and sample [1-3]. The deposition or grafting of active molecules on AFM tips enables measurements of interaction forces between chemical groups on the probe tip and molecules present on the analysed surface. This is the principle of chemical force microscopy (CFM).

The ability of chemical force microscopy to image and discriminate areas exposing different functional groups on ultra thin films has no more to be proven. Interactions between chemically modified tips and various functional groups on the surface of self-assembled monolayers (SAMs) of alkanethiols have been extensively studied [4-8]. These intermolecular forces measurements have also been used in the case of biological materials to probe interactions between ligands and receptors, antigen and antibodies, or between complementary DNA strands [9-12] Most of the CFM experiments described in the literature relate to model surfaces but rarely on real complex systems. In the case of common surfaces, such as polymers, numerous factors (roughness, morphologies, mechanical properties, ...) impede the correct interpretation of adhesion measurements. Some groups have however succeeded in characterizing adhesion properties of modified polypropylene surfaces with chemical AFM probes [13-18].

In this context, this work aims at the development of chemical force microscopy techniques for chemical and molecular recognition at the surface of polymers. The properties of the topmost surface of polymeric materials are indeed crucial in many fields such as paint adhesion or biocompatibility. These properties can be disrupted by the presence of additives on the surface. Indeed, to improve the stability and increase the lifetime of polymers, additives protecting them against thermal oxidation and UV-light are widely used. UV-light stabilisers and antioxidants are usually present in the polymer matrix at a level of 1 wt % or less. But most of the additives have several disadvantages in applications and/or processing due to their propensity to migrate towards the polymer surface, causing blooming, adhesion failure, etc. [19].

In this contribution, we present the results of a study on the ability of chemical force microscopy to detect additives on polypropylene surfaces. Gold coated AFM tips modified with self-assembled monolayers of ω-functionalised alkanethiols (octadecanethiol or 11-mercapto-1-undecanol) were used to measure adhesion forces. Before investigating polypropylene surfaces, it was necessary to determine the response of the additives towards the chemical probes. The investigated additives (Irgafos 168, Irganox 1010, Tinuvin 770, and Dastib 845) are indeed large molecules containing several chemical functionalities.

In section 2, we describe the experimental details such as the procedure used for the preparation of the alkanethiol monolayers, and the conditions for the AFM data acquisition. In section 3, we undertake a systematic investigation of the adhesion forces between tips and pure additive films. Results obtained on a polypropylene sample containing three additives are also discussed. Concluding remarks are given in section 4.

2. EXPERIMENTAL SECTION

Materials. Gold substrates were prepared by evaporating titanium (50 Å) and gold (300 Å) onto silicon wafers (100) following the procedure described elsewhere [20] The obtained substrates are polycrystalline with a (111) preferential orientation, and a typical grain size of about 200-300 Å. Commercial Si_3N_4 tips (Microlever®, ThermoMicroscopes, Sunnyvale, CA) were also coated with a 50 Å Ti adhesion layer and a 300 Å Au layer, following the same procedure. Octadecanethiol and 11-mercapto-1-undecanol (98%, Aldrich) have been used as received. Polypropylene and additives (Irgafos 168, Irganox 1010, Tinuvin 770, and Dastib 845) were received from Clariant Huningue s.a.. Irgafos 168 is an antioxidant used as a process-stabilizing agent and Irganox 1010 is an antioxidant. Tinuvin 770 and Dastib 845 are UV-light stabilizers belonging to the family of hindered amine light stabilizers (HALS). The additives are depicted in Fig. 1. Water was deionised with a Millipore system to a resistivity of

18 MΩ cm. Ethanol (99.5%, Merck-Eurolab), dichloromethane and chloroform (99.8%, Merck-Eurolab) were used as received.

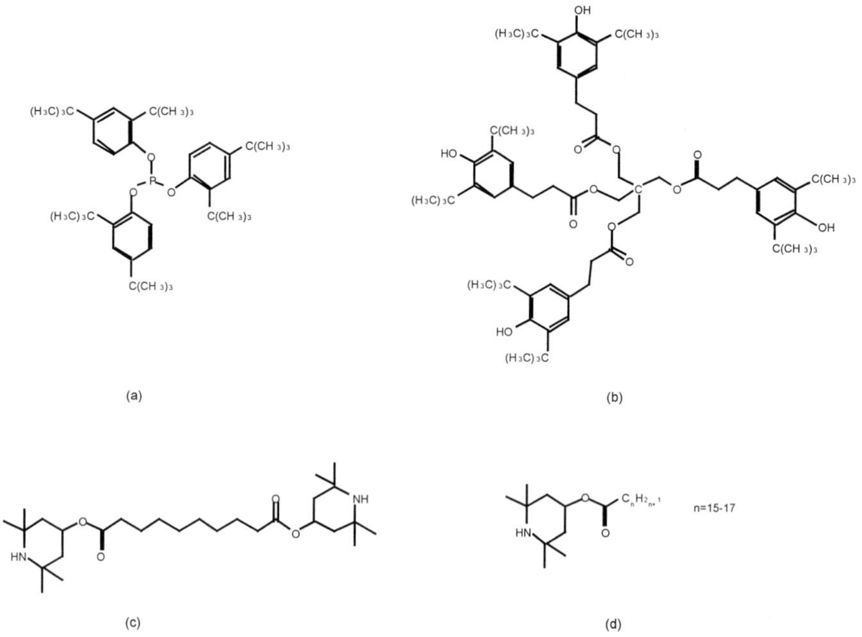

Fig. 1 Chemical structure of Irgafos 168 (a), Irganox 1010 (b), Tinuvin 770 (c), and Dastib 845 (d).

Sample Preparation. Prior to their use, the gold-coated tips and the gold substrates were cleaned to ensure a good adsorption of the thiols. They were freed from contaminants by UV/ozone treatment and the treated substrates were then immersed in ethanol to remove the gold oxide before monolayer assembly [21]. Adsorption was then realized by immersing the gold substrates for 18 h in 10^{-3} M solutions of alkanethiols using absolute ethanol as a solvent. Upon removal from the solution, the samples were rinsed twice with n-hexane and absolute ethanol, and gently dried in a nitrogen stream. The additive films were spin-coated from 10 g/l chloroform solution onto silicon wafers. Polypropylene plates were obtained by compression molding on a mirror-polished stainless steel mold. The molding conditions were the following : polymer melting at 220°C during 10 minutes, compression at 220°C during 2 minutes under a pressure of 20 MPa, and cooling during 10 minutes under a pressure of 20 MPa.

Monolayers quality. No atomic contamination could be found in the XPS spectra recorded from the thiol monolayers. The layers have also been characterized by contact angle measurements, ellipsometry, IRAS, HREELS, UPS and photoemission with synchrotron radiation. Some of these results are described in [22-25].

AFM measurements. AFM experiments were conducted with a PicoSPM equipped with a fluid cell and an environmental chamber (Molecular Imaging, Phoenix, AZ) and controlled by a Nanoscope III electronics (Digital Instruments, Santa Barbara, CA). We used silicon nitride cantilevers with a nominal spring constant of 0.1 N/m and an integrated pyramidal tip whose apex radius was typically equal to 50 nm. All AFM measurements were performed either in nitrogen atmosphere or in water rather than in air to eliminate or at least significantly reduce capillary forces. The adhesive interaction between tip and sample was determined from force vs cantilever holder displacement curves. In these measurements the deflection of the cantilever is recorded as the probe tip approaches, contacts, and is then withdrawn from the sample. The observed cantilever deflection is converted into force using the Hooke's law with the nominal cantilever spring constant. Force-displacement curves were obtained using the "Force Volume Imaging" mode. In this mode, an array of force curves over the entire probed area can be obtained. A force curve is measured at each *x-y* position in the area, and force curves from an array of *x-y* points are combined into three-dimensional array of force data. Force volume enables the investigation of the spatial distribution of the interaction force between the tip and the surface. Using a program developed under Igor Pro (Wavemetrics, Portland, OR), the force volume data were treated in order to extract from every curve the pull-off force or adhesion force and to generate adhesion maps of the probed areas. In order to eliminate possible changes that could affect the measured pull-off forces, experiments were realized with two independent tips at three different locations for each sample. For each experiment, the average value of adhesion forces was determined from 1024 force-distance curves. The reported values of adhesion force correspond to the average of the values measured for at least two independent tip-sample combinations.

3. Results and Discussion

3.1 Adhesion Force Measurements on Model Surfaces

Since common spectroscopic techniques are unable to characterize nanometer-sized tips, the quality of the functionalised tips was checked by measuring the adhesive interaction on model surfaces. The average values of adhesion obtained in water and in nitrogen atmosphere using tips and samples functionalised with SAMs terminating with either $-CH_3$ or $-OH$ groups are shown in Fig. 2. The observed trend in adhesive force agrees with the expectation that interaction between hydrophobic groups ($-CH_3/-CH_3$) will be larger in water than in N_2, whereas interaction between hydrophilic groups ($-OH/-OH$) will be lower in water than in N_2 [26, 27]. As expected, interactions forces obtained from $-CH_3$ tip on $-OH$ surface are similar to those obtained from $-OH$ tip on $-CH_3$ surface.

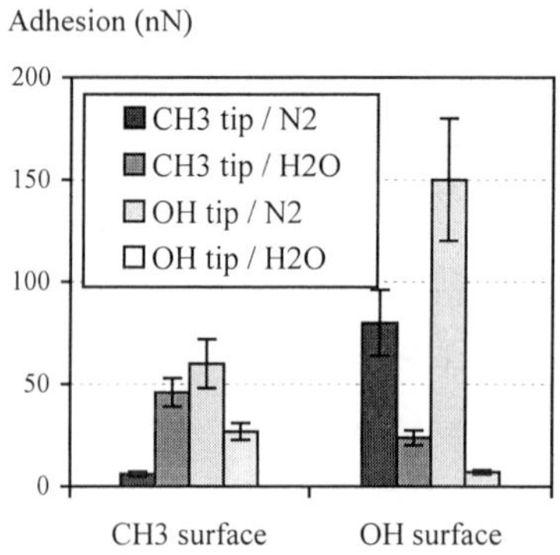

Fig. 2 Average adhesion forces (nN) obtained with $-CH_3$ and $-OH$ terminated tips on octadecanethiol and 11-mercapto-1-undecanol monolayers grafted on a Au/Si substrate. The force-distance curves were recorded in water or in nitrogen atmosphere.

These adhesion forces measurements between functionalised tips and surfaces were then used as routine operation in order to validate the grafting of functional groups on tips. It is worth noting that the standard deviation is more important in N_2 (20 to 25%) than in water (10 to 15%) due to capillary forces, which are not completely suppressed in N_2.

3.2 Adhesion Force Measurements on Additive Films

Force-distance curves were recorded on additive films (process-stabilizing agents, antioxidising agents, and UV-light stabilizers) obtained by spin-coating. Average adhesion forces are reported in Fig. 3.

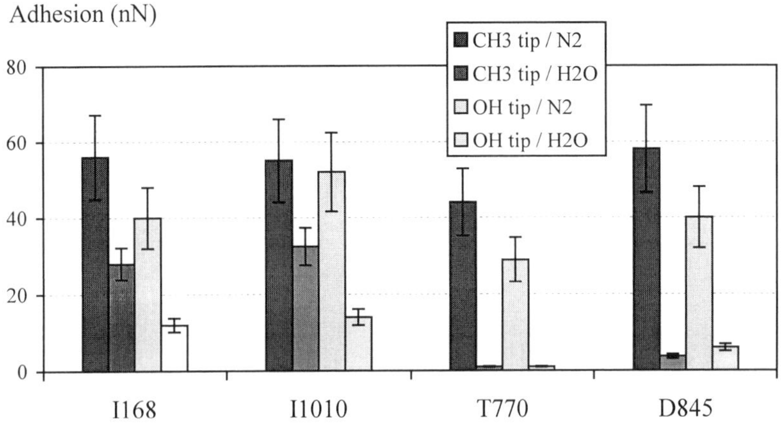

Fig. 3 Average adhesion forces (nN) obtained with $-CH_3$ and –OH terminated tips on additives spin-coated onto silicon wafers. The force-distance curves were recorded in water or in nitrogen atmosphere.

As the purpose is to detect additives on the surface of polypropylene, it was necessary to establish a fingerprint for each additive in order to be able to recognize and to localize it on the polymer surface. Four sets of measurements have been carried out for each compound : pull-off forces have been recorded with $-CH_3$ and –OH tips, in nitrogen atmosphere and in water. Each additive shows a characteristic fingerprint with respect to the tip functionality and the measurement medium. The additives can be classified according to their adhesive behaviour that can be understood when considering their chemical structure. Irgafos 168 and Irganox 1010 have a very similar behaviour. The fingerprint of Tinuvin 770 is similar to that of Dastib 845. Irgafos 168 and Irganox 1010 have a more hydrophobic behaviour than both other additives, as the adhesion obtained in water with a $-CH_3$ tip is rather high (typically around 30 nN). This behaviour is probably due to the numerous *t*-butyl groups present on these compounds. These *t*-butyl groups are indeed very bulky and probably mask the rest of the molecule. The surface is thus mainly made of $-CH_3$ groups. Tinuvin 770 and Dastib 845 have a surfactant-like chemical structure. They are made of a hydrophobic alkyl chain and a hydrophilic head. They seem to behave like a hydrophobic system in nitrogen since the observed adhesion is relatively small compared to the adhesion force measured on an –OH terminated surface (see Fig. 2). On the contrary, their behaviour in water is typical of a hydrophilic system, with low adhesion forces for both $-CH_3$ and –OH tips. They thus probably do not have the same conformation in nitrogen and in water. It is thus possible to recognize the additives from their fingerprint, or at least to recognize their family (antioxidants or UV-light stabilizers).

3.3 Adhesion Force Measurements on Polypropylene Surfaces

The results obtained on polypropylene surfaces are presented in Fig. 4.

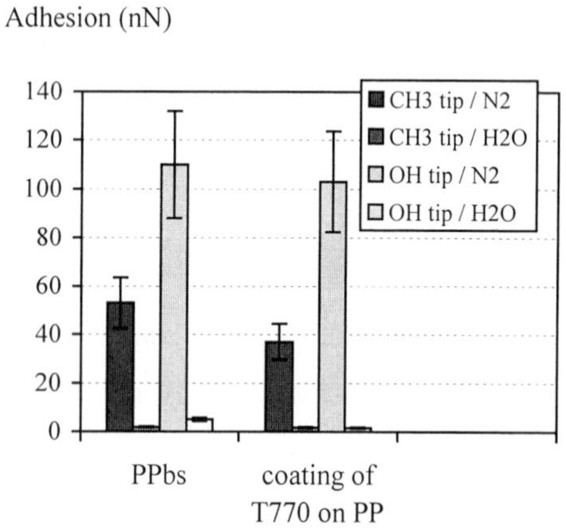

Fig. 4 Average adhesion forces (nN) obtained with $-CH_3$ and –OH terminated tips on stabilized polypropylene (0.05 wt % of Irganox 1010, 0.1 wt % of Irgafos 168 and 0.5 wt % of Tinuvin 770). The results are compared to adhesion forces obtained on polypropylene on which a film of T770 has been spin-coated. The force-distance curves were recorded in water or in nitrogen atmosphere.

PP*bs* is a compression-molded polypropylene sample containing a basic stabilization: 0.05 wt % of Irganox 1010, 0.1 wt % of Irgafos 168 and 0.5 wt % of Tinuvin 770. A pure polypropylene surface is hydrophobic and is expected to behave almost like a $-CH_3$ terminated surface. The advanced contact angle of water on a polypropylene film is indeed around 100° [13]. We should thus observe a high pull-off force when using a $-CH_3$ terminated tip in water.

On the contrary, we observe a rather small pull-off force (2 nN), i.e. one order smaller than the expected value. Moreover, the adhesion observed when using an –OH terminated tip is unexpectedly very high in N_2 and unexpectedly low in water. This adhesive behaviour in water is similar to the one observed on oxifluorinated or on UV-ozone treated polypropylene films, and is typical of a rather hydrophilic polypropylene surface [14,16]. The only compound present in the sample that could give rise to such a behavior is Tinuvin 770. Irganox 1010 and Irgafos 168 are much more hydrophobic (see Fig. 3). The surface of PP*bs* is thus probably composed of a layer of Tinuvin 770. As a comparison, we have measured the pull-off forces on a pure polypropylene surface on which a film of Tinuvin 770 was spin-coated. The behaviour of this surface (Fig. 4) is very similar to the one observed for the PP*bs* surface. This similarity shows that Tinuvin 770 is probably present on the surface of the stabilized polypropylene sample.

The presence of Tinuvin 770 on the PP*bs* surface has been confirmed by TOF-SIMS measurements [18]. But peaks corresponding to fragments belonging to both other additives (Irganox 1010 and Irgafos 168) also appear in the SIMS spectra. The three additives, initially contained in the bulk of the polymer, are thus present on the surface after material processing. However, AFM data seem to indicate that only the hydrophilic additive is detected at the outermost surface.

The results obtained from both techniques can be correlated by the analysis of adhesion maps. The adhesion map in Fig. 5 has been obtained with a $-CH_3$ tip in water. This image shows the lateral distribution of pull-off forces. The great majority of pull-off forces corresponds to a low adhesion force, around 1 nN (bright colour). This is close to the typical adhesion force measured on Tinuvin 770 (2 nN). Arrows indicate small areas presenting a higher pull-off force (10 to 20 nN). They can correspond to areas where polypropylene or the other additives (Irgafos 168 and/or Irganox 1010) are apparent. The diameter of these areas varies between 300 nm and 1.5 μm. The major part of the surface area is thus covered by an external layer of Tinuvin 770. The average adhesion forces measured by CFM thus correspond to the adhesion forces obtained on pure Tinuvin 770. On the contrary, SIMS measurements qualitatively show that the three additives are present on the surface. CFM mapping confirms the presence of the three additives and enables the characterization of their submicrometer spatial distribution. It also confirms that Tinuvin 770 is the main component of the surface.

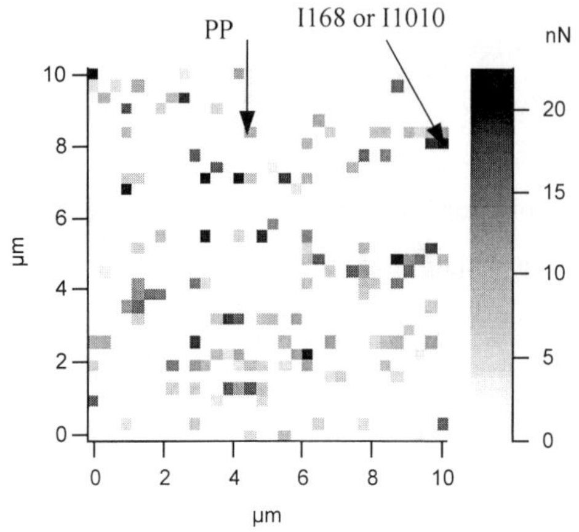

Fig. 5 Adhesion map obtained on stabilised polypropylene (0.05 wt % of Irganox 1010, 0.1 wt % of Irgafos 168 and 0.5 wt % of Tinuvin 770) with a $-CH_3$ terminated tip in water. Dark colour indicates high adhesion (10-20 nN) and bright colour indicates low adhesion (1 nN).

We have then washed the PP*bs* surface by rinsing with chloroform and dichloromethane, which are good solvents of the additives. This washed surface (Fig. 6) shows a remarkable change in adhesion from the PP*bs* surface. The average adhesion force obtained in water with a $-CH_3$ tip is equal to 15 nN. The surface behaves thus like a hydrophobic polypropylene surface.

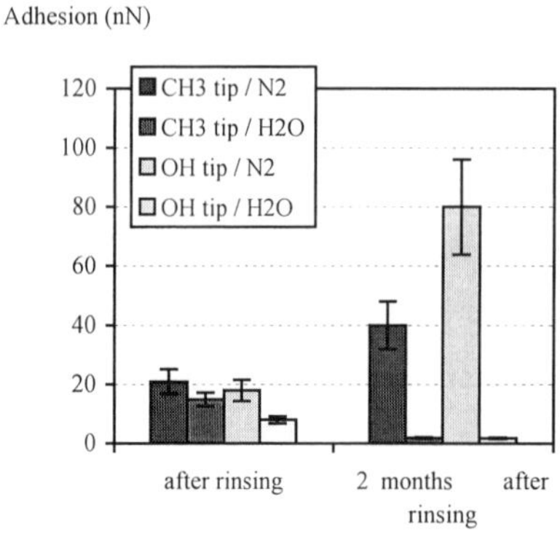

Fig. 6 Average adhesion forces (nN) obtained with $-CH_3$ and $-OH$ terminated tips on the stabilised polypropylene after rinsing with chloroform and dichloromethane, and on the same surface 2 months after the rinsing. The force-distance curves were recorded in water or in nitrogen atmosphere.

Adhesion map obtained in water with a $-CH_3$ tip is shown in Fig. 7. This image shows higher average adhesion force than in Fig. 5. The areas of small adhesion forces can be interpreted as areas where there is some Tinuvin 770 remaining on the surface.

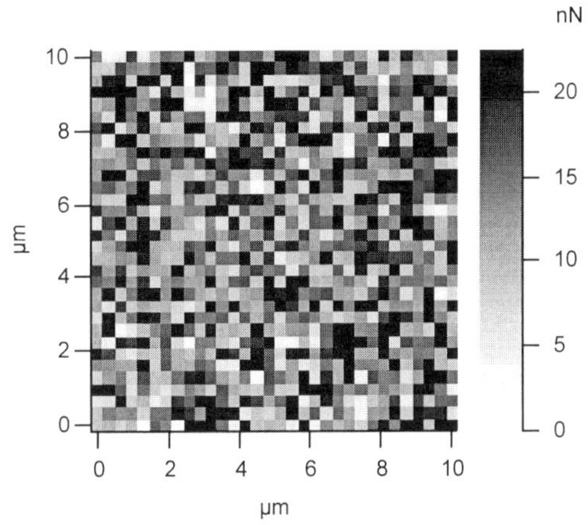

Fig. 7 Adhesion map obtained on stabilized polypropylene, after rinsing with chloroform and dichloromethane, with a $-CH_3$ terminated tip in water. Dark colour indicates high adhesion and bright colour indicates low adhesion.

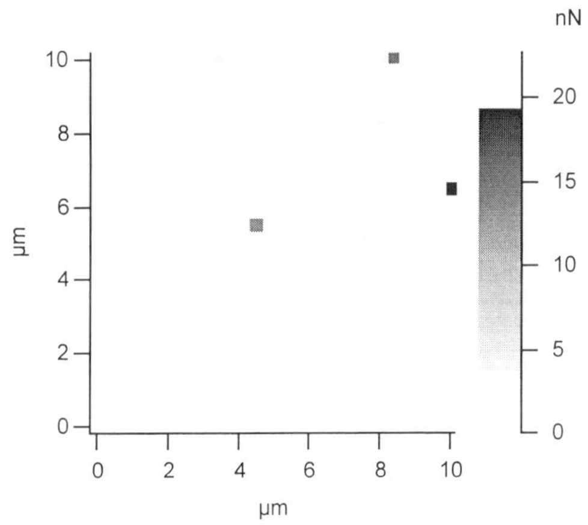

Fig. 8 Adhesion map obtained on stabilized polypropylene, 2 months after rinsing with chloroform and dichloromethane, with a $-CH_3$ terminated tip in water. Dark colour indicates high adhesion and bright colour indicates low adhesion.

This washed surface was then allowed to age in air and we have measured again adhesion forces after 2 months (Fig. 6). The behaviour of this aged surface becomes again hydrophilic and is very similar to the one observed for the PP*bs* surface. It means that an outermost layer of Tinuvin 770 has reappeared on the surface. The additives contained in the bulk of the material have migrated toward the surface. The adhesion map obtained in water with a $-CH_3$ tip (cf. Fig. 8) shows a homogeneous distribution of low adhesion forces, i.e. of Tinuvin 770.

4. CONCLUDING REMARKS

Adhesion properties of polypropylene surfaces were investigated by atomic force microscopy using chemically modified probes. These probes, obtained by grafting of $-CH_3$ or $-OH$ terminated alkanethiol monolayers, were first used to record force-distance curves on pure additives films. Four sets of measurements were carried out for each additive : pull-off forces have been recorded with $-CH_3$ and $-OH$ terminated tips, in nitrogen atmosphere and in water. We have evidenced a characteristic fingerprint for each additive with respect to the tip functionality and the measurement medium. This fingerprint enables the differentiation of the additives. The antioxidants have a more hydrophobic behaviour than the UV-light stabilizers. The results obtained on stabilized polypropylene (0.05 wt % of Irganox 1010, 0.1 wt % of Irgafos 168 and 0.5 wt % of Tinuvin 770) have shown that the extreme surface of the material is mainly composed of UV-light stabilizer (Tinuvin 770). This external layer of Tinuvin 770 can be removed by rinsing the surface with chloroform and dichloromethane, but it reappears during the aging of the material. The presence of this additive may drastically influence the adhesion properties of the polymer.

The final aim of this investigation is to characterise adhesion and aging properties of these polymers in correlation to additive migration towards the material surface. As reported in this contribution, laterally resolved chemical force microscopy enables to study distributions of chemical functional groups on the surface. These results constitute a significant step towards the ultimate aim of high-resolution chemical and molecular recognition at the surface of polymer films.

5. ACKNOWLEDGMENTS

This work was funded by the Région Wallonne (First convention n° 97 13498). The authors are grateful to Clariant Benelux for the financial support and to Clariant Huningue for providing polypropylene and additives. Stimulating discussions with Ir. P. Boudry and Ir. J.-P. Debauve (Clariant Benelux) and with Dr. J. Simonin (Clariant Huningue) were greatly appreciated. The authors also thank Prof. P. Bertrand and Ir. C. Poleunis (PCPM, UCL) for the SIMS measurements and for the gold coating on the AFM probes, and Prof. P. Rouxhet (CIFA, UCL) for access to the contact angle facilities. A.-S. D and B.N. are Post-doctoral Researcher and Research Associate of the Belgian National Fund for Scientific Research (FNRS), respectively.

6. REFERENCES

1. C.D. Frisbie, L.F. Rozsnyai, A. Noy, M.S. Wrighton and C.M. Lieber, Science, 265 (1994) 2071.
2. A. Noy, C.D. Frisbie, L.F. Rozsnyai, M.S. Wrighton and C.M. Lieber, J. Am. Chem. Soc., 117 (1995) 7943.
3. R.C. Thomas, J.E. Houston, R.M. Crooks, T. Kim and T.A. Michalske, J. Am. Chem. Soc., 117 (1995) 3830.
4. D.V. Vezenov, A. Noy, L.F. Rozsnyai and C.M. Lieber, J. Am. Chem. Soc., 119 (1997) 2006.
5. T. Han, J.M. Williams and T.P. Beebe Jr., Anal. Chim. Acta, 307 (1995) 365.
6. G. Bar, S. Rubin, A.N. Parikh, B.I. Swanson, T.A. Zawodzinski Jr. and M.-H. Whangbo, Langmuir, 13 (1997) 373.
7. H. Schönherr and G.J. Vancso, Langmuir, 13 (1997) 3769.
8. H.-X. He, W. Huang, H. Zhang, Q.G. Li, S.F.Y. Li and Z.F. Liu, Langmuir, 16 (2000) 517.
9. G.U. Lee, L.A. Chrisey and R.J. Colton, Science, 266 (1994) 771.
10. J.B.D. Green, A. Novoradovsky and G.U. Lee, Langmuir, 15 (1999) 238.
11. S.L. McGurk, R.J. Green, G.H.W. Sanders, M.C. Davies, C.J. Roberts, S.J.B. Tendler and P.M. Williams, Langmuir, 15 (1999) 5136.
12. Y.-S. Lo, N.D. Huefner, W.S. Chan, F. Stevens, J.M. Harris and T.B. Beebe Jr., Langmuir, 15 (1999) 1373.

13. H. Schönherr, Z. Hruska and G. J. Vancso, Macromolecules, 31 (1998) 3679.

14. H. Schönherr and G. J. Vancso, J. Polym. Sci., 36 (1998) 2483.

15. H. Schönherr, Z. Hruska and G. J. Vancso, Macromolecules, 33 (2000) 4532.

16. H.-Y. Nie, M.J. Walzak, B. Berno and N.S. McIntyre, Appl. Surf. Sci., 144-145 (1999) 627.

17. P.J. Eaton, P. Graham, J.R. Smith, J.D. Smart, T.G. Nevell and J. Tsibouklis, Langmuir, 16 (2000) 7887.

18. A.-S. Duwez, C. Poleunis, P. Bertrand and B. Nysten, submitted to Langmuir.

19. D. Kuila, G. Kvakovszky, M.A. Murphy, R. Vicari, M.H. Rood, K.A. Fritch, J.R. Fritch, S.T. Wellinghoff and S.F. Timmons, Chem. Mater., 11 (1999) 109.

20. Y. Golan, L. Margulis and I. Rubinstein, Surf. Sci., 264 (1992) 312.

21. H. Ron and I. Rubinstein, Langmuir, 10 (1994) 4566.

22. A.-S. Duwez, S. Di Paolo, J. Ghijsen, J. Riga, M. Deleuze, J. Delhalle, J. Phys. Chem. B, 101 (1997) 884.

23. A.-S. Duwez, L.-M. Yu, J. Riga, J. Delhalle and J.-J. Pireaux, Langmuir, 16 (2000) 6569.

24. A.-S. Duwez, L.-M. Yu, J. Riga, J. Delhalle and J.-J. Pireaux, J. Phys. Chem. B, 104 (2000) 8830.

25. A.-S. Duwez, G. Pfister-Guillouzo, J. Delhalle and J. Riga, J. Phys. Chem. B, , 104 (2000) 9029.

26. J. Israelachvili, Intermolecular and Surface Forces, Academic Press: San Diego, 1991.

27. J. Israelachvili and H. Wennerström, Nature, 379 (1996) 219.

Novel Methods to Study Interfacial Layers
D. Möbius and R. Miller (Editors)

FLUORESCENCE AND THE RELEVANT FACTORS OF ORGANIZED MOLECULAR FILMS OF A SERIES OF ATYPICAL AMPHIPHILIC β-DIKETONE RARE EARTH COMPLEXES

R.J. Zhang[1]*, K.Z. Yang[1] and J.F. Hu[2]

[1] Key Laboratory for Colloid and Interface Chemistry of Education Ministry, Shandong University, Jinan 250100, P. R. China

[2] Department of Physics, Shandong University, Jinan 250100, P. R. China

Contents

* Author to whom correspondence should be addressed. FAX: +86-531-8565167; TEL: +86-531-8564750; e-mail: zhrj@sdu.edu.cn.

Keywords: Rare earth complexes; Langmuir-Blodgett (LB) films; Fluorescence emission; Fluorescence enhancement; Fluorescence quenching.

A series of atypical amphiphilic rare earth complexes were fabricated into Langmuir-Blodgett (LB) films on a composite subphase containing two ligands and corresponding rare earth complexes. The LB films were studied by focusing on the fluorescence emission. Optimised fabrication conditions for high quality LB films of the complexes were described in detail. Influences of the ligands and the rare earth ions on the fluorescence intensity were investigated as a whole. The characteristic fluorescence peaks at 613.0, 648.0 and 545.0 nm were selected for detection of europium, samarium and terbium complexes, respectively. The influence of the ligands on the fluorescence intensity of the complexes was discussed on the viewpoint of ligand structure and the triplet energy level. For europium or samarium ion bounded by different β-diketone ligands, $Eu(TTA)_3Phen$ or $Sm(TTA)_3Phen$ emitted the most intense fluorescence in LB films. However, in terbium complexes $Tb(TTA)_3Phen$ emitted the weakest fluorescence. The influences of the rare earth ions on the fluorescence of rare earth complexes were discussed on the resonance energy levels. The influence of fluorescence emission by two compounds in adjacent LB layers were also studied.

1. Introduction

More and more attentions have been paid to rare earth complexes in recent years for their significant electronic, optical and magnetic properties both in fundamental and in practical research. Fluorescence is one of the most distinguished characters for Eu(III), Sm(III) and Tb(III) ions bounded by β-diketone ligands. Because β-diketone can absorb energy at so high a coefficiency. Then, effective intramolecular energy transfer will take place from β-diketone ligands to the ions, sharp intense emission will be obtained in colour of red for Eu(III) or Sm(III) and green in Tb(III) compounds. With the rapid development of Langmuir-Blodgett (LB) films, rare earth complexes have been fabricated into this kind of organized molecular films and the novel properties were examined. In this area, Yu et al. studied the specific intramolecular energy transfer process of an europium complex in a LB film [1], and Huang et al. [2] fabricated hemicyanine or azobenzene complexes with rare earth β-diketone groups into LB films and studied the improved nonlinear optical property.

To widen the range of film-forming functional complexes, Zhang et al. successfully fabricated $Eu(TTA)_3Phen$ and $Sm(TTA)_3Phen$ without long aliphatic chains into LB films by using a composite subphase containing ligands and corresponding rare earth complexes [3-4]. Then, fluorescence enhancement and quenching of $Eu(TTA)_3Phen$ and $Sm(TTA)_3Phen$ by several compounds in adjacent LB layers were studied [5]. Since terbium complexes can emit green fluorescence, which is more important for future application in ultrathin luminescence devices. We fabricated the $Tb(acac)_3Phen$ LB film and studied its characteristic fluorescence emission with comparison to that in solid powder and in chloroform solution [6].

A series of rare earth complexes should be studied in detail to give the answers to questions for future applications, especially for fluorescence modulation, 1) What are the factors influencing fluorescence of rare earth complexes in LB films; 2.) Can the fluorescence of a given complex be further altered in LB films? To obtain answers to these questions, in this paper, a series of atypical amphiphilic rare earth complexes are successfully fabricated into LB films. It is further indicated that the composite subphase method is valid for fabricating LB films of atypical amphiphilic rare earth complexes. Then, factors influencing the fluorescence emission of rare earth complexes, including fabrication conditions, structure and triplet energy level of ligands, resonance energy level of rare earth ions are investigated. Moreover, two compounds in adjacent LB layers are also revealed to have an effect on fluorescence of the rare earth complexes. This work is helpful both for further understanding the luminescence scheme of rare earth complexes in organized molecular films and for the potential applications in ultrathin luminescence devices.

2. Experimental details

2.1. Materials

The rare earth β-diketone complexes were synthesized by referring to the literature methods [7, 8]. Arachidic acid (AA) was a commercially available reagent. Chloroform was used without purification. Pure water for the subphase was deionised first and then doubly–distilled (pH 6.4, resistivity 18 MΩ·cm).

2.2. Procedures and apparatus

The solvent chloroform was allowed to evaporate 10 minutes before compressing the monolayers containing rare earth β-diketone complexes at the air/liquid interface. Optically-polished glass and quartz slides were used for LB film deposition.

The fluorescence spectra of the LB films were recorded on a Hitachi 850 fluorescence spectrophotometer (Japan) with the excitation bandwidth of 5 nm and the emission bandwidth of 5 nm. X-ray diffraction of the LB films was carried out on a Rigaku/rB diffractometer. UV-vis spectra were recorded on a Shimazu UV-240 (Japan) spectrophotometer.

2.3. LB film fabrication of rare earth complexes

It is easy to understand that, the precondition for the research on atypical amphiphilic rare earth complexes in Langmuir-Blodgett (LB) films is to fabricate them into LB films, which generally involves two steps: i) obtain stable monolayers, ii) successfully transfer them from monolayers to solid substrates. Since all the complex molecules in a sphere shape have no long aliphatic chains, necessary procedures should be adopted during each step. Here we would emphasis on the optimised fabrication conditions as follows.

Unfortunately the rare earth complexes themselves can not form stable monolayers at air/water interface. The π-A isotherms varied with the compression speeds. The limiting molecular areas were much smaller than the theoretical values. The reason is that the studied rare earth complexes have tendency to dissolve into water. When a composite subphase, as what mentioned in previous papers [3-6], an aqueous solution saturated by a β-diketone ligand L, the synergistic ligand Phen and the corresponding rare earth complex REL_3Phen [RE denotes Eu(III), Sm(III) or Tb(III); L denotes acac, TFA, HFA or HFA; Phen denotes 1, 10-phenanthroline] was used, the complex in subphase inhibited the dissolution of complex in monolayer into water. In this way, stable rare earth complex monolayers with reproducible isotherms could be obtained. To get highly luminescent LB films, we also add β-diketone ligand in subphase to inhibit the dissociation of complex in monolayer, otherwise only LB films with inhomogeneous fluorescence emission can be obtained.

The composite subphase mentioned above is the key factor for the stable rare earth complex monolayers at air/liquid interface. During monolayer transfer process, it is found that the monolayers are too rigid to transfer onto solid substrates. The conventional film-forming molecule AA with a long aliphatic chain was mixed with them in the spreading solution in a molar ratio of 1:1. The obtained mixed LB films emitted homogeneous intense fluorescence. So the use of a fatty acid AA is the key factor for the monolayer transfer during LB film fabrication.

A lot more works have been conducted during the monolayer transfer step to optimise the fabrication conditions. A deposition speed lower than 5 $mm \cdot min^{-1}$ was good for fabricating LB films with good periodic layered spacing. This indicates that a good orientation order of molecules of rare earth complex and AA in LB films can be well kept at a low speed. The hydrophobic solid substrate was better for the transfer of rare earth complexes than hydrophilic ones, which might be due to the effect of rare earth complexes on the mixed monolayer. This was further revealed by a drainage process of the LB films. For pure AA monolayer, after deposition of the second layer, the surface of the LB film should be hydrophobic. However, there were some water droplets left on top of the LB films transferred from the mixed monolayer. This means that certain hydrophilic parts of rare earth complex molecules in the monolayer also point to the same direction, air phase, as the hydrophobic aliphatic tails of the AA molecules at the air/liquid interface. Fortunately, after air-drying the mixed LB film for 3 minutes, the transfer ratio of following LB layers is still nearly unity. The LB films also exhibited good periodic layered structure by small angel x-ray diffraction and by fluorescence emission measurement. So, it can be deduced that the interaction between hydrophobic tail of AA and the hydrophobic solid substrate leads to the monolayer transfer for the first and following even LB layers. It can also be deduced that the interaction between AA and rare earth complex results in the transfer of rare earth complex into LB films.

Fluorescence of rare earth complex in LB films could hardly be detected, if the mixture of rare earth complex to AA was in a molar ratio larger than 1. It indicates that only when AA exists in a certain percentage can its interaction with the rare earth complex be strong enough for monolayer transfer. The question that how the two compounds interact with each other should be examined by using high sensitive characterizing methods to study the location and orientation of two kinds of molecules in monolayer at the air/liquid interface. Related work is in progress, and a periodic work will be published elsewhere [9].

In a summary for the LB film fabrication of the atypical amphiphilic rare earth complexes, the optimised experimental conditions are: a composite subphase, AA mixed with rare earth complexes in a molar ratio of 1:1, hydrophobic substrate, a deposition speed of 5$mm \cdot min^{-1}$and air-dry the LB films after the even layers. The deposition surface pressure is 20 $mN \cdot m^{-1}$, corresponding to a solid state monolayer. All experiments were carried out at room temperature ($25 \pm 1°C$).

3. Results and discussion

3. 1. Fluorescence of rare earth complex in solution

It is well known that europium complexes emit the most intense fluorescence near 610 nm. Using this wavelength for measurement, the fluorescence excitation spectra of europium complexes in solution were recorded in the range 200.0 – 400.0 nm. The maximum excitation wavelengths for different europium complexes were summarized in table 1. Using 645.0 nm and 545.0 nm for fluorescence measurement of samarium and terbium complexes, respectively, the corresponding maximum excitation wavelengths were also summarized in Table 1.

Table 1 Fluorescence result of 1.0×10^{-3} mol/l rare earth complexes in chloroform solution

Rare earth complexes	Maximum excitation wavelength (nm)	Maximum emission wavelength (nm)	Fluorescence intensity
$Eu(acac)_3Phen$	342.5	613.0	727
$Sm(acac)_3Phen$	337.5	648.0	58
$Tb(acac)_3Phen$	333.5	545.0	1146
$Eu(TFA)_3Phen$	337.5	613.0	3798
$Sm(TFA)_3Phen$	340.0	648.0	113
$Tb(TFA)_3Phen$	337.5	545.0	412
$Eu(HFA)_3Phen$	330.0	613.0	79
$Sm(HFA)_3Phen$	337.5	648.0	46
$Tb(HFA)_3Phen$	346.0	545.0	132
$Eu(TTA)_3Phen$	394.0	613.0	20843
$Sm(TTA)_3Phen$	393.5	648.0	641
$Tb(TTA)_3Phen$	390.0	545.0	7

The results show that rare earth complexes can be divided into two classes: RE-acac, TFA, HFA-Phen belongs to Class 1; while RE-TTA-Phen belongs to Class 2. The maximum excitation wavelength (λ_{ex}^{max}) for complexes in Class 1 was at about 340.0 nm. λ_{ex}^{max} of complexes in Class 2 was at about 390.0 nm. Taking terbium complexes as an example, $Tb(acac)_3Phen$, $Tb(TFA)_3Phen$, $Tb(HFA)_3Phen$ exhibited λ_{ex}^{max} at 333.5, 337.5 and 346.0 nm, respectively; λ_{ex}^{max} of $Tb(TTA)_3Phen$ was 390.0 nm. The different λ_{ex}^{max} for complexes in the two classes should be attributed to the different intramolecular energy transfer processes from ligands to rare earth ions to be discussed below.

Exciting rare earth complexes in Class 1 by the wavelength of 340.0 nm and exciting complexes in Class 2 by 390.0 nm, respectively, their fluorescence emission spectra were recorded in the range of 400.0 – 800.0 nm. Europium, samarium and terbium complexes exhibited the most intense fluorescence at the maximum emission wavelengths (λ_{em}^{max}) 613.0, 648.0 and 545.0 nm. Their fluorescence intensity at the corresponding λ_{em}^{max} was summarised in Table 1. For a given rare earth ion, the fluorescence intensity of complexes with different ligands decreased in the following order:

$$Eu(TTA)_3Phen > Eu(TFA)_3Phen > Eu(acac)_3Phen > Eu(HFA)_3Phen \quad (1)$$

$$Sm(TTA)_3Phen > Sm(TFA)_3Phen > Sm(acac)_3Phen > Sm(HFA)_3Phen \quad (2)$$

$$Tb(acac)_3Phen > Tb(TFA)_3Phen > Tb(HFA)_3Phen > Tb(TTA)_3Phen \quad (3)$$

Obviously, β-diketone ligand influences the fluorescence of europium complexes in the same order as that of samarium complexes. However, fluorescence of terbium complexes decreases in a quite different way to that of the europium and samarium complexes in chloroform solution. Moreover, the complex formed by the β-diketone ligand TTA binding Eu(III) or Sm(III) has the most intense fluorescence in europium or samarium complexes. However, $Tb(TTA)_3Phen$ emits the weakest fluorescence in terbium complexes. Here we simply show the fluorescence declining sequence, detailed discussion would be given in section 3.4-3.5.

3. 2. Fluorescence of rare earth complex in LB film

Selecting 610.0, 645.0 and 545.0 nm as the detection wavelengths for europium, samarium and terbium complexes in LB film, respectively, the fluorescence excitation spectra were recorded in the range of 200.0-400.0 nm. The maximum excitation wavelengths for different LB films were summarized in Table 2.

Table 2 Fluorescence result of rare earth complexes in mixed 18-layer LB films with AA in a molar ratio of 1:1

LB films	Maximum excitation wavelength(nm)	Maximum emission wavelength (nm)	Fluorescence intensity
$Eu(acac)_3Phen/AA(1:1)$	306.5	613.0	9
$Sm(acac)_3Phen/AA(1:1)$	324.0	648.0	7
$Tb(acac)_3Phen/AA(1:1)$	272.5	545.0	41
$Eu(TFA)_3Phen/AA(1:1)$	306.5	613.0	26
$Sm(TFA)_3Phen/AA(1:1)$	324.0	648.0	14
$Tb(TFA)_3Phen/AA(1:1)$	272.5	545.0	16
$Eu(HFA)_3Phen/AA(1:1)$	306.5	613.0	7
$Sm(HFA)_3Phen/AA(1:1)$	324.0	648.0	5
$Tb(HFA)_3Phen/AA(1:1)$	272.5	545.0	9
$Eu(TTA)_3Phen/AA(1:1)$	306.5	613.0	10800
$Sm(TTA)_3Phen/AA(1:1)$	324.0	648.0	320
$Tb(TTA)_3Phen/AA(1:1)$	272.5	545.0	2

An intense double-frequency peak appeared in each excitation spectrum since the LB film was very thin. All the complexes in Class 1 in LB film had the second largest excitation peaks near 266.0 nm, so this wavelength was selected as the excitation wavelength for the LB films, which was also selected as the excitation wavelength during the fluorescence lifetime measurement of the atypical rare earth complexes in LB films [10]. The second largest fluorescence excitation peaks for complexes in Class 2 in LB film were at about 350 nm, so it was selected as the excitation wavelength for complexes in Class 2. The obtained fluorescence emission intensity at the maximum emission wavelengths (λ_{em}^{max}) 613.0 nm for Eu(III), 648.0 nm for Sm(III), and 545.0 nm for Tb(III) complexes in LB film were summarized in Table 2. The λ_{em}^{max} is the same for the complexes in solution and LB film, which should be attributed to the stable electronic structure of rare earth ions.

Varying the β-diketone ligands binding a given rare earth ion, the fluorescence intensity of rare earth complexes in LB film decreased in the following order:

$Eu(TTA)_3Phen/AA(1:1)$>$Eu(TFA)_3Phen/AA(1:1)$>$Eu(acac)_3Phen/AA(1:1)$>$Eu(HFA)_3Phen/AA(1:1)$ (4)

$Sm(TTA)_3Phen/AA(1:1)$>$Sm(TFA)_3Phen/AA(1:1)$>$Sm(acac)_3Phen/AA(1:1)$>$Sm(HFA)_3Phen/AA(1:1)$ (5)

$Tb(acac)_3Phen/AA(1:1)$>$Tb(TFA)_3Phen/AA(1:1)$>$Tb(HFA)_3Phen/AA(1:1)$>$Tb(TTA)_3Phen/AA(1:1)$ (6)

The fluorescence intensity of rare earth complexes in LB film has the same decrease sequence as that in solution. In other words, in Eu(III) or Sm(III) complexes, the complex with the β-diketone TTA emits the most intense fluorescence in LB film. However, terbium complex with acac emits the most intense fluorescence in Tb(III) complexes.

3. 3. Influence of rare earth ion on fluorescence of complex in LB film

It is well known that, the fluorescence intensity of rare earth complexes is greatly influenced by the resonance energy level of rare earth ion. If the rare earth ions are Sc(III), Y(III), La (III) and Lu(III), no fluorescence can be detected because their resonance energy levels are very high (~ 32000 cm^{-1}), much higher than the triplet levels of ligands. When the rare earth ions are Pr(III), Nd (III), Ho(III), Er(III) Tm(III) and Yb(III), fluorescence can hardly be detected. Because energy level difference is too small, leading to high probability of non-radiative energy transfer. The rare earth ions in this paper are Eu(III), Sm(III) and Tb(III), the resonant energy levels are similar to the triplet levels of β-diketone ligands. Energy transfer can take place from ligands to rare earth ions (Fig. 1). As a result, their fluorescence is very intense and can be detected even at the ultrathin LB films [11].

Eqs. 4-6 show that, the fluorescence intensity of complexes in LB film varies with the rare earth ion. What the most distinguished is for complexes with the β-diketone ligand TTA. When the bounded rare earth ion is Eu(III) or Sm(III), Eu(TTA)$_3$Phen or Sm(TTA)$_3$Phen emits the most intense fluorescence in Eu(III) or Sm(III) complexes. However, Tb(TTA)$_3$Phen emits the weakest fluorescence in Tb(III) complexes. This result should be mainly due to different resonant energy levels of the ions. The resonance energy level $^5D_4$ of Tb(III), 21000 cm^{-1}, is much higher than $^5D_0$ of Eu(III), 17598 cm^{-1}, and $^4G_{5/2}$ of Sm(III), 17890 cm^{-1}. The triplet state of TTA is 20400 cm^{-1}, which is higher than $^5D_0$ and $^4G_{5/2}$, but lower than $^5D_4$. So energy can transfer from TTA to Eu(III) or Sm(III) at high efficiency, however only a little can transfer to Tb(III).

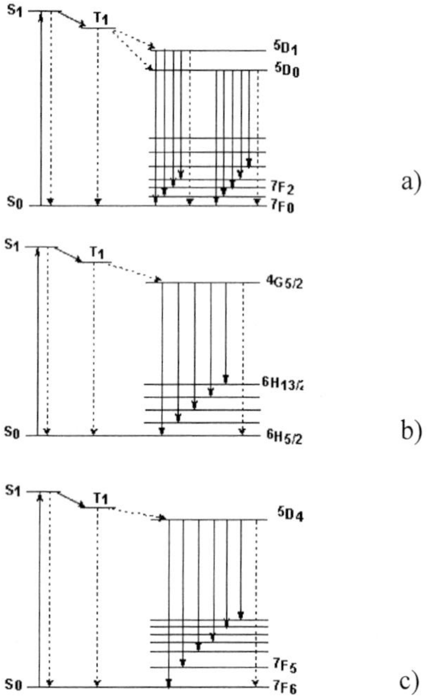

Fig. 1 Intramolecular energy transfer illustration for rare earth complexes from the ligand to (a) Eu (III), (b) Sm(III) and (c) Tb(III).

3. 4. Influence of β-diketone ligand on fluorescence of rare earth complex in LB film

The influence factors on fluorescence of rare earth complexes are very complicated, including ligand structure, matching between energy levels, thermal non-radiative deactivation, temperature, solvent and rare earth ions as discussed in Section 3.3. In this section, we only briefly discuss our results focusing on two factors. First, the ligand structure plays an important role in the fluorescence emission of rare earth complexes [12]. The β-diketone ligand can be expressed in the form of $R1COCH_2COR2$. Table 3 assigns the R1 and R2 groups for the four β-diketone ligands studied in this work. The fluorescence of the complex will be more intense if R1 is ready to donate electrons and if R2 is ready to attract electrons. Second, the higher the triplet energy level of the β-diketone ligand is, the less thermal non-radiative deactivation of resonance energy level will occur. As a result, the rare earth complex will exhibit more intense fluorescence emission [13].

Table 3 Assignment of R1 and R2 group for the β-diketone ligands represented in the form of $R1COCH_2COR2$

	acac	TFA	HFA	TTA
R1	CH_3	CH_3	CF_3	
R2	CH_3	CF_3	CF_3	CF_3

For europium complexes, the most favorable β-diketone is TTA. Because the R1 group is a thiophenyl group, which is very easy to donate electrons; while the R2 group CF_3 contains the F atom having the largest electronegativity. For the β-diketone ligand TFA, although the R2 group is the same as that in TTA, the R1 group CH_3 does not donate electrons so easily as CF_3. So the fluorescence of $Eu(TFA)_3Phen$ is weaker than that of $Eu(TTA)_3Phen$ in LB film. The R1 group of acac is the same as that of TFA, but the R2 group CH_3 only contains atoms with small electronegativity. Thus, the fluorescence of $Eu(acac)_3Phen$ is weaker than that of $Eu(TFA)_3Phen$. Although the R2 group of HFA is CF_3, the R1 group is also CF_3 without electron donating property. So $Eu(HFA)_3Phen$ emits the weakest fluorescence in europium complexes.

Sm(III) has a similar resonance energy level as Eu(III). The physical and chemical properties of the two ions are also similar [11]. So samarium complexes emit fluorescence in the same sequence as that of europium complexes with different β-diketone ligands. Tb(III) has a higher resonance energy level. The fluorescence of Tb(III) complexes decrease in a different way to that of Eu(III) and Sm(III) complexes. $Tb(acac)_3Phen$ emits the most intense fluorescence in the terbium complexes. The reason is that the triplet state of acac is 25300 cm^{-1}, much higher than the resonance energy level $^5D_4$ (21000 cm^{-1}) of Tb(III). The thermal deactivation is very limited, and the radiative energy transfer efficiency is high. The triplet state of TFA, 22800 cm^{-1}, is lower than that of acac, so the fluorescence of $Tb(TFA)_3Phen$ is weaker than that of $Tb(acac)_3Phen$ in LB film. The triplet state of TFA is higher than that of HFA, 22200 cm^{-1}. Moreover, the R1 group of TFA, CH_3, is relatively easier to donate electrons than the R1 group of HFA, CF_3. So $Tb(TFA)_3Phen$ emits more intense fluorescence than $Tb(HFA)_3Phen$. For $Tb(TTA)_3Phen$, since the $^5D_4$ resonance level of Tb(III) is higher than the triplet state of TTA, energy transfer efficiency from TTA to Tb(III) is limited. As a result, $Tb(TTA)_3Phen$ emits the weakest fluorescence in terbium complexes in LB film.

3.5. Influence of compounds in adjacent layer on fluorescence of rare earth complex in LB film

In previous Section 3.4, it is revealed that ligand influences fluorescence of rare earth complex in LB film. Question may arise as i) what the fluorescence intensity will be, if there exists a compound which can absorb energy in the range of triplet state of the ligand. ii) whether the fluorescence of a complex can be further changed in LB film? To answer these questions, alternate LB films were purposely fabricated benefiting from the structure design of the NIMA 2022 trough. At the air/liquid interfaces on the two relatively separated compartments, different spreading solutions were spread, and nine LB films with the alternative components of ABAB...AB were obtained. Here, A denotes $Eu(TTA)_3Phen$/AA (1:1), $Sm(TTA)_3Phen$/AA (1:1), and $Tb(TTA)_3Phen$/AA (1:1), respectively; B denotes SB/AA (1:4), FC/AA (1:4) and AA. The LB films with pure AA in alternative layer were used for reference. Previous experiment has proved that AA has no influence on fluorescence of rare earth complexes [5]. SB is a long-chain Schiff base and FC is a ferrocene derivative, which were studied in detail in a published paper [5]. The fluorescence of europium or samarium complex in LB film is enhanced by SB and quenched by FC as studied before [5]. However, as shown in Fig. 2, the fluorescence of the terbium complex is enhanced both by SB and by FC in adjacent layer. Opposite influence of FC on the fluorescence of terbium complex can be due to the different triplet energy levels of TTA and acac. Fig. 3 shows the absorption spectra of SB and FC in LB film. The two compounds can both absorb energy at the excitation wavelength of 266.0 nm.

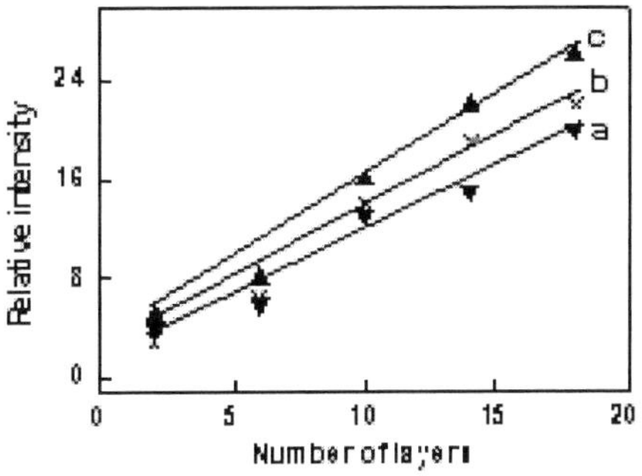

Fig. 2 Relationship between fluorescence intensity and the number of layers for LB films in the form of ABAB...AB; (a) A= $Tb(acac)_3Phen$ /AA (1:1), B=AA; (b) A= $Tb(acac)_3Phen$ /AA (1:1), B=FC/AA(1:4); (c) A= $Tb(acac)_3Phen$ /AA (1:1), B=SB/AA(1:4).

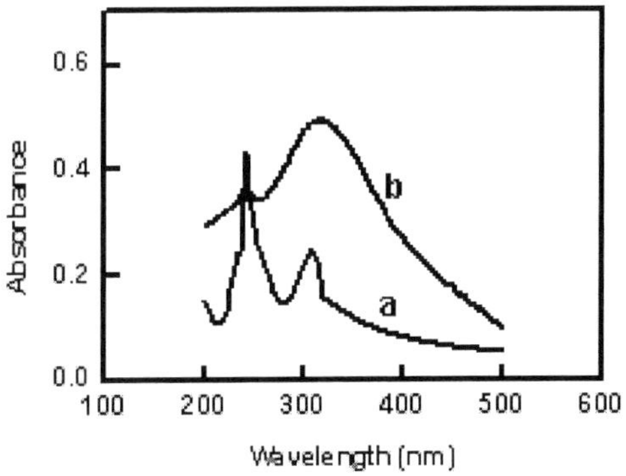

Fig. 3 UV-vis spectra of LB films of (a) SB/AA (1:4) and (b) FC/AA (1:4).

For SB, energy might transfer from it to Eu(III), Sm(III) and Tb(III). There seems no absorbance at the wavelengths corresponding to triplet state of TTA, 20400 cm^{-1} (equal to 490nm) and that of acac, 25300 cm^{-1} (equal to 395nm). So SB enhances fluorescence of all the complexes. However, for europium and samarium complexes, although energy might transfer from FC to them, energy transfer from complexes to FC should be dominant, resulting in the observed fluorescence quenching. On the contrary, energy transfer from FC to $Tb(acac)_3Phen$ is dominant, resulting in fluorescence enhancement of $Tb(acac)_3Phen$ by FC. The complicated intermolecular energy transfer scheme including fluorescence yield and energy transfer efficiency is in progress.

The above results show that, fluorescence from rare earth complexes can be further altered by certain molecules in adjacent layer of LB films, which is helpful for their future application in luminescence devices involving fluorescence modulation.

4. ACKNOWLEDGEMENT

The authors are grateful for the financial support of the Climbing Program (A State Fundamental Research Key Project).

5. REFERENCES

1. A.C. Yu, L.M. Ying, X.S. Zhao, W.S. Xia and C.H. Huang, Prog. Nat. Sci., 7 (1997) 692.
2. H.Li, C.H. Huang, X.S. Zhao, X.M. Xie, L.G. Xu and T.K. Li, Langmuir, 10 (1994) 3794.
3. R. J. Zhang, H. G. Liu, K. Z. Yang, Z. K. Si, G. Y Zhu and H. W. Zhang, Thin Solid Films, 295 (1997) 228.
4. R.J. Zhang, H.G. Liu, C.R. Zhang, K.Z. Yang, Z.K. σ and G.Y. Zhu, Chin. Chem. Lett., 6 (1995) 627.
5. R.J. Zhang, H.G. Liu, C.R. Zhang, K.Z. Yang, G.Y. Zhu and H.W. Zhang, Thin Solid Films, 302 (1997) 223.
6. R.J. Zhang and K.Z. Yang, Langmuir, 13 (1997) 7141.
7. L.R. Melby, N.J. Rose, E. Abramson and J.C. Caris, J. Am. Chem. Soc., 86 (1964) 5117.
8. H. Bauer, J. Blanc and D.L. Ross, J. Am. Chem. Soc., 86 (1964) 5125.
9. R. J. Zhang, P. Krüger, B. Kohlstrunk and M. Lösche, Submitted to J. Phys. Chem.
10. R. J. Zhang, K. Z. Yang, A. C. Yu and X. S. Zhao, Thin Solid Films 363 (2000) 275.
11. T. Moller, The Chemistry of the Lanthanides, Pergamon Press, 1973.
12. H. G. Huang, K. Hiraki and Y. Nishikawa, Nippon Kagaku Kaishi, 1 (1981) 66.
13. S. Sato and M. Wada, Bull. Chem. Soc. Jpn., 43 (1970) 1955.

Novel Methods to Study Interfacial Layers
D. Möbius and R. Miller (Editors)

Electroluminescent device using PbBr-based layered perovskite having self-organized organic-inorganic quantum-well structure

Masanao Era,[1,2] Tetsuya Ano[1] and Mitsuharu Noto[1]

[1]Department of Chemistry and Applied Chemistry, Faculty of Science and Engineering, Saga University, Honjo 1, Saga-shi, Saga 840-8502, Japan

[2]Core Research for Evolutional Science and Technology (CREST), Japan Science and Technology Corporation, Saga 840-8502, Japan

Contents

PbBr-based layered perovskites exhibit sharp and intense exciton emission around 410 nm owing to their self-organized quantum-well structure. We prepared heterostructure electroluminescent devices combined with the PbBr-based layered perovskite as emissive material and organic carrier transporting materials: indium tin oxide (ITO) /PbBr-based layered perovskite (CHEPbBr4)/ electron-transporting oxadiazole (OXD7)/AlLi and ITO/Copper phthalocyanine (CuPc)/CHEPbBr4/OXD7/AlLi. When the devices were driven at 100K, sharp electroluminescence corresponding well to exciton emission of CHEPbBr4 was observed. Further, comparison between voltage-current density-EL intensity characteristics of the devices demonstrated that employment of CuPc layer as hole-injection and transporting layer provides large lowering driving voltage and enhancement of EL efficiency of the heterostructure EL device.

1. Introduction

Lead halide-based layered perovskites $(RNH_3)_2PbX_4$ self-organize a multiple quantum-well structure where a semiconductor layer consisting of two-dimensional network of corner-sharing octahedral PbX_6 and organic ammonium RNH_3 layer are alternately piled up [1]. Owing to the self-organized quantum-well structure, they form stable exciton with a large binding energy of several hundred meV, and exhibit attractive optical properties due to the exciton, efficient emission and optical nonlinearity and so on [1-3].

PbBr-based layered perovskite possesses exciton emission in violet region (the emission is peaking around 410 nm). In addition, the layered perovskite possesses excellent film processability; one can prepare optically high-quality thin film of the PbBr-based layered perovskite by using simple spin-coating, dipping, vacuum-deposition and the Langmuir-Blodgett technique [4-9]. The attractive properties of the PbBr-based layered perovskite motivate us to apply them as thin film emissive material of light emitting device in violet region.

To attain electroluminescence from the PbBr-based layered perovskite, we employ the heterostructure device where emissive layer of the PbBr-based layered perovskite are combined with organic carrier-transporting layers. In 1990, Adachi et. al. demonstrated that carriers and excitons to be confined within emissive layer (EML) in organic double heterostructure (DH) device consisting of electron-transporting layer (ETL), EML and hole-transporting layer (HTL). As a result of the confinement effect, highly efficient electroluminescence (EL) was attained even in the DH device with very thin emissive layer (5 nm thick) [10]. The carrier and exciton confinement in the organic DH device is expected to provide a new approach to construct EL devices with a variety of emissive materials including inorganic materials; based on the confinement, one can drive the emissive materials sandwiched between ETL and HTL by injected carriers [11].

In this work, we fabricated heterostructure electroluminescent devices by the combination of the PbBr-based layered perovskite as emissive material and organic carrier transporting materials, electron-transporting oxadiazole and hole-transporting copper phthalocyanine, and evaluated their electroluminescent properties.

2. EXPERIMENTS

PbBr-based layered perovskite having cyclohexenylehtyl ammonium layer (CHEPbBr4) was employed as emissive material (Fig. 1). The crystal sample of CHEPbBr4 was prepared by solvent-evaporation from acetone-nitromethan solution where stoichiometric amount of $PbBr_2$ and cyclohexenylethyl ammonium bromide were dissolved. Film samples of CHEPbBr4 were spin-coated from the DMF solution of the crystal sample (conc. = 10 mg/ml).

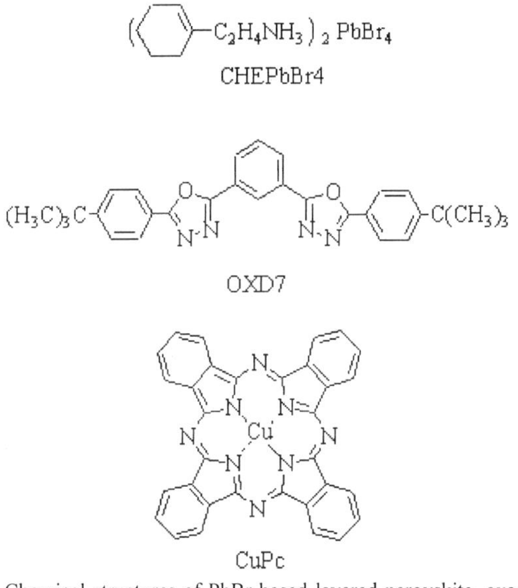

Fig. 1 Chemical structures of PbBr-based layered perovskite, oxadiazole and copper phthalocyanine employed in this study.

Using the CHEPbBr4 spin-coated film, we fabricated two types of EL device: a single heterostructure device (SHD) combined with CHEPbBr4 and an electron-transporting oxadiazole (OXD7) and double heterostructure device (DHD) with CHEPbBr4, OXD7 and a hole-transporting phthalocyanine (CuPc). For SHD, CHEPbBr4 layer (thickness is about 20 nm) was firstly spin-coated on indium tin oxide (ITO)-coated glass substrates. Then, OXD7 layer (thickness of 100 nm) and AlLi cathode (thickness is more than 100 nm) were successively vacuum-deposited on the substrates at 10^{-5} torr. For DHD, CuPc layer (thickness of 400 nm) was firstly vacuum-deposited on ITO substrates. Then, spin-coating of CHEPbBr4 layer (thickness of about 20 nm) and vacuum deposition of OXD7 layer (thickness of 500 nm) and AlLi (thickness is more than 100 nm) were successively carried out.

3. RESULTS AND DISCUSSION

Fig. 2 shows absorption and photoluminescence (PL) spectra of a spin-coated film of CHEPbBr4. Sharp and intense absorption and PL due to exciton, which are characteristic of PbBr-based layered perovskite, are observed around 400 nm. The spectral features demonstrate that PbBr-based layered perovskite structure is surely formed in the spin-coated film.

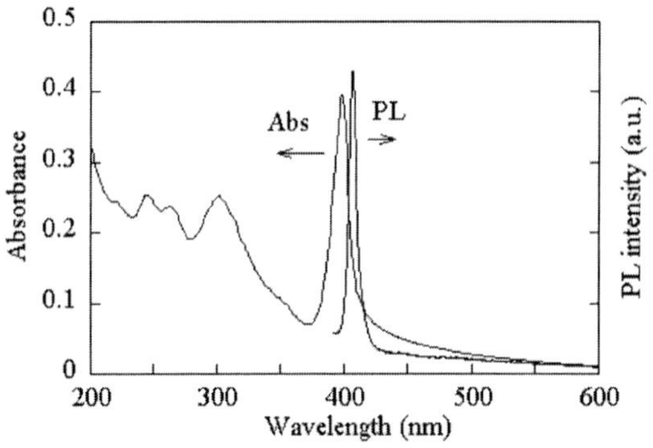

Fig. 2 Absorption and photoluminescence spectra of spin-coated film of PbBr-based layered perovskite CHEPbBr4.

Fig. 3 is comparison between X-ray diffraction profiles of film and crystal samples of CHEPbBr4. Only diffraction peaks corresponding to the layer structure of CHEPbBr4, (00n) series, are observed in the film sample while not only (00n) peaks but (100) or (010) peak is observed around 2 theta of 15 degrees in the crystal sample. The profiles demonstrate that the layer structure is oriented parallel to the film plane in the CHEPbBr4 film. From the (00n) peaks, the layer spacing was calculated to be 1.7 nm.

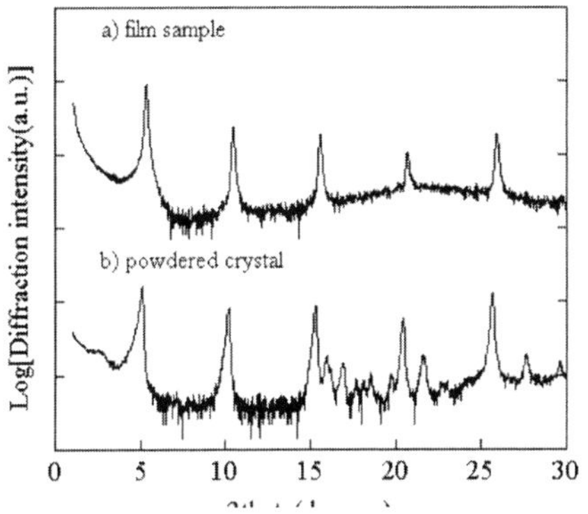

Fig. 3 X-ray diffraction profiles of film sample (a) and crystal sample (b) of PbBr-based layered perovskite CHEPbBr4.

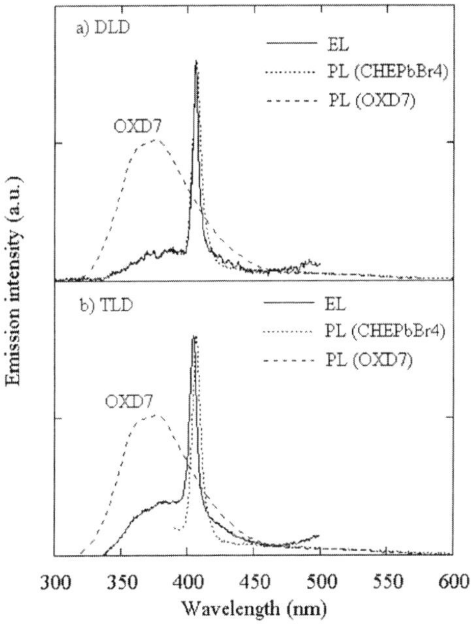

Fig. 4 Electroluminescent spectra of heterostructure device using PbBr-based layered perovskite: a) single heterostructure device and b) double heterostructure device.

When the heterostructure EL devices were driven at 100 K, sharp EL due to exciton emission of CHEPbBr4 was observed. EL spectra of the heterostructure devices are shown in Fig. 4.

EL spectra are in good agreement with PL spectrum of CHEPbBr4 film, demonstrating that carrier-recombination and radiative decay mainly occurred within the CHEPbBr4 emissive layer. Beside the EL due to the exciton, however, small and broaden emission peaking around 380 nm is observed. The broaden emission corresponds well to PL of OXD7 layer. The emission reveals that emission region was slightly expanded to the OXD7 layer. The emission due to OXD7 in the DHD is somewhat stronger than that of the SHD; degree of expansion of emissive region to OXD7 layer in the DHD is large compared with that of the SHD.

Current density-voltage characteristics of the SHD and DHD are compared in Fig. 5. The characteristics clearly demonstrate that insertion of CuPc layer largely lowered drive voltage to inject carriers into the device; drive voltage to inject a current of 100 $mAcm^{-2}$ was 19.8V for DHD and 72 V for SHD although device thickness was almost same with each other. Further, current density-EL intensity characteristics of the heterostructure devices demonstrated that EL of the DHD was highly efficient in comparison with that of SHD (Fig. 6). At a current density of about 150$mAcm^{-2}$, EL intensity of DHD was about 30 times large that of SHD.

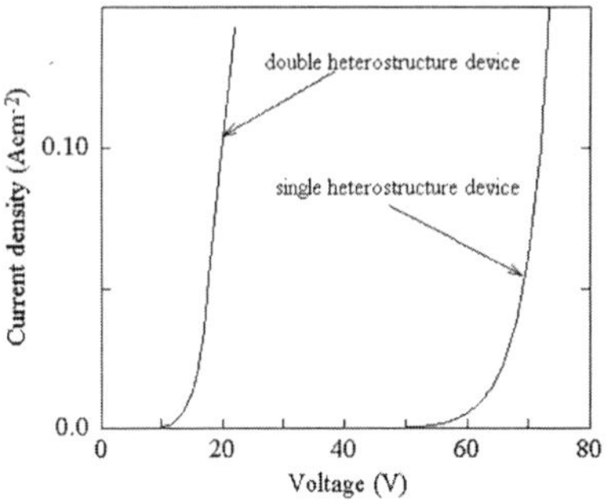

Fig. 5 Current density-voltage characteristics of heterostructure devices using PbBr-based layered perovskite.

The energy diagram of SHD and DHD is shown in Fig. 7. In SHD, there are large potential barriers of more than 1.2 eV for both hole-injection at ITO/CHEPbBr4 interface and electron-injection at AlLi/OXD7 interface.

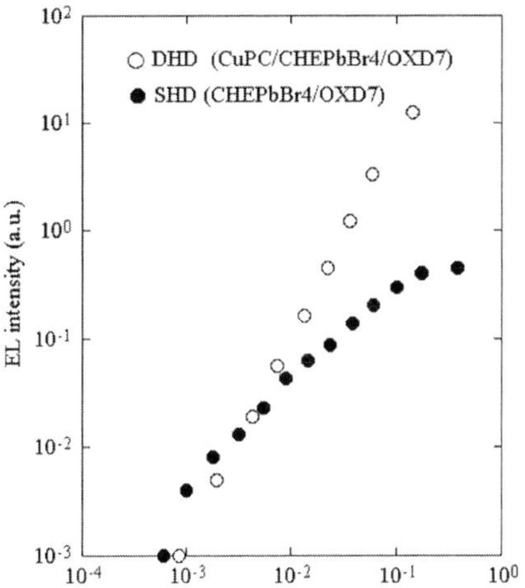

Fig. 6 Current density-EL intensity of heterostructure devices using PbBr-based layered perovskite: filled circles; single heterostructure device and open circles; double heterostructure device.

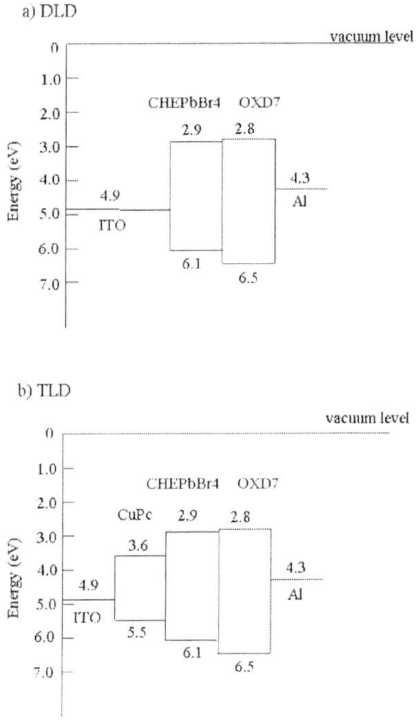

Fig. 7 Energy diagram of heterostructure device using PbBr-based layered perovskite: a) single heterostructure device and b) double heterostructure device.

The large potential barriers are most likely to inhibit carrier injection and enlarge driving voltage of the SHD. On the other hand, insertion of CuPc layer lowers potential barrier for hole-injection in DHD. The low potential barrier makes it possible to inject large amount of holes to device at relatively low driving voltage. Further, most part of injected holes is expected to confine at the interface of CHEPbBr4/OXD7 because of small hole-transporting capability of OXD7 and potential barrier for hole-injection at the interface.

The carrier confinement enlarges effective electric field applied to the OXD7/AlLi interface and increase electron-injection. As a result, it is supposed that driving voltage is lowered and EL efficiency is enhanced in the DHD. In other words, employment of CuPc as hole-injection and hole-transporting layer provides good device performance of the DHD.

4. Conclusion

Sharp electroluminescence due to exciton emission of PbBr-based layered perovskite was successfully observed in the heterostructure devices combined with PbBr-based layered perovskite as emissive material and organic carrier-transporting materials. In addition, lowering of drive voltage and enhancement of EL efficiency was attained in the double heterostructure device emplying copper phthalocyanine CuPc as hole-injection and hole-transporting layer.

5. Acknowledgement

This work was supported by the Core Research for Evolutional Science and Technology Program form the Japan Science, and Technology Corporation (CREST/JST), and Grant-in-Aid for Scientific Research from the Ministry of Education, Science, Sports and Culture of Japan (12650016).

6. References

1. T. Ishihara, In „Optical Properties of Low-Dimensional Materials“, T.Ogawa and Y.Kanemitsu (Eds.), World Scientific, Singapore, 1996, p289.
2. T. Hattori, T. Taira, M. Era, T. Tsutsui and S. Saito, Chem. Phys. Lett. 254 (1996) 103.

3. T. Kondo, A. Azuma, T. Yuasa and R. Ito, Solid State Commun. 105 (1998) 253.

4. C.-Q. Xu, S. Fukuta, H. Sakakura. T. Kondo, R. Ito, Y. Takahashi and K. Kumata, Solid State Commun. 77 (1991) 923.

5. K. Liang, D.B. Mitzi and M.T. Prikas, Chem. Mater. 10 (1998) 403.

6. D.B. Mitizi, M.T. Prikas and K. Chondroudis, Chem. Mater. 11 (1999) 542.

7. M. Era, T. Hattori, T. Taira and T. Tsutsui, Chem. Mater. 9 (1997) 8.

8. M. Era, N. Kakiyama, T. Ano and M. Nagano, Trans. Mater. Res. Soc. Jpn., 24 (1999) 509.

9. M. Era and S. Oka, Thin Solid Films, 376 (2000) 232.

10. C. Adachi, T. Tsutsui and S. Saito, Appl. Phys. Lett., 57 (1990) 531.

11. M. Era, C. Adachi, T. Tsutsui, and S. Satio, Chem. Phys. Lett., 178 (1991) 488.

12. M. Era, T. Tsutsui, K. Takehara, K. Isomura, and H. Taniguchi, Thin Solid Films, 363 (2000) 229.

Novel Methods to Study Interfacial Layers
D. Möbius and R. Miller (Editors)

THE USE OF LB INSULATING LAYERS TO IMPROVE THE EFFICIENCY OF LIGHT EMITTING DIODES BASED ON EVAPORATED MOLECULAR FILMS

G. Y. Jung, C. Pearson and M. C. Petty

Centre for Molecular and Nanoscale Electronics and School of Engineering, University of Durham, South Road, Durham DH1 3LE, UK.

Contents

Keywords: Langmuir-Blodgett film, light emitting diode, Alq, TPD.

We report on the use of insulating Langmuir-Blodgett films of arachidic acid (AA) to improve the quantum efficiency of light emitting devices based on evaporated molecular films of 8-hydroxyquinoline aluminium (Alq) and N,N'-diphenyl-N,N'-(3-methylphenyl)-1,1'-biphenyl-4,4'-diamine (TPD). Devices based on indium-tin oxide (ITO)/TPD/Alq/Al structures were found to emit green visible electroluminescence (peak at 517 nm) when a positive bias was applied to the ITO electrode. The external quantum efficiency was approximately 0.13 %. The incorporation of an arachidic acid LB film between the Alq and the Al electrode was found to increase both the forward current and light output of the LED and to produce a threefold increase in efficiency – to 0.4 %.

1. INTRODUCTION

Since the discovery of electroluminescence (EL) in poly(phenylene vinylene)(PPV), the use of organic materials in light emitting devices (LEDs) has received increasing attention [1]. There is now intense interest in the application of low molecular weight materials and organic polymers [2,3] to a variety of display technologies [4].

Many techniques have been applied in attempts to optimise the performance of organic LEDs. Electron and hole transporting films have been deposited between the cathode and the emitting layer and the anode and the emitting layer, respectively, to improve and balance the injection of carriers [5]. The use of low work function cathode materials such as calcium [6] or Mg/Ag [7] has led to high efficiency devices. These metals are unstable on exposure to air, however.

Bilayer heterojunction devices consisting of aromatic diamine hole transporting layers and 8-hydroxyquinoline aluminium (Alq) [8] or dye-doped Alq [9] emissive layers sandwiched between indium tin oxide (ITO) and Mg/Ag electrodes have been found to exhibit high external efficiency, luminous efficiency and brightness.

Electron injection has been improved by inserting an inorganic insulating layer such as AlO_x [10], SiO_2 [11] or LiF [12] between the cathode and the emissive material. Recently, we have demonstrated how insulating Langmuir-Blodgett (LB) films may be used to improve the efficiency of LEDs based on poly-(2-methoxy,5-(2'ethylhexyloxy)-*p*-phenylenevinylene (MEH-PPV) [13]. In this work, we describe the use of multilayer LB films of arachidic acid (AA), sandwiched between the emissive layer and the aluminium cathode, to enhance the operation of N,N'-diphenyl-N,N'-bis(3-methylphenyl)-1,1'-biphenyl-4,4'-diamine (TPD)/Alq bilayer devices.

2. EXPERIMENTAL

The materials used in this investigation are shown in Fig. 1. The Alq, TPD and AA compounds were purchased from the Sigma Aldrich Co. Ltd. and used without further purification. The anode was indium-tin oxide (ITO)-coated glass from the Samsung Co., with a sheet resistance of 32 $\Omega\ \square^{-1}$. This was patterned into stripes (2 mm wide) with a photoresist pen and, after drying, the exposed ITO was etched using zinc powder dissolved in dilute (50%) hydrochloric acid. Following removal of the resist, the substrate was cleaned by sonication in ultrapure water, acetone and isopropyl alcohol for 30 minutes each and finally dried in a flow of dry nitrogen.

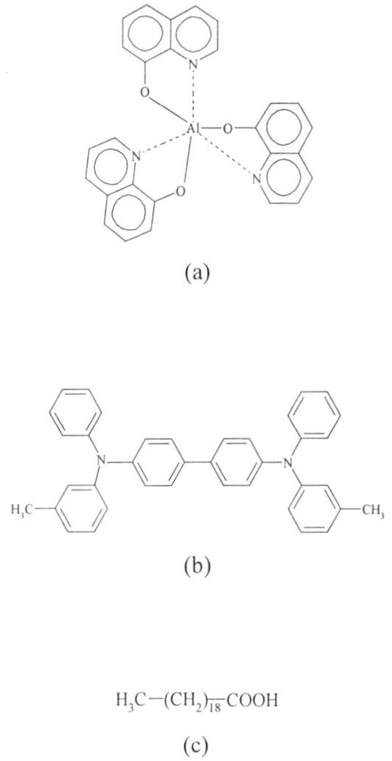

Fig. 1 Materials used in this work: (a) 8-hydroxyquinoline aluminium (Alq); (b) N,N'-diphenyl-N,N'-(3-methylphenyl)-1,1'-biphenyl-4,4'-diamine (TPD); (c) arachidic acid (AA).

The Alq and TPD were deposited by thermal evaporation in a purpose built vacuum chamber at a pressure of approximately 10^{-6} mbar. Each material was placed in a glass crucible, which was heated to 180 °C for TPD and 220 °C in the case of Alq; the corresponding deposition rates were 0.4 nm s^{-1} for TPD and 0.1 nm s^{-1} for Alq. The thickness of the films was measured by a Tencor Instruments Alpha-step 200 stylus profilometer: 50 nm for TPD and 55 nm for Alq.

Arachidic acid LB films were built-up at a surface pressure of 32 mN m^{-1} using a constant perimeter barrier trough located in a microelectronics clean room. The subphase used was ultrapure water purified by reverse osmosis, carbon filtration, two stages of deionisation and UV sterilisation at a temperature of 20 ± 1 °C and pH 5.8 ± 0.2. Y-type deposition was noted with an average transfer ratio of 0.9 (upstrokes and downstrokes).

The device structure was completed by vacuum evaporation of aluminium stripes (1 mm wide, 150 nm thick) onto the organic layers at a pressure of about 10^{-6} mbar. The active area of an LED device was 2 mm^2, defined by the overlap of the ITO and cathode electrode. A schematic diagram of the device structure used in this work is given in Fig. 2.

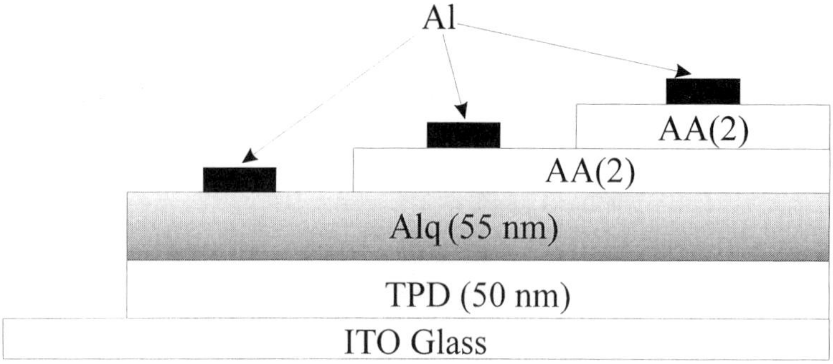

Fig. 2 Schematic diagram showing light-emitting device structure used in this work.

All electrical measurements were undertaken in a vacuum chamber (10^{-1} mbar). Biases were supplied, and device currents were measured, by a Keithey 2400 source meter using a linear staircase step of 0.5 V with a 2 s delay between measurements. The EL devices were mounted over a large area silicon photodiode. Not all the light from the LED (i.e. scattered light) was collected by the photodiode and so the efficiencies quoted in this paper are lower limits. The photocurrents generated were recorded using a Keithley 485 digital picoammeter. For quantum efficiency measurements, the light power was calculated with typical conversion factor of 0.26

Amp / Watt at 517 nm, the EL peak of the TPD/Alq device. EL spectra were recorded using a Jobin Yvon-Spex Instruments S.A. Spectrum One CCD detector.

The current and light output from the LED device were measured simultaneously. The current versus voltage characteristics were found to be strongly dependent on the bias history of the device. For example, an irreproducible and spiky behaviour was recorded during the first scan. This behaviour lessened but was still evident during the second voltage scan. All curves shown in this paper are scans taken after reproducible characteristics had been obtained (usually following the third voltage scan).

3. Results and Discussion

ITO/TPD/Alq/Al 'reference' devices exhibited rectifying electrical characteristics, as reported by others for similar EL structures [8]. Green electroluminescence was clearly visible under ambient lighting conditions when a positive bias was applied to the ITO electrode. Incorporation of the fatty acid LB layer did not significantly change the colour of the EL. Fig. 3 compares the normalised EL output of ITO/TPD/Alq/Al and ITO/TPD/Alq/AA/Al devices. The peak output for the LB film-containing device (511 nm) is very similar to that of the reference structure (517 nm) suggesting that most of the light output originates from the Alq layer [14].

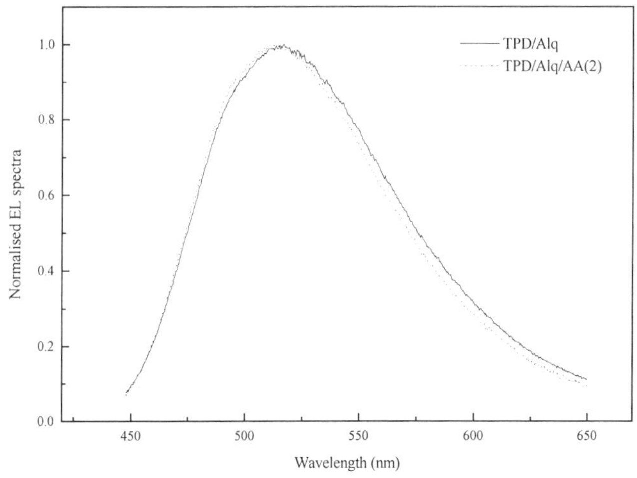

Fig. 3 Normalised EL spectra for ITO/TPD/Alq/Al and ITO/TPD/Alq/AA(2 LB cycles).

Fig. 4 shows the current versus voltage behaviour (Fig 4(a)) and light output versus voltage behaviour (Fig. 4(b)) for LED devices in which different numbers of AA layers were sandwiched between the Alq and the top Al electrode. The maximum luminous intensity for a device containing 4 layers (2 LB deposition cycles) of arachidic acid was estimated as 7000 cd m^{-2} at a bias voltage of 18 V.

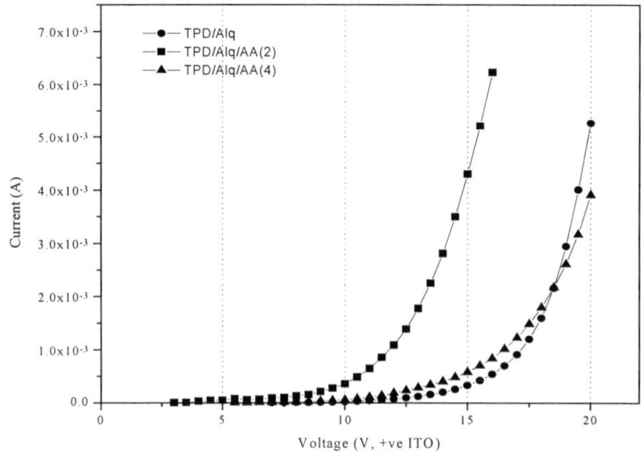

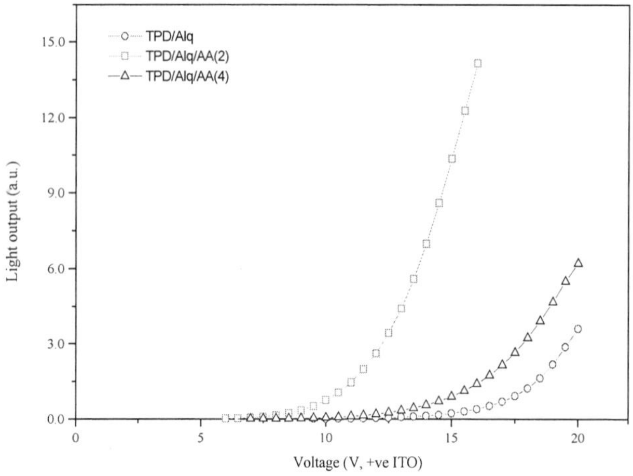

Fig. 4 (a) Current versus voltage characteristics for ITO/TPD/Alq/AA devices containing different thicknesses of fatty acid. (b) Corresponding light output versus voltage characteristics. The number of LB deposition cycles is shown for each device: λ (0); ν (2); σ (4) for the current; open symbols for the light output.

It is evident that the incorporation of 4 layers (2 LB dipping cycles) of the fatty acid has produced an increase in both the current and the light output of the EL structure. At the same time, the EL versus voltage curve is displaced to lower biases. The EL structure incorporating 8 LB layers (4 dipping cycles) exhibits similar current voltage characteristic to the reference device and a slightly enhanced light output. These results contrast with data published by Kim et al. for ITO/spin-coated MEH-PPV/poly(methyl methacrylate) LB film/Al structures [15]. Here, insertion of the LB film resulted in a current decrease at all bias voltages. The effect on the quantum efficiency of our EL structures is shown in Fig. 5. The devices incorporating the LB films both have greater efficiencies over the reference structure, with the 4 LB layer structure exhibiting an approximately threefold increase in efficiency .

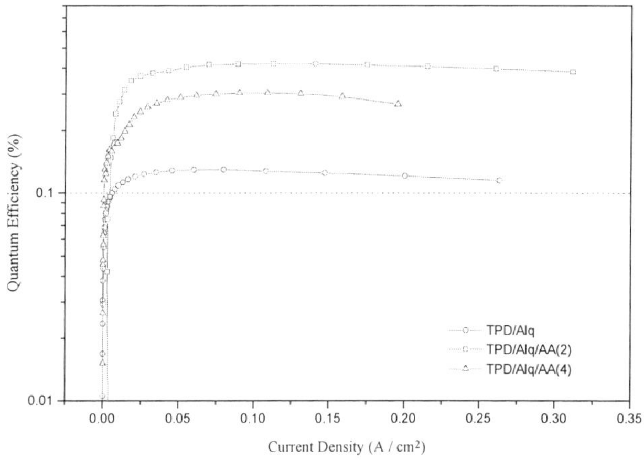

Fig. 5 Quantum efficiency versus current density for the devices whose photoelectrical behaviour is shown in Fig. 3.

In our previous publication we have discussed the possible effects of introducing an insulating layer into an organic EL device [13]. We therefore suggest the following as one possible explanation for our results. For the ITO/TPD/Alq/Al reference structure, the electron current injected from the Al into the Alq is limited by a barrier of about 1 eV at the Al/Alq interface [16]. A relatively large forward applied voltage is therefore required to enable electrons to tunnel through this barrier so that EL can be observed. A significant difference for the devices incorporating the AA layers is that the Al Fermi level is no longer 'tied' to the energy band structure of the Alq. In particular, as the applied voltage is increased, the Al Fermi level can move with respect to the LUMO level in the Alq aligning filled electron states in the metal with vacant states in the organic material. As long as the fatty acid layer is transparent to electrons (from our results, this is clearly the case for devices with 4 LB layers) the electron current, and hence the EL efficiency, increases. The precise mechanism of conduction through the arachidic acid film is unclear. The film thicknesses used in this work are too high to support simple quantum mechanical tunnelling. However, our previous work on metal/LB/inorganic semiconductor devices revealed that relatively thick fatty acid films could support large currents [17].

An alternative explanation is that the presence of the fatty acid layer effectively lowers the work function of the cathode. This would occur if the LB layer incorporated a positive charge. However, our previous work on fatty acid LB layers suggests that the deposited layers are negatively charged [18]. There are two further effects that should also be considered. First, the insulating layer will act as a physical barrier, preventing any chemical interaction between Al and Alq reducing penetration of the metal into the organic film and impeding the diffusion of oxygen/water vapour. The presence of the insulating fatty acid layer between the polymer and the top electrode will also separate the light emission zone from the metal [19]. Some (or all) of the enhanced light output may result from the reduction in nonradiative quenching effects.

4. Conclusions

We have studied the effects of incorporating fatty acid Langmuir-Blodgett films into dual layer organic light emitting structures based on 8-hydroxyquinoline aluminium (Alq) and N,N'-diphenyl-N,N'-(3-methylphenyl)-1,1'-biphenyl-4,4'-diamine (TPD). The insertion of arachidic acid layers between the Alq emissive layer and an aluminium cathode was found to produce a threefold increase in the quantum efficiency. The precise mechanism for this enhancement is currently subject to further study.

5. References

1. J.H. Burroughes, D.D.C. Bradley, A.R. Brown, R.N. Marks, K. Mackay, R.H. Friend, P.L. Burns and A.B. Holmes, Nature, 347 (1990) 359.

2. S.A. VanSlyke, C.H. Chen and C.W. Tang, Appl. Phys. Lett., 69 (1996) 2160.

3. Y. Cao, I.D. Parker, G. Yu, C. Zhang and A.J. Heeger, Nature, 397 (1999) 414.

4. J. Kido, Phys. World, 12 (1999) 414.

5. K.H. Choi, D.H. Hwang, H.M. Lee, L.M. Do and T.H. Zyung, Synthetic Metals, 96 (1998) 123.

6. S. Karg, M. Meier and W. Riess, J. Appl. Phys., 82 (1997) 1951.

7. X.Z. Jiang, Y.Q. Liu, S.G. Liu, W.F. Qiu, X.Q. Song and D.B. Zhu, Synthetic Metals, 91 (1997) 253.

8. C.W. Tang and S.A. VanSlyke, Appl. Phys. Lett., 51 (1987) 913.

9 C.W. Tang, S.A. VanSlyke and C.H. Chen, J. Appl. Phys., 65 (1989) 3610.

10. M. Meier, M. Cölle, S. Karg, E. Buchwald, J. Gmeiner, W. Rieß and M. Schwoerer, Mol. Cryst. Liq. Cryst., 283 (1996) 197.

11. I.D. Parker and H.H. Kim, Appl. Phys. Lett., 64 (1994) 1774.

12. M. Matsumura, K. Furukawa and Y. Jinde, Thin Solid Films, 331 (1998) 96.

13. G.Y. Jung, C. Pearson, L.E. Horsburgh, I.D. W. Samuel, A.P. Monkman and M.C. Petty, J. Phys. D: Appl. Phys., 33 (2000) 1029.

14. J. Kalinowski, J. Phys. D: Appl. Phys., 32 (1999) R179.

15. Y.-E. Kim, H. Park and J.-J Kim, Appl. Phys. Lett., 69 (1996) 599.

16. M. Matsumura and Y. Jinde, Appl. Phys. Lett., 73 (1998) 2872.

17. M.C. Petty, J. Batey and G.G. Roberts, IEE Proc., 132 (1985) 133.

18. N.J. Evans, M.C Petty and G.G. Roberts, Thin Solid Films, 160 (1988) 177.

19. M. Matsumara, K. Furukawa and Y. Jinde, Thin Solid Films, 331 (1998) 96.

Novel Methods to Study Interfacial Layers
D. Möbius and R. Miller (Editors)

LIGHT EMITTING EFFICIENCIES IN ORGANIC LIGHT EMITTING DIODES (OLEDs)

S.T.Lim[1], J.H.Moon[1], J.I.Won[2], M.S.Kwon[1], D.M.Shin[1]

[1] Department of Chemical Engineering, Hong-Ik University, 121-791, Seoul, Korea, gcoxcomb@wow1.hongik.ac.kr

[2] Hyundai Electronics Industries Co., Ltd, Cheongju, Korea

Contents

Keywords; organic light emitting diodes(OLEDs), buffer layer, interface, energy barrier, electron-hole pair

The interface between the organic layer and metallic anode of an OLED is crucial to the stability and the performance of the device. The α-septithiophene (α-7T) has been used as the buffer layer at the interface between the ITO electrode and the hole transport layer (HTL). The insertion of α-7T layer lowered the operating voltage and improved the external power efficiency. Moreover EL emission was not saturated up to 1600mA/cm^2. The Maximum EL intensity was over 17000cd/m^2 and the maximum external power efficiency at 2000cd/m^2 is 6.41m/W and that at 100cd/m^2 is 9.34lm/W. The EL intensity and external power efficiency of OLEDs depend on the thickness of the α-7T layer.

1. INTRODUCTION

Organic light emitting diodes (OLEDs) have drawn a lot of attention for its practical display applications. In many numerical model and experimental reports, it has been found out that the device performance efficiency strongly depends on the height of interfacial energy barrier [1, 2, 3]. Several methods have been proposed to improve the hole-electron balance in OLEDs [4, 5]. Insertion of a buffer layer between the metallic anode and the hole transport layer (HTL) is what can control the hole concentration in the emission region [6, 7, 8]. The buffer layer lowers the energy barrier at the anodic-metal/organic layer interface and induces uniform injection of the hole carriers from the metal to the organic layer. Since the highest occupied molecular orbital (HOMO) of hole transporting materials in OLEDs is located below the work function of the anode electrode, it is required that buffer material should have an intermediate value of HOMO, from 4.8eV to 5.5eV. A lowering of the energy barrier at the interface anodic-metal/organic layer in OLEDs plays a key role in the performance of the devices [9]. It has been reported that OLEDs using oligothiophene as a buffer material show improved device performance [10]. Oligothiophene has relatively low ionisation potential (IP) (5.0 – 5.2eV) and high hole mobility (as high as $0.03cm^2/Vs$) [11]. Schön and his co-workers have observed that the single crystal of α-sexithiophene (α-6T) showed a high room temperature mobility over $0.2cm^2/Vs$ in space charge limited current (SCLC) region [12]. Silicon monoxide (SiO) has been frequently used to improve the performances of OLEDs. The effects of a SiO layer, at the interface between the organic layer and the aluminum electrode, were investigated and a strong dependence of the diode efficiency on the thickness of SiO layer was observed [13]. In this work, α-septithiophene was used to act as a buffer material, and the effect of the SiO layer on the device performance efficiency was investigated.

2. EXPERIMENTAL

2.1. Sample preparation

Four kinds of OLEDs were fabricated to investigate the performance of the buffer layer and the SiO layer. Fig. 1 shows the molecular structure of the organic materials used in the devices. ITO-coated glass with sheet resistance of 10Ω/□ was used as the substrate which is commercially available from Samsung Corning Co. Ltd. N,N'-diphenyl-N,N'-(3-

methylphenyl)-1-1'-biphenyl-4,4'-diamine (TPD) was used as HTL and 8-Hydroxyquinoline aluminium (Alq_3) was purified by train sublimation technique and used as emitting layer (EML) and electron transport layer (ETL). Lithium fluoride was layered on top of the ETL followed by aluminium layer of 2000Å. α-septithiophene was synthesised and purified by a train-sublimation technique in our laboratory. Before deposition of the organic layer, the ITO substrates were pretreated with a wet (aquaregia) process [14]. Film thicknesses of the organic layer were measured using a Plasmon SD series ellipsometer.

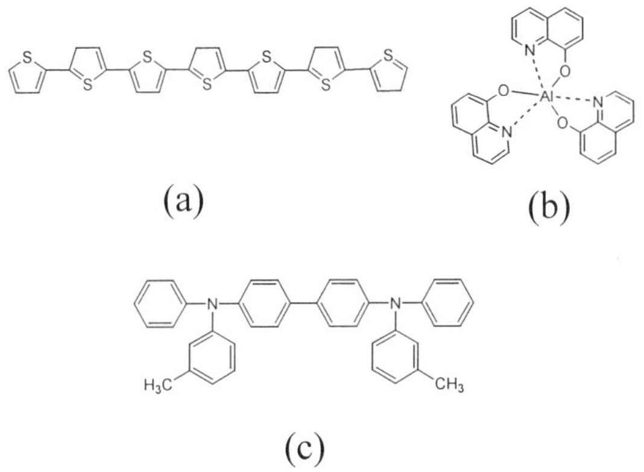

Fig. 1 Molecular structure of (a) α-7T; (b) TPD; (c) Alq_3

To investigate the effect of α-septithiophene as buffer material, three kinds of devices, device I, device II and device III were fabricated. The device structures are as follows:

device I has a structure of ITO/TPD(400Å)/Alq_3(600Å)/LiF/Al;

device II ITO/α-7T(200Å)/TPD (400Å)/ Alq_3(600Å)/ LiF/Al;

device III ITO/α-7T(30Å)/TPD(400Å) /Alq_3(600Å)/LiF/Al.

To investigate the effect of controlling the hole injection to HTL the device IV was fabricated with the device structure of ITO/α-7T(30Å)/SiO(50Å)/TPD(400Å)/Alq_3(600Å)/LiF/Al. Although the SiO layer generally shows hole blocking and insulating properties, the charge carrier may be transported through thin SiO solid films up to 10nm by tunnelling processes [15]. All the organic and inorganic layers were deposited via conventional thermal evaporation under the pressure of 10^{-5} Torr. The deposition rates of the organic compounds were 1.0 ± 0.2Å/s. ITO was used as a hole-injecting and LiF/Al as an electron-injecting contact. The insertion of a thin insulating layer of LiF between the emission layer and the cathode can decrease the effective work-function of the cathode. The LiF layer can also reduce some chemical reactivity of that interface.[16]

2.2. Sample characterisation

The samples were characterised under steady state condition. All the electrical properties and the emission spectra were measured at room temperature under ambient atmosphere. Steady state current measurements on samples were performed using the high voltage Keithley Source Measurement Unit SMU 236. The EL spectra were obtained using the Perkin-Elmer LS50B.

3. Results and Discussion

As shown in Fig. 2(a), the devices I and IV showed similar current density-applied voltage (J-V) characteristics. The maximum EL intensity obtained from the device I was two times higher than that obtained from device IV (Fig. 2. (b)). From device III, we observed the highest EL intensity of over 17000cd/m^2.

The EL intensity-current density (lm-J) and external power efficiency –current density (η-J) characteristics are shown in Fig. 3. The turn-on voltages were about 10V for devices I and IV and about 4V for devices II and III. This clearly indicates that the α-7T lowers the turn-on voltage. At all the current density, the EL intensity of device I compares with that of device IV. This may indicate that the current density is not the dominant factor to improve the device performance. The EL intensity was saturated above 600mA/cm^2 for device I and IV (cf. Fig. 3 (a)), while this is not the case for the device II and III.

Staudigel and his co-workers have performed numerical simulation [17]. According to their report about a three-layer system, the holes are accumulated before the highest energy barrier (in our case α-7T and TPD), whereas the electrons are stored behind the interface between the second HTL and the ETL (in our case TPD and Alq_3). The number of holes injected from the hole-injecting layer is increased due to the lowered energy barrier at the interface between the organic layer and the anode. However, the reduced hole velocity which is induced by drift mobility within the hole transporting layer may contribute to the balanced number of the electrons and holes. The balanced number of electrons and holes may contribute to the sharp exciton distribution near the interface of HTL and ETL, and so to the improved device performance.

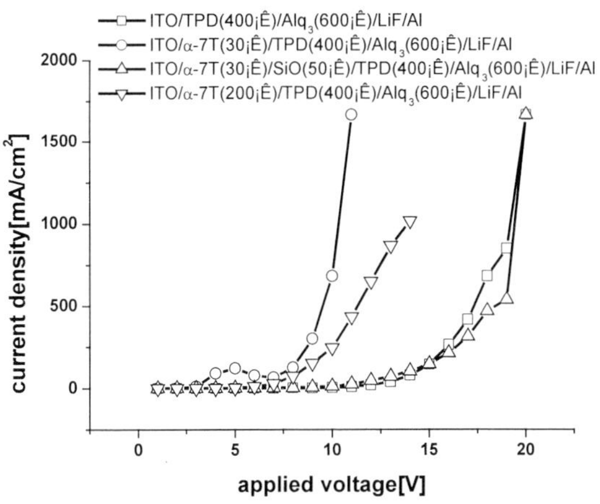

a)

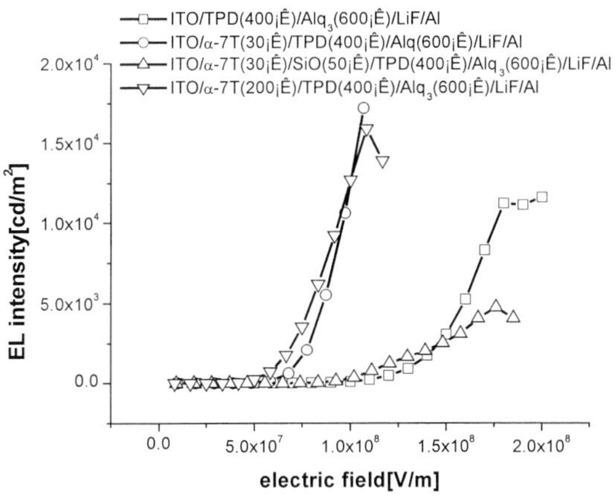

b)

Fig. 2 (a) Current density (J) vs. applied voltage (J-V); (b) Electroluminescence (EL) intensity vs. applied electric field (lm-F)

The big contrast between Staudigel's results and ours is the increase of the device efficiency, although the current density is increased, which usually tend to reduce the quantum efficiency. No energy barriers for the electrons between TPD and α-7T may not only facilitate the charge transport but also the built-up of the sharp recombination zone, which increases the luminescence efficiency. Therefore, the devices II and III did not show a decreased current density. The low energy barrier at the anode/HTL interface is a very effective source for bulk excitons, and yields high device luminance. The low energy barrier between anode and HTL in our case is in contrast with Staudigel's simulation that was based on a low HTL/ETL energy barrier (cf. Fig. 4).

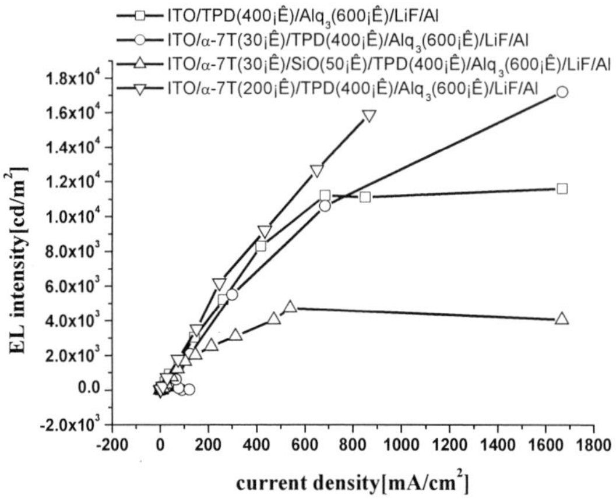

a)

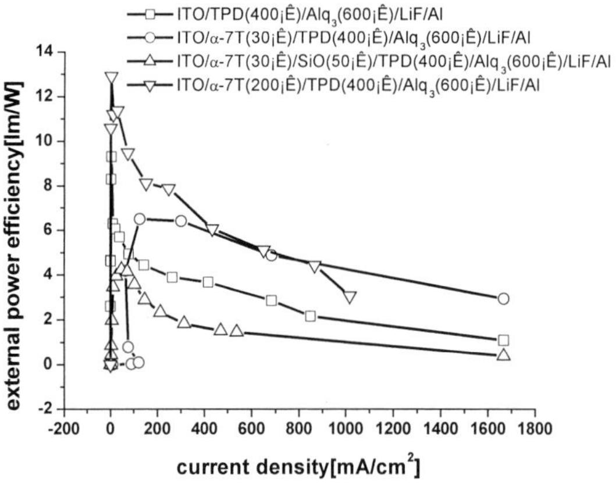

b)

Fig. 3 (a) EL intensity-current density(lm-J); (b) external power efficiency-current density(η-J)

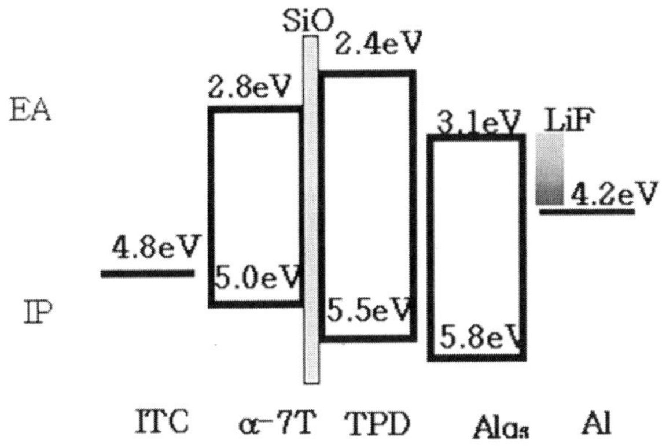

Fig. 4 Energy band diagram of materials. Ionisation potentials(IP) and electron affinities (EA) are literature values for TPD, Alq_3 [7] IP and EA of α-7T were measured via cyclovoltammetry (CV) technique.

Fig. 5 shows the uniformity and smooth surface of α-7T that may contribute to the uniform injection of holes at the interface α-7T/TPD and improve the device efficiency. With the SiO layer, the device IV showed a similar current density over the range of applied voltage but showed decreased EL intensity with respect to the current density in comparison with device I (cf. Fig. 2(b)). Since the SiO layer implements the height for the holes, the numbers of holes and electrons are out balanced a the HTL/ETL interface, and lowered the device efficiency.

In conclusion, the use of α-7T as a buffer material has increased the current density and external power efficiency of OLEDs. The energy barrier for the hole injection at the interface of the hole injecting layer/HTL is one of the important factors to dominate the device performance of OLEDs. The low energy barrier for hole injection from the anode to the organic layer is also one of the important factors to dominate the device efficiency. The improved surface morphology of the hole injecting layer, α-7T layer, can also contribute to the improvement of the device performance, Fig. 5. The hole injection from the anode to the organic layer, the transport from the organic layer to organic layer to recombine with electrons should be controlled in a way to lower the energy barrier and balance the number of electrons and holes to recombine in the emission zone.

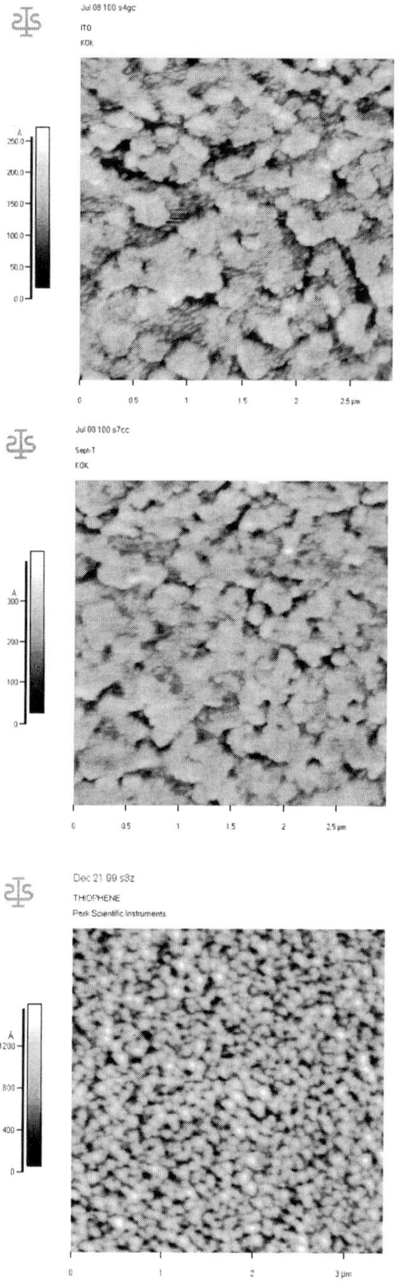

a)

b)

c)

Fig. 5 AFM images of (a) ITO surface on glass substrate (b) α-7T on ITO (30Å) (c) α-7T on ITO (400Å)

4. ACKNOWLEDGEMENTS

The Brain Korea 21 supported this work.

5. REFERENCES

1. D.V. Khramtchenkov, H. Bassler and V.I. Arkhipov, J. Appl. Phys., **79**(2000) 9283
2. D.V. Khramtchenkov, V.I. Arkhipov and H. Bassler, J. Appl. Phys., **81**(2000) 6954
3. Y.-H. Tak and H. Bassler, J. Appl. Phys., **81**(2000), 6963
4. M. Ottmar, D. Hohnholz, A. Wedel and M. Hanack, Synthetic Metals, **105**(1999) 145
5. M. Uchida, C. Adachi, T. Koyama and Y. Taniguchi, J. Appl. Phys., **86**(1999) 1680
6. C. Hosokawa, H. Higashi and T. Kusimoto, Appl. Phys. Lett., **62**(1993) 3238
7. J. Kido and Y. Iizumi, Appl. Phys. Lett., **73**(1998) 2721
8. Y. Kurosawa, N. Tada, Y. Ohmori and K. Yoshino, Jpn. J. Appl. Phys., **37**(1998), L872
9. S.C. Veenstra, U. Stalmach, V.V. Krasnikov, H.T. Jonk nan, A. Heeres and G.A. Sawatzky, Appl. Phys. Lett., **76**(2000), 2253
10. D.M. Shin, S.T. Lim, J.S. Choi and J.S. Kim, Thin Solid Films, **363**(2000) 268
11. A. Ddabalapur, L. Torsi and H.E. Katz, Science, **268** (1995) 270
12. J.H. Schön, Ch.Kloc, R.A.Laudise and B.Batlogg, Physical Review B, **58**(1998) 12952
13. Y. Ohmori, Y. Kurosaka, N. Tada, A. Fujii and K. Yoshino, Jpn. J. Appl. Phys., **36**(1997) L1022
14. J.S. Kim, R.H. Friend and F. Cacialli, Appl. Phys. Lett., **74**(1999) 3084
15. Y. Ohmori, Y. Kurosara, N. Tada, A. Fujii and K. Yoshino, Jpn. J. Appl. Phys., **36** (1997) L1022
16. J. Staudigel, M. Stobel, F. Steuber and J. Simmerer, J. Appl. Phys., **86**(1999) 3895

Novel Methods to Study Interfacial Layers
D. Möbius and R. Miller (Editors)

LIGHT-EMITTING DEVICES BASED ON SEQUENTIALLY ADSORBED LAYER-BY-LAYER SELF-ASSEMBLED FILMS OF ALIZARIN VIOLET

Sharmistha Das and Amlan J. Pal*

Indian Association for the Cultivation of Science, Department of Solid State Physics, Jadavpur, Calcutta 700 032, India

Contents

Keywords: Charge injection, electroluminescence, layer-by-layer self-assembly, light-emitting diodes, organic semiconductors

*E-mail: sspajp@mahendra.iacs.res.in

Layer-by-layer self-assembled films of a small organic molecule have been deposited by physi-adsorbing the molecules with a polycation. Light-emitting devices have been fabricated based on layer-by-layer self-assembled films. Alizarin violet mixed with poly(allylamine hydrochloride) and poly(acrylic acid) have been used as polycation and polyanion, respectively. The pH of the polyanion has been varied which in turn controlled the film morphology, and device characteristics based on such films have been compared. Charge injection mechanisms have been studied in the devices, and barrier heights for charge carriers have been calculated. The results showed that lower turn-on current and higher light output can be obtained from the devices based on self-assembled films as compared to devices based on spin-cast films.

1. Introduction

Alternate physi-adsorption of oppositely charged polymers, the so called layer-by-layer self-assembly (SA), or electrostatic self-assembly is an interesting field of research in recent years [1-4]. With a simple preparative procedure, large area films of uniform thickness are possible. The method is based on electrostatic adsorption of alternate layers of polyanion and polycation. A polycation can be adsorbed onto a negatively charged glass surface, efficiently reversing the net surface charge. A subsequent dip in a polyanion solution adsorbs a second layer onto the substrate and again reverses the net surface charge. Repetition of this dipping process results in a multilayer nanocomposite organic film consisting of the optically or electrically functional materials of interest.

Past work with weak polyelectrolytes, such as poly(allylamine hydrochloride) (PAH) and poly(acrylic acid) (PAA) multilayers has shown that bilayer composition, wettability of the surface, layer interpenetration thickness of the individual polyelectrolyte layers vary as a function of pH [5]. Since the SA method relies on the attractive interaction of oppositely charges, small number of charged groups are not favourable to the formation of homogeneous SA films. This restricted the use of small organic molecules (with a couple of charged groups) in this method of film deposition.

Most of the light-emitting devices (LEDs) based on conjugated organics are fabricated with spin-cast or vacuum-evaporated films [6]. Several devices based on such layer-by-layer self-assembled films have recently been demonstrated [7]. In this article, we have demonstrated a simple method to obtain SA films of small organic molecules and aimed to make a comparative study of electrical characteristics of devices based on layer-by-layer self-assembly process and conventional spin-cast technique. The semiconductor used in this work has been alizarin violet. Devices based on spin-cast films of alizarin violet have been found to yield electroluminescence (EL) in a wide range of concentration of the active material [8]. PAH with physi-adsorbed alizarin violet (in controlled concentrations) and PAA have been used as polycation and polyanion, respectively, to deposit layer-by-layer self-assembled films.

2. Experimental

The dye used in this work was alizarin violet. PAH (MW = 70 000) and the dye was purchased from Aldrich Chemical Co. and PAA (MW = 90 000) was obtained from Polyscience as 25 % aqueous solution. All the chemicals were used without further purification. The chemical structure of alizarin violet is shown in Fig. 1. The polyelectrolyte deposition baths were prepared with 10^{-2} M (based on repeat units) aqueous solutions using 18.2 MΩ Millipore water. The solutions of alizarin violet and PAH were mixed in a controlled way so that SO_3^- ions of alizarin violet get attached to NH_3^+ ions of PAH. The low concentration of alizarin in PAH allowed 80 % of the NH_3^+ ions of PAH to take part in the adsorption process during layer-by-layer deposition. Use of organic molecule alone as anion generally results in material loss during washing [9, 10].

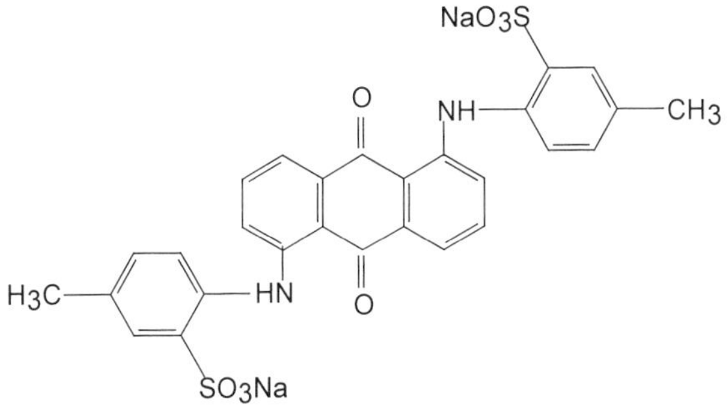

Fig. 1 Molecular structure of alizarin violet.

PAA was used as polyanion, and mixed solution of PAH and alizarin (10:1) was used as polycation. The pH of the each solution was adjusted with either HCl or NaOH. In this work, we have varied the pH of the polyanion solution, which will control the free charge density of the absorbing polyelectrolyte, to get different morphology in the self-assembled films. Three different pH levels of PAA were 3, 3.3 (or 3.5), and 4, whereas the pH of polycation system was kept at 7.5. Thoroughly cleaned substrates were immersed in the polycation solution for 15 min followed by rinsing in three water baths for 2, 1, and 1 min, respectively. The rinsing washes off the surplus polymers attached to the surface. The substrates were then immersed in polyanion solution for 15 min followed by the same rinsing protocol. This resulted in one bilayer of self-assembled film, and the whole sequence was repeated for a desired number of times of get a desired number of bilayer films. The films were then oven-dried in vacuum at 80 °C for 6 hours.

To fabricate the LEDs, layer-by-layer self-assembled (SA) films were deposited on indium tin oxide (ITO) coated glass substrates. Aluminium as top electrode was vacuum-evaporated at a pressure below 10^{-5} Torr. Device testing was computer-automated using GPIB interface and performed in nitrogen environment. DC voltage was supplied by a Yokogawa 7651 dc voltage source. The device current was measured by using a Keithley 617 electrometer. The luminous output was measured by Electron Tubes 9256B photomultiplier coupled with a Hewlett-Packard 34401A multimeter. The thickness of the film was measured by a Planer Products Limited SF 101 Surfometer.

3. RESULTS AND DISCUSSION

Fig. 2 shows the current-voltage characteristics of LEDs based on SA films of alizarin violet deposited at three different pHs of polyanion. The rectification ratio of the devices depended on the pH level. Current-voltage characteristics showed two distinct regions: ohmic behaviour in the low field region with current proportional to voltage, and space-charge limited conduction (SCLC) behaviour at higher voltages.

To study charge injection mechanisms, we have tried to fit Richardson-Schottky thermionic emission and Fowler-Nordheim tunnelling mechanisms. We have found that under forward bias, the temperature-independent Fowler-Nordheim (FN) tunnelling mechanism is applicable, which presumes tunnelling of charge carriers directly into the bands of the semiconductor. According to the model, the current density (J) is related to the applied field (F) as [11,12]:

$$J = \left(J_0 F^2 / \phi\right) \exp\left[-\kappa \phi^{3/2} / F\right] \qquad (1)$$

where J_0 is a prefactor, ϕ is the energy barrier that controls hole injection, and κ is a constant. The FN plot, which is represented by $\log\left(J/F^2\right)$ versus $1/F$, should therefore show linear behaviour with a negative slope.

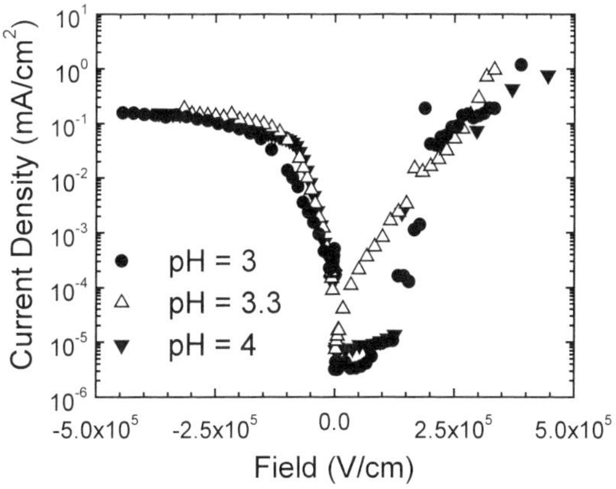

Fig. 2 Device current-field characteristics of LEDs based on layer-by-layer self-assembled films of alizarin violet/PAH and PAA, deposited at three different pHs of the polyanion (PAA).

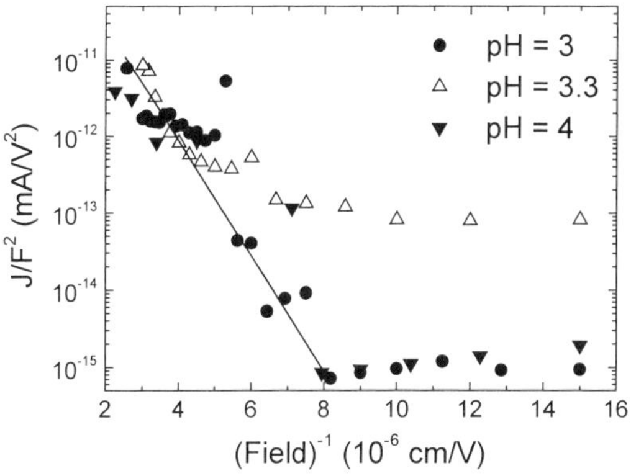

Fig. 3 Fowler Nordheim plots [*log(J/F²)* vs. *1/F*] under forward bias for the same structures as described in Fig. 2. The line is the best fit of the model to the experimental points above 10^5 V/cm.

Figure 3 shows the FN plots for the same structures as shown in Fig. 2 under forward bias. The plots show linear behaviour at higher fields. This suggests that FN tunnelling mechanism is applicable in these devices. At low fields, the deviation from the linear behaviour probably indicates additional contribution in current due to thermionic emission. From the slope of the plot, we have calculated a hole barrier height of 0.1 eV, which did not show any dependence on the pH of the polyanion.

Under reverse bias, the FN tunnelling mechanism did not apply. On the other hand, the current-density was found to follow Richardson-Schottky thermionic emission model, where tunnelling through the barrier is ignored and field-induced barrier lowering is taken into consideration. The current density at a temperature T is given by [13]:

$$J = AT^2 \exp\left[-\left(\phi - \beta F^{1/2}\right)/kT\right] \qquad (2)$$

where A is a constant, β relates to the relative dielectric constant (ε_r) of the emitting material as $\left(e^3/4\pi\varepsilon_0\varepsilon_r\right)^{1/2}$. A plot of $log(J)$ versus $F^{1/2}$ would therefore give a straight line. Fig. 4 shows such plots for devices based on SA films deposited at different polyanion pHs. The plots fit to straight lines, which is in agreement with the concept of thermionic emission. The extrapolated linear region of $log(J)$ versus $F^{1/2}$ plots towards zero field resulted a hole-barrier height of around 0.75 eV under reverse bias direction, which was independent of polyanion pH. This shows that the barrier for charge carriers is in general higher in reverse bias than in the forward bias and hence could make tunnelling mechanism less suitable for charge injection in the former case.

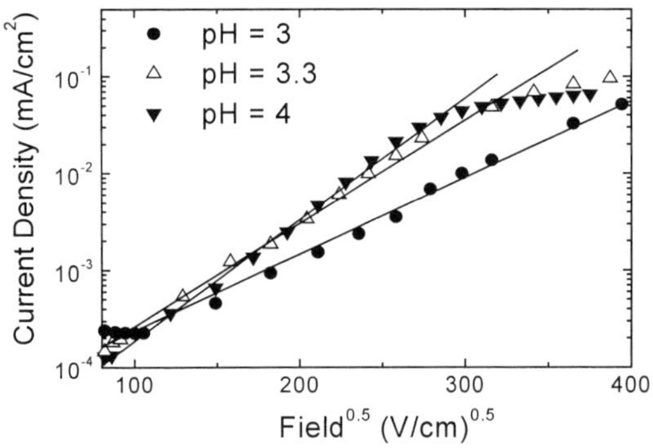

Fig. 4 Thermionic emission plots [$log\ (J)$ vs. $F^{1/2}$] under reverse bias for the same structures as described in Fig. 2. The lines are best fits of the model to the experimental points in each case.

We have studied luminance versus current density plots in devices based on SA films deposited at different polyanion pHs. Luminance has been detected only under forward bias directions. Luminance versus current density plots for devices based on different SA films are presented in Fig. 5. Similar plot for devices with spin-cast film of alizarin violet has also been shown in the figure for comparison. Each of the plots shows a sharp turn-on followed by a linear behaviour in the log-log scale. The turn-on current is generally determined by the amount of electron injection required for light emission in these devices. The turn-on current and the luminance from the LEDs at any current density depended on the polyanion pH used in SA film deposition. The figure shows that the LEDs based on SA films (deposited at a suitable polyanion pH) can yield higher luminance than the devices based on spin cast films. The turn-on currents in such devices are also lower as compared to the spin-cast counterpart.

Above the turn-on current, the linear behaviour of luminance-current density plots in the log-log scale can be explained in terms of FN tunnelling model. Considering the FN mechanism to be valid for both electron and hole injections, a relationship between luminance (L) and current density (J) can be established as [14]:

$$\log(L) = D \log(J) + \text{constant} \tag{3}$$

where $D = (\phi_e/\phi_h)^{3/2}$, with ϕ_e and ϕ_h being barrier the heights for electrons and holes, respectively. The equation explains the linear behaviour of *log(L)* versus *log(J)* plots in Fig. 5.

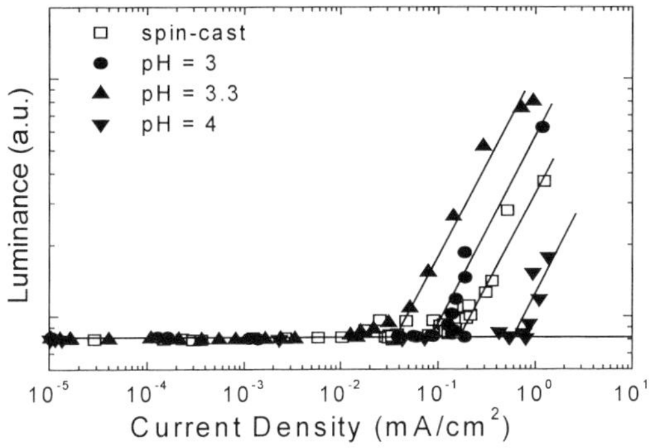

Fig. 5 Luminance-diode current characteristics under forward bias of the same LEDs as described in Fig. 2. The horizontal line represents the background signal level and the other lines are the best fits to the points above turn-on current for each case.

The slope of the plots (= 0.8) was independent of pH of the polyanion, and the equation therefore suggests that the barrier heights (or, their ratio) remained the same in the devices. The FN plots also predicted a constant hole barrier heights in these devices.

4. Conclusion

In conclusion, we have shown that layer-by-layer sequentially adsorbed films of conjugated dyes can be used as an emitting layer in LEDs. Alizarin violet, mixed with a polymer has been used as polycation. The pH of the polyanion has been varied to get layer-by-layer self-assembled films of different morphology. Current-voltage characteristics showed ohmic and space-charge limited conduction behaviours in the low and high field regions, respectively. Applicability of Richardson-Schottky thermionic emission and Fowler-Nordheim tunnelling mechanism have been tested for charge injection mechanisms in the devices. A hole-barrier height of 0.1 eV was calculated from the FN plots under forward bias direction. Luminance was observed only under forward bias direction. The turn-on current and the luminance level at any particular device current depended on the pH of the polyanion used in SA film deposition. We have shown that devices based on SA films can yield higher luminance level than the devices based on spin cast films. The turn-on current in devices based on SA films can also be lower than that in the spin-cast devices.

5. Acknowledgements

The authors thank Dr. Partha Chowdhury of Indian Association for the Cultivation of Science for his cooperation in using Surfometer to measure the thickness of the emitting layer. This work was supported by Department of Science and Technology, Government of India (Project No. SP/S2/M-11/94), and Department of Atomic Energy, Government of India (Project No. 37/16/96-R&D-II).

6. References

1. G. Decher, J.-D. Hong and J. Schmitt, Thin Solid Films, 210/211 (1992) 831.

2. G. Decher, Science, 277 (1997) 1232.

3. P. Bertrand, A. Jonas, A. Laschewsky and R. Legras, Macromol. Rapid Comm., 21 (2000) 319.

4. S.S. Shiratori and M.F. Rubner, Macromolecules, 33 (2000) 4213.

5. D. Yoo, S.S. Shiratori and M.F. Rubner, Macromolecules, 31 (1998) 4309.

6. P.A. Lane and D.D.C. Bradley (Eds.). Proc. 2nd International Conference on Electroluminescence of Molecular Materials and Related Phenomena, Sheffield, 15-18 May 1999, Synthetic Metals, Elsevier, Amsterdam, Vol. 111-112 (2000).

7. J.W. Baur, S. Kim, P. B. Balanda, J. R. Reynolds and M. F. Rubner, Adv. Mat., 10 (1998) 1452.

8. S. Das, A. Chowdhury, S. Roy and A.J. Pal, Phys. Stat. Sol. A., 178 (2000) 811.

9. X. Zhang, M.L. Gao, X.X. Kong, Y.P. Sun and J.C. Shen, J. Chem. Soc., Chem. Comm., (1994) 1055.

10. M.R. Linford, M. Auch and H. Möhwald, J. Am. Chem. Soc., 120 (1998) 178.

11. R.H. Fowler and L. Nordheim, Proc. R. Soc. (London), Ser. A, 119 (1928) 173.

12. A.J. Heeger, I. D. Parker, Y. Yang, Synth. Met. 67 (1994) 23.

13. H. Vestweber, J. Pommerehne, R. Sander, R.F. Mahrt, A. Greiner, W. Heitz and H. Bässler, Synth. Met., 68 (1995) 263.

14. T. Östergård, J. Paloheimo, A.J. Pal and H. Stubb, Synth. Met., 88 (1997) 171.

Novel Methods to Study Interfacial Layers
D. Möbius and R. Miller (Editors)

STRUCTURAL PROPERTIES AND INTERACTIONS OF THIN FILMS AT THE AIR-LIQUID INTERFACE EXPLORED BY SYNCHROTRON X-RAY SCATTERING

Torben R. Jensen[1,3] and Kristian Kjaer[2,*]

[1] Department of Chemistry, University of Aarhus, DK-8000 Århus C, Denmark.

[2] Materials Research Department, Risø National Laboratory, DK-4000 Roskilde, Denmark

Contents

3) Work initiated while at Risø National Laboratory.

*) Address correspondence to Kristian.Kjaer@Risoe.DK

Key words: grazing-incidence X-ray diffraction, specular X-ray reflectivity, liquid surface X-ray scattering, advanced materials, thin films, lipid structure, lipid-protein interactions.

Monolayers of lipids have been studied for more than a century. During the past decade new insight into this area has been obtained due to the development of surface sensitive X-ray scattering methods using synchrotron radiation: Grazing-incidence X-ray diffraction (GIXD) and specular X-ray reflectivity (XR). These novel methods provide direct microscopic information about the systems in question and allow *in situ* investigations at the air-liquid interface. The present review outlines the underlying theory for surface sensitive X-ray scattering and describes a dedicated liquid surface diffractometer and the sample requirements for this relatively new X-ray scattering technique. The (simple) relation between the data (measured in reciprocal space) and the molecular model (in direct space) is discussed. We show how the first data analysis can be done with a pocket calculator and allude to more detailed approaches. Some recent results are presented with special emphasis on biologically relevant systems including lipid and protein structures and lipid/protein interactions at interfaces. Generally, these investigations have impact within the fields of chemistry, physics, materials science, molecular engineering and biology.

1. Introduction

A wide variety of thin films have been studied during the past approximately one hundred years, mainly by macroscopic methods [1 - 3]. For the past decade X-ray scattering measurements of thin films assembled on a liquid surface have been available at some synchrotron radiation facilities [4]. This novel method providing structural information on a molecular scale has had a deep impact on the research field, providing much new structural information [5 - 23] yet still has a significant potential for exploring both complex natural biological systems and advanced artificial nano-sized assemblies. This review presents an introduction to the surface sensitive X-ray scattering methods, grazing-incidence X-ray diffraction (GIXD) and specular X-ray reflectivity (XR), along with selected results and a discussion of perspectives for future development and investigations.

As shown by Pockels, Langmuir [24 - 26] and others, amphiphilic molecules, including lipids, can be stabilised as monomolecular layers at the air-water interface. Also, some proteins can self-assemble – on their own or with lipids - at the air-water interface to form monolayers, *e.g.*, lipase [27], bacterial surface-layer proteins [28, 29], or trans-membrane proteins such as bacteriorhodopsin [30]. Transfer of monomolecular layers to solid support by the Langmuir-

Blodgett (L-B) technique [31 - 33], resulting in mono-, bi- or multilayers allows characterisation of the resulting L-B films by other techniques, *e.g.*, atomic force microscopy (AFM), X-ray scattering [34] and many others. The present review will be restricted to thin films at liquid surfaces.

A biological membrane can be considered as two coupled monomolecular layers of lipid. Thus, a lipid monolayer assembled on a liquid interface can serve as an important model system for investigation of structure and function of biological membranes as well as of the interactions of the membrane with other lipids, proteins, steroids or ions [35]. Therefore, important information can be extracted by X-ray studies of a monolayer on a liquid surface. Furthermore, the X-ray scattering methods allow direct *in situ* investigations under near physiological conditions.

X-ray reflectivity studies are related to the observation of the colours of the rainbow in the light reflected from an oil film on a wet road. The phenomenon is due to the interference of light scattered from the air/oil and oil/water interfaces, the thickness of the oil films being of the same order of magnitude as the wavelength of visible light ($\lambda \approx 5000$ Å). This suggests that X-rays ($\lambda \approx 1$ Å) can be reflected from a monomolecular layer to provide information about the molecular packing. A laterally crystalline organisation of the molecules gives rise to X-ray diffraction in the same way that visible light is scattered by a 2D grating (*e.g.*, a compact-disc). To study the extremely small amount of matter present in a monomolecular layer, very intense X-ray radiation is needed. Fortunately, synchrotron sources dedicated to the production of X-rays have become available during the last decades allowing important applications within a broad range of research fields, such as, biology, medicine, chemistry and physics [36]. The first X-ray scattering experiments on monolayers on liquid surfaces were conducted in the late eighties [5, 6, 37, 38].

The continuously growing interest in nanometer scale molecular engineering within chemistry and materials science [39 - 41], along with interest in complex biological systems, provides strong incentives for the further development of surface sensitive X-ray scattering. Combination of this technique with other powerful complementary methods such as AFM and MD also capable of providing information on a molecular scale [42 - 47] provides a great potential for deeper insight.

The following section outlines the underlying theory for X-ray scattering methods along with a description of the practical techniques for investigations *in situ* at the liquid surface by GIXD and/or XR. Subsequently, some recent results are reviewed, *e.g.,* advanced materials and lipid and protein structure as well as their interactions at the air-water interface. This paper reviews a quite young research field that has, still, much potential yet to be explored for the investigation of a variety of complex systems. We will attempt to illustrate the general possibilities offered by these novel methods.

2. THEORY AND METHODS

2.1 The materials

The molecules investigated on a liquid surface diffractometer are usually insoluble and amphiphilic [48], having a hydrophobic moiety (typically a hydrocarbon chain with at least 13 carbons), traditionally denoted the *tail*, and a hydrophilic *head* that can be neutral or charged depending on *p*H or on the concentration of other ions in the subphase. More complex molecules can also be studied, *e.g.,* proteins, which can form a monomolecular layer by adsorption to the air-water interface. Interactions or chemical reactions can be studied *in situ* by injecting macromolecules or a catalyst under a given surface layer initially assembled at the air/liquid interface.

2.2 Synchrotron X-ray radiation and Liquid surface diffraction

The advent of synchrotron X-ray sources providing intense, well-collimated beams has allowed the development of surface-specific X-ray scattering methods. The methods described here are relatively new; the first X-ray scattering experiments using a liquid surface diffractometer were performed in the latter half of the eighties [5, 6, 37, 38]. Excellent surface sensitivity can be achieved, allowing detailed structural investigations of thin films assembled on a liquid surface. Instruments should allow *in situ* investigation of phase transitions and reactions by manipulation of pressure and temperature. Furthermore, *in vivo*-like conditions can be arranged for studies of biological systems.

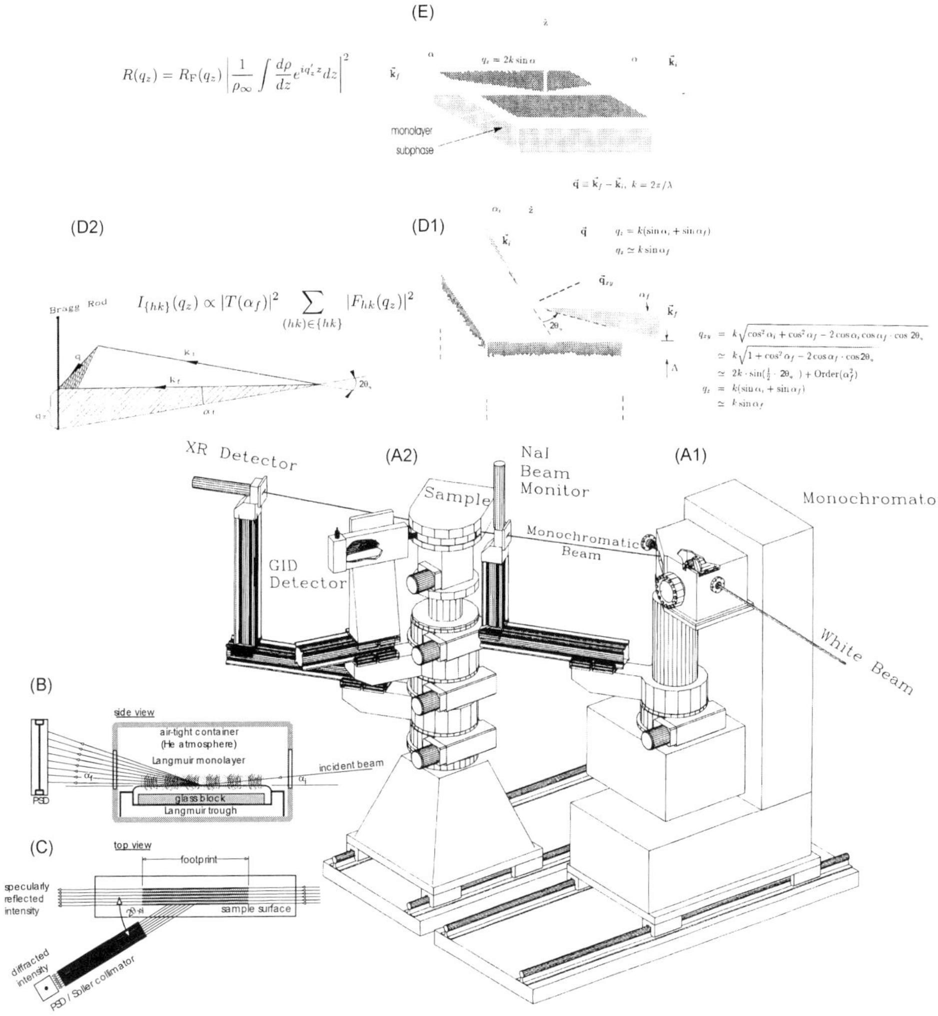

Figure 1 (**A**) Overview of the liquid surface diffractometer at the undulator beamline BW1 at the synchrotron radiation facility HASYLAB, Hamburg, showing (**A1**) the monochromator and (**A2**) sample towers. (**B**) Side view of the sample cell. (**C**) Top view showing the X-ray footprint on the sample surface and the crossed-beam area for diffraction. (**D1**) Geometry for grazing-incidence X-ray diffraction (GIXD) from a liquid surface and (**D2**) detail showing the scattering triangle in *reciprocal space* and illustrating how the Bragg rod pattern is probed by variation of α_f. (**E**) Geometry for specular X-ray reflection (XR) from a liquid surface.

In a synchrotron or in a storage ring fed from a synchrotron, electrons (or positrons) are centripetally accelerated in a closed orbit. These accelerated charged particles, with a velocity close to the speed of light, emit electromagnetic radiation with a wavelength range extending into the X-ray region [49]. An example is the storage ring DORIS III at HASYLAB in Hamburg, Germany, with positrons orbiting a circumference of 289.2 m [50, 51]. At the undulator source BW1 [52, 53] the initial white beam power is *ca.* 1000 W but the beam that finally illuminates the sample is a monochromatic beam of a selected wavelength in the range 1.2 to 1.6 Å with a power of *ca.* 0.3 mW. The radiation is well collimated, allowing a large source to sample distance (here 36 m). X-ray scattering experiments on liquid surfaces are possible using the diffractometer [54 - 56] shown in Figure 1A. The X-ray beam illuminates a monolayer area of, *e.g.*, 2×50 mm^2 (*cf.* Figure 1C), which contains as little as *ca.* 10^{15} lipid molecules or *ca.* 10^{13} molecules of, *e.g.*, *HLL* lipase (a *ca.* 30 kDa protein [57]). Although such very small amounts of material can be probed using the intense X-ray radiation from a synchrotron source, the total sample area has to be considerably larger to allow for manipulation of surface pressure.

The sample cell in the BW1 liquid surface diffractometer (Figure 1B) is a Teflon-made Langmuir film balance placed in a gas-tight container with windows transparent to the X-rays. The sample cell allows variation of temperature (typically 2°C to 35°C) and surface pressure of the monolayer during the X-ray scattering experiment. A glass plate is placed in the trough under the X-ray footprint area in order to reduce the subphase depth to *ca.* 0.3 mm and thus suppress mechanically excited long-wavelength waves on the liquid surface [58]. In order to reduce background scattering from the gas phase (and possibly also beam damage), the atmosphere is often helium gas, saturated by water vapour to reduce evaporation of the subphase. Lipids, peptides or proteins can assemble at the air-water interface and form 2D-crystalline monomolecular layers.

2.3 Surface sensitivity by grazing incidence of X-rays

The enhanced surface sensitivity obtained by grazing-incidence of X-rays on an air/bulk interface can be understood by standard textbook optics [59]. Only the essentials of the theory are outlined here, for details see [8, 13, 15, 16, 18, 60]. For X-rays of wavelength $\lambda \sim 1$ Å the refractive index, *n,* of matter is slightly less than unity [15, 60]:

$$n = 1 - \delta - i\beta \qquad (1)$$

where $\delta = 2\pi\rho r_0/k^2$, $k = 2\pi/\lambda$ is the X-ray wavenumber, ρ the electron density and $r_0 = 2.82 \cdot 10^{-15}$ m the classical electron radius. Typically δ is of the order 10^{-6} to 10^{-5} in condensed matter and only *ca.* 10^{-9} in air. The term β is related to the linear absorption coefficient, μ, by $\beta = \mu/(2k)$. For $\lambda \sim 1$ Å, $\beta << \delta$. Snell's law relates the incident grazing angle, α_i, to the angle between the surface and the refracted wave, α':

$$n \cos(\alpha') = \cos(\alpha_i) \tag{2}$$

A refractive index less than unity implies that below a certain critical angle, $\alpha_i = \alpha_c$, total external reflection of X-rays will occur. Writing $\alpha_c \equiv (2\delta)^{1/2}$ (this identity will be justified below), we see that for small angles and neglecting absorption ($\beta = 0$), Snell's law can be written

$$\alpha_i^2 = \alpha_c^2 + (\alpha')^2 \tag{3}$$

For an ideal planar interface the reflectivity is given by Fresnel's law, which in the small-angle limit becomes

$$R_F = |(\alpha_i - \alpha')/(\alpha_i + \alpha')|^2 \rightarrow (\alpha_c/2\alpha_i)^4 \text{ for } \alpha_i >> \alpha_c \tag{4}$$

For $\alpha_i < \alpha_c$, α' is imaginary and total reflection occurs. This justifies defining the critical angle α_c as $(2\delta)^{1/2}$. For an air-water interface, at $\lambda = 1.30$ Å, $\alpha_c = 0.13°$. At total reflection, below the interface an *evanescent* wave travels parallel to the interface, its intensity decreasing exponentially with depth. For an incident angle $\alpha_i = 0.85\ \alpha_c$, the beam illuminates only the top *ca.* 90 Å beneath the water surface, thus enhancing the surface sensitivity and reducing the background scattering from the subphase.

In the following, two X-ray scattering geometries are discussed, *cf.* Figure 1. The lateral structure of the monolayer is probed by *grazing-incidence X-ray diffraction* (GIXD), *cf.* Figure 1B, 1C and 1D. To gain information about the vertical structure of a monomolecular layer at the air-water interface, vertical scattering is required, *i.e.* measurement of the *specular X-ray reflectivity* (XR) for $\alpha_i = \alpha_f$ and $2\theta_{xy} = 0$, *cf.* Figure 1E, discussed further below.

2.4 *Grazing-incidence X-ray diffraction*

Grazing-incidence X-ray diffraction (GIXD) is performed with a constant incident glancing angle, α_i, less than the critical angle, α_c [61, 62], while the diffracted intensity is recorded as function of the horizontal and vertical angles, $2\theta_{xy} \neq 0$ and $\alpha_f \geq 0$, respectively. The directions of the incident and scattered X-rays are conveniently given by the wave vectors, $\boldsymbol{k}_i$ and $\boldsymbol{k}_f$, where $| \boldsymbol{k}_i | = | \boldsymbol{k}_f | = k = 2\pi/\lambda$. The scattering process is characterised by the scattering vector, $\boldsymbol{q} = \boldsymbol{k}_f - \boldsymbol{k}_i$, which can be separated into its horizontal and vertical components, $\boldsymbol{q}_{xy}$ and q_z, respectively, as shown in Figure 1D. The domain of the scattering vector $\boldsymbol{q}$ and the wave vectors $\boldsymbol{k}_i$ and $\boldsymbol{k}_f$ is called *reciprocal space*.

A short diversion into bulk (3D) crystallography is in order here. Consider a 3D crystal, built by repetition of identical unit cells on a 3D lattice defined by the primitive vectors $\boldsymbol{a}$, $\boldsymbol{b}$ and $\boldsymbol{c}$. Due to constructive interference of the scattering from all the unit cells of the crystal, Bragg diffraction occurs at points $\boldsymbol{q} = \boldsymbol{q}_{hkl}$ satisfying the three Laue conditions

$$\boldsymbol{q}\cdot\boldsymbol{a} = 2\pi h, \quad \boldsymbol{q}\cdot\boldsymbol{b} = 2\pi k, \quad \boldsymbol{q}\cdot\boldsymbol{c} = 2\pi l \tag{5}$$

where h, k, l are integers called Miller indices. The set of points $\{\boldsymbol{q}_{hkl}\}$ is called the reciprocal lattice. It is spanned by three reciprocal lattice vectors $\boldsymbol{a}^*$, $\boldsymbol{b}^*$ and $\boldsymbol{c}^*$:

$$\boldsymbol{q} \equiv \boldsymbol{q}_{hkl} = h\,\boldsymbol{a}^* + k\,\boldsymbol{b}^* + l\,\boldsymbol{c}^*, \quad h, k, l\text{: integers} \tag{6}$$

As can be seen from eqs. (5) and (6), $\boldsymbol{a}^*$ has the properties: $\boldsymbol{a}^*\cdot\boldsymbol{a} = 2\pi$, $\boldsymbol{a}^*\cdot\boldsymbol{b} = 0$, $\boldsymbol{a}^*\cdot\boldsymbol{c} = 0$; and similarly for $\boldsymbol{b}^*$ and $\boldsymbol{c}^*$. The crystal plane spacing d is given by $d = 2\pi/|\boldsymbol{q}|$.

Returning to monolayers, diffraction can occur if the molecules are packed with 2D-'crystalline' order of sufficiently long range. With a crystalline repeat of a unit cell along only two primitive vectors $\boldsymbol{a}$ and $\boldsymbol{b}$ (*in* the monolayer plane), and no repeat out of the plane, only two Laue conditions apply:

$$\boldsymbol{q}\cdot\boldsymbol{a} = 2\pi h, \quad \boldsymbol{q}\cdot\boldsymbol{b} = 2\pi k \tag{7}$$

Then, in reciprocal space, the diffracted intensity occurs where the horizontal scattering vector component, $\boldsymbol{q}_{xy}$, coincides with a reciprocal lattice vector $\boldsymbol{q}_{hk}$ satisfying *eq.* (7), while the q_z - component is not similarly restricted. Thus, instead of reciprocal lattice *points* (or Bragg points), *eq.* (6), the diffracted intensity is extended along Bragg *rods* defined by $\boldsymbol{q} = (\boldsymbol{q}_{xy}; q_z)$ where the q_z - component is unrestricted and

$$\boldsymbol{q}_{xy} \equiv \boldsymbol{q}_{hk} = h\,\boldsymbol{a}^* + k\,\boldsymbol{b}^*, \quad h, k\text{: integers} \tag{8}$$

Here the two reciprocal lattice vectors $\boldsymbol{a}^*$ and $\boldsymbol{b}^*$ are parallel to the monolayer plane. $\boldsymbol{a}^*$ is orthogonal to $\boldsymbol{b}$ and $\boldsymbol{a}^*\cdot\boldsymbol{a} = 2\pi$; and similarly for $\boldsymbol{b}^*$. Thus, $|\boldsymbol{a}^*| = 2\pi/(a\sin(\gamma))$ and $|\boldsymbol{b}^*| = 2\pi/(b\sin(\gamma))$, where γ is the angle between $\boldsymbol{a}$ and $\boldsymbol{b}$. Figure 1D shows the geometry for GIXD in terms of the vertical incidence angle, $\alpha_i \approx 0$, and the exit angle, α_f, of the X-rays and the horizontal scattering angle, $2\theta_{xy}$. The angle α_f between horizon and the diffracted beam determines the vertical component, q_z, of the scattering vector [16]. In contrast to 3D crystallography, where often thousands of Bragg reflections are recorded, with monolayers usually only some few Bragg rods can be measured. This is due to the very limited amount of matter intercepting the X-ray beam and to various kinds of crystalline disorder. As a result, crystal structure solution using methods directly adapted from conventional 3D crystallography are possible to a limited extent only. Conversely, it must be noted that each Bragg rod pattern contains much more information than just a single Bragg point from a 3D single crystal. Therefore important structural information *can* be extracted from the Bragg rods and in some cases these data allow the refinement of, *e.g.*, a molecular model defined in terms of a few rigid bodies. This can provide a detailed structural model for the molecular crystalline arrangement in the monolayer. Along the (h,k) Bragg rod, the diffracted intensity is given by:

$$I(h,k,q_z) = |F(h,k,q_z)|^2 \cdot e^{-\left(q_{hk}^2 u_{\text{hor}} + q_z^2 u_z\right)} \cdot |T(\alpha_f)|^2 \tag{9}$$

Here the most important factor is the structure factor $F(h,k,q_z)$, which is the Fourier transform of the electron density of the molecules in the unit cell,

$$F(h,k,q_z) = \int_{\boldsymbol{r} \in \text{Unit Cell}} \rho(\boldsymbol{r}) e^{i(\boldsymbol{q}_{hk} \cdot \boldsymbol{r} + q_z z)} d^3\boldsymbol{r} \tag{10}$$

We can also write F in terms of the constituent atoms with form factors f_j [63]:

$$F(h,k,q_z) = \sum_{j \in \text{Unit Cell}} f_j e^{i(\boldsymbol{q}_{hk} \cdot \boldsymbol{r}_j + q_z z_j)} \tag{11}$$

Notice that while F is a complex number, only the absolute square, $|F|^2$, appears in *eq.* (9). This gives rise to the well-known *phase problem* of X-ray crystallography. The exponential in *eq.* (9), the Debye-Waller factor, involves the terms u_{hor} and u_z, which are the mean-square atomic displacements in the horizontal and vertical directions, respectively. The factor $|T(\alpha_f)|^2$ describes interference of X-rays diffracted upwards with X-rays diffracted down and subsequently reflected back up by the interface. The factor $|T(\alpha_f)|^2$ equals unity except near $\alpha_f = \alpha_c$ where it peaks sharply (the Yoneda-Vineyard peak) [65, 66]. It is convenient for deducing the zero-point of the α_f scale, which often covers a range of 0 to 10°, but otherwise

$|T(\alpha_f)|^2$ is unimportant for our purposes. So far it has not been possible to prepare a monomolecular layer that was a *single* 2D-crystal. The films usually consist of a large number of 2D-crystalline domains, each with a different orientation around the surface normal, *i.e.* a 2D powder. Therefore, the horizontal components, q_x and q_y, can be measured only in their combination, $q_{xy} = |\,\mathbf{q}_{xy}\,| = (q_x^2 + q_y^2)^{½}$, not individually, and the measured intensity is a sum over Bragg rods (h, k) which have the same $|\mathbf{q}_{xy}|$. Thus, denoting this set of reflections $\{h, k\}$, the measured signal is

$$I_{\{hk\}}(q_z)_{\text{meas}} = \sum_{(hk)\in\{hk\}} I(h,k,q_z) \tag{12}$$

In terms of the vertical incidence and exit angles, α_i and α_f, and $2\theta_{xy}$, the angle between the horizontal projections of the incident and diffracted beams (Figure 1), the vertical and horizontal scattering vector components are given [16] by

$$q_z = (2\pi/\lambda)\,[\sin(\alpha_i) + \sin(\alpha_f)] \approx (2\pi/\lambda)\sin(\alpha_f), \tag{13}$$

$$q_{xy} = (2\pi/\lambda)\,[\cos^2(\alpha_i) + \cos^2(\alpha_f) - 2\cos(\alpha_i)\cos(\alpha_f)\cos(2\theta_{xy})]^{1/2}$$

$$\approx (2\pi/\lambda)\,[1 + \cos^2(\alpha_f) - 2\cos(\alpha_f)\cos(2\theta_{xy})]^{1/2} \approx (4\pi/\lambda)\sin(2\theta_{xy}/2) + \text{Order}(\alpha_f^2) \tag{14}$$

and analogous to the plane spacing in 3D, the repeat (line-)spacing is $d = 2\pi/q_{xy}$. The Bragg peaks can be indexed by Miller indexes, h and k, and then the d-spacing (or q_{xy}) can be used to determine the lattice parameters, a, b and γ (the angle between the vectors $\mathbf{a}$ and $\mathbf{b}$) as well as the area of the unit cell, $A_{\text{Cell}} = a\,b\sin(\gamma)$, using

$$d = 2\pi/q_{xy} = [h^2/a^2 + k^2/b^2 - 2\,(hk/ab)\cos\gamma]^{-½}\sin\gamma \tag{15}$$

In some cases the quantity $d_{\min} = 2\pi/q_{\max}$ is referred to as the (real space) *resolution* of a diffraction data set [67].

In 3D, seven crystal systems and 232 space groups are possible for the bulk crystalline symmetry. For the two-dimensional case this reduces to only four crystal systems and 17 plane groups [68]. An example [27] of a diffraction data set for the acylglycerol, 1,2-dipalmitoyl-*sn*-glycerol (*L*-1,2-DPG) is given in Figure 2 in four different projections. Figure 2B shows a surface plot of the diffracted intensity as a function of q_{xy} and q_z, whereas Figure 2C is the corresponding contour plot. Projection of the measured intensity onto the q_{xy} or the q_z axis visualises the data as a *Bragg peak* (Figure 2A) or a *Bragg rod* (Figure 2D), respectively. The single Bragg peak observed in the low-order region of the 2D powder diffractogram (Figure 2A) is indexed as the six coinciding reflections, $\{h,k\}$ = {(1,0), (0,1), (-1,1), (-1,0), (0,-1), (1,-1)}, leading to $a = b = 4.80$ Å, $\gamma = 120°$ and $A_{\mathrm{Cell}} = 20.0$ Å^2. This area will accommodate only one alkyl chain, not both chains of the DPG molecule, so the lattice (a, b, γ) deduced must describe (close) packing of the alkyl *chains*. No lower-order reflections were observable, indicating that despite being bound to the alkyl chains, the glycerol backbones are too disordered (*i.e.*, have too large mean-square displacements) to be observed by diffraction. This is a common finding in diffraction studies of monolayers of lipids with multiple alkyl chains. In addition to the hexagonal lattice of this example, the other possible crystal systems are explored in Table 1.

Let us now, by way of example, analyse Figure 2 in more detail, using as little mathematics as possible and the numerics of only a pocket calculator [16]. A finite size of the crystalline domains gives rise to broadening of the Bragg peaks. Indeed, in Figure 2A the width of the peak, $\mathrm{FWHM_{meas}}(q_{xy})$ = 0.014 Å^{-1} (Full Width at Half-Maximum height), exceeds the instrumental resolution [67], $\mathrm{FWHM_{resol}}(q_{xy})$ = 0.008 Å^{-1}. Correcting for the latter by *eq.* (16) (which is strictly valid for Gaussian shaped peaks) one arrives at the intrinsic peak width, $\mathrm{FWHM_{intrinsic}}(q_{xy})$ = 0.011 Å^{-1}:

$$\mathrm{FWHM_{intrinsic}}(q_{xy}) = [\mathrm{FWHM_{meas}}(q_{xy})^2 - \mathrm{FWHM_{resol}}(q_{xy})^2]^{1/2} \tag{16}$$

A simple model assumes the monolayer to consist of 2D crystallites that are perfect and have a finite size, L_{xy}. Then, by the Scherrer formula [69],

$$L_{xy} \approx 0.9\ [2\pi/\ \mathrm{FWHM_{intrinsic}}(q_{xy})] \tag{17}$$

we find $L_{xy} \approx 500$ Å. Yet more can be learned by analysis of the *shape* of the peak [70, 71].

In the compressed DPG monolayer, the alkyl chains can be considered to be all-*trans* and straight. Then by a reasoning similar to above the thickness of the diffracting layer can be estimated using $L_z \approx 0.9\ (2\pi/\Delta q_z)$, where Δq_z is the FWHM of the Bragg rod. This relation can be derived by insertion in *eqs.* (9) to (12) of a straight alkyl chain - or, representing it in sufficient detail, a pencil-shaped cloud of electrons [16].

Table 1 The crystal systems for molecular packing in two-dimensions.

Crystal system	*Unit cell parameters* (1)	*d-spacing*	*Low-order reflections* (2)
Hexagonal	$a = b, \gamma = 120°$	$d = a\,[(h^2+k^2+hk)\,]^{-1/2}\,3^{1/2}/2$	{(1,0), (0,1), (-1,1), (-1,0), (0,-1), (1,-1)}
Rectangular (centred) (3)	$a \neq b, \gamma = 90°$	$d = [(h^2/a^2+k^2/b^2)\,]^{-1/2}$	{(1,1), (1,-1), (-1,1), (-1,-1)} and {(0,2), (0,-2)}
Rectangular	$a \neq b$ and $\gamma = 90°$	$d = [(h^2/a^2+k^2/b^2)\,]^{-1/2}$	{(1,1), (1,-1), (-1,1), (-1,-1)} and {(0,2), (0,-2)} (4)
Square	$a = b$ and $\gamma = 90°$	$d = [(h^2+k^2)/a^2\,]^{-1/2}$	{(1,0), (-1,0), (0,1), (0,-1)} and {(1,1), (1,-1), (-1,1), (-1,-1)}
Oblique	$a \neq b$ and $\gamma \neq 90°$	*Eq.* (15)	{(1,0), (-1,0)}, {(0,1), (0,-1)} and {(-1,1), (1,-1)}

Footnotes:

(1) The unit cell area can be calculated from $A_{Cell} = a\,b\,\sin(\gamma)$.

(2) Individual reflections are denoted (h,k) and coinciding reflections are denoted $\{(h,k), (h,k), \ldots\}$.

(3) A hexagonal cell (H) can be transformed to a centred rectangular (R) unit cell: $a_R = a_H$, $b_R = (3^{1/2})\,a_H$, $\gamma_R = 90°$. Likewise, a hexagonal cell, upon compression along a symmetry axis, leading to a distorted-hexagonal cell with $a_{DH} = b_{DH}$, $\gamma_{DH} \neq 120°$, is best described by transformation to centred rectangular unit cell: a_R, $b_R \neq (3^{1/2})\,a_R$, $\gamma_R = 90°$.

(4) Most rectangular-type packing assumed by amphiphilic monolayers (including the herringbone packing [19] of hydrocarbon chains) are 'almost' centred, so that the {(1,0), (-1,0)} or {(0,1), (0,-1)} reflections, even when allowed by the 2D-crystal symmetry, are too weak to be detected.

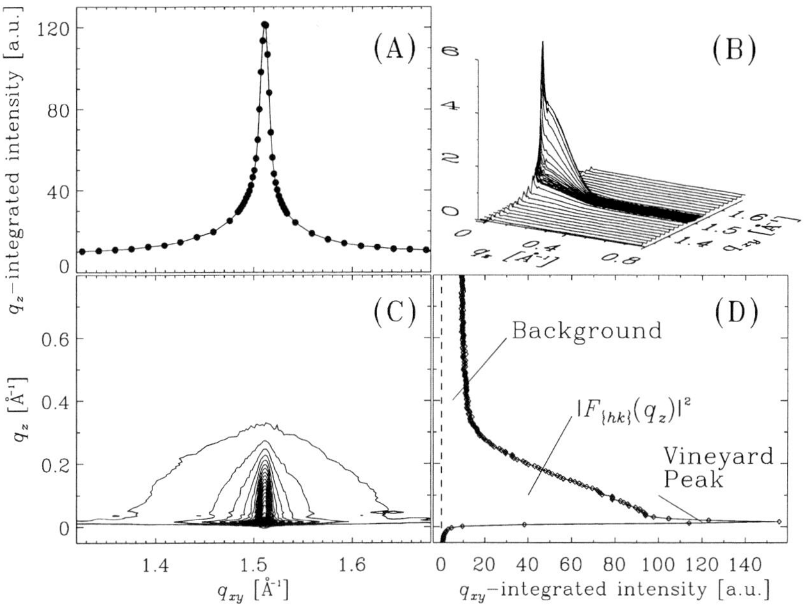

Figure 2 Grazing-incidence diffraction data for a Langmuir film of 1,2-dipalmitoyl-*sn*-glycerol (DPG) on an aqueous subphase (T = 20°C and Π = 29 mN/m). (**B**) shows a surface plot of the diffracted intensity as a function of q_{xy} and q_z, whereas (**C**) is a contour plot. Projection of the measured intensity onto the q_{xy} or the q_z axis visualises the data as a Bragg *peak* (**A**) or a Bragg *rod* (**D**), respectively.

The structure factor, $|F(\boldsymbol{q})|^2$, of such a long, straight scatterer will be large only near the plane where $\boldsymbol{q}$ is orthogonal to the long axis of the chain: only near this plane will all parts of the chain scatter in phase, and as $\boldsymbol{q}$ deviates from the plane of maximal structure factor, the intensity will decrease according to a bell-shaped curve of width Δq_z [8]. In Figure 2D we see (on top of a uniform background level) such a broad bell-shaped curve – due to the structure factor $|F(\boldsymbol{q})|^2$, and on top of that the sharp Vineyard-Yoneda peak. The $|F(\boldsymbol{q})|^2$-signal peaks at the horizon (at q_z = 0) and below the horizon the signal is zero due to absorption of the X-rays by the subphase. Measuring with a ruler (the vertical instrumental resolution is negligible in this case) and applying a factor of 2 to account for the hidden half of the bell-shaped curve below the horizon, we find $\Delta q_z = 2 \cdot 0.155$ Å^{-1}, leading to $L_z \approx 18$ Å in fair agreement with the expected length, *ca.* 19 Å, of the $C_{15}H_{31}$ chain (*cf. eq.* 21 below).

We can also derive the orientation of these straight alkyl chains. Along the Bragg rods, the bell-shaped maxima occur where the plane of maximal structure factor intercepts the rods. In

Figure 2D the maximum of each of the six coinciding Bragg rods occurs at the horizon, so the plane must be horizontal, and the alkyl chains, accordingly, vertical. We have now arrived at a model of the DPG monolayer: Only the alkyl chains are visible by X-ray diffraction. They are straight and have their long axes parallel and vertical, organised 4.80 Å apart in a hexagonal lattice, each occupying an area of 20.0 $Å^2$, in 2D-crystalline patches at least 500 Å wide (*i.e.*, the correlation of alkyl chain positions decays over *ca.* L_{xy} = 500 Å. The condensed patches can possibly be larger than this).

Modelling in further detail, warranted for more complex systems (provided that sufficiently precise and detailed data are obtained) would require the use [72 - 74] of computer programs [75 - 78] for modelling the molecule and for computing the resulting structure factor ($|F|^2$ in *eqs*. 9, 10 and 11).

In the next simplest case we encounter alkyl chains which are straight and parallel, but possibly not vertical. Then the structure factor plane will intercept each Bragg rod (h,k) at a (possibly nonzero) height, $q_{z,hk}$, and from those of the maximum positions, $q_{z,hk}$, that are at or above the horizon (hence observable) there will be enough information to determine the orientation of the plane, and of the long chain axes. Specifically, we find [16] that if the long axes are tilted from the surface normal by the angle t, in a particular azimuthal direction given by the horizontal unit vector $\hat{\boldsymbol{e}}$ then the relation

$$(\boldsymbol{q}_{hk} \cdot \hat{\boldsymbol{e}}) \cdot \tan(t) = q_{z,hk} \tag{18}$$

must hold for all the observable $q_{z,hk}$. Note that the vectors $\boldsymbol{q}_{hk}$ are given (*cf. eq.* 8) by the lattice ($\boldsymbol{a}$, $\boldsymbol{b}$, γ) derived in turn from the Bragg peak positions, $|\boldsymbol{q}_{hk}|$, in the 2D-powder diffractogram, so that *eq.* (18) has only two unknowns: The tilt angle, t, and the azimuthal tilt direction. Thus, from the set of ($|\boldsymbol{q}_{hk}|$, $q_{z,hk}$) data, the tilt angle and the tilt direction with respect to the 2D-crystalline lattice can be deduced.

Consider, for example, the data in Figure 3 for a compressed monolayer of (racemic) MPG (1-Monopalmitoyl-rac-glycerol) at 20°C [79, 80]. Two peaks are apparent in the 2D-powder diffractogram, so the lattice is not hexagonal: If a hexagonal lattice ($a_H = b_H$, $\gamma_H = 120°$) is distorted along a symmetry direction, giving $a_{DH} = b_{DH}$, $\gamma_{DH} \neq 120°$, two peaks result instead of the single peak of the hexagonal lattice. Such a cell is best described by transformation (*cf.* table 1 and Figure 4**A**) to a centred rectangular unit cell: $a_R \neq b_R$, $\gamma_R = 90°$ with $Z = 2$ alkyl chains per unit cell (at the corner and at the centre). The two peaks are then $\{(\pm1,\pm1)\}$ and $\{(0,\pm2)\}$, *cf.* Table 1. The peak at $|\boldsymbol{q}_{xy}| = |\boldsymbol{q}_{hk}| \sim 1.44\text{Å}^{-1}$ has $q_{z,hk} = 0.48\ \text{Å}^{-1} > 0$ so by *eq.* (18) the tilt angle t is not zero. The peak at $|\boldsymbol{q}_{xy}| = |\boldsymbol{q}_{hk}| \sim 1.49\ \text{Å}^{-1}$ has $q_{z,hk} = 0$ so by *eq.* (18) the tilt must be in a direction $\hat{\boldsymbol{e}}$ orthogonal to this $\boldsymbol{q}_{hk}$-vector. $\hat{\boldsymbol{e}}$ cannot be orthogonal to all of the four vectors $\{(h,k)\} = \{(\pm1,\pm1)\}$ so the peak at 1.49 Å^{-1} must in stead be the $\{(0,\pm2)\}$ and that at 1.44 Å^{-1} must then be the $\{(\pm1,\pm1)\}$. With this indexing we find (*cf.* Table 1) $a_R = 5.07\text{Å}$, $b_R = 8.42\text{Å}$, $\gamma_R = 90°$ and $A_{Cell} = 42.7\text{Å}^2$ per $Z = 2$ alkyl chains. Being orthogonal to $\boldsymbol{q}_{0,\pm2}$, the tilt direction $\hat{\boldsymbol{e}}$ is then along the nearest-neighbour direction $\boldsymbol{a}_R$. From *eqs.* (18) and (7) we now find $0.48\ \text{Å}^{-1}/\tan(t) = q_{z,11}/\tan(t) = \boldsymbol{q}_{11} \cdot \hat{\boldsymbol{e}} = \boldsymbol{q}_{11} \cdot (\boldsymbol{a}_R/a_R) = 2\pi/a_R = 2\pi/(5.07\text{Å})$, whence $t = 21°$. In conclusion, the chains are tilted by 21°, in a nearest-neighbour ($\boldsymbol{a}_R$) direction, and their lattice is centred rectangular which can be described as distorted-hexagonal, being slightly expanded in the nearest-neighbour (NN) $\boldsymbol{a}_R$ direction.

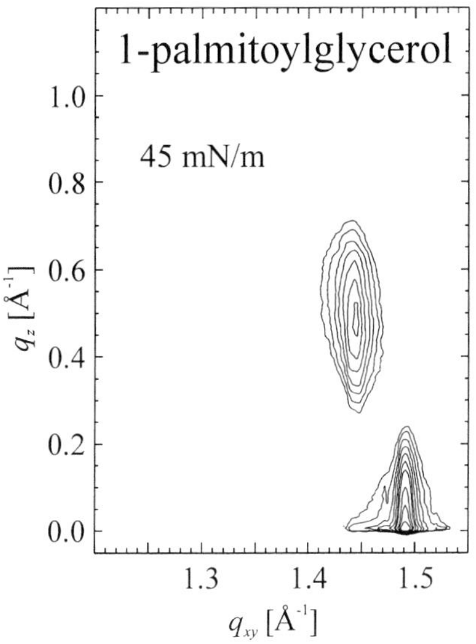

Figure 3 GIXD pattern of a compressed monolayer of (racemic) MPG (1-Monopalmitoyl-rac-glycerol) at 20°C. Adapted from ref. [80].

The reasoning is explained also in Figure 4, which shows (**A**, **B**) a top view and a side view of a lattice of long, linear molecules with bulky head groups. The molecules tilt in the (Nearest-Neighbour) $\boldsymbol{a}_R$ direction, and the lattice is expanded in the $\boldsymbol{a}_R$ direction. (**C**) shows a top view of the corresponding reciprocal lattice and (**D**) is a side view of it, showing the plane of maximal structure factor that is orthogonal to the long molecular axes in (**B**). (**E**) is the resulting 2D powder pattern that indeed is similar to Figure 3.

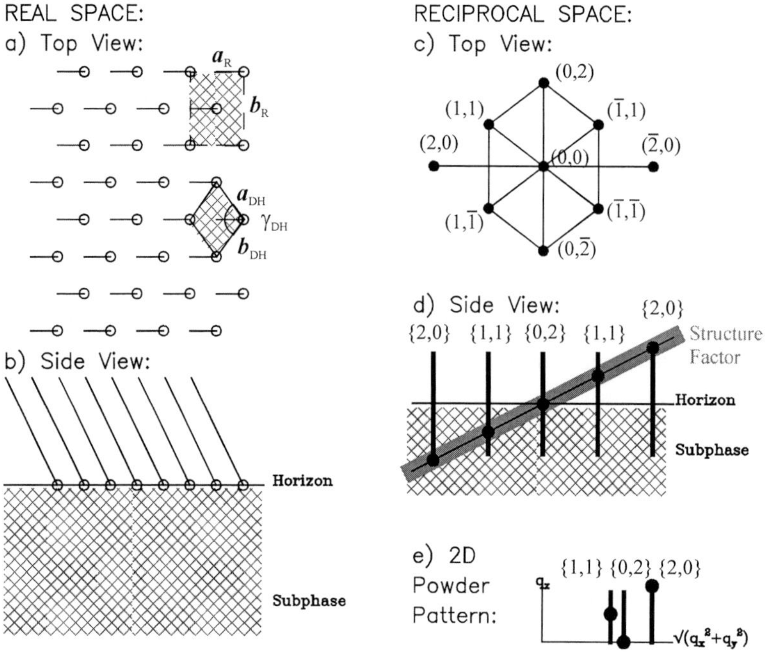

Figure 4 Top view (**A**) and side view (**B**) of a lattice of long, linear molecules with bulky head groups. The molecules tilt in the (Nearest-Neighbour) $\boldsymbol{a}_R$ direction, and the lattice is expanded in the $\boldsymbol{a}_R$ direction. (**C**) shows a top view of the corresponding reciprocal lattice and (**D**) is a side view of it, showing the plane of maximal structure factor that is orthogonal to the long molecular axes in (**B**). (**E**) is the resulting 2D powder pattern. (**A**) also illustrates the relation between the lattices ($a_{DH} = b_{DH}$, $\gamma_{DH} \neq 120°$) and ($a_R \neq b_R$, $\gamma_R = 90°$). Adapted from ref. [16].

In other systems we may encounter chain tilts in the next-nearest-neighbour (NNN) direction or in general (non-symmetry) directions.

In many cases (such as the MPG monolayer) the head group requires a larger area than the cross section of the alkyl chain(s). Then, even at high monolayer pressure, the chains can be tilted in order to maximise the chain-chain interactions while accommodating the head groups in the lattice. The cross sectional area per alkyl chain, A_o, can be calculated as

$$A_o = (A_{Cell} / Z)\cos(t) \qquad (19)$$

where Z is the number of alkyl chains per unit cell. For MPG we find $A_o = (A_{Cell} / 2) \cos(t) = 19.8$ Å^2, a value in the range of so-called rotator phases (typically, $A_o = 20 - 21$ Å^2 [19, 20]).

The name rotator phase should not be taken too literally. It indicates a phase in which the alkyl chains pack in a lattice that is nearly hexagonal when viewed along the long chain axis, such as would be the case for freely spinning alkyl chains. At lower temperatures and/or longer alkyl chains, and in particular for single chain lipids, closer chain packing is observed. Indeed, at least eight structurally distinct condensed phases were found for nonchiral single chain substances [19] and a number of phase transitions have been investigated using GIXD [20]. In some cases hexagonal or nearly hexagonal packing were observed, often at higher temperatures corresponding to the so-called rotator phases. At lower temperatures one often encounters alkyl chain packing in a more dense rectangular unit cell containing two alkyl chains related by glide symmetry (so-called herringbone and pseudo-herringbone packing) [19] with *projected* areas down to $A_o(minimal) \approx 18.5$ Å^2, as illustrated in Figure 5.

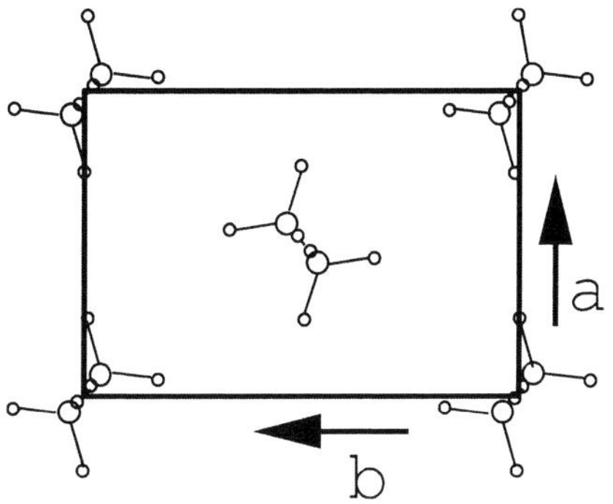

Figure 5 *Herringbone* packing of (*in casu*) long *n*-alkanes $C_{36}H_{74}$, as viewed along the long molecular axes; adapted from ref. [101].

Many biological molecules are chiral, *i.e.*, they exist in two forms that can be superimposed by mirroring but not by rotation or translation. In solution, a 50:50 mixture of enantiomers will tend to form a racemic mixture in order to maximise the entropy of mixing. In a monolayer segregation of enantiomers into chiral-resolved 2D-crystallites can occur if interactions between individual enantiomers are significant [81]. The tendency of long aliphatic chains to pack with a herringbone motif in a centred rectangular lattice with glide plane symmetry creates a bias towards racemic crystalline domains. In order to induce separation into chiral domains the glide symmetry and the possible formation of solid solutions between the two enantiomers must be prevented [82, 83]. These resolved enantiomers would then pack in an oblique lattice.

2.5 *Specular X-ray reflectivity (XR)*

To gain information about the vertical (laterally averaged) structure of monomolecular layers or interfaces a purely vertical scattering vector is required: Specular X-ray reflection (XR). Then the incident and reflected angles are the same, $\alpha_i = \alpha_f = \alpha$ and $2\theta_{xy} = 0$ (Figure 1E), and the scattering vector is $q_z = 2\ k \sin(\alpha) = (4\pi/\lambda) \sin(\alpha)$. The X-ray reflectivity, $R(q_z)$, can be measured by a NaI scintillation detector moving on the α_f arc. Typically, $0 \leq \alpha \leq 5°$. Given the electron density profile, $\rho(z)$, across the interface, there are essentially two different methods for calculating the reflectivity $R(q_z)$. It can be calculated *dynamically* by matching the electromagnetic wave fields at each interface [84], or *kinematically* using the so-called 'master formula for reflectivity',

$$R(q_z) = R_F(q_z) \left| \frac{1}{\rho_\infty} \int \frac{d\rho}{dz} e^{iq'_z z} dz \right|^2 \qquad (20)$$

Though derived by approximation [85 - 87], the 'master formula', *eq.* (20), is sufficiently accurate [8] when used with thin films such as Langmuir monolayers. Important, however, is the use of the refraction corrected scattering vector, $q'_z = (4\pi/\lambda) \sin(\alpha')$, as indicated, *cf.* *eq.* (3). The Fresnel reflectivity, $R_F(q_z)$, is calculated from standard optics (*eq.* 4, *cf.* [59]) for a perfectly sharp interface between air and pure subphase (of average electron density ρ_∞). The root-mean-square roughness, σ, for a *bare* water surface is *ca.* 3 Å. It is due to thermally excited microscopic capillary waves on the surface [58]. Inserting in *eq.* (20) a diffuse interface that is smeared out vertically by the root-mean-square roughness, σ, yields a factor $\exp(-q_z^2\sigma^2)$, equivalent to the Debye-Waller factor familiar from 3D crystallography. In a reflectivity study of monolayers of arachidic acid, $C_{19}H_{39}CO_2H$, it was found that σ, generated by capillary waves, is given by the surface tension for relatively low surface pressures, as for the bare water surface. At higher surface pressures, above a threshold value, the monolayer aquires a stiffness against bending, resulting in a 20% decrease of σ [88, 89]. The liquid-liquid interface of hexane-water was found to have a roughness $\sigma = 3.3 \pm 0.3$ Å [90]. The master formula, *eq.* (20), states that the ratio between the measured reflectivity and the Fresnel reflectivity is the absolute square of the Fourier transform of the normalised gradient of the electron density across the interface. Equation 20 (in parallel with *eq.* 9) states that only the modulus squared,

$R(q_z) = |r(q_z)|^2$, not the phase, of a complex function r is measured by XR (or GIXD) and one is thus faced with the usual *phase problem* of X-ray crystallography. Nevertheless, the measured normalised reflectivity, $R(q_z)/R_F(q_z)$, can usually be inverted to yield the laterally averaged electron density, $\rho(z)$, of the structure as a function of the vertical z co-ordinate. Note that the (dimensionless) reflectivity, $R(q_z)$, can be measured on an absolute scale and that *eq.* (20) contains no unknown factors. In consequence the electron density profile $\rho(z)$ can be derived on an absolute scale, as well.

The inversion of $R(q_z)/R_F(q_z)$ to yield $\rho(z)$ can be performed either by a model independent method [91 - 93] or using an explicit molecular or layered model of the interface [15, 16]. It can be difficult to model complex systems such as monolayers of pure proteins or proteins adsorbed to a lipid monomolecular layer at the air-water interface. Then the model-free method is useful, *e.g.*, writing $\rho(z)$ quite generally in terms of cubic-spline functions and least-squares fitting such a profile for agreement between the corresponding model reflectivity and the measured data, subject only to a smoothness constraint [91 - 93]. Although this approach does not directly lead to a molecular interpretation of the interface structure, it is well suited for generating electron density profiles that agree with the observed reflectivity data. Further, such an approach can provide new ideas for *models* based on the molecular structure. Figure 6 shows an example of an electron density profile $\rho(z)$ (shown as a black curve across the interface) extracted from measured XR data $R(q_z)$. The electron density model was least-squares fitted to the observed data, subject only to the restriction of an upper limit on the thickness of the surface film, while the smoothness of the $\rho(z)$ model, its deviation from an assumed average density and the width of the air-film interface entered as weighted terms in the χ^2 minimised by the fitting algorithm [91 - 93].

The structural model sketched in Figure 6 was based [28, 29] on the electron density profile and the available knowledge of the lipid, dipalmitoylphosphatidylethanolamine (DPPE) and of the surface layer protein from *B. sphaericus* CCM2177.

For a known, simple material at an interface, another strategy for generating a structural model is to generate $\rho(z)$ as a sum of contributions, $\rho_i(z)$, from individual atoms at heights z_i. Each atom (or pseudo atom, say, CH_2) in an aliphatic chain is modelled by, *e.g.*, a Gaussian-shaped $\rho_i(z)$ function (of root-mean-square width σ) to smear out the 8 electrons as shown in Figure 7.

As the reflectivity curve $R(q_z)$ does not contain enough information to *deduce* the positions, z_i, of the individual atoms, this strategy requires that an atomic model can be generated and parametrically varied for agreement with the observed $R(q_z)$; this can be cumbersome for large molecules such as proteins or lipid-protein systems The model should include all atoms or molecules contributing to the electron density also, *e.g.,* bound water molecules.

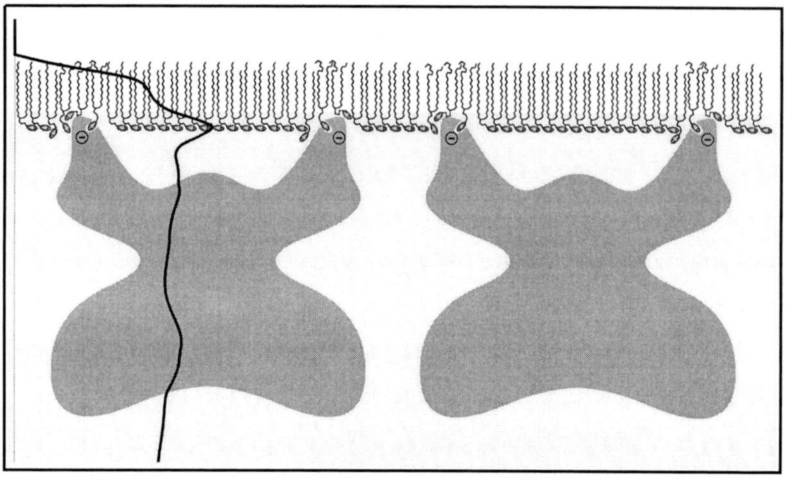

Figure 6 Artist's view of lipid/protein interaction derived from the electron density model from XR measurements and available structural data. The lipid monolayer is dipalmitoyl phosphatidyl ethanolamine (DPPE) and the protein is a membrane surface layer protein from *B. sphaericus* (from ref. [28]). The structural model sketched was based [28, 29] on the electron density profile (black line) inverted from measured reflectivity data as described in the text.

Finally, an electron density model can be built from a stack of homogeneous slabs. Then the fitting parameters are the densities and thicknesses of the slabs. Or, the parameters to be least squares refined can be the slab thicknesses, the number of electrons per molecule in each slab and the area per molecule, while the slab densities are calculated from these parameters. In either case a *common* roughness, σ, of the slab interfaces, or *individual* inter-slab roughnesses, σ_i, are additional variables to be refined [15, 16]. Such a slab model is shown by a dotted line in Figure 7 and the continuous electron density profile $\rho(z)$ obtained after applying the roughness, σ, is represented by a dashed line (almost coincident with the solid line).

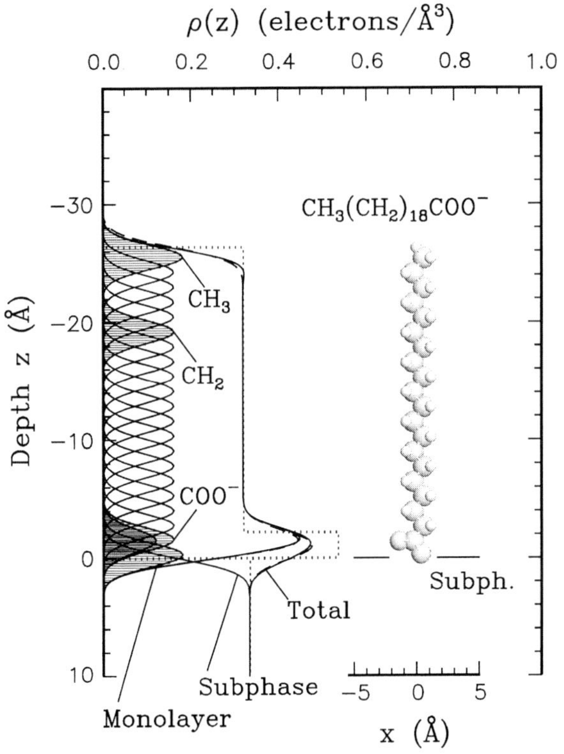

Figure 7 The electron density profile $\rho(z)$ (left) of a close-packed monolayer of arachidic acid (right). The electron density profile $\rho(z)$ can be constructed *either* as the sum of densities of individual atoms (or pseudo atoms, *e.g.*, CH_2), represented by Gaussian-shaped $\rho_i(z)$ functions of root-mean-square width σ (full lines) *or* as the sum of two slabs of constant densities (dotted lines). In either case the subphase contributes to the total $\rho(z)$. The electron density due to the slabs is smeared by a roughness σ, giving the dashed lines that are barely distinguishable from the full line. Typically σ is of order 3 Å but for display purposes, the Figure was constructed using σ = 1 Å). Adapted from ref. [16].

This method is often fruitful for "simple" monolayers of known molecules where the model can consist of typically two to five slabs: one section for the tail and one or more for the head group region of the molecule. Just two such slabs can suffice for representing a fatty acid or an alcohol. The number of electrons in the tail can often be kept fixed at the calculated value and the refined number of electrons in the head group can give information about co-ordination to water molecules. In the special case of lipids with all-*trans* hydrocarbon tails, comparison of the tail slab thickness, L_z, with the calculated length of the CH chain can give an estimate of the tilt angle, t, of the lipid tails. The maximal length of a saturated CH chain can be estimated [8] by

$$L_{CH} \approx (m + 9/8)\ 1.265\ \text{Å} \tag{21}$$

where m is the number of CH_2 groups and the terminal CH_3 effectively contributes 9/8. The factor 1.265Å is given by Tanford [94], *cf.* [95], and agrees well with 3D crystal structures. In such cases the tilt angle, t, can be calculated from

$$\cos(t) = L_z / L_{CH} \qquad (22)$$

Some simple rules can be derived for preliminary quantitative analysis of reflectivity data. For a two-slab model of, *e.g.*, a monolayer of close packed fatty acid, an acylglycerol or a phospholipid, consisting of a thin head group slab of higher density in-between a tail slab and subphase of nearly equal density (such as in Figure 7), it can be shown [16] that the first few minima in the observed $R(q_z)/R_F(q_z)$ data occur at q_z-values where

$$q'_z \approx (\tfrac{1}{2} + p)\, \pi / (L_T + \tfrac{1}{2}L_H), \qquad p = 1, 3, 5, \ldots \qquad (23)$$

Here the thicknesses of the head and tail regions are denoted L_H and L_T. *Eq.* (23) can be useful for a first estimation of the layer thickness for monomolecular layers. For an example, see Figure 9C below where the arrows mark the positions calculated from *eq.* (23). Other better known rules for simpler cases are given in [16].

Recently Lösche *et al.* have argued that, given reflectivity data of sufficient quality, the complex head groups of, *e.g.*, phospholipids should be modelled in further detail. Specifically, instead of the single head group box of, *e.g.*, Figure 7, they [96, 97] have proposed the use of distribution functions for the sub-moieties of the head group and the associated water, stabilising the least-squares fit by volumetric constraints.

The discussion given so far applies to XR from surfaces covered by homogenous thin films. The interpretation of reflectivity data from heterogeneous monomolecular layers is more difficult. Consider, *e.g.*, a surface with randomly distributed domains of lipid and protein or domains of two lipids with different reflectivity, say, $R_1(q_z) = | r_1(q_z)|^2$ and $R_2(q_z) = | r_2(q_z)|^2$, where $r_i(q_z)$ are complex functions describing the amplitudes of the reflectivity, *cf. eq.* (20).

Then, for the case of macroscopically large homogeneous patches of each of the constituents the patches will reflect X-rays independently of one-another and thus the measured reflectivity will be the (*incoherent*) average of the intensities:

$$R(q_z) = \psi\, R_1(q_z) + (1-\psi)\, R_2(q_z) = \psi\, | r_1(q_z)|^2 + (1-\psi)\, | r_2(q_z)|^2, \qquad (24)$$

ψ being the fraction of the total area occupied by constituent *no.* 1. For a microscopically homogeneous mixture of the constituents, conversely, the total electron density will *locally* be the sum of contributions from each of the constituents and from the subphase,

$$\rho(z) = \varphi\, \rho_1(z) + (1\text{-}\varphi)\, \rho_2(z) + \rho_{Sub}(z) = \varphi\, [\rho_1(z) + \rho_{Sub}(z)] + (1\text{-}\varphi)\, [\rho_2(z) + \rho_{Sub}(z)], \qquad (25)$$

φ being the fraction present of constituent *no.* 1. Insertion into *eq.* (20) then yields the average of the amplitudes, *i.e.* the *coherently* averaged reflectivity:

$$R(q_z) = |\varphi\, r_1(q_z) + (1\text{-}\varphi)\, r_2(q_z)|^2 \qquad (26)$$

The incoherent (or the coherent) average applies when the domains are much larger (or much smaller) than the coherence area of the reflectivity experiment. The coherence area depends on the ***q***-resolution [67] and is typically of order micrometers along the X-ray beam by tens of nanometers in the perpendicular direction [8].

In either case, however, the reflectivity is *an average* over the entire surface in the X-ray footprint area. By contrast, a grazing-incidence X-ray diffraction (GIXD) experiment will probe only domains with 2D-crystalline order (and for each Bragg rod there will be contributions only from those few domains of the 2D powder which are azimuthally oriented to diffract) while the rest of the sample will contribute to the background intensity only. In this respect, and in the nature of the structural information that they contribute, GIXD and XR are complementary.

2.6 Other Sources of Radiation. Other methods for nano-scale imaging

Structural information on the Ångström scale can be obtained by scattering or imaging techniques with radiation of X-rays, neutrons or electrons, with a wavelength of the order 1 Å. Atomic Force Microscopy can also provide information on the Ångström-scale, but is more important for the nanometer range. Most other techniques give information on a micrometer or larger scale.

X-rays are scattered by the atomic electrons and therefore the scattering amplitude increases with the atomic number. By contrast, neutrons are scattered by the atomic nuclei and there is a large difference in the scattering amplitudes of, for example, hydrogen, H, and deuterium, D. Neutrons can be useful for the investigation of lipid-protein interactions using, *e.g.,* deuterated lipid and heavy water (D_2O) to increase the contrast to protein adsorbed at the lipid-water interface [98 - 100]. Due to the low brightness of the available neutron sources, for Langmuir monolayers neutrons are useful for reflectivity measurements only. Electron diffraction and electron microscopy are important techniques. A drawback is that artefacts may be caused by the need to transfer the sample to a solid support and subject it to the high vacuum of the electron microscope.

2.7 *Theory - conclusions*

In the following we will summarise the similarities and differences between grazing-incidence X-ray diffraction (GIXD) and specular X-ray reflectivity (XR). Diffraction occurs when the lateral scattering vector, $\boldsymbol{q}_{xy}$, coincides with a 2D reciprocal lattice vector, $\boldsymbol{q}_{hk}$, and thus the diffracted radiation is concentrated in Bragg rods. In GIXD there is no interference between scattering from the subphase and from the monomolecular layer so the subphase only contributes a flat background, which can be subtracted from the measured intensity. The purpose of the use of grazing incidence is to minimise the background level by illuminating only a few nanometers deep. Specular X-ray reflectivity (XR) is characterised by $\boldsymbol{q}_{xy} = 0$. Thus, GIXD and XR measure different parts of the structure factor (the Fourier transformed electron density) of the monomolecular layer. XR measures the projection of the electron density onto the *z*-axis and also includes contribution from the subphase.

In a number of investigations, lipids, fatty acids and their derivatives were observed to have a laterally well ordered packing of the aliphatic chains whereas the polar head were laterally less well ordered. This disorder partly stems from co-ordination with water in the subphase and results in a large Debye-Waller factor for the head group. Thus, in most cases only the tail region contributes to diffraction and the polar head group contributes to the background. In XR experiments the lateral order is irrelevant and both the tail, head and the subphase contribute to measurable modulations of the electron density across the surface [8].

Heterogeneous monomolecular layers are an important class of samples relevant for, *e.g.*, biological systems. Examples are mixtures of different lipids or lipids and proteins *etc.*, which can form monomolecular layers built from domains of, say, a lipid in a 2D-crystalline ordered phase immersed in a liquid phase, or crystalline domains of protein among islands of disordered material. A GIXD experiment will probe the crystalline domains only and the rest of the sample will contribute to the background intensity. XR, by contrast, will provide information about an average structure projected onto the vertical *z* axis. As discussed above, the appropriate type of averaging depends on the lateral scale of inhomogeneity as compared to the coherence length of the X-ray experiment. The examples discussed in the following illustrate how GIXD and XR provide complementary information about thin films on liquid surfaces.

3. LIPID AND PROTEIN STRUCTURE AT THE AIR-WATER INTERFACE INVESTIGATED USING X-RAY SCATTERING

Since their initial development in the latter half of the eighties the surface sensitive X-ray scattering methods described here have been applied to a wide variety of systems assembled at the air-water interface. An overview of selected results on molecular structural characteristics of bio-related materials is presented in this chapter. A biological membrane is a mixture of a number of different compounds, *e.g.*, fatty acids, acylglycerols, phospholipids and proteins, as discussed in the following subsections. Surface pressure-area measurements provide phase diagrams revealing the position of phase boundaries depending on temperature, pressure and molecular chain length but give only very limited information on the structure of the phases. Prior to the development of surface sensitive X-ray scattering techniques speculations about the structure and behaviour of molecules in the monolayer state were sometimes based on knowledge about molecular arrangements in 3D. However, because of the environmental differences between the 2D and 3D states, such analogous packing does not always occur.

3.1 Alkanes, alcohols, fatty acids and their salts

Structural data extracted by grazing-incidence X-ray diffraction for a few selected materials at the air-water interface are summarised in Table 2. As an interesting special case we note that the formation of stable monolayers at the air water interface is not limited to amphiphilic molecules; also some purely hydrophobic materials have been observed to spontaneously form 2D-crystalline monolayers, *e.g.*, long n-alkanes C_mH_{2m+2}, $m \geq 36$ [101, 102]. These long alkanes are packed with the long molecular axes vertical and in a rectangular lattice with two molecules per cell forming a herringbone pattern, cf. Figure 5. However, for shorter alkanes, $m = 28, 29$, multilayer films of 3-4 molecular layers, and for $m = 23, 24$, thicker multi-layers occur [101].

Long chain alcohols have been the subjects of careful investigation using X-ray scattering [103 - 113]. The 1-alcohols form monomolecular layers when spread at the air-water interface, whereas bi- or multi-layers were observed for α,ω-docosanediol [103 - 105]. Octadecanol, $C_{18}H_{37}OH$, has been carefully investigated by GIXD [70, 71, 111]. At constant surface pressure of 18 mN/m a single Bragg peak (indicating hexagonal packing of the vertical aliphatic chains) is observed above 8°C. Below 8°C two diffraction peaks are observed showing that the vertical molecules are packed in a centred rectangular lattice with two molecules per cell, their carbon backbones forming a herringbone pattern when viewed along the long molecular axes [70]. At low surface pressure, the 1-alcohols are tilted towards their next-nearest neighbours (NNN-tilt), the tilt angle varying with pressure and chain length [14, 17, 106, 111, 114].

The long chain fatty acids, $C_mH_{2m+1}COOH$, show a rich variety of phases and phase transitions that have been studied structurally by GIXD [9 – 12, 20, 113, 115 - 119]. Docosanoic acid ($m = 21$) and heneicosanoic acid ($m = 20$) were selected as representatives of the homologous class since most of the phases are observable in the temperature range 5 to 30°C [10, 11, 113]. Eicosanoic (arachidic) acid ($m = 19$) was investigated using both GIXD and XR and was found to exhibit a continuous change of tilt angle of the aliphatic chains from 33 to 0° for increasing surface pressure in the range *ca.* 1 to 25.6 mN/m [9]. Fatty acids and alcohols, *e.g.*, heneicosanol and heneicosanoic acid, were found by GIXD [120, 121] to be fully miscible at the air-water interface.

Interacting with a monolayer of charged lipids, ions can substantially alter the organisation, as shown by a study of uncompressed arachidic acid on a dilute $CdCl_2$(*aq.*) subphase at elevated *p*H (ammonia). The large number of relatively intense and sharp Bragg peaks in the 2D powder pattern shown in Figure 8 result from the self-assembly of fatty acid in co-ordination with metal ions, forming a super-structure relative to that of the close-packed molecules.

The three peaks assigned integer Miller indices attest mainly to the packing of the molecules while the remaining peaks, with fractional indices, are direct evidence of the super-structure of metal ions [122, 123]. Initially, Cd^{2+} was assumed to co-ordinate to the arachidate anions in the ratio 1:2, but a detailed X-ray reflectivity study revealed that the cadmium: arachidate ratio was close to 1:1 due to the presence of $CdOH^+$ ions in a layer under the arachidate [123]. Also, the presence in the monolayer of ammonia complexed to the cadmium ions was deduced [123].

Table 2 Crystallographic data for monolayers of selected alkanes, fatty acids and alcohols investigated at the air/water interface using grazing-incidence X-ray diffraction.

Compound	***Physical conditions***			***Unit cell parameters*** (1)				***Molecular parameters***		(1)	***Ref.***
	T / °C	Π / (mN/m)	*p*H	a / Å	b / Å	γ / °	(A_{Cell}/Z) / Å^2	t / °	tilt direction	A_o / Å^2	
n-alkane (2)											
$C_{36}H_{74}$	5	~0	neutral	5.0	7.4	90	18.5	0	-	18.5	[101]
1-alcohols (2)											
$C_{23}H_{47}OH$	5	~0	neutral	5.00	7.56	90	18.9	9.5	NNN	18.6	[107]
$C_{30}H_{61}OH$	5	~0	neutral	4.99	7.49	90	18.7	7.7	NNN	18.5	[107]
$C_{31}H_{63}OH$	5	~0	neutral	4.99	7.53	90	18.8	10.1	NNN	18.5	[107]
Fatty acids (2)											
$C_{15}H_{31}COOH$	24	10	2	5.259	8.474	90	22.3	25	NN	20.2	[166]
$C_{15}H_{31}COOH$	24	18	2	4.865	8.690	90	21.1	16	NNN	20.3	[166]
$C_{15}H_{31}COOH$	24	25	2	4.830	8.366	90	20.2	0	-	20.2	[166]
$C_{29}H_{59}COOH$	5	~0	neutral	5.53	7.44	90	20.6	26.5	NN	18.4	[12]
Arachidic acid on $CdCl_2$ (3)											
$C_{19}H_{39}COOH$	9	~0	8.8	4.60	8.37	93.4	19.2	11	Non-symmetry	18.9	[123]

Footnotes:

(1) The area per alkyl chain (*i.e.*, per molecule) is given.
(2) Subphase: water.
(3) Subphase: 10^{-3} M $CdCl_2$ (*aq*) solution.

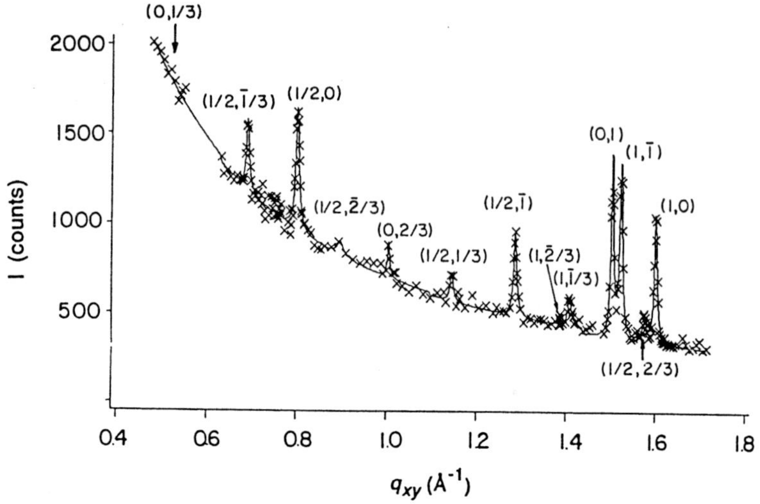

Figure 8 Bragg peaks measured by GIXD on a self-assembled uncompressed monolayer of arachidic acid on $CdCl_2$ (10^{-3} M, *p*H = 8.8 (ammonia), T = 9°C), from ref. [122, 123]. The corresponding Bragg rods (not shown) show that the integer order peaks result *mainly* from the close-packed molecules while *only* the metal ions, laterally ordered in a *thin* layer, contribute to the fractional order peaks.

The shorter myristic acid (m = 13) did not self-assemble to give rise to measurable diffraction when uncompressed on pure water; however, the diffraction signal was measurable when the subphase contained ions, *e.g.*, $CdCl_2$ (*aq*) [124].

Intense X-ray diffraction was also observed for metal ion grid complexes formed *in situ* at the air-water interface from Ag^+ and Co^{2+} ions with various ligands. Molecular self-assembly at the air-water interface can provide novel materials not available from bulk solution [125, 126].

In the following two sections we discuss acylglycerols and then phospholipids, which have larger head groups and possibly more than one alkyl chain per head group. Larger head groups often introduce tilt of hydrocarbon tails in order to maximise the hydrophobic contact between the hydrocarbon chains. The properties of lipids are important for a variety of biological processes and are therefore a key to a deeper understanding of them.

3.2 Acylglycerols

A number of different acylglycerols are found in biomembranes and contribute to, *e.g.*, storage and generation of metabolic energy and maintenance of the barrier properties of membranes. Some of these acylglycerols have saturated hydrocarbon chains, *e.g.*, palmitate, or *cis* unsaturated chains, *e.g.*, oleoate. Such compounds are important for the regulation of physical and chemical properties, such as the fluidity, of some biological membranes owing to their very different molecular chain packing. Previously structural information on acylglycerols was based on assumed analogies between 3D bulk crystal structures and 2D membranes and on spectroscopic methods [127, 128]. In order to gain detailed *in situ* structural knowledge on crystalline and non-crystalline acylglycerols assembled in monomolecular layers at the air-water interface, dipalmitoylglycerol, DPG, and monooleoylglycerol, MOG, have been investigated by X-ray scattering [27, 45 - 47] as discussed in the following, and also with other methods, *e.g.*, molecular modelling [45 -47, 129] and pressure-area isotherms [130 - 132].

Monomolecular layers of the pure lipids DPG and MOG were investigated by GIXD and some results are shown in Figure 9**A**, 9**B** as Bragg peaks. MOG diffracts only weakly and the Bragg peak could be described as a broad hump just above the background. (The background is similar to the signal obtained with a bare water surface, shown as a dashed line in Figure 9**A**, 9**B**.) Apparently, in the MOG monolayer the *cis* unsaturated bond in the hydrocarbon chain disrupts ordering of the alkyl chains. By contrast, dipalmitoylglycerol gives a single intense and sharp diffraction peak (see also Figure 2, which shows the same DPG data in different projections). From the data we deduce that the alkyl chains of DPG are packed in a hexagonal lattice with lattice constant a = 4.80 Å corresponding to an area per hydrocarbon chain, A_0 = 20.0 Å^2. The extracted GIXD data for DPG, MOG and selected acylglycerols are given in Table 3. The widths of the Bragg peaks (Figure 9**A**, 9**B**) reveal a large difference between the estimated sizes of the crystalline patches of the DPG and MOG monolayers: L_{xy} = 400 Å and L_{xy} = 21 Å, respectively.

Specular X-ray reflectivity (XR) data for monolayers of the pure lipids DPG and MOG are shown in Figure 9**C**. Two-slab models were refined to the measured data (points) and the resulting calculated reflectivity curves are shown as lines in Figure 9**C**. The parameters of the slab models are given in Table 4 and plots of the corresponding electron densities, $\rho(z)$, are shown in Figure 9**D**, 9**E**.

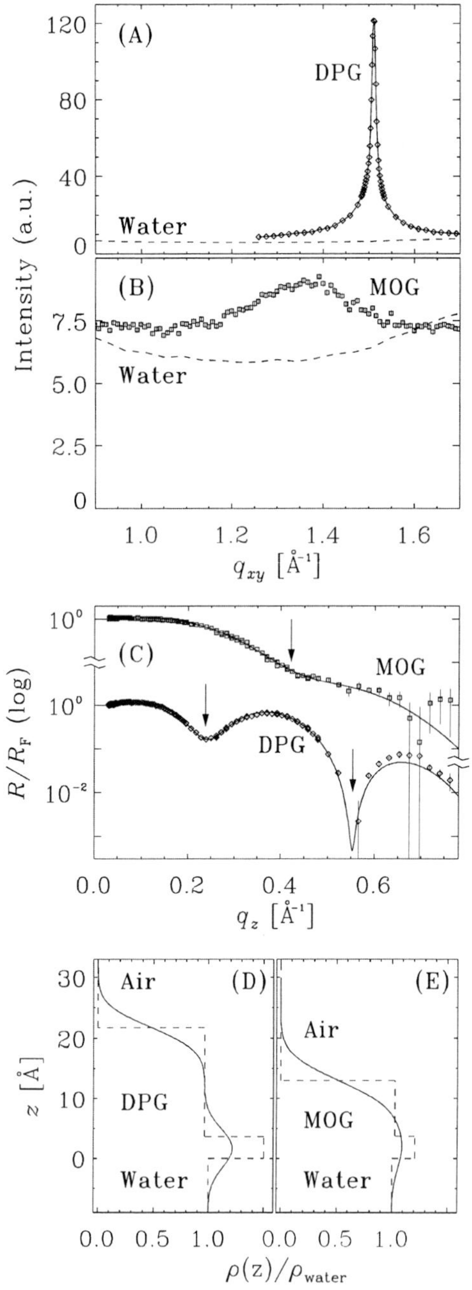

Figure 9 Bragg peaks of (**A**) 1,2-dipalmitoyl-*sn*-glycerol (*L*-1,2-DPG) and (**B**) racemic monooleoylglycerol (rac-MOG) measured by grazing-incidence X-ray diffraction. The X-ray scattering from a pure water surface is shown (dashed curve) for comparison. (**C**) Specular X-ray reflectivity normalised by the Fresnel reflectivity, $R(q_z)/R_F(q_z)$, for monomolecular layers of MOG and DPG. Calculated reflectivities (corresponding to least-squares fitted electron density slab models) are shown as full lines. For clarity, the MOG data and model are offset vertically. (**D**, **E**) Electron density profiles normalized by the electron density of water, $\rho(z)/\rho_{water}$, *versus* depth, z, corresponding to the reflectivity models shown in (**C**). (**D** DPG; **E**, MOG). In **C**, the arrows mark the minimum positions as calculated from *eq.* (23). Monomolecular layers were prepared on a subphase of water (T = 20°C) compressed to a pressure of Π = 29 and 35 mN/m for the GIXD and Π = 16 and 15 mN/m for the XR measurement for DPG and MOG, respectively; adapted from ref. [27].

Table 3 Crystallographic data for monolayers of selected lipids investigated at the air/water interface using grazing-incidence X-ray diffraction.

Compound	***Physical conditions***		***Unit cell parameters***			(3)	***Molecular parameters***		(3)	***Ref.***
	T / °C	Π / (mN/m)	a / Å	b / Å	γ / °	(A_{Cell}/Z) / $Å^2$	t / °	tilt direction	A_o / $Å^2$	
rac-MPG (1)	5	2.5	5.170	5.170	113.7	24.6	36.3	NN	19.8	[133]
rac (1)	5	20	5. 055	5.055	115.8	23.0	30.5	NN	19.8	[133]
rac (1)	5	40	5.070	5.070	122.0	21.8		NNN	19.8	[133]
rac/*L* (2)	20	5.5	5.678	8.577	90	24.4	34	NN	20.2	[79]
rac/*L* (2)	20	15	5.458	8.543	90	23.3	31	NN	20.0	[79]
rac/*L* (2)	20	30	5.224	8.473	90	22.1	26	NN	19.9	[79]
rac/*L* (2)	20	44	5.072	8.423	90	21.4	21	NN	20.0	[79]
L-1,2-DPG (4)	20	5	4.83	4.83	120	20.2	14		19.6	[45]
	20	40	4.74	4.74	120	19.5	≤ 2		19.5	[45]
L-α-DPPE	20	10	5.10	8.48	91.5	21.6	24		19.8	[138]
	20	38	4.79	8.30	90.0	19.9	0		19.9	[138]
DL-α-DPPE	20	10	5.10	8.46	90.0	21.6	22		20.0	[138]
	20	38	4.78	8.28	90.0	19.8	0		19.8	[138]
L-α-DPPC	15	30	5.15	5.04	115.5	23.4	30		20.3	[139]
	15	45	5.08	4.93	117.2	22.3	25		20.2	[139]
DL-α-DPPC	15	30	5.15	4.94	116.4	22.8	27		20.3	[139]
	15	45	5.07	4.95	116.9	22.4	25		20.3	[139]
L-diT-DPPC	15	20	4.92	4.92	120	21.0	0		21.0	[146]
	15	33	4.91	4.91	120	20.9	0		20.9	[146]

Footnotes:

(1) Here a lattice with one chain per cell and $a = b$ is given (recalculated from ref. [133]). The crystal symmetry can then also be expressed in terms of a centred rectangular unit cell (*cf.* Table 1).

(2) Within the experimental accuracy the racemic mixture and the *L* enantiomer gave identical GIXD data at 20°C [79].

(3) The area *per chain* is given.

(4) 1,2-dipalmitoyl-*sn*-glycerol; a.k.a. *L*-1,2-Dipalmitin

In the two models the hydrocarbon chains of MOG and DPG have electron densities that are similar within the uncertainty, but due to the smaller thickness of the tail slab of MOG and (mainly) to the smaller contrast of the MOG head slab, distinct head and tail regions are not apparent in the $\rho(z)$ profile for MOG (Figure 9**E**, full line), contrary to the case for DPG (Figure 9**D**, full line). Accordingly, the observed specular reflectivity data for DPG exhibit more pronounced interferences than do the MOG data (Figure 9**C**). Glycerols with a single saturated hydrocarbon chain reach zero chain tilt only at the very highest pressures (if at all) [80, 133], presumably in order to preserve close contact between the hydrocarbon chains despite the comparatively large glycerol head group.

A fluorescence microscopy study of 1-hexadecyl-glycerol revealed spiral shaped crystalline patches turning clockwise and counter clockwise for the *S*- and *R*-enantiomer, respectively. The racemate showed spirals in both directions [134]. Despite these observations, which seemed to indicate separation of the enantiomers in different patches, with GIXD an oblique lattice was observed for the *S* enantiomer but a centred rectangular cell for the racemic mixture [135], indicating that the individual 2D-crystals are racemic at the microscopic level (a not straightforward relationship between domain shapes and chirality in structure has been observed in other similar systems, as well [136]). While in the pressure and temperature range 15-18 mN/m and 5 to 20°C the racemate has a first order phase transition from NN-tilt to NNN-tilt, the enantiomer exhibits a continuous change of tilt direction and lattice distortion through non-symmetry directions, resulting in the above-mentioned oblique structure [135]. The effects of chirality decrease with increasing temperature [136]. Qualitatively, this observation can be interpreted as an increase in rotational disorder of the tail backbone with increasing temperature, which effectively makes an enantiomer achiral by rotation about its long axis thereby reducing the chiral interaction [136].

By contrast to 1-hexadecyl-glycerol, for the closely related molecule 1-monopalmitoyl-glycerol (*cf.* Table 3), both (Π-A) isotherms and GIXD data are identical (at 20°C) for the *L* enantiomer and the racemic mixture, revealing that the influence of the chiral centre in the head group is too weak to influence the tail lattice. Only tilt in the NN direction occurs at 20°C [79, 80].

At these temperatures, complete miscibility of the two molecules (1-hexadecyl-glycerol and 1-monopalmitoyl-glycerol) was observed and the temperature for the NN-tilt to NNN-tilt phase transition changed linearly with composition [80]. These results suggest that the presence of the carbonyl group stabilises the NN-tilted phase [80].

Table 4 Structural data for monolayers of selected lipid monolayers at the surface of water (neutral *p*H) extracted from specular X-ray and neutron reflectivity data by modelling in terms of two-slab models.[(1)]

Compound	*Physical conditions*		*Model parameters*								*Ref.*
	T / °C	Π / (mN/m)	σ / Å	A / Å^2	L_H / Å	E_H [(2)]	ρ_H / Å^{-3}	L_T / Å	E_T [(2)]	ρ_T / Å^{-3}	
L-1,2-DPG [(3)]	20.3	16	3.05	41.7	3.64	76	0.50	18.05	242	0.32	[27]
rac-MOG [(4)]	20.3	15	3.50	42	3.7	63	0.40	9.3	135	0.34	[27]
L-α-DLPE [(5)]	15	5	3.09	77.2	11.80	377	0.413	8.00	178	0.287	[137]
	15	37	3.79	42.0	7.72	170	0.523	13.28	178	0.318	[137]
L-α-DMPE [(5)]	21	5	2.85	73.7	11.11	337	0.410	9.88	210	0.287	[137]
	21	35	3.88	40.0	7.55	155	0.512	16.04	210	0.326	[137]
L-α-DPPC [(6)]	18	42	3.6	49.1	9.3	204 [(7)]	0.447	16.0	242	0.307	[148]

Footnote:

(1) The electron density of the subphase is $\rho_{water} = 0.334$ Å^{-3}.

(2) E_T (known *a priori*) is the total number of electron in the tail moieties and E_H is the corresponding number for the head group, but including hydration.

(3) 1,2-dipalmitoyl-*sn*-glycerol; a.k.a. *L*-1,2-Dipalmitin.

(4) rac-MOG, 1-Monooleoyl-rac-glycerol.

(5) The number of electrons in the PE head group (phosphatidyl ethanolamine) is larger than the calculated value of 140 due to incorporation of water molecules in this slab, *viz.* 24 and 3.0, resp. 20 and 1.5 water molecules at low and high pressure for DLPE and DMPE, respectively.

(6) *L*-α-DPPC: The model parameters were simultaneously refined to four data sets, two X-ray reflectivities and two neutron reflectivities. For the neutron experiments deuterated lipid and subphases of both pure H_2O and pure D_2O were used.

(7) This corresponds to a hydration of (4 ± 2.5) water molecules per lipid.

3.3 Phospholipids

The phospholipids *L*-α-dimyristoyl phosphatidic acid, DMPA, and *L*-α-dipalmitoyl phosphatidyl choline, DPPC, were among the first Langmuir monolayer systems to be structurally characterised by surface sensitive X-ray scattering at the air-water interface in the late eighties [5, 38]. Since then a number of other phospholipids have been investigated by GIXD and XR as shown in Tables 3 and 4. Phospholipids with relatively small head groups have been studied carefully using GIXD and XR, *e.g.*, the dimyristoyl derivatives of phosphatic acid, PA, and phosphatidylethanolamine, PE, denoted DMPA and DMPE [37, 137, 138]. On compression, DMPE exhibits a phase transition from an NN-tilted chain structure to a hexagonal lattice of vertical chains. At low pressure, an oblique chain lattice was seen for the *L* enantiomer and a centred rectangular lattice for the racemic mixture (*DL*) (Table 3) [138]. Adding three methyl groups to the terminal amino group of PE gives phosphatidylcholine, PC, which has a head group as larger than the cross sectional area of the two aliphatic chains. Thus, in order to maximise the hydrophobic interactions while adapting to the head group lattice, the alkyl chains are highly tilted even at high pressure, see Table 3 [38, 139, 140].

The influence of the head group size on the chain packing is illustrated in a GIXD study of DPPC and DPPE in contact with the *n*-alkane dodecane. At surface pressures higher than *ca.* 8 and 5 mN/m DPPC and DPPE, respectively, formed non-tilted phases. For the DPPE monolayer, but not for the DPPC monolayer, the alkane can be squeezed out at elevated surface pressure [141]. In monolayers of DPPC, an oblique structure is observed for both the racemic mixture and the enantiomer [139]. Chirality of phospholipids has a strong influence on the functional properties of these molecules in a biological membrane.

Natural phospholipids often have one or two unsaturated bonds in the aliphatic chains, creating disorder in the molecular packing and reducing phase transition temperatures. An alternative way to create disorder is to attach methyl or ethyl groups to the aliphatic tails, as carefully studied by Brezesinski *et al.* using GIXD. Such short alkyl branches close to the head group have a weak effect on the molecular packing [142 - 145]. A long branch may act as an additional tail, and triple and quadruple chain phosphatidyl cholines have been investigated [146, 147]. Triple chain phosphatidylcholine lipids have alkyl structures similar to those of double chain phospholipids with a smaller head. Chiral structures of triple chain compounds were found with chirality corresponding to their head groups [147]. *L*-α-DMPE and *L*-α-dilauroyl phosphatidyl ethanolamine (*L*-α-DLPE) were investigated by XR and the results for the fluid phase (low pressure) and for the solid phase (high pressure) are given in Table 4 [137]. At high pressure the molecular area corresponds to two upright alkyl chains. The number of electrons associated with the PE head group is higher than the calculated 'dry' value of 140, revealing hydration. In agreement with *eq.* (24), the XR data (figure 8b of [137]) recorded for the fluid/solid coexistence region exhibited an *isosbestic* point.

For a compressed monolayer of DPPC, contrast variation was employed. Three data sets: XR data and neutron reflection data for two different contrasts (deuterated DPPC-d62 on both H_2O and D_2O) were jointly refined to yield a model with parameters better determined than from any of the data sets separately [148].

3.4 Amino acids, peptides and proteins at the air-water interface

The transformation of racemic chemistry to chiral biology remain an unsolved mystery of nature, but the spontaneous segregation of a racemic mixture into enantiomers in 2D and 3D might have played an important role [33]. To address this problem the molecular structures (at the air-water interface) of racemic and enantiomerically resolved monolayers have been explored, *e.g.*, α-amino acids with a long alkyl chain residue. A brief overview of pure derivatives of amino acids, peptides and proteins characterised at the air-water interface is presented.

In order to elucidate the interaction between solute and surface layer Wolf *et al.* investigated the molecular structure of palmitoyl-lysine at the air-water interface using GIXD and XR [149]. Such a monolayer can serve as template and induce oriented nucleation and growth of α-amino acid crystals, *e.g.*, when deposited on a subphase containing glycine [150]. Weissbuch *et al.* improved and extended the investigation to other similar α-amino acids with a long alkyl chain residue having an extra amide group, designed to promote, *via* hydrogen bonding, translational packing only [82]. A racemic mixture was found to spontaneously separate into crystalline domains of opposite handedness, whereas racemic mixtures of α-amino acids without the extra amide moiety formed structures containing symmetry-related right- and left-handed molecule [82]. Soluble (α-amino acid:Cu) complexes were seen by GIXD to bind enantioselectively to a chiral monolayers of palmitoyl-(*S*)-lysine [151].

The *D*-enantiomer of N-docosyl-leucine 2D-crystallises in an oblique unit cell, as expected. However, GIXD revealed a *different* oblique unit cell for the racemic mixture suggesting miscibility of the two enantiomers [152]. For myristoyl alanine monolayers, GIXD data indicated that a racemic mixture separated into 2D-crystalline islands of opposite chirality [153].

X-ray monolayer techniques are important for investigation of peptides and their interactions with lipids, *e.g.*, by relating the structure of the pure peptides adsorbed at the air water interface to their physical and chemical properties. The cyclic peptides valinomycin and nonactin, investigated by Rapaport *et al.* [154], are ionophores capable of selectively carrying ions across natural and artificial membranes and are highly selective to binding of certain alkali ions. The unbound cyclic peptides do not order at the air-water interface but the complexed forms did order and revealed a tendency in some cases to form films of seven to eight layers thick, as deduced by GIXD [154]. GIXD also proved valuable for structural studies assisting the design of artificial nano-tubes (built by self-assembly of cyclic peptides in Langmuir films), which were investigated with a view to, *e.g.*, transport of solute components through the tube across a film [155].

Recently, it was demonstrated that X-ray diffraction data could be recorded from 2D-crystals of proteins at the air-liquid interface. GIXD from the water soluble protein streptavidin attached to a lipid monolayer revealed a rectangular protein structure with lattice parameters 84 and 85 Å [156], in accordance with 3D structural data. X-ray diffraction patterns from as few as 10^{13} protein molecules in a single layer of purple membrane were recorded by Verclass *et al.* [30]. The protein crystallised in a 2D-hexagonal lattice, a = 61.3 Å, and diffraction up to the order (h,k) = (4,3) was observed, corresponding to a real space *resolution* [67] of $d_{\min} = 2\pi/q_{\max} \cong 9$ Å. The protein monolayer has a thickness of *ca.* 50 Å. GIXD has thus been shown to be a promising method for obtaining structural information on membrane proteins under near physiological condition [30]. Several proteins, *e.g.*, glucose oxidase, alcohol dehydrogenase and urease were however found to denaturate at the air-water interface, as XR measurements revealed layer thicknesses in the range 8 to 14 Å [157].

3.5 Probing the interaction of dissolved macromolecules with a monomolecular layer

Biological cells contain a variety of macromolecular structures in an aqueous phase and their interaction with lipid membranes play a critical role in many processes. This has been part of the motivation for investigations of polymer-lipid interactions [158 - 160]. Coulomb interactions can be screened upon binding polyelectrolyte to a monolayer of, *e.g.*, phospholipid [161, 162].

Adsorption of phospholipase A_2 (PLA_2) to a monolayer of *D*-DPPC, studied by GIXD, revealed a dramatic change in the lipid structure (the *D* enantiomer was selected in order to suppress hydrolysis and focus on the adsorption process) [163]. The alkyl chains of a pure DPPC monolayer are highly tilted at all pressures [139], *cf.* Table 3, but upon PLA_2 adsorption the tilt angle decreased, *e.g.*, from 29 to 8° at *ca.* 30 mN/m [163]. Unspecific adsorption of serum albumin was used as a control and GIXD showed that this caused no change in alkyl chain packing at higher pressures [163].

The interaction between *Humicola Lanuginosa* lipase (*HLL*) [57] and sodium dodecyl sulphate (SDS) has been investigated measuring neutron reflectivity from deuterated SDS dissolved in a solution of protonated lipase in D_2O. The lipase was found to adsorb readily at the air-water interface but SDS can displace lipase from the interface at higher SDS concentrations [164]. Also, the interaction of dissolved *HLL* with monolayers of MOG and DPG was studied by GIXD and XR [27].

The coupling of bacterial surface (S)-layer proteins from *Bacillus sphaericus* and *B. coagulans* to a monolayer of the lipid dipalmitoylphosphatidylethanolamine, DPPE, was studied by GIXD combined with X-ray and neutron reflectivity [28, 29]. Comparison of the lipid structure before and after protein adsorption reveals minimal reorganisation of the lipid alkyl chains but major rearrangements of the head groups. XR suggest that amino acid side chains intercalate the lipid head groups at least to the phosphate moiety. Remarkably, it proved possible to measure the diffraction pattern from the S-layer protein lattice under a lipid monolayer. The GIXD data showed that S-proteins from *Bacillus sphaericus* crystallised in a square lattice, $a = b = 130.5$ Å [28]. By contrast, those from *B. coagulans* were organised in an oblique lattice when adsorbed under DPPE, $a = 99.5$, $b = 76$ Å and $\gamma = 80°$ [29].

Cholesterol crystallisation is thought to be the first step in the formation of gallstones in the human biliary system and the process of cholesterol nucleation remains incompletely understood. GIXD revealed a phase transition from a monolayer to a highly crystalline rectangular bilayer phase [165]. The presence of the phospholipid DPPC in the cholesterol film inhibited cholesterol crystallisation [165]. AFM provided complementary information on the thickness and morphology of the cholesterol films transferred to a solid support: The cholesterol monolayer thickness was 13 ± 2 Å and in the bilayer phase the presence of elongated faceted crystallites of pure cholesterol about 10 layers thick could be observed [165].

4. CONCLUDING REMARKS

This review has attempted to present an overview of the surface sensitive X-ray scattering methods X-ray specular reflection (XR) and grazing-incidence X-ray diffraction (GIXD) and, by way of examples, to summarise some selected results gained by the application of these techniques to Langmuir film systems. Because X-ray scattering provides direct microscopic structural information about the systems under study these novel methods have already had a considerable impact on our present understanding of the molecular structure and phase diagrams of many Langmuir monolayers. Some of the more important advances include direct measurements of the structures of lipids and other molecules and studies of interacting lipid-macromolecule aggregates.

Further development of the surface sensitive X-ray scattering techniques will probably allow further advances in the characterisation of a variety of complex or interacting systems. This is of considerable importance also in view of the widespread interest in structural information on complex nano-sized systems with numerous potential applications. There is a significant potential to be further explored in the combination of information from several powerful and somewhat complementary methods like MD and AFM.

ACKNOWLEDGEMENTS

We thank the European Community and the Danish National Research Council for financial support under the IHP (Contract HPRI-CT-1999-00040) and DANSYNC programmes, respectively. We are grateful to many colleagues for inspiring collaboration, resulting in a number of the papers cited. Of the experimental work reviewed here, much was done in HASYLAB at DESY, Hamburg, Germany and we are grateful for access under the IHP programme and for excellent support by HASYLAB staff. Finally, we are indebted to Prof. T. Bjørnholm and Dr. G. Brezesinski for useful discussions.

5. REFERENCES

1. G. Gaines, *Insoluble Monolayers at the Liquid-Gas Interface*, Interscience, New York, 1966.
2. D.M. Small, *The Physical Chemistry of Lipids*, Handbook of Lipid Research, Plenum Press, New York, Vol. 4, 1986.
3. A.-F. Mingotaud, C. Mingotaud, and L.K. Patterson, *Handbook of monolayers*, Academic Press, Inc., 1993.
4. See articles in: *Synchrotron Radiation News*, 12(2)(1999), edited by B. Ocko.
5. K. Kjaer, J. Als-Nielsen, C.A. Helm, L.A. Laxhuber, H. Möhwald, *Phys. Rev. Lett.* 58(1987)2224.
6. P. Dutta, J.B. Peng, B. Lin, J.B. Ketterson, M. Prakash, P. Georgopoulos, S. Ehrlich, *Phys. Rev. Lett.*, 58(1987)2228
7. S.G. Wolf, M. Lahav, L. Leiserowitz, M. Deutsch, K. Kjaer, J. Als-Nielsen, *Nature (London)*, 328(1987)63.
8. J. Als-Nielsen, K. Kjaer, X-ray reflectivity and diffraction studies of liquid surfaces and surfactant monolayers, In: (Eds.) T. Riste, D. Sherrington, Phase transitions in soft condensed matter, *Proc. Nato Adv. Study Institute*, Geilo, Norway, April 4-14, Plenum Press, New York, 1989, pp. 113.
9. K. Kjaer, J. Als-Nielsen, C.A. Helm, P. Tippman-Krayer, H. Möhwald, *J. Phys. Chem.*, 93(1989)3200.
10. B. Lin, M.C. Shih, T.M. Bohanon, G.E. Ice, P. Dutta, *Phys. Rev. Lett.*, 65(1990)191.
11. R.M. Kenn, C. Böhm, A.M. Bibo, I.R. Peterson, H. Möhwald, K. Kjaer, J. Als-Nielsen, *J. Phys. Chem.*, 95(1991)2092.
12. F. Leveiller, D. Jacquemain, L. Leiserowitz, J. Als-Nielsen and K. Kjaer, *J. Phys. Chem.*, 96(12)(1992)10380
13. D. Jacquemain, S.G. Wolf, F. Leveiller, M. Deutsch, K. Kjaer, J. Als-Nielsen, M. Lahav, L. Leiserowitz, *Angew. Chem. Int. Ed. Engl.*, 31(1992)130.

14. J.F. Legrand, A. Renault, O. Konovalov, E. Chevigny, J. Als-Nielsen, G. Grübel, B. Berge, *Thin Solid Films*, 248(1994)95.

15. J. Als-Nielsen, D. Jacquemain, K. Kjaer, F. Leveiller, M. Lahav, L. Leiserowitz, *Phys. Rep.*, 246(1994)251.

16. K. Kjaer, *Physica B*, 198(1994)100.

17. B. Berge, O. Konovalov, J. Lajzerowicz, A. Renault, J.P. Rieu, M. Vallade, J. Als-Nielsen, G. Grübel, J.F. Legrand, *Phys. Rev. Lett.*, 73(12)(1994)1652.

18. I. Weissbuch, R. Popovitz-Biro, M. Lahav, L. Leiserowitz, K. Kjaer, J. Als-Nielsen, Molecular self-assembly into crystals at air-liquid interfaces, In: (Eds.) I. Prigogine, S. Rice, *Advances in Chemical Physics*, John Wiley, New York, Vol. 102 (1997) pp. 39.

19. I. Kuzmenko, V.M. Kaganer, L. Leiserowitz, *Langmuir*, 14(1998)3882.

20. V.M. Kaganer, H. Möhwald, P. Dutta, *Rev. Mod. Phys.*, 71(3)(1999)779.

21. H. Rapaport, I. Kuzmenko, M. Berfeld, K. Kjaer, J. Als-Nielsen, R. Popovitz-Biro, I. Weissbuch, M. Lahav, L. Leiserowitz, *Phys. Chem. B*, 104(2000)1399.

22. I. Kuzmenko, H. Rapaport, K. Kjaer, J. Als-Nielsen, I. Weissbuch, M. Lahav, and L. Leiserowitz, *Chem. Rev.* 101(2001), 1659.

23. T.R. Jensen, K. Balashev, T. Bjørnholm, K. Kjaer, *Biochimie*, 83(2001)399.

24. A. Pockels, *Nature (London)*, 43(1891)437.

25. I. Langmuir, *J. Am. Chem. Soc.*, 39(1917)1848.

26. I. Langmuir, *Trans. Faraday Soc.*, 15(1920)62.

27. T.R. Jensen, K. Kjær, P.B. Howes, A. Svendsen, K. Balashev, N. Reitzel, T. Bjørnholm, In: (Eds.) G. Kokotos, V. Constantinou-Kokotou, *Lipases and lipids: Structure, function and biotechnological applications*. Crete University Press, Rethymnon, 2000, pp. 127.

28. M. Weygand, B. Wetzer, D. Pum, U.B. Sleytr, N. Cuvillier, K. Kjaer, P.B. Howes, M. Lösche, *Biophys., J.* 76(1999)458.

29. M. Weygand, M. Schalke, P.B. Howes, K. Kjaer, J. Friedmann, B. Wetzer, D. Pum, U.B. Sleytr and M. Lösche, *J. Mater. Chem.*, 10(2000)141.

30. S.A.W. Verclas, P.B. Howes, K. Kjaer, A. Wurlitzer, M. Weygand, G. Büldt, N.A. Dencher and M. Lösche, *J. Mol. Biol.*, 287(1999)837.

31. K.B. Blodgett, *J. Am. Chem. Soc.*, 56(1934)495.

32. K.B. Blodgett, *J. Am. Chem. Soc.*, 57(1935)1007.

33. M.C. Petty, *Langmuir-Blodgett films An Introduction*, Cambridge University Press, 1996.

34. R. Feidenhans'l, *Surf. Sci. Rep.*, 10(1989)105.

35. H. Brockman, *Curr. Opinion Struc. Biol.*, 9(1999)438.

36. W. Clegg, *J. Chem. Soc., Dalton Trans.*, (2000)3223.

37. C.A. Helm, H. Möhwald, K. Kjaer and J. Als-Nielsen, *Biophys. J.*, 52(1987)381.

38. C.A. Helm, H. Möhwald, K. Kjaer and J. Als-Nielsen, *Europhys. Lett.*, 4(1987)697.

39. I. Kuzmenko, R. Buller, W.G. Bouwman, K. Kjaer, J. Als-Nielsen, M. Lahav, L. Leiserowitz, *Science*, 274(1996)2046.

40. J. Garnaes, N.B. Larsen, T. Bjørnholm, M. Jørgensen, K. Kjaer, J.Als-Nielsen, J.F. Jørgensen, J.A. Zasadzinski, *Science*, 264(1994)1301.

41. T. Bjørnholm, T. Hassenkam, N. Reitzel, *J. Mater. Chem.*, 9(1999)1975.

42. L.K. Nielsen, T. Bjørnholm, O. Mouritsen, *Nature (London)*, 352(2000)404.

43. K. Balashev, T.R. Jensen, K. Kjaer, T. Bjørnholm, *Biochimie*, 83(2001)387.

44. M.Ø. Jensen, G.H. Peters *et al*, work in progress.

45. G.H. Peters, S. Toxvaerd, N.B. Larsen, T. Bjørnholm, K. Schaumburg, K. Kjaer, *Nature Struct. Biol.*, 2(1995)395.

46. G.H. Peters, N.B. Larsen, T. Bjørnholm, S. Toxvaerd, K. Schaumburg, K. Kjaer, *Phys. Rev. E*, 57(1998)3153.

47. G.H. Peters, S. Toxvaerd, N.B. Larsen, T. Bjørnholm, K. Schaumburg, K. Kjaer, *Il Nuovo Cimento*, 16(9)(1994)1479.

48. G. Roberts, *Langmuir-Blodgett films*, Plenum Press, New York, 1990.

49. D.E. Eastman, Y. Farge, E.-E. Koch. (Eds.), *Handbook on Synchrotron Radiation*, North Holland, Amsterdam, Vol. 1, 1983.

50. W. Brefeld, P. Gürtler, In: *Handbook on Synchrotron Radiation*, (Eds.) S. Ebashi, M. Koch, E. Rubenstein, North Holland, Amsterdam, Vol. 4, 1991, ch. 7, pp. 269.

51. See also http://www-hasylab.desy.de/facility/doris/parameters.htm

52. R. Frahm, J. Weigelt, G. Meyer, G. Materlik, *Rev. Sci. Instrum.*, 66(2)(1995)1677.

53. See also http://www-hasylab.desy.de/facility/experimental_stations/stations/BW1_X-Ray_Undulator_Beamline.htm

54. Built at Risø National Laboratory, Denmark, and available at beamline BW1, in HASYLAB at DESY, Hamburg, Germany. *Cf.* K. Kjaer, in: *Experimental Stations at HASYLAB*, (1994)88.

55. See also http://www-hasylab.desy.de/facility/experimental_stations /stations/BW1_Horizontal_Scattering_Diffractometer.htm

56. J. Als-Nielsen, In: *Handbook on Synchrotron Radiation*, (Eds.) G. Brown, D.E. Moncton, North Holland, Amsterdam, Vol. 3, 1991, pp. 471.

57. *HLL* is also known as *thermomyces lanuginosa* lipase (*TLL*) and is marketed by Novozymes, Inc. under the brand name *Lipolase* as an additive to washing powder.

58. A. Braslau, M. Deutsch, P.S. Pershan, A.H. Weiss, J. Als-Nielsen, J. Bohr, *Phys. Rev. Lett.*, 54(1985)114.

59. M. Born, E. Wolf, *Principles of Optics*, Pergamon Press, Oxford, 1984.

60. J. Als-Nielsen, D. McMorrow, Elements of Modern X-ray Physics, Wiley, New York, 2001, ch. 3.

61. W.C. Marra, P. Eisenberger, A.Y. Cho, *J. Appl. Phys.*, 50(1979)6927.

62. P. Eisenberger, W.C. Marra, *Phys. Rev. Lett.*, 46(1981)1081.

63. f_j is the atomic form factor of atom *no. j*. Atomic form factors are usually tabulated as $f_j = f_j(\sin(\theta)/\lambda)$, where $\sin(\theta)/\lambda = |\boldsymbol{q}|/4\pi$ 64..

64. U. Shmueli (Ed.), *International tables for crystallography*, Kluwer Academic Publishers, Dordrecht, Vol. B, 1993.

65. Y. Yoneda, *Phys. Rev.*, 131(1963)2010.

66. G.H. Vineyard, *Phys. Rev. B*, 26(8)(1982)4146.

67. Some care is called for when using the phrase *resolution* since in different communities it evokes either of the following two notions: **A.** Clearly, from, *e.g.*, *eq.* (15), the *shortest* real space distance that can be resolved by a diffraction experiment is inversely related to the accessible range of q. Indeed, in a subset of the crystallography community it is customary to refer to the quantity $d_{\min} = 2\pi/q_{\max}$ as *the resolution.* **B.** Conversely, the *instrumental resolution*, or $\boldsymbol{q}$-resolution is inversely related to the *longest* real space separation that can give rise to interferences in an X-ray scattering experiment. This concept is at the basis of the discussion of *eq.* (24) *vs. eq.* (26) and for *eq.* (17).

68. T. Hahn (Ed.), *International tables for crystallography*, Kluwer Academic Publishers, Dordrecht, Vol. A, 1992.

69. A. Guinier, *X-Ray Diffraction*, Freeman, 1963, ch. 5.5.

70. V.M. Kaganer, G. Brezesinski, H. Möhwald, P.B. Howes, K. Kjaer, *Phys. Rev. Lett.*, 81(26)(1998)5864.

71. V.M. Kaganer, G. Brezesinski, H. Möhwald, P.B. Howes, K. Kjaer, *Phys. Rev. E*, 59(2)(1999)2141.

72. N. Reitzel, D.R. Greve, K. Kjær, P.B. Howes, M. Jayaraman, S. Savoy, R.D. McCullough, J.T. McDevitt; T. Bjørnholm, *J. Am. Chem. Soc.*, 122(2000)5788.

73. I. Weissbuch, P.N.W. Baxter, I. Kuzmenko, H. Cohen, S. Cohen, K. Kjær, P.B. Howes, J. Als-Nielsen, J.M. Lehn, L. Leiserowitz, M. Lahav, *Chem. Eur. J.*, 6(2000)725.

74. H. Rapaport, I. Kuzmenko, S. Lafont, K. Kjaer, P.B. Howes, J. Als-Nielsen, M. Lahav and L. Leiserowitz, *Biophys. J.*, (2001), submitted.

75. P.B. Howes, POW, a version of E. Vlieg's ROD program 76. modified for the modelling of 2D-powder diffractograms., unpublished (1998).

76. E. Vlieg, ROD, a computer program for the analysis of single crystal surface X-ray diffraction data. See http://www.esrf.fr/computing/scientific/joint_projects/ANA-ROD/index.html

77. The CERIUS^2 computational package (Molecular Simulations Inc., San Diego. CA).

78. G.M. Sheldrick, SHELX-97 Program for Crystal Structure Determination, University of Göttingen, Germany (1997).

79. G. Brezesinski, E. Scalas, B. Struth, H. Möhwald, F. Bringezu, U. Gehlert, G. Weidemann, D. Vollhardt, *J. Phys. Chem.*, 99(1995)8758.

80. C. DeWolf, F. Bringezu, G. Brezesinski, H. Möhwald, P.B. Howes, K. Kjaer, *Physica B*, 248(1998)199.

81. D. Andelman, *J. Am. Chem. Soc.*, 111(1989)6536.

82. I. Weissbuch, M. Berfeld, W. Bouwman, K. Kjaer, J. Als-Nielsen, M. Lahav, L. Leiserowitz, *J. Am. Chem. Soc.*, 119(1997)933.

83. I. Kuzmenko, I. Weissbuch, E. Gurovich, L. Leiserowitz, M. Lahav, *Chirality*, 10(1998)415.

84. L.G. Parratt, *Phys. Rev.*, 95(1954)359.

85. E.S. Wu, W.W. Webb, *Phys. Rev. A*, 8(1973)2077.

86. J. Als-Nielsen, *Z. Phys. B*, 61(1985)411.

87. J. Als-Nielsen, In: Topics in Current Physics, (Eds.) W. Schommers, P. Blanckenhagen, Springer, Berlin, Vol. 2, ch. 5, 1986.

88. J. Daillant, L. Bosio, J.J. Benattar, J. Menunier, *Europhys. Lett.*, 8(1989)453.

89. J. Daillant, L. Bosio, B. Harzallah, J.J. Benattar, *Phys. II* (Paris), 1(1991)149.

90. D.M. Mitrinovic, Z. Zhang, S.M. Williams, Z. Huang, M.L. Schlossman, *J. Phys.Chem. B*, 103(1999)1779.

91. J.S. Pedersen, *J. Appl. Crystallogr.*, 25(1992)129.

92. I.W. Hamley and J.S. Pedersen, *J. Appl. Crystallogr.*, 27(1994)29.

93. J.S. Pedersen and I.W. Hamley, *J. Appl. Crystallogr.*, 27(1994)36.

94. C. Tanford, *The Hydrophobic Effect*, Wiley, New York, (1980) ch. 6.

95. J. Israelachvili, *Intermolecular and surface forces*, Academic Press, (1992) ch. 17.3.

96. M. Schalke, M. Lösche, *Adv. Colloid and Interface Sci.*, 88(2000)243.

97. M. Schalke, P. Krüger, M. Weygand, M. Lösche, *Biochim. Biophys. Acta*, 1464(2000)113.

98. D. Vaknin, K. Kjaer, H. Ringsdorf, R. Blankenburg, M. Piepenstock, A. Diederich, M. Lösche, *Langmuir*, 9(1993)1171.

99. M. Lösche, M. Piepenstock, A. Diederich, T. Grünwald, K. Kjaer, D. Vaknin, *Biophys. J.*, 65(1993)2160.

100. M. Lösche, C. Erdelen, E. Rump, H. Ringsdorf, K. Kjaer, D. Vaknin, *Thin Solid Films*, 242(1994)112.

101. S.P. Weinbach, I. Weissbuch, K. Kjaer, W.G. Bouwman, J. Als-Nielsen, M. Lahav, L. Leiserowitz, *Adv. Mater.*, 10(1995)857.

102. R. PopovitzBiro, R. Edgar, J. Majewski, S. Cohen, L. Margulis, K. Kjaer, J. Als-Nielsen, L. Leiserowitz, M. Lahav, *Croatica Chemica Acta*, 69(2)(1996)689.

103. J. Majewski, R. Edgar, R. Popovitz-Biro, K. Kjaer, W.G. Bouwman, J. Als-Nielsen, M. Lahav, L. Leiserowitz, *Angew. Chem. Int. Ed.*, 34(1995)649.

104. R. Popovitz-Biro, J. Majewski, L. Margulis, S. Cohen, L. Leiserowitz, M. Lahav, *Adv. Mater.*, 6(1994)956.

105. M.C. Shih, T.M. Bohanon, J.M. Mikrut, P. Zschack, P. Dutta, *J. Chem. Phys.*, 97(1992)4485.

106. C. Lautz, Th.M. Fischer, M. Weygand, M. Lösche, P.B. Howes, K. Kjaer, *J. Chem. Phys.*, 108(1998)4640.

107. J.-L. Wang, F. Leveiller, D. Jacquemain, K. Kjaer, J. Als-Nielsen, M. Lahav, L. Leiserowitz, *J. Am. Chem. Soc.*, 116(1994)1192.

108. G. Weidemann, G. Brezesinski, D. Vollhardt, C. DeWolf, H. Möhwald, *Langmuir*, 15(1999)2901.

109. R. Popovitz-Biro, J. Majewski, L. Margulis, S. Cohen, L. Leiserowitz, M. Lahav, *J. Phys. Chem.*, 98(1994)4970.

110. S.W. Barton, B.N. Thomas, E.B. Flom, S.A. Rice, B. Lin, J.B. Peng, J.B. Ketterson, P. Dutta, *J. Chem. Phys.*, 89(1988)2257.

111. G. Brezesinski, V.M. Kaganer, H. Möhwald, P.B. Howes, *J. Chem. Phys.*, 109(5)(1998)2006.

112. M.C. Shih, T.M. Bohanon, J.M. Mikrut, P. Zschack, P. Dutta, *J. Chem. Phys.*, 97(1992)2303.

113. T.M. Bohanon, B. Lin, M.C. Shih, G.E. Ice, P. Dutta, *Phys. Rev. B*, 41(1990)4846.

114. J. Majewski, R. Popovitz-Biro, W.G. Bouwman, K. Kjaer, J. Als-Nielsen, M. Lahav, L. Leiserowitz, *Chem. Eur. J.*, 1(1995)304.

115. P. Tippman-Krayer, H. Möhwald, *Langmuir*, 7(1991)2303.

116. V.M. Kaganer, I.R Peterson., R.M. Kenn, M.C. Shih, M. Durbin, P. Dutta, *J. Chem. Phys.*, 102(1995)9412.

117. M.K. Durbin, A. Malik, R. Ghaskadvi, M.C. Shih, P. Zschack, P. Dutta, *J. Phys. Chem.*, 98(1994)1753.

118. M.C. Shih, T.M. Bohanon, J.M. Mikrut, P. Zschack, P. Dutta, *Phys. Rev. A*, 45(1992)5734.

119. M.L. Schlossman, D.K. Schwartz, P.S. Pershan, E.H. Kawamoto, G.J. Kellogg, S. Lee, *Phys. Rev. Lett.*, 66(1991)1599.

120. M.K. Durbin, M.C. Shih, A. Malik, P. Zschack, P. Dutta, *Colloids and Surfaces A*, 102(1995)173.

121. M.C. Shih, M.K. Durbin, A. Malik, P. Zschack, P. Dutta, *J. Chem. Phys.*, 101(1994)9132.

122. F. Leveiller, D. Jacquemain, M. Lahav, L. Leiserowitz, M. Deutsch, K. Kjaer, J. Als-Nielsen, *Science*, 252(1991)1532.

123. F. Leveiller, C. Böhm, D. Jacquemain, H. Möhwald, L. Leiserowitz, K. Kjaer, J. Als-Nielsen, *Langmuir*, 10(1994)819.

124. C. Böhm, F. Leveiller, D. Jacquemain, H. Möhwald, K. Kjaer, J. Als-Nielsen, I. Weissbuch, L. Leiserowitz, *Langmuir*, 10(1994)830.

125. I. Weissbuch, P.N.W. Baxter, S. Cohen, H. Cohen, K. Kjaer, P.B. Howes, J. Als-Nielsen, G.S. Hanan, U.S. Schubert, J.-M. Lehn, L. Leiserowitz, M. Lahav, *J. Am. Chem. Soc.*, 120(1998)4850.

126. I. Weissbuch, P.N.W. Baxter, I. Kuzmenko, H. Cohen, S. Cohen, K. Kjaer, P.B. Howes, J. Als-Nielsen, J.-M. Lehn, L. Leiserowitz, M. Lahav, *Chem. Eur. J.*, 6(2000)725.

127. I. Pascher, *Curr. Opinion Struc. Biol.*, 6(4)(1996)439.

128. I. Pascher, M. Lundmark, P.-G. Nyholm, S. Sundell, *Biochim. Biophys. Acta*, 1113(1992)339.

129. G.H. Peters, S. Toxvaerd, A. Svendsen, O.H. Olsen, *J. Chem. Phys.*, 100(8)(1994)5996.

130. D.A. Fahey, D.M. Small, *Biochem.*, 25(1986)4468.

131. D.A. Fahey, D.M. Small, *Langmuir*, 4(1988)589.

132. J.M.R. Patino, C.C. Sanchez, M.R.R. Nino, *Langmuir*, 15(1999)2484.

133. C. DeWolf, G. Brezesinski, H. Möhwald, K. Kjaer, P.B. Howes, *J. Phys. Chem. B*, 102(1998)3238.

134. G. Brezesinski, R. Rietz, K. Kjaer, W.G. Bouwman, H. Möhwald, *Il Nuovo Cimento*, 16(9)(1994)1487.

135. E. Scalas, G. Brezesinski, H. Möhwald, V.M. Kaganer, W.G. Bouwman, K. Kjaer, *Thin Solid Films*, 284-285(1996)56.

136. E. Scalas, G. Brezesinski, V.M. Kaganer, H. Möhwald, *Phys. Rev. E*, 58(2)(1998)2172.

137. C.A. Helm, P. Tippmann-Krayer, H. Möhwald, J. Als-Nielsen, K. Kjaer, *Biophys. J.*, 60(1991)1457.

138. C. Böhm, H. Möhwald, L. Leiserowitz, J. Als-Nielsen, K. Kjaer, *Biophys. J.*, 64(1993)553.

139. G. Brezesinski, A. Dietrich, B. Struth, C. Böhm, W.G. Bouwman, K. Kjaer, H. Möhwald, *Chem. Phys. Lipids*, 76(1995)145.

140. D. Vaknin, K. Kjaer, J. Als-Nielsen, M. Lösche, *Makromol. Chem., Macromol. Symp.* 46(1991)383.

141. G. Brezesinski, M. Thoma, B. Struth, H. Möhwald, *J. Phys. Chem.*, 100(1996)3126.

142. F. Bringezu, G. Brezesinski, H. Möhwald, *Chem. Phys. Lipids*, 94(1998)251.

143. G. Brezesinski, B. Dobner, B. Elsner, H. Möhwald, *Pharmazie*, 52(1997)703.

144. F. Bringezu, G. Brezesinski, P. Nuhn, H. Möhwald, *Thin Solid Films*, 327-329(1998)28.

145. G. Brezesinski, F. Bringezu, G. Weidemann, P.B. Howes, K. Kjaer, H. Möhwald, *Thin Solid Films*, 327-329(1998)256.

146. G. Brezesinski, A. Dietrich, B. Dobner, H. Möhwald, *Progr. Colloid Polym. Sci.*, 98(1995)255.

147. F. Bringezu, G. Brezesinski, P. Nuhn, H. Möhwald, *Biophys. J.*, 70(1996)1789.

148. D. Vaknin, K. Kjaer, J. Als-Nielsen, M. Lösche, *Biophys. J.*, 59(1991)1325.

149. S.G. Wolf, M. Lahav, L. Leiserowitz, M. Deutsch, K. Kjaer, J. Als-Nielsen, *Nature (London)*, 328(1987)63.

150. E.M. Landau, M. Levanon, L. Leiserowitz, M. Lahav, J. Sagiv, *Nature (London)*, 318(1985)353.

151. M. Berfeld, I. Kuzmenko, I. Weissbuch, H. Cohen, P.B. Howes, K. Kjaer, J. Als-Nielsen, L. Leiserowitz, M. Lahav, *J. Phys. Chem. B*, 103(1999)6891.

152. R. Reitz, W. Rettig, G. Brezesinski, H. Möhwald, *Pharmazie*, 52(1997)701.

153. P. Nassoy, M. Goldmann, O. Bouloussa, F. Rondelez, *Phys. Rev. Lett.*, 75(1995)457.

154. H. Rapaport, I. Kuzmenko, K. Kjaer, P.B. Howes, W. Bouwman, J. Als-Nielsen, L. Leiserowitz, M. Lahav, *J. Am. Chem. Soc.*, 119(1997)11211.

155. H. Rapaport, H.S. Kim, K. Kjaer, P.B. Howes, S. Cohen, J. Als-Nielsen, M.R. Ghadiri, L. Leiserowitz, M. Lahav, *J. Am. Chem. Soc.*, 121(1999)1186.

156. H. Haas, G. Brezesinski, H. Möhwald, *Biophys. J.*, 68(1995)312.

157. D. Gidalevitz, Z. Huang, S.A. Rice, *Proc. Natl. Acad. Sci. USA*, 96(1999)2608.

158. J. Majewski, T.L. Kuhl, K. Kjaer, M.C. Gerstenberg, J. Als-Nielsen, J.N. Israelachvili, G.S. Smith, *J. Am. Chem. Soc.*, 120(1998)1469.

159. T.L. Kuhl, J. Majewski, P.B. Howes, K. Kjaer, A. von Nahmen, K.Y.C. Lee, B. Ocko, J.N. Israelachvili, G.S. Smith, *J. Am. Chem. Soc.*, 121(1999)7682.

160. P. Lehmann, D.G. Kurth, G. Brezesinski, C. Symietz, *Chem. Eur. J.*, 7(8)(2001)1646.

161. K. de Meijere, G. Brezesinski, H. Möhwald, *Macromolecules*, 30(1997)2337.

162. K. de Meijere, G. Brezesinski, K. Kjaer, H. Möhwald, *Langmuir*, 14(1998)4204.

163. U. Dahmen-Levison, G. Brezesinski, H. Möhwald, *Thin Solid Films*, 327-329(1998)616.

164. L-.T. Lee, B.K. Jha, M. Malmsten, K. Holmberg, *J. Phys. Chem. B*, 103(1999)7489.

165. S. Lafont, H. Rapaport, G.J. Sömjen, A. Renault, P.B. Howes, K. Kjaer, J. Als-Nielsen, L. Leiserowitz, M. Lahav, *J. Phys. Chem. B*, 102(1998)761.

166. G. Weidemann, G. Brezesinski, D. Vollhardt, F. Bringezu, K. de Meijere, H. Möhwald, *J. Phys. Chem.*, 102(1998)148.

6. Symbols and Abbreviations

AFM	atomic force microscopy
LB	Langmuir-Blodgett
MD	molecular dynamics (simulation)
T	temperature
Π	lateral surface pressure
diT-DPPC	1,2-di(2-tetradecyl-palmitoyl) phosphatidylcholine
DPPC	dipalmitoyl phosphatidyl choline
DPG	dipalmitoyl glycerol
DLPE	dilauroyl phosphatidyl ethanolamine
DMPE	dimyristoyl phosphatidyl ethanolamine
MOG	monooleoylglycerol
MPG	monopalmitoylglycerol
PA	phosphatic acid
PE	phosphatidylethanolamine
m	a natural number, as in: C_mH_{2m+2}
SR	synchrotron radiation
BW1	name of a synchrotron radiation beamline
λ	wavelength
$\boldsymbol{k}_i$, $\boldsymbol{k}_f$,	wave vectors in reciprocal space
k	$= \lvert \boldsymbol{k}_i \rvert = \lvert \boldsymbol{k}_f \rvert = 2\pi/\lambda$. X-ray wavenumber
r_o	classical electron radius
$\boldsymbol{r}$	position vector in real space
XR	X-ray specular reflectivity
β	imaginary part of refractive index; related to the linear absorption coefficient
δ	deviation from unity of real part of refractive index

μ	linear absorption coefficient
n	refractive index
α_c	critical angle for total reflection
α_i	angle between the surface and the incident wave
α'	angle between the surface and the refracted wave
α_f	angle between the surface and the outgoing reflected or diffracted ray
$2\theta_{xy}$	horizontal part of scattering angle
2θ	total scattering angle composed of α_i , α_f and $2\theta_{xy}$
R_F	Fresnel reflectivity calculated for an ideal (sharp and planar) interface between vacuum and a subphase of (constant) electron density ρ_∞
$R(q_z)$	X-ray reflectivity
$r(q_z)$	reflectivity amplitude function
q_z	$= (4\pi/\lambda)\sin(\alpha)$. Value of scattering vector for reflectivity ($\alpha_i = \alpha_f = \alpha$ and $2\theta_{xy} = 0$)
q'_z	$= (4\pi/\lambda)\sin(\alpha')$. Value of refraction corrected scattering vector for reflectivity.
p	an odd number in *eq.* (23)
$\rho(z)$	electron density profile across the interface (laterally averaged),
$\rho_{Sub}(z)$	contribution by the subphase to $\rho(z)$
ρ_∞	average electron density of the subphase
L_H, L_T	heights of head and tail regions in a slab model of a molecular monolayer
E_H, E_T	numbers of electrons in head and tail regions
ρ_H, ρ_T	electron densities of slabs representing head and tail regions
3D	three-dimensional
$\boldsymbol{a}$, $\boldsymbol{b}$, $\boldsymbol{c}$	primitive vectors of 3D lattice
$\boldsymbol{a}^*$, $\boldsymbol{b}^*$, $\boldsymbol{c}^*$	basis vectors of 3D reciprocal lattice
h, k, l	Miller indices with respect to 3D reciprocal lattice
2D	two-dimensional
GIXD	grazing incidence X-ray diffraction
$\boldsymbol{q}$	$= \boldsymbol{k}_f - \boldsymbol{k}_i$, scattering vector in reciprocal space
q	$= \lvert \boldsymbol{q} \rvert$
$\boldsymbol{q}_{xy}$, q_z	horizontal and vertical components of the scattering vector
q_{xy}	$= \lvert \boldsymbol{q}_{xy} \rvert$
q_z	$\approx (2\pi/\lambda)\sin(\alpha_f)$. Value of vertical component of scattering vector for GIXD ($\alpha_i \approx 0$).

$\boldsymbol{a}, \boldsymbol{b}$	primitive vectors of 2D lattice (*in* the monolayer plane)
a	$= \lvert\boldsymbol{a}\rvert$
b	$= \lvert\boldsymbol{b}\rvert$
γ	angle between the primitive vectors $\boldsymbol{a}, \boldsymbol{b}$
A_{Cell}	area of unit cell
Z	number of alkyl chains per unit cell
t	tilt angle (from vertical) of alkyl chains
A_{o}	area *per alkyl chain* projected on a plane orthogonal to the long chain axis
NN	nearest-neighbour direction
NNN	next-nearest-neighbour direction
DH	distorted hexagonal
H	hexagonal
R	rectangular
$\boldsymbol{a}^*, \boldsymbol{b}^*$	basis vectors of 2D reciprocal lattice (*in* the monolayer plane)
a^*	$= \lvert\boldsymbol{a}^*\rvert = 2\pi/(a \sin(\gamma))$
b^*	$= \lvert\boldsymbol{b}^*\rvert = 2\pi/(b \sin(\gamma))$
h, k	Miller indices with respect to 2D reciprocal lattice
ρ	electron density
$\rho(\boldsymbol{r})$	3D-variation of electron density due to molecules in unit cell
F	structure factor
f_j	atomic form factor of atom *no. j*.

Novel Methods to Study Interfacial Layers
D. Möbius and R. Miller (Editors)

In and Out Plane X-ray Diffraction of Cellulose LB Films

Shin-ichi Kimura[1,2]*, Masahiko Kitagawa[2], Hiroyuki Kusano[1] and Hiroshi Kobayashi[2]

[1]Research Institute of Technology, Tottori Prefecture, Wakabadai-minami, Tottori 689-1112, Japan;

[2]Department of Electrical and Electronic Engineering, Tottori Univ., Koyama, Tottori 680-8552, Japan

Contents

* e-mail:shin@pref.tottori.jp tel:+81-(857)-38-6200 fax:+81-(857)-38-6210

Keyword: Cellulose, Langmuir-Blodgett film, Indium tin oxide, X-ray diffraction analysis

We have investigated in and out plane X-ray diffraction from the palmitoyl cellulose Langmuir-Blodgett (PC-LB) films prepared on indium-tin-oxide (ITO) substrate by the horizontal lifting method. In the in plane analysis, the peak intensity of diffracted X-ray from PC-LB film decreased in accordance with the increase in the incident angle ω, in contrast to the increase of the diffraction intensities from ITO substrate. The average distance between the palmitoyl chains in PC-LB film have been determined to be 4.20Å, which indicates the regular structure of PC-LB glucose units on the ITO substrate. The molecular structure has also been discussed based on MM2 molecular dynamic calculation. In the out-of-plane analysis, the diffraction from ITO substrate dominated in the grazing incident alignment, therefore the diffraction from PC-LB films were so weak to be detected.

1. Introduction

Structurally controlled organic thin films are expected for the application in functional sensors and electronic devices [1-5]. Langmuir-Blodgett (LB) technique is one of the most promising methods for preparing mono-molecular layered structure in atomic scale. It is also inevitable that the physical and electronic properties of such advanced thin film are only defined after the detailed characterization of the structure of those films.

We have reported the structure analysis of palmitoyl cellulose Langmuir-Blodgett (PC-LB) film by using transmission electron microscope (TEM) [6], scanning probe microscope (SPM) [7] and Fourier transform-infrared spectroscopy [7]. In addition, we have reported the electrical properties of the PC-LB film on the indium tin oxide (ITO) substrate [2]. The average grain size in the PC-LB film was about 200nm. The monolayer thickness was about 22 Å.

Recently, it has become possible to analyse the fine surface structure of organic thin films by using X-ray total reflection [8, 9]. All incidents X-ray reflected on the sample surface can be detected when incident angle of X-ray is less than a critical angle. Further, when incident angle is larger than the critical angle, the intensities decrease proportional to (incident angle)$^{-4}$. By analysing in and out plane X-ray diffraction patterns, the periodic structure perpendicular to the sample surface and that parallel to the film surface are estimated.

In this study, we have investigated the structure of PC-LB films on the ITO substrate by using the in and out plane X-ray diffraction method. The size of crystallites and grains in the PC-LB film and the distance between palmitoyl side chains of PC's are determined and discussed together with the consideration by the structural calculation based on the molecular dynamic calculation.

2. EXPERIMENTAL

Fig.1 shows the schematic diagram to prepare the PC-LB films on the ITO substrate by the horizontal lifting method. A monolayer of stearic acid was deposited on the ITO substrate at first and then X-type layers of palmitoyl cellulose LB film were built up at a trough temperature of 290 K. The degree of substitution (DS) of the palmitoyl cellulose have been reported elsewhere.

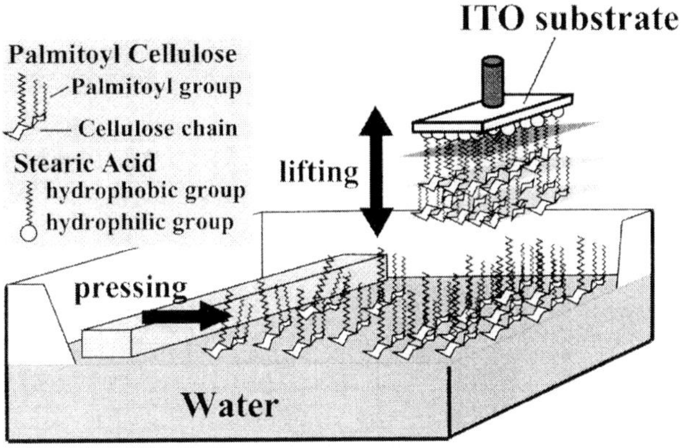

Fig. 1 Schematic diagram to prepare the palmitoyl cellulose Langmuir-Blodgett film on the ITO substrate by the horizontal lifting method.

In and out-plane X-ray diffraction pattern were recorded on a X-ray diffractometer equipped with a 4-axes goniometer (Rigaku ATX-G). CuKα radiation from copper rotating anode was used for the experiment. Incident angles (ω) for in plane geometry were between 0.14° and 0.36°, those for out of plane were between 0° and 5°. In-plane diffraction pattern is recorded as a result of the periodic structure of the direction perpendicular to the sample surface, out-plane diffraction pattern by structure of the direction parallel to the sample surface.

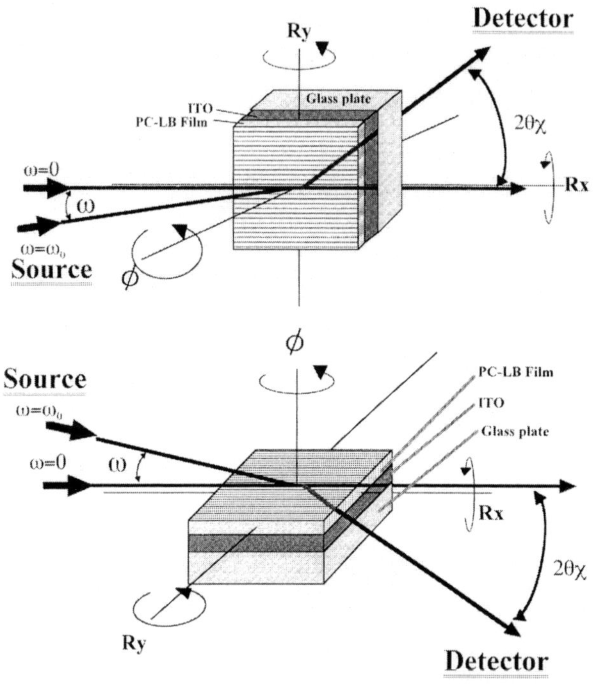

Fig. 2 Geometry for in plane X-ray diffraction measurement of palmitoyl cellulose Langmuir-Blodegett film on the ITO substrate.

Fig. 2 shows the geometry for in plane X-ray diffraction measurement of the PC-LB film on the ITO substrate. Molecular dynamic calculation has been carried out by CAChe (MM2).

3. Results and Discussion

Fig. 3 shows In-plane X-ray diffraction pattern of the PC-LB film on the ITO substrate. The peak intensity of $2\theta\chi = 21.1^\circ$ (Peak A) at $\omega=0.14^\circ$ was larger than that at $\omega=0.20^\circ$. To the contrary, the peaks of $2\theta\chi = 30.7^\circ$ (Peak B), 35.5° (Peak C) appeared at $\omega=0.20^\circ$ almost disappeared at $\omega=0.14^\circ$. The interplanar spacings of these two peaks (B,C) correspond to those of ITO(222), ITO(004).

The peak intensity of A decreased with the increase of incident angle. Contrary, the peak intensities of ITO substrate increased with increase incident angle. Consequently, the peak A is due to the diffraction peak of PC-LB film which appeared at the small incident angle of X-ray (ω=0.14°) .

The interplanar spacing of the peak A of PC-LB film was 4.20Å. This value was attributed to a distance between the palmitoyl chains in PC molecules perpendicular to the glucose units. Palmitoyl chains was long acyl group (16 carbon) and bonded to the hydroxyl group of glucose unit.

On the other hands, the size of apparent crystallite can be estimated from the peak broadening. Hence, the size of crystallite can be estimated by the Scherrer equation ($D=0.9\lambda/(\beta\cos\theta)$), where D is the size of crystallite, λ is the wave length of CuKα, β is the half width of the peak and θ is the Bragg angle. From the half width of the peak of $2\theta\chi$= 21.1°, the size apparent of crystallite in the PC-LB film was about 15Å.

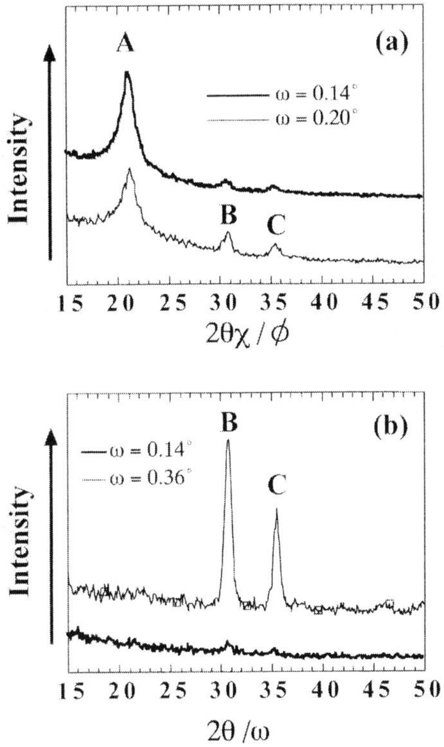

Fig. 3 In plane X-ray diffraction pattern of palmitoyl cellulose Langmuir-Blodegett film on the ITO substrate.

Fig. 3 (b) shows in plane X-ray diffraction pattern of the stearic acid LB (SA-LB) film on the ITO substrate. The peaks of $2\theta\chi$=30.7° (peak B), 35.5° (Peak C) due to the ITO substrate appeared. However, any peaks due to SA-LB film did not appear. Therefore, the structure of the direction of perpendicular to the SA-LB film surface was considered to be not well aligned.

Fig. 4 shows the calculated results of the model of palmitoyl cellulose molecular structure by CAChe (MM2). Fig. 4 (a) indicated ideal structure of PC molecule at most stable state in energy. Palmitoyl chains were on either side of cellulose main chain. Total energy of this structure by MM2 was 115.7 kcal/mol. On the other hands, when PC molecules spread on water surface, the stable conformation of palmitoyl cellulose on water surface may be alignment of palmitoyl chains opposite to water surface. Fig. 4 (b) shows the metastable structure of PC molecules. Palmitoyl chains are considered to be aligned on the upper half plane of each glucose units. The total energy was 464.5 kcal/mol which is rather larger than 115.7 kcal/mol for the most stable state, which indicated metastable state exist in the formation of PC-LB films probably owing to hydrophilic and hydrophobic structures. The average distance between palmitoyl chains was 4.6Å. This distance well corresponds to the results obtained from the in plane X-ray diffraction analysis.

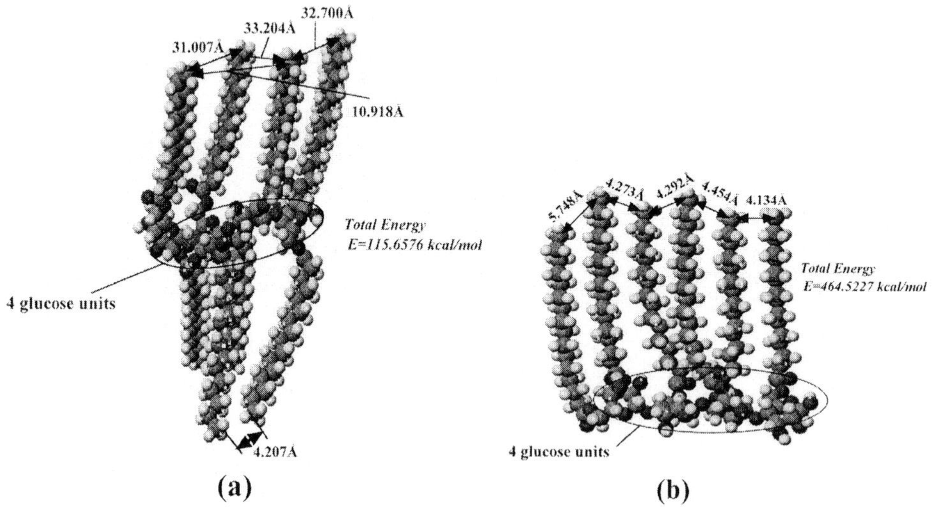

Fig. 4 Calculated results of the model of palmitoyl cellulose by CAChe (MM2).

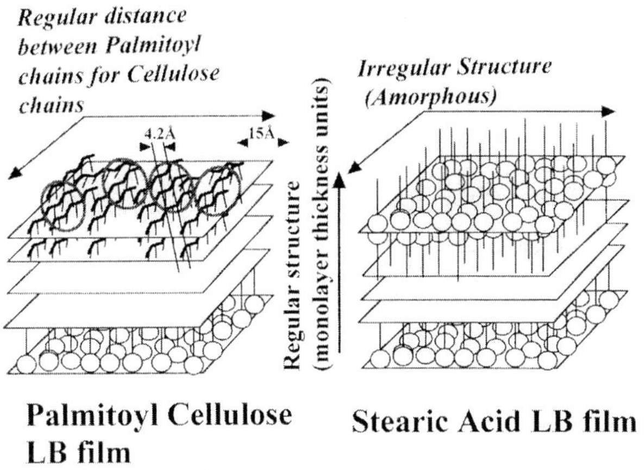

Fig. 5 Schematic model of (a) palmitoyl cellulose Langmuir-Blodgett films and (b) stearic acid Langmuir-Blodgett films on the ITO substrate.

Fig.5 shows schematic models of PC-LB film and SA-LB film on the ITO substrate. Both structures of PC-LB film and SA-LB film are regular in the direction parallel to film surface. However, the structure of SA-LB film in the direction perpendicular to the film surface seems to be irregularly aligned. In PC-LB film, the distance between palmitoyl chains keep some same distance. The conformation of palmitoyl chains is most probably determined due to cellulose main chains.

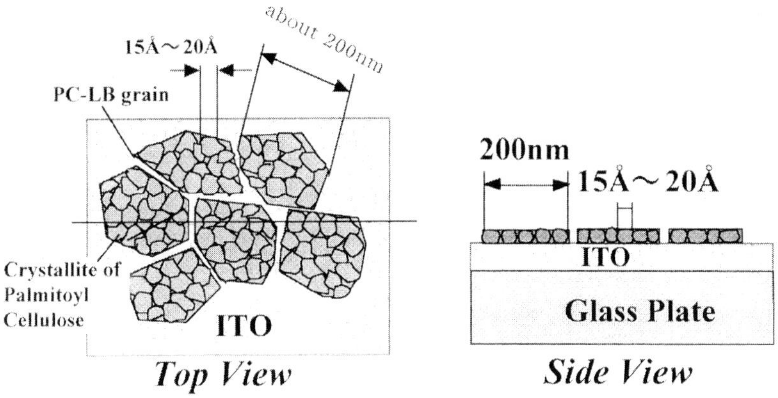

Fig. 6 Schematic model of grains in palmitoyl cellulose Langmuir-Blodgett films on the ITO substrate.

Fig.6 illustrates the model of grains in the PC-LB film. We have reported the grain size was estimated 200nm by using TEM[6] and SPM[7]. Therefore, the structure of 200nm size grains in the PC-LB film is considered to be aggregation of many 15Å size crystallites with 4.2Å distance between palmitoyl chains.

Fig.7 (a) shows out-plane X-ray diffraction pattern from the PC-LB films on the ITO substrates. The intensities of all X-ray diffractions oscillate similarly. Therefore, periodic structures were considered to exist in the samples. The thickness of the periodic structure was derived from Fourier-transform of the X-ray diffraction.

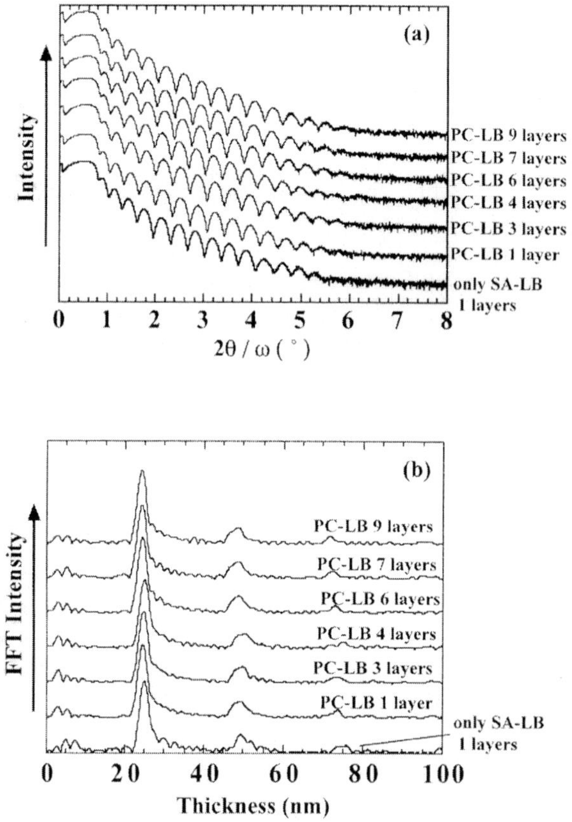

Fig. 7 Out plane X-ray diffraction pattern of palmitoyl cellulose Langmuir-Blodgett film on the ITO substrate. (a) Diffraction patterns (b) Fourier transform results

Fig.7 (b) shows the Fourier transform of the results of out plane X-ray diffractions. Several peaks were obtained at 25nm, 50nm and 75nm in all samples. In addition, all curves were same. Therefore, these results does not seem to depend on the layer number of PC-LB films. From these results, the diffraction from ITO was detected and the thickness was estimated about 250Å which coincide well with ITO film thickness used in this study. However, any diffractions of PC-LB films and SA-LB films were so weak to be not detected.

4. Conclusion

We analysed the structure of PC-LB film on the ITO substrate by using in and out plane X-ray diffraction patterns. The periodic structure in the direction perpendicular to the film surface was clarified.

The molecular structure of the PC-LB film was considered on the base of molecular dynamic calculations.

5. Acknowledgements

We are indebted to Prof. S. Tanaka of Tottori University for valuable advice, and H. Kimoto, A.Kaneda, K.Sato of Research Institute of Technology, Tottori prefecture for advice and support of this work. We would also like to express gratitude to K. Inaba, and A. Takano of X-ray Research Laboratory, Rigaku Corporation for the advice of the technique using X-ray diffractometer.

6. References

1. H. Kusano, S Kimura, M Kitagawa and H Kobayashi, Thin Solid Films, 1997, 295, 53.
2. S. Kimura, S Kimura, M Kitagawa and H Kobayashi, Polymers for Advanced Technologies, to be published.
3. M Sugi. F L Carter. R E Siatkowski and H Wohltjen (Eds.). Molecular Electronic Devices, 1988, 441. Elsevier, Amsterdam.
4. K Sakai. H Matsuda. H Kawada. K Eguchi and T Nakagiri. Appl Phys Lett, 1988, 53, 1274.

5. K Yano. M Kyogaku. R Kuroda. Y Shimada. S Shido. H Matsuda. K Takimoto. O Albrecht. K Eguchi and T Nakagiri. Appl Phys Lett, 1996, 68, 188.

6. S Kimura. H Kusano. M Kitagawa and H Kobayashi. ,Appl. Surf. Sci., 1999, 142, 579.

7. S Kimura. H Kusano. M Kitagawa and H Kobayashi. ,Appl. Surf. Sci., 1999, 142, 585

8. T. Ikeda, H. Fujioka, S. Hayakawa, K. Ono, M. Oshima, M. Yoshimoto, H. Maruta, H. Koinuma, K. Inaba, R. Matsuo, J. Crystal Growth, 2000, 208, 395.

9. H. Miyata, K. Kuroda, J. Am. Chem. Soc., 1999, 121, 7618.

Novel Methods to Study Interfacial Layers
D. Möbius and R. Miller (Editors)

ISOMERICALLY-FEATURED AGGREGATE IMAGES OF PHENOL-FORMALDEHYDE MONOLAYERS BY BAM AND SMM

Jin Un Kim[a], Burm-Jong Lee[a,*], Jung-Hyuk Im[b], Jae-Ho Kim[b], Hoon-Kyu Shin[c] and Young-Soo Kwon[c]

[a]*Department of Chemistry, Inje University, Kimhae 621-749, South Korea*
[b] *Department of Applied Chemistry, Ajou University, Suwon 442-749, South Korea*
[c]*Department of Electrical Eng., Dong-A University, Pusan 604-714, South Korea*

Contents

Keywords: Phenol-formaldehyde resin; Monolayer; LB film; Condensation; Brewster angle microscopy; scanning Maxwell-stress microscopy

*Corresponding author. Tel.: +82 55 320 3223; Fax.: +82 55 321 9718; e-mail: chemlbj@ijnc.inje.ac.kr

Brewster angle microscopy (BAM) and scanning Maxwell-stress microscopy (SMM) were applied to investigate the characteristic aggregate patterns of reactive phenolic resin monolayers. For the two-dimensional condensation reaction in the monolayer at the air-water interface, amphiphilic phenols of structural isomers, *o*-, *m*- and *p*-hexadecoxyphenol (*o*-, *m*- and *p*-HP), were synthesized by the reaction of catechol, resorcinol, and hydroquinone, respectively, with 1-iodohexadecane. Monolayers of the HPs were spread on aqueous 1% formaldehyde subphases. The surface pressure-area isotherms revealed characteristic monolayer phases and collapse pressures according to the structural isomers at the air-water interface. The optical images of the HP-formaldehyde monolayers were monitored by BAM. The noticeable image was not observed until the monolayer collapse began. The images similar to dots, entangled threads, and blood vessels were respectively produced from *o*-, *m*-, and *p*-HP-formaldehyde monolayers after monolayer collapse. The HP-formaldehyde monolayers were transferred onto silicon wafers, and the morphologies of LB monolayers were observed by SMM. The SMM images have also supported the characteristic aggregate patterns of isomeric HP-formaldehyde monolayers.

1. INTRODUCTION

Considerable progress has been made in the structural characterization of monolayer at the air-water interface, utilizing techniques such as fluorescence microscopy [1, 2], Brewster angle microscopy [3, 4], X-ray and neutron methods [5-8], ellipsometric measurements [9], infrared reflection-absorption spectroscopy [10], and so forth. In particular, with the development of Brewster angle microscopy (BAM), a sensitive and effective method has become available for studying the organization in spread monolayers at the air-water interface at the microscopic level. BAM provides information on the morphology of amphiphilic monolayer [11-16], including the inner structure of condensed domains and phase transitions in monolayer [17-20], the orientational order of the monolayer domains [19, 21, 22], deformation [23-25] and relaxation phenomena [23, 26-28] in monolayer domains caused by compression-expansion cycles or by interfacial flow. As the light intensity at each point in the BAM image depends on the local thickness and monolayer optical properties, this technique can be used to determine the thickness of film regions, even when the optical properties of the film (refractive index) are unknown [29], if a model for the observed morphology is adopted [19].

The scanning probe technique introduced by Binnig and Rohrer [30] about 15 years ago now forms a novel area of microscopic analytical techniques. The scanning force microscope (SFM), represented by the atomic force microscope (AFM), typifies the class of scanning probe techniques based on detecting a minute force between the sample and the scanning tip [31] via the deflection of a small cantilever. Since the SFM does not necessarily require electrical conductivity of the sample as the scanning tunneling microscope (STM) does and is operable in both aqueous and atmospheric environments, it is extremely suitable for the study of organic and biological systems. Furthermore, it has been recognized and demonstrated that SFMs can be used for microscopic observations of not only short-range repulsive forces as in the AFM but also long-range forces such as van der Waals forces and electric and magnetic forces as well [31-33]. The scanning Maxwell-stress microscopy (SMM) is a versatile electric force microscope, which, unlike its predecessors [34-40], relies only on the harmonic analysis of AC-voltage-induced oscillations of the cantilever under non-resonant conditions [41-43].

We describe herein the adoption of BAM and SMM to investigate the isomerically-featured monolayer system of a thermosetting resin. Although the thermosetting resins such as phenolic resin have been employed for various plastic applications for a long time, the LB film fabrication and their characterization of the commercially important resins have been seldom reported [44,45]. Besides the preparation of mechanically stable LB films, the two-dimensional network formation of the resins can give an understanding about network structures in the bulk resins, which are not still fully clarified. Particularly, we report characteristic aggregate images of two-dimensional phenol-formaldehyde polymers which are prepared at the air-water interface. Three kinds of structural isomers of amphiphilic phenols were examined on monolayer behaviour by surface pressure-area isotherms and Brewster angle microscopy (BAM), and on the characteristic images of network LB films by scanning Maxwell-stress microscopy (SMM).

2. SCANNING MAXWELL-STRESS MICROSCOPY

2.1. Principles

The SMM detects, just like the AFM in the non-contact mode, the electric force exerted on the scanning tip via the inflection of a tiny cantilever and displays the two-dimensional (2D) distribution of the force field on the sample surface (Fig.1).

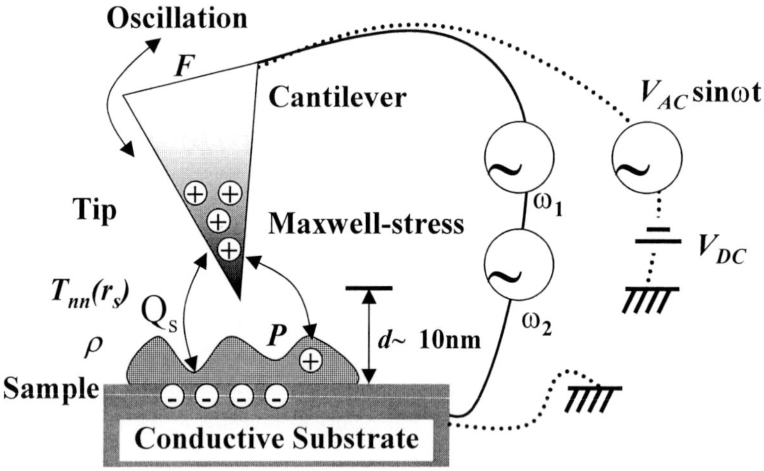

Fig. 1 Principles of scanning Maxwell-stress microscopy.

As well documented in literatures, the electromagnetic force ***F*** that the tip feels is given by the integral of the Maxwell stress tensor $\boldsymbol{T}_{ij}$ over the entire tip surface,

$$F_i = \oint_{\partial tip} T_{ij} dS_i \tag{1}$$

In the absence of a magnetic held the Maxwell stress tensor is given in a vacuum by

$$T_{ij} = E_i D_j - \frac{1}{2}\delta_{ij} E \cdot D \tag{2}$$

where ***E*** and ***D*** are the electric field and electrical displacement respectively. Since the electric field is always perpendicular to a conductive surface, the normal component of the Maxwell stress tensor at a point $\boldsymbol{r}_S$ on the tip surface reduces to

$$T_{nn}(r_s) = \frac{1}{2}\varepsilon(r_s)E(r_s) \cdot E(r_s) = \frac{1}{2}\rho_t^2(r_s) \tag{3}$$

where $\rho_t(\boldsymbol{r}_s)$ denotes the surface charge density on the tip. And,

$$T_{nn}(r_s) = D^2 / 2\varepsilon_0(r_s), \qquad D = \rho_t(r_s) \tag{4}$$

Thus, tip force $\boldsymbol{F}$ is

$$F = \oint_{\partial tip} \frac{1}{2\varepsilon_0} D^2 \cos\theta \; dS = \oint_{\partial tip} \frac{1}{2\varepsilon_0} \rho_t^2 \cos\theta \; dS \tag{5}$$

Expression (4) $\boldsymbol{D}$ can be written as follows:

$$D = \rho_t = \frac{\varepsilon_0 V_t - P - \rho \; d_c \varepsilon_0 / \varepsilon}{d - (1 - \varepsilon_0 / \varepsilon) d_0} \tag{6}$$

where d and ε are, respectively, the distance of tip and substrate and dielectric constant, d_0 and ρ are, respectively, the thickness of dielectric layer and the charge density, and $\boldsymbol{S}$ and $\boldsymbol{P}$ denote the area of electrode and polarization.

Then, tip force $\boldsymbol{F}$ is

$$F = \frac{1}{2}\left(\frac{V_t - P / \varepsilon_0 - \rho \; d_c}{d - (1 - \varepsilon_0 / \varepsilon) d_0}\right)^2 \varepsilon_0 S \tag{7}$$

If we apply a d.c. biased alternating voltage

$$V_t = V_{DC} + V_{AC} \sin(\omega t) \tag{8}$$

to the tip, there appear three force components on the tip, i.e. the d.c. component $\boldsymbol{T}_{DC}$, the fundamental component T_ω and the second-harmonic component $T_{2\omega}$. Specifically, the latter two oscillating components are written as

$$T_\omega = \frac{1}{\varepsilon}\left(\int \rho_t(r_s)\right) c_t(r_s) V_{AC} \sin(\omega t) \tag{9}$$

$$T_{2\omega} = \frac{1}{2} V_{AC}^2 \; c_t^2(r_s) \sin^2(\omega t) \tag{10}$$

Expression (9) shows that T_ω is due to the interaction between the oscillating charge on the tip induced by the applied a.c. voltage and the charge and polarization in the sample and on other conductors. On the other hand, $T_{2\omega}$ represents a purely capacitive force.

The SMM detects both the oscillating force components

$$F_{\omega} = \oint_{\partial tip} T_{\omega} \cos\theta \; dS_s, \qquad\qquad F_{2\omega} = \oint_{\partial tip} T_{2\omega} \cos\theta \; dS_s \tag{11}$$

Thus,

$$F_{\omega} = \frac{V_{DC} - P/\varepsilon_0 - \rho \, d_c}{[d - (1 - \varepsilon_0/\varepsilon)d_0]^2} \varepsilon_0 S V_{AC} \sin \omega t \tag{12}$$

$$F_{2\omega} = -\frac{1}{4[d - (1 - \varepsilon_0/\varepsilon)d_0]^2} \varepsilon_0 S V_{AC}^2 s \cos 2\omega t \tag{13}$$

via the deflection of the cantilever, where 8 denotes the angle between the surface normal of the tip and the z axis, along which the force is detected. Then it recovers information on the distribution of charge and potential and that of dielectric constant and topography of the sample from F_{ω} and $F_{2\omega}$ respectively.

The SMM controls the tip-sample separation in such a way that $F_{2\omega}$ is fixed at a given value, which is the unique feature of the SMM compared with previous electric force microscopes. Also, the resultant tip position gives rise to an image reflecting the topography and dielectric constant of the sample. As Eqs. (10) and (11) show, fixing $F_{2\omega}$ constant is equivalent to following the contour of equal average squared capacitance or, in other words, the contour of equal capacitance gradient, $\partial C_t/\partial_z$, with $C_t = \boldsymbol{c}_t(\boldsymbol{r}_s)\, \mathrm{d}\boldsymbol{S}_j$ being the total capacitance of the tip [34]. Therefore the apparent topography observed by the SMM is not in general identical to the geometrical shape of the surface, but a mixture of the topography and the dielectric properties of the medium. When the sample is electrically conductive, however, the SMM gives the true topography as demonstrated in the section 2.2. Since the capacitance scales as $1/d$, where d is the separation between the tip and the counter-conductor, the tip-sample interaction responsible for topography measurements in the SMM can be estimated to be roughly proportional to $1/d^2$.

2.2. Basic performance

To examine the performance of the SMM, we prepared a test sample by means of photolithography. The sample is comprised of tungsten strips of 1.0×1.0μm wide and 500 nm thick fabricated on a silicon wafer by means of reactive ion etching (RIE). Fig. 2 respectively shows the surface potential and topography images of the test sample respectively, observed simultaneously by the SMM with the Kelvin feedback loop closed. The surface potential image in particular indicates that the potential on the tungsten strips is about 20 mV both higher and lower than on the silicon surface. This potential difference between tungsten and silicon approximately agrees with that expected from their work function values found in the literature.

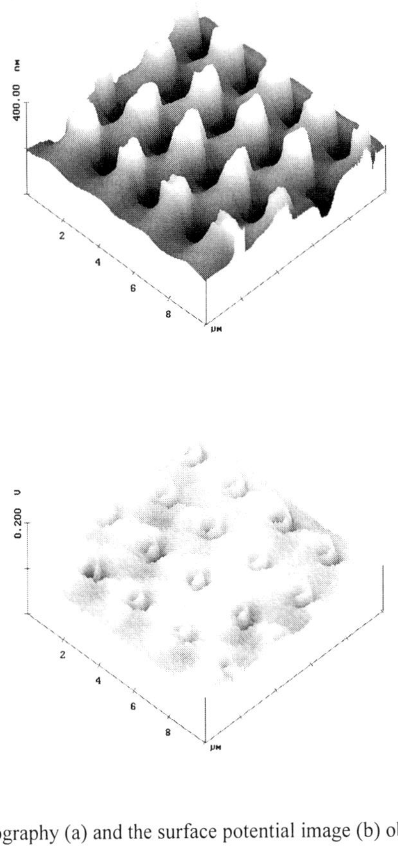

Fig. 2 Topography (a) and the surface potential image (b) observed from the test sample by the SMM.

A marked difference in spatial resolution between the surface potential and topography images is also quite obvious. In the topography image the roughness on the silicon surface, which was presumably caused by the RIE process, is visible just as revealed by the SEM picture. In the surface potential image, on the other hand, although the periodic modulation of potential is relatively deep at the edge of the tungsten strips, it becomes shallower as one moves into the deep region of the tungsten strips. As indicated in section 2.1., this difference results from the differing range of tip-sample interaction involved in the respective images. In the case of surface potential mapping the interaction is inversely proportional to observations, which clearly indicates that both the radius of the tip apex and overall shape of the tip have a significant influence on the resulting spatial resolution in the case of SMM owing to the long-range nature of the relevant interaction.

3. EXPERIMENTAL

3.1. Materials

Amphiphilic phenols, *o*-, *m*-, and *p*-hexadecoxyphenol (*o*-, *m*-, and *p*-HP) were respectively synthesised by the reaction of catechol, resorcinol, and hydroquinone with 1-iodohexadecane as described before [44].

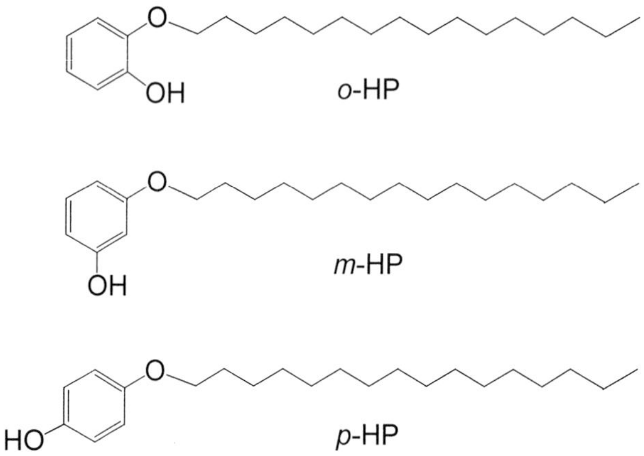

Fig. 3 Molecular structures of the *o*-, *m*- and *p*-HP.

o-HP: white powder; FT-IR (KBr, cm^{-1}), 3400 (O-H) 3100 (aromatic C-H), 2950-2850 (aliphatic C-H), 1600, 1500, 1470 (aromatic C=C), 1250, 1230 (C-O); ^{1}H-NMR (chloroform-d, ppm) 0.92 (t, 3H), 1.11-1.80 (broad s, 28H), 4.04 (t, 2H), 5.72 (s, phenol ring 1H), 6.8-6.95 (m, phenol ring 3H). *m*-HP: white powder; FT-IR (KBr, cm^{-1}), 3450 (O-H), 3100 (aromatic C-H), 2950-2850 (aliphatic C-H), 1600, 1500, 1470 (aromatic C=C), 1200,1180, 1150 (C-O); ^{1}H-NMR (chloroform-d, ppm) 0.92 (t, 3H), 1.11-1.80 (broad s, 28H), 3.91 (t, 2H), 6.40 (s, phenol ring 2H), 6.52 (d, phenol ring 1H), 7.13 (t, phenol ring 1H). *p*-HP: white powder; FT-IR (KBr, cm^{-1}), 3450 (O-H), 3100 (aromatic C-H), 2950-2850 (aliphatic C-H), 1520, 1470 (aromatic C=C), 1250 (C-O); ^{1}H-NMR (chloroform-d, ppm) 0.92 (t, 3H), 1.11-1.80 (broad s, 28H), 3.85 (t, 2H), 6.78 (s, phenol ring 4H). Formaldehyde was purchased from Yakuri Pure Chemicals Co. as 37 wt.% aq. solution and used as-diluted to 1 wt.%.

3.2. Surface pressure-area (π-A) isotherm of monolayer and LB deposition

A film balance system NLE-LB200-MWC (Nippon Laser and Electronics LAB, Moving Wall Method, trough size, 80×585 mm) was used for measuring surface pressure as a function of molecular area. Isotherms were taken at a compression rate of 50mm/sec and at ambient temperature. Chloroform was employed as spreading solvent. Monolayers were spread on pure water subphase or on 1 wt.% aq. formaldehyde subphase and incubated for 10 min before starting the compression. The deposition of monolayer was performed in the vertical mode. The transfer onto solid substrates was carried out at a surface pressure of 30mN/m and at a lifter speed of 50mm/min. The employed substrate was silicon wafer plates (Mitsubishi co.) for the measurement of SMM.

3.3. Brewster angle microscopy (BAM)

A commercial Brewster angle microscope miniBAM, manufactured by Nanofilm Technologie GmbH (Göttingen-Germany), was used to study the morphology of the monolayer. The Brewster angle microscope was positioned over the film balance on a specially designed frame structure, which makes it possible to move the Brewster angle microscope easily along the length of the film balance. The location of the Brewster angle microscope along the film balance makes it possible to visualize any inhomogeneity in the overall film.

3.4. Scanning Maxwell-stress microscopy (SMM)

Fig. 4 shows the schematic diagram of the setup of SMM. A DC-biased AC voltage $V_t = V_{DC} + V_{AC} \sin\omega t$ is applied to a conductive AFM tip, which is normally held 20~30 nm away from the sample surface by monitoring the oscillating electric force on the tip with the second harmonic frequency 2ω.

There also appears a fundamental force component F_ω, which is proportional to the surface potential difference between the sample and the tip. We scan the tip over the surface typically at the rate of 0.1 Hz/line, while the tip-surface distance is controlled to give a constant $F_{2\omega}$. The actual SMM system consists of a commercial AFM (Nanoscope, Digital Instruments) and a double-lock-in amplifier to selectively detect the oscillating electric force signals. We used AFM cantilevers with sharpened tips purchased from Olympus (OMCL-RC-800-PSA) with a spring constant of k=0.37 N/m. The AC voltage was 3 V_{pp} at 7.2 kHz, which should be chosen well below the resonance frequency of the cantilever. The resultant oscillation was about 2 nm or so at the closest approach of the tip to the surface. The lateral resolution was 10 nm with the potential sensitively of 1 mV.

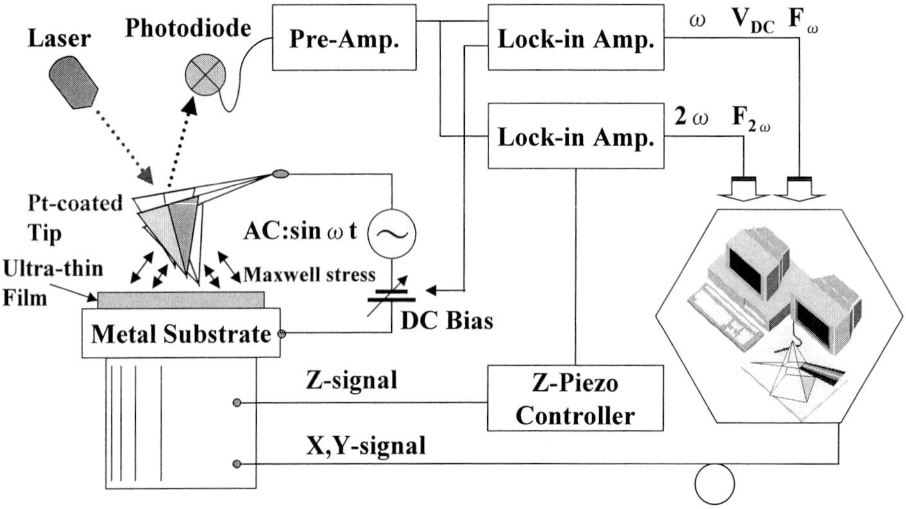

Fig. 4 Schematic diagram of the SMM.

4. Results and Discussion

4.1. Surface pressure-area (π-A) isotherm of monolayer

Chloroform solutions (1 mM) of the amphiphilic phenols, *o*-HP, *m*-HP, and *p*-HP, were spread on pure water or on 1 wt.% aq. formaldehyde. Fig. 5 shows the π-A isotherms of *o*-, *m*-, and *p*-HP monolayers on the aq. formaldehyde at pH 3.5. The monolayers of *m*- and *p*-HP showed the collapse pressure of over 35mN/m, while the *o*-HP monolayer revealed low collapse pressure of ca. 15mN/m. The collapse of *o*-HP monolayer at the relatively low surface pressure can be understood owing to the relatively low hydrophilicity of the phenol moiety of *o*-HP due to the intramolecular hydrogen bond. The isomers also showed different monolayer phases, i.e., *p*-HP monolayer showed relatively the condensed phase, while those of *o*- and *m*-HP revealed the expanded phase. In particular, *o*-HP monolayer developed positive surface pressure upon the monolayer spreading. From the above result, we estimate that the hydroxy group of *p*-HP is directed effectively toward inside the water subphase and the hydrophobic chains are packed most closely. Interestingly, a spontaneous increase of surface pressure was found in the monolayers on the aq. formaldehyde subphase. The increase of surface pressure continued to about 2 hr after the monolayer spreading and reached the plateau at surface pressure of ca. 8 mN/m in case of *p*-HP. This result indicates that, as known in bulk phenolic polymers, the formation of methylolphenols and network structures through their condensation reaction proceeds partly at the air-water interface.

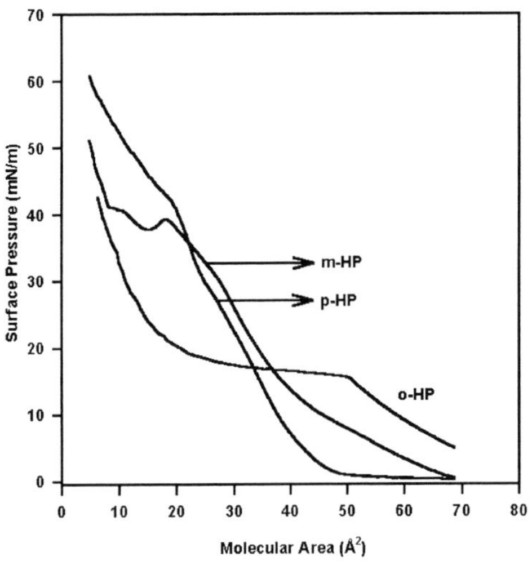

Fig. 5 Surface pressure area (π-A) isotherms of *o*-, *m*-, and *p*-HP on aq. 1% formaldehyde at pH=3.5.

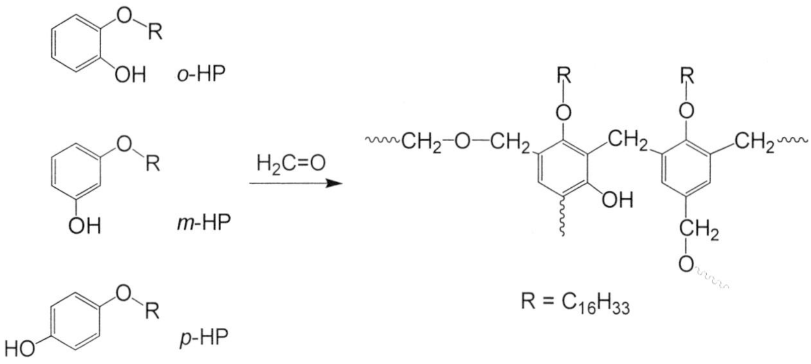

Fig. 6 An illustration of network structure from phenol-formaldehyde condensation reaction.

4.2. BAM images of molecular aggregates at the air-water interface

The images of the HP monolayers on 1 wt.% aq. formaldehyde subphase were observed by BAM. Any noticeable morphology of *o*-, *m*- and *p*-HP domains was not observed by BAM at the surface pressure lower than collapse pressure. However, very different images were monitored according to the structural isomers when the surface pressure reached around the collapse point. The BAM images are shown in Figure 7-9.

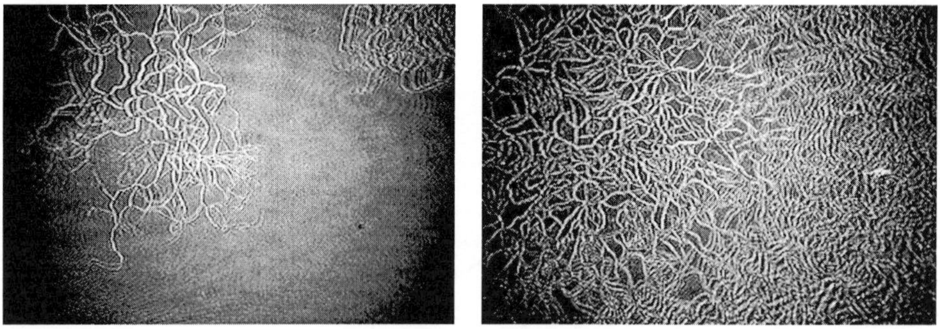

Fig. 7 BAM images of the *o*-HP monolayer spread on aq. 1 wt.% formaldehyde: (a) at 16 mN/m of surface pressure and (b) at 18 mN/m of surface pressure.

Fig. 7 shows the thread-like images, which are observed at the surface pressures of 16 mN/m (7a) and 18 mN/m (7b). As estimated from the π-A isotherm, *o*-HP monolayer shows relatively low collapse pressure. Therefore, the image appeared at the surface pressure of 16 mN/m can be understood as collapsed aggregate structure. The particular aggregate structure is supposed to depend on the molecular structure of *o*-HP, the extent of condensation reaction with formaldehyde, the structures of the methylolphenols, and partly networked intermediates. However, the molecular architecture for the thread-like images is not clear at the molecular level. In cases of *m*-HP and *p*-HP monolayers, the characteristic images appeared at higher surface pressure than that of *o*-HP monolayer. It is understandable because the π-A isotherms show higher collapse pressure in case of *m*-HP and *p*-HP monolayers.

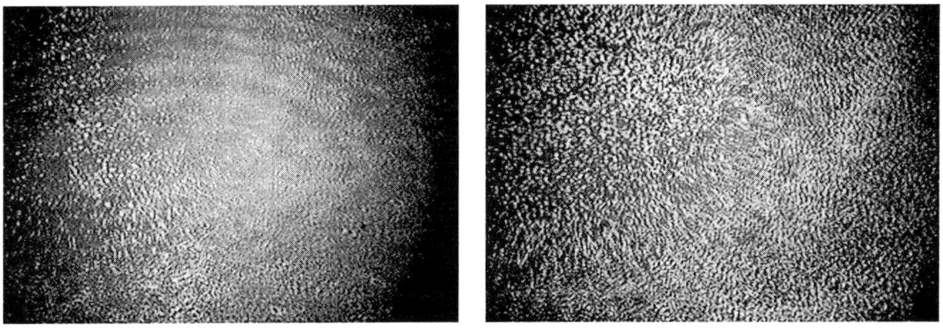

Fig. 8 BAM images of the *m*-HP monolayer spread on aq. 1 wt.% formaldehyde: (a) at 38 mN/m of surface pressure and (b) at 40 mN/m of surface pressure.

Figure 8 shows the dots-like images from *m*-HP monolayer, which was observed at the surface pressures of 38 mN/m (8a) and 40 mN/m (8b). Meanwhile, Figure 9 shows images similar to blood vessels, which was observed from *p*-HP monolayer at the surface pressures of 35 mN/m (9a) and 45 mN/m (9b). As mentioned above, the statement from one molecule to characteristic aggregate structure is very difficult at this point. However, when considered together with the π-A isotherm and the spontaneous increase of surface pressure, the images of *p*-HP system seem to be strongly influenced by methylolphenols and network intermediates.

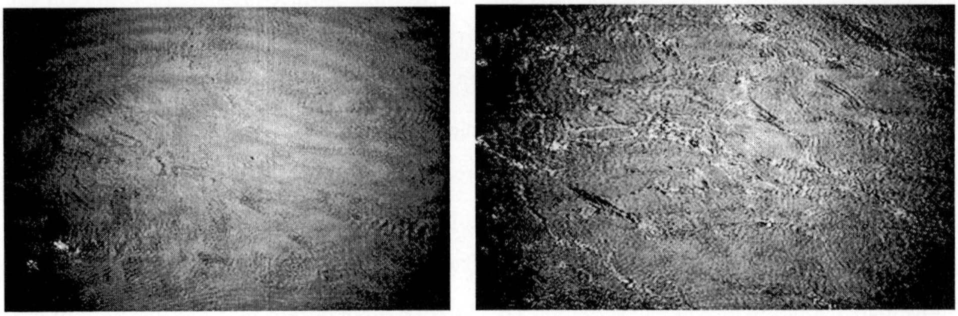

Fig. 9 BAM images of the *p*-HP monolayer spread on aq. 1 wt.% formaldehyde: (a) at 35 mN/m of surface pressure and (b) at 45 mN/m of surface pressure.

4.3. SMM images of LB monolayer films deposited on solid substrate

Figs. 10-12 show the SMM topography and surface potential images of the HP LB films deposited on silicon wafer substrates after monolayer collapse. As mentioned in Introduction, the SMM is a variant of AFM operated in the non-contact mode, which can image the distribution of surface charge and potential over ultra-thin films with a nanometer scale resolution [10]. SMM allows simultaneous observation of electrically imaged topography and surface potential based only on the harmonic analysis of forced oscillations of the cantilever driven by an external AC-voltage. As for the potential image, the potential distribution of substrate was homogeneous over the substrate surface before the LB film was deposited. The spatial distribution of the surface potential was created after the monolayer formation on the Si wafer, because the surface potential would be different between the monolayer film and the substrate. The difference of the surface dipole moments is due to the different kinds of functional moieties in the molecule [11]. Together with this intramolecular contribution on the surface dipole moments, the aggregated structure of the molecules can create the change of surface potential.

The SMM images of the *o*-HP-formaldehyde LB film shown in Figure 10 could be evaluated together with the thread-like BAM images. When considered the image scale and topographic height, the SMM image was estimated as beginning stage for much entangled thread-like BAM images. Although the surface potential image is well matched with the topographic image at large, different details are partly seen. This image difference tells that the same topographic structure can be differently viewed when the aggregate was built by different kinds of molecules. In case of *m*-HP-formaldehyde system as shown in Figure 11, the SMM image makes clear that one dot-like image by BAM is composed of many smaller aggregates. And, the smaller aggregates are seen as grown up to 100 nm high.

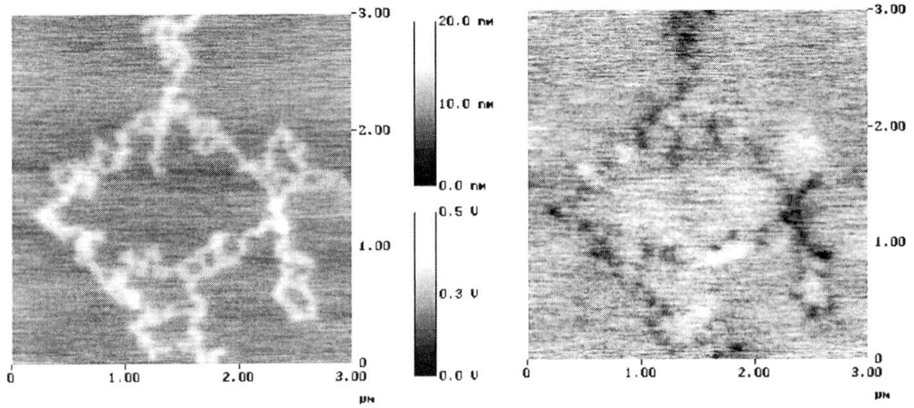

Fig. 10 The SMM images of the *o*-HP LB monolayer film on silicon wafer substrate: (a) topography and (b) surface potential of the *o*-HP LB film deposited from aq. 1 wt.% formaldehyde subphase.

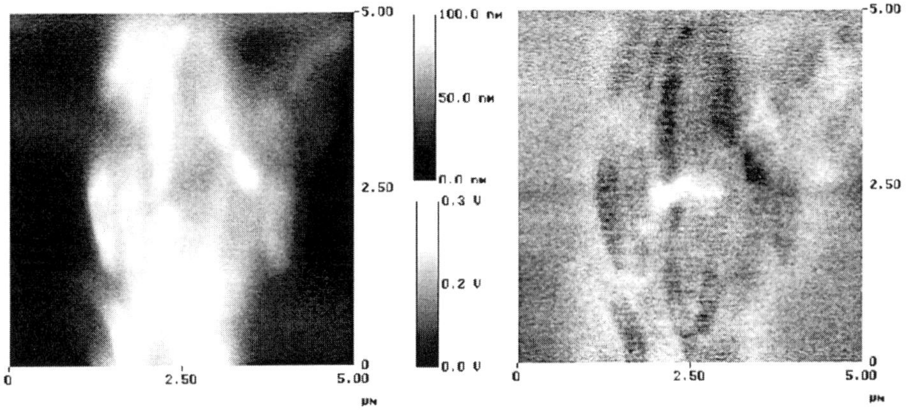

Fig. 11 The SMM images of the *m*-HP LB monolayer film on silicon wafer substrate: (a) topography and (b) surface potential of the *m*-HP LB film deposited from aq. 1 wt.% formaldehyde subphase.

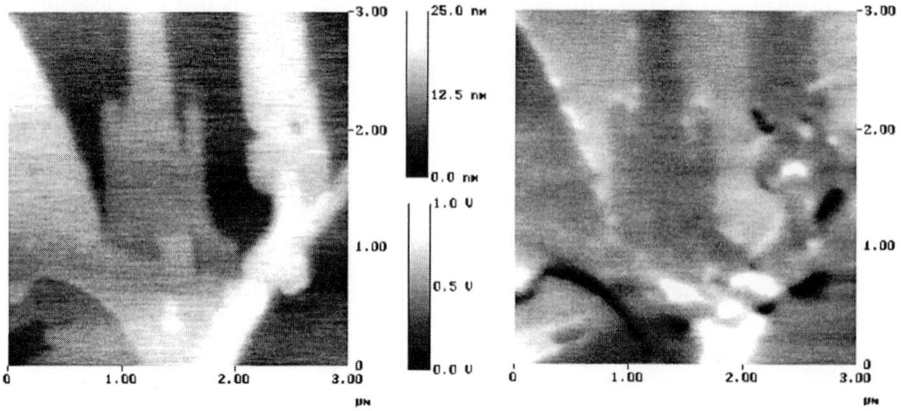

Fig. 12 The SMM images of the *p*-HP LB monolayer film on silicon wafer substrate: (a) topography and (b) surface potential of the *p*-HP LB film deposited from aq. 1 wt.% formaldehyde subphase.

Figure 12 shows the SMM images of *p*-HP-formaldehyde aggregates. The image is also different from the images by *o*-HP-formaldehyde or *m*-HP-formaldehyde LB films. The *p*-HP-formaldehyde monolayer at the air-water interface has shown most reactive from π-A isotherm and spontaneous increase of surface pressure. Therefore, much condensation intermediates are thought to be related with the formation of the aggregate structure of the *p*-HP-formaldehyde LB film.

5. Conclusion

We have clearly demonstrated the different aggregate structures of isomeric phenols-formaldehyde monolayers by BAM and SMM. The synthetic amphiphilic phenols of structural isomers, *o*-, *m*- and *p*-HP, showed characteristic monolayer properties on the aq. formaldehyde subphase. The aggregate patterns are supposed to give a clue to understand the microscopic morphology of the commercially important phenolic resins. The micron-sized domains observed by SMM can be related to the surface or interior morphology of the bulk phenolic resins.

6. Acknowledgment

The authors wish to thank Dr. H. Yokoyama for invaluable assistance.

7. References

1. H. Möhwald, Annu. Rev. Phys. Chem., 41 (1990) 441.
2. H.M. McConnell, Annu. Rev. Phys. Chem., 42 (1991) 171.
3. S. Hénon and J. Meunier, Rev. Sci. Instrum., 62 (1991) 936.
4. D. Hönig and D. Möbius, J. Phys. Chem., 95 (1991) 4590.
5. J. Als-Nielsen and K.K. Kjaer, In Phase Transition in Soft and Condensed Matter, T. Riste and D. Sherrington (Eds.), NATO Ser. B, Vol. 211, Plenum Press, New York, 1989, pp 251-321.
6. J. Als-Nielsen and H. Möhwald, In Handbook on Synchroton Radiation, S. Ebashi, M. Koch and E. Rubenstein (Eds.), Elsevier/North-Holland, Amsterdam, 1991, pp 1-53.
7. R.M. Kenn, C. Böhm, A.M. Bibo, I.R. Peterson and H. Möhwald, J. Phys. Chem., 95 (1991) 2092.
8. M. Paudler, J. Ruths and H. Riegler, Langmuir, 8 (1992) 184.
9. R. Reiter, H. Motschmann, H. Orendi, A. Nemetz and W. Knoll, Langmuir, 8 (1992) 1784.
10. R. Mendelsohn, J.W. Brauner and A. Gericke, Annu. Rev. Phys. Chem., 46 (1995) 305.
11. P.J. Werkman, A.J. Schouten, M.A. Noordegraaf, P. Kimkes and E.J.R. Sudhölter, Langmuir, 14 (1998) 157.
12. U. Gehlert, G. Weidemann, D. Vollhardt, G. Brezesinski, R. Wagner and H. Möhwald, Langmuir,14 (1998) 2112.
13. R. Deschenaux, S. Megert, C. Zumbrunn, J. Ketterer and R. Steiger, Langmuir, 13 (1997) 2363.

14. M.A.C. Stuart, R.A.J. Wegh, J.M. Jroon, E.J.R. Sudhölter, Langmuir, 12 (1996) 2863.

15. D. Vollhardt, Adv. Colloid Interface Sci., 64 (1996) 143.

16. D. Hönig, G.A. Overbeck and D. Möbius, Adv. Mater., 4 (1993) 419.

17. D. Vollhardt and V.J. Melzer, J. Phys. Chem. B, 101 (1997) 3370.

18. V. Melzer and D. Vollhardt, Phys. Rev. Lett., 76 (1996) 3770.

19. G.A. Overbeck, D. Hönig and D. Möbius, Thin Solid Films, 242 (1994) 231.

20. G.A. Overbeck and D. Möbius, J. Phys. Chem., 97 (1993) 7999.

21. G. Brezesinski, E. Scalas, B. Struth, H. Möhwald, G. Bringezu, U. Gehlert, G. Weidemann and D. Vollhardt, J. Phys. Chem., 99 (1995) 8755.

22. G.A. Overbeck, D. Hönig, L. Wolthaus, M. Gnade, D. Möbius, Thin Solid Films, 242 (1994) 26.

23. J. Läuger, C.R. Robertson, C.W. Frank and G.G. Fuller, Langmuir, 12 (1996) 5630.

24. M.C. Friedenberg, G.G. Fuller, C.W. Frank and C.R. Robertson, Langmuir, 12 (1996) 1594.

25. E.K. Mann, S. Hénon, D Langevin and J. Meunier, Phys. Rev., E 51 (1995) 5708.

26. U. Gehlert and D. Vollhardt, Langmuir, 13 (1997) 277.

27. C. Lautz, Th.M. Fischer and J. Kildea, J. Chem. Phys., 106 (1997) 7448.

28. T. Reda, H. Hermel and H.D. Höltje, Langmuir, 12 (1996) 6452.

29. M.N.G. de Mul and J.A. Mann, jr., Langmuir, 14 (1998) 2455.

30. G. Binnig, H. Rohrer, Ch. Gerber and E. Weibel, Appl. Phys. Lett., 40 (1982)178.

31. D. Rugar and P. Hansma, Phys. Today, 43 (1990) 23.

32. W.A. Ducker, T.J. Senden and R.M. Pashley, Nature, 353 (1991) 239.

33. D. Sarid, Scanning Force Microscopy, Oxford University Press, Oxford, 1991.

34. Y. Martin, D.W. Abraham and H.K. Wickramasinghe, Appl. Phys. Lett., 52 (1988) 1103.

35. J.E. Stern, B.D. Terris, H.J. Mamin and D. Rugar, Appl. Phys. Lett., 52 (1988) 2717.

36. B.D. Terris, J.E. Stern, D. Rugar and H.J. Mamin, Phys. Rev. Lett., 63 (1989) 2669.

37. F. Saurenbach, D. Wollmanm, B.D. Terris and A.F. Diaz, Langmuir, 8 (1992) 1199.

38. M. Nonnenmacher, J. Greschher, O. Wolter and R.J. Vac. Sci. Technol., B9 (1991) 1358.

39. J.M.R. Weaver and D.W. Abraham, J. Vac. Sci. Technol., B9 (1991) 1559.

40. M. Nonnenmacher, M.P. O'Boyle and H.K. Wickramasinghe, Appl. Phys. Lett., 58 (1991) 2921.

41. H. Yokoyama, M.J. Jeffery and T. Inoue, Jpn. J. Appl. Phys., 32 (1993) L1845.

42. M. Fujihira, H. Kawate and M. Yasutake, Chem. Lett., (1992) 2223.

43. H. Yokoyama and T. Inoue, Thin Solid Films, 242 (1994) 33.

44. J.U. Kim, B.-J. Lee and Y.-S. Kwon, Thin Solid Films, 327-329 (1998) 486.

45. J. U. Kim, B.-J. Lee and Y.-S. Kwon, Bull. Korean Chem. Soc., 18 (1997) 1056.

Novel Methods to Study Interfacial Layers
D. Möbius and R. Miller (Editors)

NMR METHODS FOR STUDIES OF ORGANIC ADSORPTION LAYERS

Monika Schönhoff

Max-Planck-Institute of Colloids and Interfaces, D-14424 Potsdam/Golm

1. Introduction

1.1. *Scope of this Review and Related Fields*

This review provides an overview of the possibilities to apply NMR methods in the investigation of organic adsorption layers at solid interfaces, focussing on adsorption from aqueous solutions. In adsorption studies of compounds such as surfactants, polymers or lipids, standard methods of investigation are for example optical reflection techniques, infrared spectroscopy and X-ray or neutron reflectivity. These reveal pure structural parameters such as layer thickness, adsorbed amount, surface roughness or molecular orientations. The variety of local dynamic information achievable from various NMR methods makes it an especially attractive goal to apply magnetic resonance techniques to interfaces as well, however, with a low sensitivity and motional aspects as the main limitations, investigations of large organic molecules in interaction with interfaces have been rare for a long time. NMR methods have only in the past decade started to become a tool to obtain unique information on adsorption layers, since instrumental developments led to increased sensitivity making studies on adsorption layers feasible. Furthermore, specific methodologies for surfaces have been developed, and finally the preparation of adsorption layers on colloidal particles has advanced to a stage where even complex composite layers can be handled.

Still, NMR investigations of adsorption layers are few seen in contrast to the wealth of different compounds and surface structures investigated with other, truly monolayer sensitive techniques. They are not yet introduced as a standard characterisation into the field of adsorption research, and very often the question is asked by experimentalists, as to which kind of information can be obtained by an NMR experiment. The answer to this question is not straightforward, since the type of NMR experiments applicable to a particular adsorption layer will depend strongly on the structure and dynamics of the layer. Parameters determined in different NMR techniques include for example motional correlation times, orientational order parameters, or bound fractions of segments. Adsorption layers can exhibit a large range of molecular mobility and structures, and can in many cases not unambiguously be described by a solid or liquid state. The most important task in NMR adsorption layer studies is thus to choose suitable methods according to the system under investigation, and the underlying principles and basics of NMR are described in this article. Aimed at experimentalists in the field of organic adsorption layers, the purpose is to provide an overview of the current possibilities and achievements of NMR in surfactant, lipid and polymer films. The focus is on methodological aspects in order to clarify the question of applicability of different kinds of NMR techniques in dependence on the widely varying conditions in different classes of adsorption layers.

The review is limited to the field of NMR applications to interfaces which form model systems for layers at planar solid/liquid interfaces, and it focuses on colloidal templates as substrates for adsorption from aqueous solution. Some studies of grafted polymers and chemisorbed alkyl chains, as well as investigations of dried adsorption layers are included to provide a broader range of methodologies. A previous review by Blum from 1994 had summarised the existing investigations of polymer and surfactant monolayers up to then, also including studies of smaller moieties like coupling agents [1]. Another review by Griffiths and Cosgrove had summarised the 1H work of the Bristol group on polymer adsorption [2].

Here, special emphasis is put on more recent developments, including complex films, and a particular focus lies on introducing the reader to the principles governing the applicability of different NMR methods in dependence of the mobility of the system. The first part contains some general considerations of spin resonance in adsorption layers. The second part provides the basics of the main NMR methods, which have been introduced to the investigation of adsorption layers. Selected aspects relevant for interfaces are explained, while for detailed NMR theories or technical aspects the reader is referred to standard textbooks. In the third part, an overview of recent experimental studies is given, where a range of examples are described from a methodological viewpoint to demonstrate the application of the basic principles explained in the second part.

1.2. Chances and Challenges of NMR Applications to Interfaces

The Motivation for performing NMR investigations at interfaces is given by the tremendously large range of magnetic resonance methods successfully employed to volume phases of organic compounds, such as in solutions, melts, liquid crystalline, or glassy systems. Especially concerning dynamic aspects and local molecular information, magnetic resonance has provided a large number of different methods, each leading to unique information. Knowledge obtained for volume systems of organic materials is thus manifold and ranges from the analysis of conformations (trans, gauche), bond angles or interatomic distances, to motional information like diffusion, mobility of segments, and furthermore even to information on the mesoscopic scale like aggregation, local order in liquid crystalline systems, hydration and interaction with small molecules. This variety in combination with the local site-selectivity implied by investigating a specific nucleus, are the main advantages of NMR methods.

The major complications are given by the low sensitivity and by motional aspects governing the choice of applicable methods: The mobility of surface attached large molecules is generally significantly reduced, leading to spectra which are inhomogeneously broadened, however, in most cases suitable narrowing techniques can be applied. On the other hand, segments in adsorption layers can exhibit fast segmental motions and lead to narrow liquid spectra. In practice, many layers show a gradient of mobility and are thus covering the range from solid to liquid materials within nanometers. Thus, in the study of adsorption layers, the otherwise separated fields of high resolution and solid state NMR meet. Appropriate methods from either field have to be chosen to address the questions of interest.

Sensitivity is a major issue, since investigations of materials at interfaces are limited by the intrinsically low sensitivity of NMR combined with the small amount of compound provided by a monolayer. The low sensitivity is due to the relatively small spacing of the energy levels of nuclear spins, ΔE. For spin I = ½ nuclei the spin energy levels split up in a magnetic field $\boldsymbol{B} = B_0\, \boldsymbol{z}$ according to the Hamiltonian of the Zeeman interaction

$$H_Z = -\mu_Z B_0 = -\gamma\hbar B_0 I_Z, \qquad (1)$$

where γ is the gyromagnetic ratio of the nucleus, μ_Z is the magnetic moment of a spin, and I_Z can assume the values ½ or -½. Transitions between these states are in pulsed NMR induced by irradiation with a resonant radiofrequency pulse of the Larmor frequency $\omega_L = \gamma B_0$.

In thermal equilibrium the spin population N_β in the upper state I_Z = -½, is given by a Boltzmann distribution: $N_\beta = N_\alpha \exp(-\Delta E/kT)$, leading to an excess fraction of spins in the ground state given by

$$\frac{N_\alpha - N_\beta}{N_\alpha + N_\beta} \approx \frac{\Delta E}{2kT} = \frac{\gamma\hbar B_0}{2kT} \qquad (2)$$

This fraction can be very low, i.e. about 10^{-5} to 10^{-4} at room temperature and a field of 9.4 T, so that the number of spins which can be excited and detected, is extremely small. This is the reason why NMR, in contrast to UV or IR spectroscopy, is not sensitive to a single monolayer.

A crucial requirement is therefore a suitable design of the adsorption geometry, maximising the surface area per volume, which can be realised by colloidal dispersions or porous materials, as described in the following chapter.

1.3. Design of Systems: NMR on Colloidal Materials and Related Fields

A broad field of application of NMR methods is given by adsorption studies of small solvent or gas molecules to porous materials, and since rather small pore sizes of ~ nm are available, sensitivity is no major problem. Experiments of porous materials have a long tradition, being driven by a large number of practical applications, e.g. in catalysis, chromatography, geology or oil recovery. For monitoring covalently attached groups at the surface of porous materials in order to study surface chemical aspects, magic angle spinning is applicable. Porous materials saturated with solvent give liquid NMR signals of the surface attached solvent

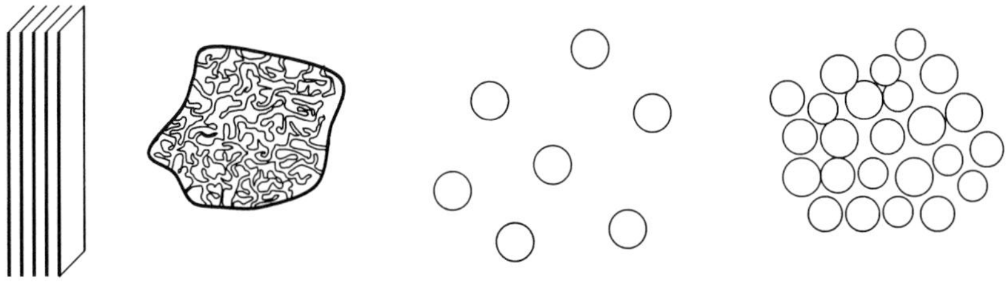

Fig 1. Geometries for the formation of an internal solid surface in volume samples: Stacked planar glass slides, porous materials, nonporous colloidal particle dispersions, and densely packed colloidal particles.

molecules, which are in exchange with free solvent. Though the high surface areas of porous materials are attractive, the length scales and geometrical requirements for adsorbed molecules of larger dimensions to interfaces are very different: Polymeric or long chain amphiphilic molecules require a minimum pore size for controlled monolayer formation. A number of investigations in porous materials were performed using comparatively large pore volumes filled with for example liquid crystalline phases of amphiphiles [3], polymers [4, 5], or surfactants in mesophase silicates [6]. In these systems confined geometry effects on volume materials are studied. For the investigation of defined monolayers, however, or even of multilayer build-up, the convex geometry is preferable. The convex interface of colloidal particles is more easily accessible to the adsorbing species, and better methods exist for additional structural characterisation of the layers, e.g. light scattering.

The choice of a suitable particle size for NMR investigations is a compromise between two opposed requirements: On the one hand, controlled layer formation requires large particles, since a model surface should have a radius *R*, which is large enough not to influence the properties of the adsorption layer as compared to planar substrates. In particular for polymer adsorption *R* should be large compared to the molecular dimensions (i.e. the radius of gyration in solution, R_g, of typically some tens of nm). This is a very critical aspect especially for the controlled formation of multilayers. On the other hand, achieving sufficient sensitivity requires small particles. In practice, the optimum choice of particles is a compromise of the arguments above, and largely depending on the system, the method and the nucleus: The majority of investigations of monolayers has been performed on small polydisperse silica of radii in the range of 5- 20 nm, i.e. not large compared to typical polymer dimensions. Though in the case of $R \approx R_g$ the curvature can be expected to strongly influence the adsorption, and for $R < R_g$ it is critical to speak of a defined layer, small particles are sufficient to at least serve as model systems for the influence of adsorption on the dynamics of a polymer chain. On the other hand, with comparatively sensitive methods, particles with a radius as large as some 100 nm have been employed (see chapters 3.1.2, 3.4.3, and 3.5.2), and even 2H 2-dimensional spectra were taken on 320 nm particles (see chapter 3.3). As the most extreme case, even planar substrates were applied, where a dense packing of 50 thin silica slides provided a 2H spectrum of perdeuterated layers of 80Å thickness within several days experiment time [7].

In a colloidal dispersion, the geometric parameters of the system determining the sensitivity are the particle radius and the packing density. Additional influence is imposed by the NMR parameters such as the gyromagnetic ratio and the natural abundance (or labelled fraction, respectively) of the nucleus employed, and most importantly the width of the lines, which can vary over orders of magnitude leading to correspondingly decreased intensities for motionally restricted molecules.

In a number of layer systems, experiments are not directly limited by the small total number of spins, but rather by the overlap of the signal of a few surface spins with a large signal of spins in the volume. Even with suitable materials, e.g. 1H-free particles and deuterated solvent chosen for 1H investigations of interfacial material, in practice still suppression techniques have to be employed to reduce the comparatively large signal of residual volume protons.

A related field of research where NMR investigations have yielded a large variety of information are 'soft' colloidal volume systems, so-called association colloids, composed of amphiphilics (surfactants, lipids) or copolymers, e.g. block-copolymers. Structures such as micellar solutions, microemulsions, or liquid crystalline phases possess internal interfaces where the arrangement of molecules can be governed by similar principles as at a solid/liquid interface. NMR methods have very successfully been employed in this area, as has been reviewed [8, 9], and some ideas and approaches can be transferred to liquid interfacial layers in colloidal particle dispersions.

Electron spin resonance (ESR) has an even longer standing history in the investigation of adsorption layers than NMR methods, due to the higher sensitivity. Its disadvantage is to require a spin probe, that can disturb the local molecular environment, which is a critical issue on molecular dimensions in monolayers. ESR and NMR methods can be considered complementary sources of information, and some of the very early NMR studies of grafted polymers at interfaces were in fact conducted complementarily to ESR [10]. A review of the application of ESR techniques to polymers at solid interfaces is given by Hommel [11].

2. Basics of NMR in Colloid and Interface Science

The most important principles in NMR investigation of adsorption layers at solid/liquid interfaces are the averaging of interactions and the relaxation of spins as measures of segmental mobility, while only in a few systems spectral shifts were analysed. As mentioned before, the motional state of surface adsorbed molecules can range from solid to liquid type behaviour, determining the NMR methods applicable to the system. The difference between the solid or liquid state in NMR is established by anisotropic interactions which cause broad spectra for solids, whereas in liquids molecular motions are averaging this anisotropy in time and are causing narrow spectra.

In this chapter these interactions and the relaxation of spins are described with respect to their relevance in adsorption layer studies. A detailed account of the foundations of NMR, practical aspects of Fourier transform (FT) NMR methods, and descriptions of pulse sequences, however, can be found in textbooks of liquid or solid state NMR, respectively.

2.1. Spectroscopic Studies: Chemical Shifts and Interactions

2.1.1. 1H and 13C NMR

Nuclei. The nuclei best accessible to NMR investigations of organic materials are 1H, 13C, and 2H, the latter with appropriate isotope labelling. Due to their large gyromagnetic ratio and natural abundance, 1H measurements are most sensitive, see table 1, and generally cause sufficient signals in adsorption layer studies.

13C studies have, in spite of the low natural abundance (~ 1%) and the lower γ, in combination leading to an absolute sensitivity of about 3 orders of magnitude lower as compared to 1H, become the method of choice for investigations of purely solid layers. The reason is a larger range of chemical shifts, as described further below. 2H investigations can only be performed with isotope labelling, an advantage in this case is that a selective introduction of the 2H label to the functional group of interest leads to very localised dynamic information.

Nucleus	*I*	*Natural abundance (%)*	γ $(10^7 radT^{-1}s^{-1})$
1H	½	99.98	26.75
13C	½	1.11	6.73
2H	1	0.015	4.11

Table 1: Properties of the spins of interest in organic materials.

Interactions. In addition to the Zeeman interaction H_Z, which is defining the spectrometer frequency ω_0, a number of additional interactions determine the spectrum, and the resulting Hamiltonian becomes:

$$H = H_Z + H_Q + H_D + H_{CS} + H_J . \qquad (3)$$

H_Q, H_D, H_{CS} and H_J describe the quadrupolar, dipolar, chemical shift and indirect electron coupled interactions, respectively, and are listed according to their typical magnitude. The Zeeman term was given by equ. (1) and is not influenced by the local environment. Quadrupolar interactions occur for nuclei with $I > ½$ only and are described in the next chapter for the case of 2H with $I = 1$, thus the dominant interaction for 1H and 13C is given by H_D and H_{CS}. The dipole-dipole interaction between spins is a strongly anisotropic interaction. Similarly, the chemical shift depends on the orientation of the molecule with respect to B_0, this part is the so-called chemical shift anisotropy (CSA). In a static sample without any molecular motions both interactions are causing broad solid state spectra, where the resonance frequency depends on the angle formed with B_0. The linewidth then corresponds to the typical strength of the interaction, Δ_D or Δ_{CSA}, which can be as large as several 10 kHz. In addition to anisotropic interactions, the lineshape can be affected by a variation of the magnetic susceptibility at the interface between two different materials. This leads to local gradients in $\boldsymbol{B}$, causing line broadening.

Liquid state. In the case of rapid isotropic motions being present, the anisotropic interactions can be averaged. This is the case if the correlation time of the motions, τ_c, is

$$\tau_c << 1/\Delta_D, 1/\Delta_{CSA} \qquad (4)$$

and thus the isotropic motions are fast enough to average the anisotropic interaction. In this case the resonances for a specific functional group occur at an averaged frequency which is determined by the isotropic contribution of the chemical shift. The spectrum becomes a liquid-state or high resolution spectrum, covering only the range of the isotropic chemical shifts.

For studies of adsorption layers the fact whether or not correlation times are fast enough to average the interactions can lead to conclusions about the mobility. Pure spectral investigations in the sense of detecting changes in the chemical shifts due to adsorption are limited to a few systems, e.g. solid layers. The reason is that in samples of known composition, not the analysis of chemical shifts of different functional groups is of interest, but shift changes due to intermolecular interactions, which are considerably smaller. For 1H spectra the total chemical shift range for different functional groups is about 10 ppm, with changes due to different local chain arrangements being considerably smaller.

The advantage of 13C investigations over 1H lies in the large total chemical shift range of about 200 ppm, which makes the spectra very sensitive to the local environment. Dipolar interactions can be suppressed by working with 1H decoupling techniques. In ordered alkyl chains of amphiphiles shift changes due to adsorption could be resolved, which are due to a change of the fraction of trans conformations. In swollen polymer films, the direct molecular environment is too heterogeneous to allow the detection of small chemical shifts differences.

Solid state. In a number of adsorption layers, particularly at the solid-air interface, the organic compound is in the solid state, i.e. spectra are significantly broadened compared to solution due to both dipolar interaction and chemical shift anisotropy (CSA). The Hamiltonians describing both interactions contain the anisotropy term $3cos^2\vartheta-1$, where ϑ is the angle between the relevant internuclear vector and the field B_0. Narrowing techniques can be applied to average these interactions. The most common procedure today is magic-angle-spinning (MAS), where the sample rotates about an axis inclined to B_0 by $\vartheta_M = 54.7°$, the so-called 'magic angle' (see Fig. 2).

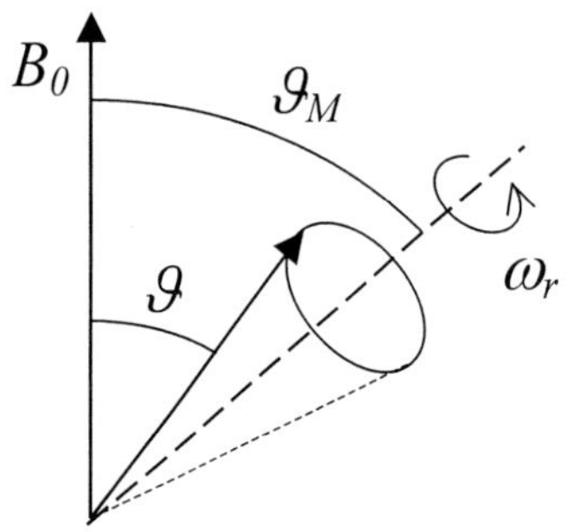

Fig. 2: Time averaging of an internuclear vector due to rotation of the sample about the magic angle.

The Hamiltonian then becomes a periodic function of time, and provided sufficiently fast rotation, the time average of any angle ϑ defining the interaction strength would become $<\vartheta> = \vartheta_M$, resulting in $<3cos^2\vartheta_M-1>=0$, and thus leading to zero interaction and narrow high resolution resonances. Typical interaction strengths are about 25 kHz for 13C CSA or for 1H-13C dipolar coupling, and about 50 kHz for 1H-1H dipolar interactions. Organic materials often provide partial mobility, e.g. by rotations of mobile functional groups, so that the motional average giving the remaining interaction, $\Delta_m = \langle H \rangle_{motions}$, can be substantially less than the above values, and thus easier to average by rotation in MAS. The required frequency of rotation is $\omega_r >> \Delta_m/\hbar$. Standard equipment today provides rotation speeds of some kHz, and the most advanced probes allow up to 30 kHz. In cases where MAS does not lead to sufficient line narrowing, it can be further combined with multi-pulse sequences, which are averaging the dipolar interaction only.

In adsorption layer studies MAS techniques have been applied to dried samples only, i.e. adsorption layers at the solid/air interface or chemisorbed alkyl chains. One reason is that fast spinning is difficult to accomplish in dispersed colloidal solutions, since densely packed and mechanically stable samples are required. Here, similar as in liquid state studies of layers at the solid/liquid interface, spectral investigations provide information rather by 13C spectra than by

1H spectra due to the wider range of chemical shifts. MAS spectra of 13C are mostly acquired under cross-polarisation (CP-MAS), i.e. the dipolar interaction is employed to transfer magnetisation from excited 1H nuclei to the 13C nuclei and thus enhance the signal.

Since dipolar coupling to protons broadens the 13C resonances, it can be useful to reduce it, which is achieved by 1H decoupling applied during acquisition. Decoupling consists of the resonant irradiation and thus saturation of the protons, and causes the dipolar 13C-1H interaction to average and vanish.

A useful method to correlate 1H and 13C signals is wide-line separation: In a two-dimensional (2D) spectrum high-resolution 13C CP MAS spectra are taken in one dimension, and correlated to the 1H wide-line spectra of the protons attached to each resolvable carbon atom. This enables the correlation between the 1H mobility, as detected by the linewidth, and the 13C structure viewed by the chemical shifts.

2.1.2. 2H NMR Static and Liquid Spectra

2H NMR has proven an extremely useful method and acquired a long standing tradition in the investigation of partially ordered volume phases, such as lipid and surfactant mesophases (see e.g. [12]). It thus appears a suitable method in the investigation of adsorption layers, which form partially ordered systems as well. In contrast to spin ½ nuclei, the dominant interaction determining the spin properties, i.e. both the resonance frequencies and the relaxation mechanism, is the quadrupolar term H_Q. H_Q describes the interaction between the quadrupolar 2H nucleus and the electric field gradient (EFG) present in the C-2H bond. This is a *local* interaction, depending only on the motion and orientation of the particular C-2H bond. In contrast to this, for dipolar interactions dominating proton spectra, complex sums of dipolar interactions of spin pairs have to be calculated and the resonance frequency and relaxation of each 1H nucleus will depend on the parameters of all surrounding protons. 2H spectra and relaxation rates are therefore far easier to describe quantitatively, which is shown in this chapter.

The energy levels of a 2H nucleus with spin $I = 1$ will in an axially symmetric field gradient split up into states with $I_Z = -1, 0, 1$, leading to a doublet of transitions separated by

$$\Delta \nu_Q(\vartheta) = \frac{3}{2}\chi\frac{1}{2}\left(3\cos^2\vartheta - 1\right), \qquad (5)$$

where χ is the quadrupolar coupling constant, and ϑ the angle between the C-2H bond and the magnetic field B_0. Due to the anisotropy of the interaction, in polycrystalline samples a superposition of the resonance frequencies for all orientations ϑ is observed. Fig. 3a) shows the typical 'Pake pattern' shaped spectrum arising from this superposition for a macroscopically isotropic sample ('Powder spectrum'). The dashed lines indicate each of the two transitions with $\Delta I_Z = 1$, and the frequency distribution of each is determined by the angular dependence in equ. (5).

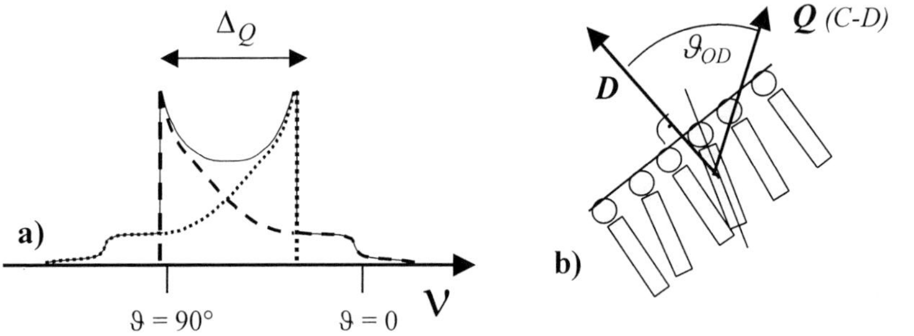

Fig. 3 a) 2H NMR static spectrum 'Pake pattern' for the case of an axially symmetric EFG. The symmetric lineshape is a result of the superposition of two transitions, indicated by the dashed and the dotted lines. The angles correspond to the dashed spectrum. b) Definition of vectors and angles in an amphiphilic aggregate.

Even in ordered phases the angle ϑ can be time dependent due to fast local motions, such as bond rotations, and the time average of equ. (5) has to be taken into account. If these motions are fast and isotropic, the spectrum will be completely averaged and collapse to a narrow liquid resonance. The interesting case is *partial* averaging by anisotropic fast motions, as it occurs in ordered chains. In fig. 3b) the case where motions relative to the director axis ***D*** are partly averaged by a correlation time of fast motions, τ_f, is considered. The angle ϑ is then split up into an angle ϑ_{QD} between the C-2H bond and the surface director ***D*** and ϑ_{DB} between ***D*** and ***B***. Anisotropic fast motions lead to partial averaging of ϑ_{QD} and representation of the time average by the order parameter $S_{QD} = \frac{1}{2}\left\langle 3\cos^2 \vartheta_{QD} - 1 \right\rangle$. The director ***D*** is not averaged with respect to B_0, and thus (5) becomes

$$\Delta\nu_{DB}(\vartheta)=\frac{3}{2}\chi\frac{1}{2}\left(3\cos^2\vartheta_{DB}-1\right)S_{QD}\cdot \tag{6}$$

The Pake pattern spectral shape is therefore preserved, but the 2H spectrum partly collapses to a narrower splitting due to the factor S_{QD}. The spacing of the maxima, the quadrupolar splitting Δ_Q, is then given by

$$\Delta_Q=\Delta\nu_Q(90°)=\frac{3}{n}\chi S_{QD} \tag{7}.$$

By an order parameter $S:=S_{QD}$ thus the residual anisotropy, which is not averaged by fast local motions, is characterised. This is the main information obtained for partially ordered systems, and can be considered a characterisation of the mobility of the C-2H bond, which on the other hand describes the chain packing order in oriented aggregates.

In practice, the situation can be more complicated, since motions around interdependent axes might lead to complex motional patterns, or the EFG may have no axial symmetry. The resulting 2H wide/line spectrum then can exhibit a modified shape, from which further information about the geometry of motions can be obtained. 2H wide-line spectra thus contain a wealth of information about the time scales and geometries of local bond motions in systems with partial motional averaging. Order parameters are established in characterising liquid crystalline phases, particularly for lipids. Typical values of S in aggregated alkyl chains of volume phases are on the order of 0.1- 0.2.

In addition to the fast local motions, also slow motional modes with a correlation time τ_s might be present, which average ϑ_{DB} in time. In volume systems this might be rotational tumbling of aggregates, or diffusion. The criterion for averaging is $\tau_S^{-1} >> \Delta\nu_Q$, and in this case the time average of ϑ_{DB} leads to a complete averaging of the quadrupolar interaction, and an isotropic (liquid state) Lorentzian line is obtained.

The principle of fast and slow motional averaging (also described in the next chapter) applies similarly to protons and dipolar interactions. 1H wide-line spectra are generally without structure, due to the superposition of a large number of dipolar interactions. This makes 2H wide-line investigations far more attractive than those of protons in spite of the lower sensitivity and necessity for isotopic labelling.

2.2. *Relaxation and Dynamics*

2.2.1. Spin Relaxation and Molecular Motion

Following the application of a radio frequency pulse, which is spectrally broad enough to excite all resonance frequencies of a particular nucleus, the system is in an excited state and the net magnetisation vector $\boldsymbol{M}$ is no longer parallel to $\boldsymbol{B_0}$. The process of equilibration with the environment is called 'relaxation' of the spin system. Transitions between spin states which are leading to relaxation are induced by local magnetic fields at the positions of the spins, which are fluctuating with an appropriate frequency. These random fields are for example induced by random molecular tumbling, rotations about bonds or other local motions. This is the fundamental basis of employing relaxation rates as measures of local molecular mobility.

Relaxation consists of two different mechanisms, namely spin-spin and spin-lattice relaxation. Spin-lattice relaxation describes the equilibration of the spins with the surroundings by a transfer of energy. A typical NMR experiment starts with the excitation of spins such that the magnetisation $\boldsymbol{M} = M_0 z$ is switched by a so-called 90° pulse, producing a magnetisation vector precessing in the x-y plane (see Fig. 4a, $t = 0$).

The decay of the magnetisation vector $\boldsymbol{M}$ to the equilibrium value $\boldsymbol{M} = M_0 z$ following the excitation is described by

$$\frac{dM_z}{dt} = -\frac{M_z - M_0}{T_1} \tag{8}$$

with T_1 the spin-lattice or longitudinal relaxation time and $R_1 = T_1^{-1}$ the corresponding rate. Furthermore, spin-spin interactions can cause a loss of coherence of single spins, such that they are not precessing in phase any more about $\boldsymbol{B_0}$. Since $\boldsymbol{M}$ is the vector sum of the individual magnetic moments of the nuclei, such coherence loss is leading to a decay of the precessing x-y component of $\boldsymbol{M}$ (see Fig. 4, $t \approx T_2$). This is the spin-spin or transverse relaxation described by

$$\frac{dM_x}{dt} = -\frac{M_x}{T_2}, \qquad \frac{dM_y}{dt} = -\frac{M_y}{T_2} \tag{9}$$

with T_2 the spin-spin relaxation time and $R_2 = T_2^{-1}$. It is generally $T_2 \leq T_1$, and the decay processes are described in Fig. 4 for the case of $T_2 << T_1$.

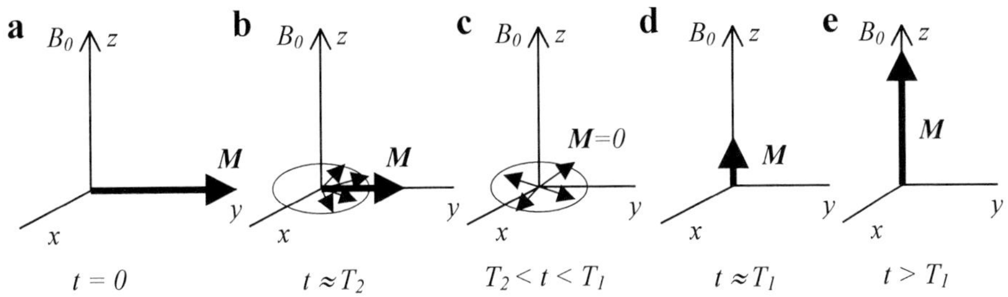

Fig. 4: The vector sum of the single spin magnetic moments, ***M***, is decaying due to coherence loss of the spins (b,c), in the case of $T_2 << T_1$ the magnetisation detected in the x-y-plane can decrease to zero (c), before T_1 relaxation becomes effective and M is increasing to the equilibrium value parallel to B_0.

In an NMR-experiment magnetisation is detected in the x-y-plane only, therefore in liquids T_2 could be measured by the time constant T_2* of the signal decay following an excitation pulse. In many cases, however, the FID decay constant, and correspondingly the linewidth are influenced by magnetic field inhomogeneities, which cause an additional FID decrease and the total linewidth (FWHM) becomes:

$$\Delta\nu = \frac{1}{\pi}\frac{1}{T_2^*} = \frac{1}{T_2} + f\left\{\frac{\partial B_0}{\partial x}\right\} \tag{10}$$

In cases of sufficiently short T_2 the linewidth is determined by relaxation, and a homogeneous Lorentzian lineshape occurs, provided that all inhomogeneous broadening interactions, such as H_D or H_{CSA} are time averaged.

Spin- spin relaxation is usually determined applying spin echoes (Hahn-Echo, CPMG method), which delete other influences such as B_0 inhomogeneity. The spin-lattice relaxation time T_1 is measured by detecting the relaxation after inverting M_0 by a 180° pulse (inversion recovery method).

The connection between spin relaxation and molecular motions is given by a description of motion by a motional correlation function. The autocorrelation function $g(\tau)$ contains the time dependent fluctuations $f(t)$ in the average

$$g(\tau) = \left\langle f(t) f(t+\tau) \right\rangle \; . \tag{11}$$

Fluctuations in dependence of frequency are given by the spectral density function $J(\omega)$, which is the Fourier transform of the correlation function

$$J(\omega) = \int_{-\infty}^{\infty} g(\tau) e^{-i\omega\tau} d\tau \tag{12}$$

In the case of 2H, where it is $I=1$ and the relaxation is dominated by the interaction of the quadrupole moment of the nucleus with an axially symmetric EFG, the relaxation rates can be expressed as:

$$R_1 = \frac{1}{T_1} = \frac{3\pi^2}{40} \chi^2 \left[2J(\omega_0) + 8J(2\omega_0)\right] \tag{13}$$

, (14)

where χ is the quadrupolar coupling constant. These equations relate experimentally determined relaxation rates to the spectral density function at the Larmor frequency ω_0 and $2\omega_0$. R_2 is furthermore dependent on the contribution at zero frequency and thus probes very slow motions, whereas R_1 is determined by fast motions only, since typical spectrometer frequencies are several hundred MHz. The interpretation of relaxation rates in terms of motional correlation times involves suitable models providing a functional dependence of $J(\omega)$. The simplest one is given by completely random motions in an isotropic system, for example rotational tumbling of small spherical molecules, described by an exponential decay of the correlation function

$$g(\tau) = \exp(-\tau/\tau_c) \tag{15}$$

leading to

$$J(\omega) = \frac{\tau_c}{1 + (\omega\tau_c)^2} \tag{16}$$

In motional narrowing, i.e. if $\tau_c\omega_0 << 1$, the simple expression

$$R_2 = R_1 = \frac{3\pi^2}{4} \chi^2 \tau_c \tag{17}$$

is obtained. Thus longer correlation times lead to faster relaxation, a result which might seem counterintuitive, but is due to the normalisation of $g(\tau)$ and $J(\omega)$.

An often employed model for the interpretation of NMR relaxation data, which was developed for amphiphilic molecules, is based on a separation of the molecular motions into two well separated time regimes, and has been termed 'two-step model' [13]: Local anisotropic motions, such as bond rotations, are 'fast motions' determining the longitudinal relaxation rate R_1. Isotropic motions, which are averaging the residual anisotropy, such as aggregate rotation or diffusion, occur on much slower time scales, described by the slow motion correlation time τ_S.

The transverse relaxation rate R_2 is dominated by these 'slow motions' due to the term $J(0)$. The spectral density function can be written as

(18)

Under the assumption of $\omega\tau_S >> 1$, a simplified expression for the relaxation rate difference $\Delta R = R_2 - R_1$ as a function of the correlation time is obtained:

$$\Delta R = R_2 - R_1 = \frac{9\pi^2}{20}\chi^2 S^2 \tau_s, \qquad (19)$$

and it becomes obvious that predominantly R_2 (which is often $R_2 \approx \Delta R$) probes the slow motions, as argued above.

For polymers, the local motional heterogeneity is represented by the use of a broad distribution of correlation times instead of a single τ_C. A typical approach is a log-normal distribution of τ_c in $J(\omega)$, from which a mean τ_c and width of the distribution can be extracted by comparison to experimental R_2 data [14]. On the other hand, polymers are suitably described by a separation into fast and slow segmental motions as well, and for example a Hall-Helfland model employing two rate constants for both, respectively, can be applied [15]. By any such model distribution applied to two relaxation rate values R_1 and R_2 determined at a given frequency ω_0, only two model parameters can be extracted. More detailed information is achievable in relaxation experiments where $J(\omega)$ is probed in dependence of frequency, which is called relaxometry [16]. This has, however, not yet been applied to polymeric or amphiphilic monolayers.

In contrast to 2H, relaxation is far more complex to describe for 1H or 13C: With spin ½ nuclei, the main relaxation mechanism is dipolar relaxation, where again complex sums over all interacting pairs of spins complicate the expressions for R_1 and R_2. 1H relaxation data determined for adsorption systems are interpreted qualitatively as a measure for mobility, since the proportionality to the spectral density function still holds.

2.2.2. Exchange and Solvent Studies

In heterogeneous systems, where molecules can occupy different sites with differing properties, the time scale of exchange between the sites is a substantial parameter. In the investigation of organic adsorption layers, different sites are often established by adsorbed versus non-adsorbed molecules, or by free solvent molecules in contrast to solvent bound in the adsorption layer. In such cases the parameters measured by NMR depend on the time scale of the exchange of molecules between the sites, τ_{ex}, with respect to the time scale of the NMR experiment, τ_{NMR}, which is given for example by the delay between pulses in echo experiments, by the spacing of gradient pulses in diffusion experiments, or by the inverse of the strength of the residual interaction, e.g. $1/\Delta_Q$ in wide-line 2H experiments. Typical values of τ_{NMR} can range from μs to some 100 ms, depending on the type of experiment.

Fig. 5: Spectral lines of a molecular species exchanging between two sites, A and B, in the case of slow exchange (solid lines). In fast exchange a single resonance frequency occurs as a weighted average of ω_A and ω_B (dashed line).

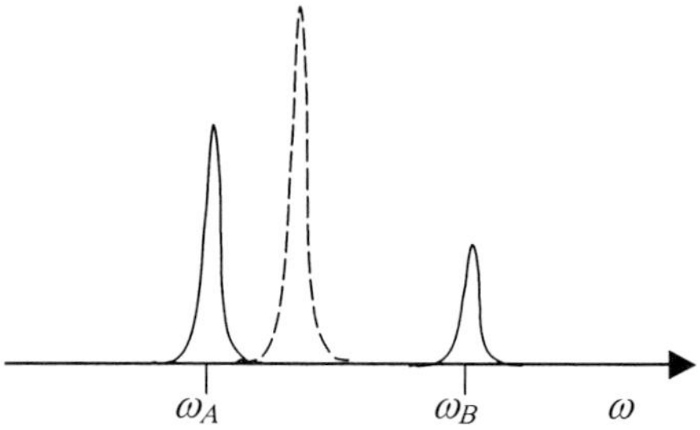

Fig. 5 explains the principle of exchange by giving a simple example for a spectrum of a nucleus which is present in two sites with different chemical shifts. In slow exchange two resonances are observed (solid lines), and the averaged spectrum (dashed line) is obtained for the case of fast exchange with $\tau_{ex} << |\omega_A - \omega_B|^{-1}$.

In adsorption systems the time scale of exchange is often determined by the diffusion coefficient of the molecular species and the binding energies: Polymers mostly exhibit slow adsorption/desorption kinetics, and can be considered irreversibly bound to the surface for the duration of any NMR experiment, i.e. it is $\tau_{NMR} << \tau_{ex}$. Here, the regime of slow exchange with a superposition of the NMR properties from both sites is established. For small solvent molecules with faster diffusion in solution and lower binding energies to the surface on the other hand, it is $\tau_{ex} << \tau_{NMR}$ for any experimental NMR time scale. In this regime of 'fast exchange' the NMR parameters are averaged quantities from both sites, weighted by the fraction $f_{A,B}$ of molecules in the respective site, A or B. For example in fast exchange a relaxation experiment will result in one relaxation rate only, which is given by:

$$R_2 = f_A R_{2A} + f_B R_{2B} \tag{20}$$

The regime of fast exchange has become the basis for a series of investigations of the solvent dynamics in colloidal dispersions, where direct observation of the adsorbed species does not lead to satisfying results. The averaged relaxation rate serves as an indirect measure for the relaxation rate and fraction of the solvent (i.e. water) immobilised in the adsorption layer. The water immobilisation is thus determined by the product of bound amounts f_B and their dynamics, R_{2B}. However, not only immobilisation, but also additional effects such as relaxation induced by diffusion close to susceptibility discontinuities, or cross relaxation can further increase the relaxation rates.

To facilitate the evaluation of relaxation rates in composite systems, a specific relaxation rate R_{2sp}, normalised on R_2^0 of free water, has been introduced [17], and is defined as:

$$R_{2sp} = \left(R_2 - R_2^0\right) / R_2^0 \qquad (21)$$

R_{2sp} monitors the relative change of the relaxation rate due to the presence of water binding sites. It can be shown that R_{2sp} is an additive quantity in samples containing several water binding components. In adsorption samples, the quantity ΔR_{2sp} is calculated from the experimentally determined R_{2sp} by

$$\Delta R_{2sp} = R_{2sp} - \left(R_{2sp}^{CP} + R_{2sp}^{Pol}\right). \qquad (22)$$

Here, the water immobilisation of a noninteracting system, given by the specific relaxation rates of the colloidal particle solution, $R_{2sp}{}^{CP}$, and of the polymer solution, $R_{2sp}{}^{Pol}$, is subtracted from the experimentally determined value R_{2sp}. ΔR_{2sp} serves as the parameter characterising additional relaxation effects due to adsorption. Such experiments on the solvent molecules circumvent problems of lacking sensitivity and are particularly useful for polymers in cases of strong binding, leading to immobile segments, and strong hydration at the same time.

In some adsorption systems, as an additional method, pulsed field gradient (PFG) diffusion measurements were applied to dilute colloidal dispersions, these methods are described in detail elsewhere [18, 19]. Briefly, by this method the self-diffusion coefficients for all spectrally resolvable liquid resonances can be obtained. Usually the method is applied to protons, and can in coated colloid dispersions monitor mobile species such as the solvent or free surfactants.

3. Adsorption Systems investigated by various NMR Methods

3.1. Surfactant Adsorption from Solution

The adsorption of amphiphilic molecules to interfaces is not only governed by the molecular interaction with the interface, but also by the collective aggregation of the amphiphiles and can thus be understood in terms of surface aggregates. Surface aggregates exhibit a large variety of shapes and structures, such as monolayers, bilayers, adsorbed spherical micelles or hemimicelles, or large anisotropic rod- or disc-shaped micelles. The structures depend on the nature of the surface and the solvent, on the spontaneous curvature of the surfactant, and can vary with the degree of surface coverage.

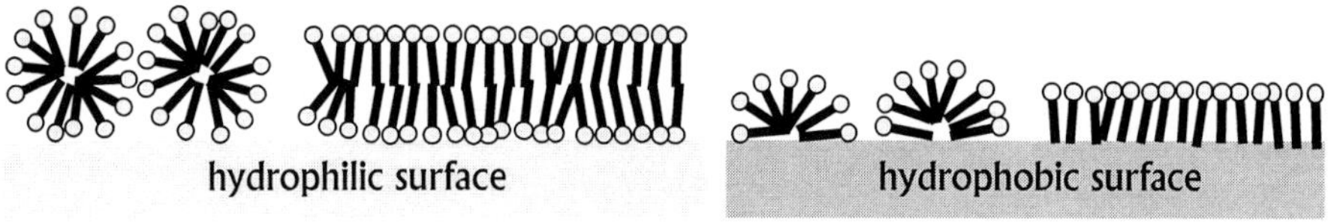

Fig. 6: Surface aggregates of amphiphilic molecules in aqueous solution: spherical micelles, bilayer, hemimicelles and monolayer.

Since the structure is anisotropic on the molecular scale, in general residual interactions are present, leading to broad spectra. The shape and the size of surface aggregates is determining whether an isotropic motional mode is fast enough to average the anisotropy in amphiphilic aggregates, and in a number of investigations (mainly involving 2H NMR), the spectral shape was interpreted in terms of such isotropic motions and their correlation times.

2H NMR was very successful in the investigation of surfactant adsorption, and provides the majority of all surfactant studies. A few 13C spectral studies reveal information about trans conformations of the alkyl chain, and furthermore the dynamics of adsorption can be monitored by 1H self-diffusion. Other approaches which will not be further described here involved investigations of the counterions, e.g. 23Na relaxation in latex particles/ ionic surfactant mixtures [20], and 7Li DQF spectra of clay suspensions containing nonionic surfactants [21]. The counterion behaviour is rather complex and direct implications for adsorbed surfactant layers are difficult to obtain.

3.1.1. SPECTRAL STUDIES OF 1H AND 13C

Pure spectral 1H or 13C studies of physisorbed surfactants are rare. Chemical shift changes at the interface could be detected in few cases for 13C spectra. 1H spectra even of fairly mobile species at the interface are often considerably broadened due to surface immobilisation, and less mobile segments closer to the surface are not detectable at all in liquid spectra. This solid-liquid contrast was made use of in a 1H study: Haggerty et al. used a simple integration of the ethyleneoxide 1H signal of the headgroup of the nonionic surfactant $C_{16}EO_x$, $x \approx 20$, on hydrophobic PS latex to monitor the adsorption: The signal decrease in a surfactant solution due to the addition of particles is assigned to immobilised segments at the surface. Flat adsorption at low coverage led to no 1H signal, while at full monolayer coverage one third of the bound EO-chains was not detected due to motional restriction [22]. The immobilisation of the ethyleneoxide groups in layers is also manifested in the decreased relaxation times of $T_2 \approx 4$ ms as opposed to about 300 ms in solution [23].

13C shifts. In two 13C investigations of ionic surfactants on charged colloids changes of the chemical shift due to adsorption were detected: 13C shifts in the hydrocarbon chains increased to up to 1.5 ppm compared to monomeric surfactant in solution. The shift change is generally towards higher shifts, indicating deshielding of the nuclei, which was attributed to a larger fraction of trans conformations in the alkyl chain in the adsorbed state. Compared to 13C shifts in micelles, the shifts of adsorbed surfactants are however extremely small, and it was concluded that the chain conformation in adsorption layers was rather similar to that in micelles [24].

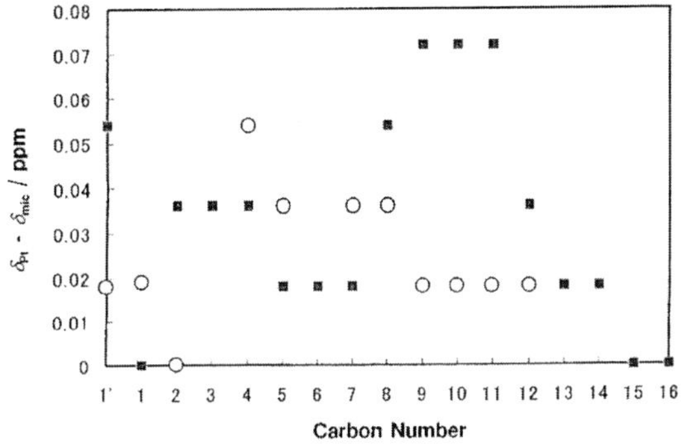

Fig. 7: Chemical shift changes due to adsorption for $C_{12}TAB$ (open symbols), and $C_{16}TAC$ (filled symbols). Figure taken from [25] with permission.

Larger, detectable shift changes compared to micelles were found for hydrophobic adsorption of C_nTAB on small (2 nm) Pt nanoparticles [25]. Shift changes were always towards higher frequency (see Fig. 7), and thus attributed to an increased fraction of trans conformations in adsorbed chains. The fact that the adsorption on Pt particles takes place with the headgroups pointing outwards, and the fact that the particle radius is extremely small seems to enhance close packing and trans conformations as compared to micellar volume aggregates. Fig. 7 also shows that trans conformations are enforced mainly in the centre of the chain [25].

Adsorption dynamics. 1H NMR was employed beyond spectral studies by performing diffusion measurements: Applying 1H PFG-NMR diffusion experiments to surfactants adsorbed to latex particles in dilute dispersions, a method was developed for the investigation of surfactant adsorption dynamics [23, 26]. Since surfactant molecules were occurring in two sites, i.e. in solution and as adsorbed surfactant, each site exhibited a different diffusion coefficient and was distinguished in a PFG experiment. This offered a convenient way to vary the relevant experimental time scale, which is determined by the spacing of the gradient pulses Δ.

The decay curves in Fig. 8 exhibit a biexponential character for small Δ, this indicates a superposition of signals from two species, which are in slow exchange and have different diffusion coefficients. The two coefficients are D_A arising from free surfactant in solution, and D_B from the slow diffusion of particle bound surfactant. With increasing experiment time $\tau_{NMR} = \Delta$, the curves become approximately monoexponential. This is indicative of a diffusion coefficient determined by the weighted average $D = f_A D_A + f_B D_B$ as it is the case in the regime of fast exchange. The echo decays of the 1H signal at varying Δ thus cover the intermediate range between slow and fast exchange. In a model, taking into account the exchange of surfactant between the surface bound site and the free solution site, this region of intermediate exchange was described. A quantitative analysis resulted in the time scales, i.e. the average residence times of a molecule in the free or bound site, which determine the rates of adsorption and desorption.

For the non-ionic surfactant $C_{12}(EO)_5$ the resulting residence time on the particle, $\tau_B = 13.4$ ms, was consistent with purely diffusion controlled adsorption, while the kinetics of incorporation into the adsorption layer is fast compared to this time scale. At higher concentrations additional contributions of micelles to the exchange rate were found to increase the exchange rate [23]. PFG-NMR experiments can probe surfactant exchange dynamics at a time scale as fast as the ms range, which had not been achieved by any other method at solid/liquid interfaces.

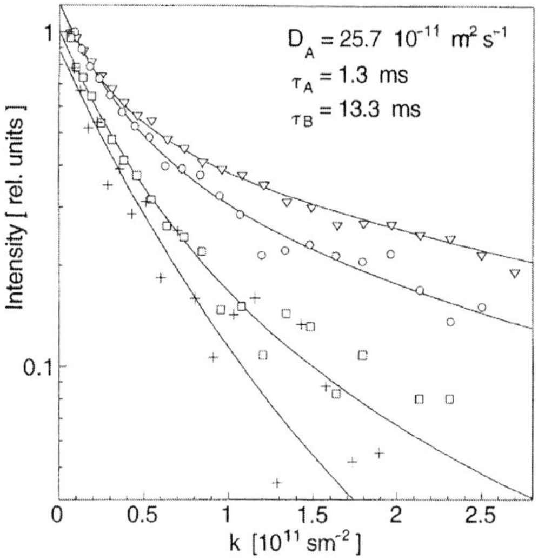

Fig. 8: PFG echo decays in dependence of gradient spacing Δ for nonionic surfactant $C_{12}(EO)_5$: (∇): Δ = 15 ms; (O): Δ = 25 ms; (□): Δ = 50 ms; (+): Δ = 100 ms. The echo intensity is displayed in dependence of the experimental parameter k, such that D can be derived from the slope according *to* $I=I_0 exp(-kD)$. Solid lines are the results of fits to all decay curves, obtained by a two-site model in the region of intermediate exchange, and yielding the parameters given in the figure. Figure taken from [23] with permission.

3.1.2. 2H NMR STUDIES

Already in early NMR studies of surfactant adsorption layers, 2H investigations had been the method of choice [27-30], and a number of investigations has evolved in the meantime. The disadvantages over 1H studies, i.e. lower sensitivity and the necessity of labelling, are overcompensated by the achievement of quantitative results: In the case of no isotropic motional mode averaging the quadrupolar interaction, wide line 2H spectra resulted in order parameters, while for surfactants in the isotropically averaged state linewidths and relaxation rates can be evaluated.

Several compounds, among them ionic, zwitterionic and non-ionic surfactants, have been studied on hydrophilic or hydrophobic particles. All investigations reported were performed on surfactants physisorbed to colloidal particles and investigated in concentrated (centrifuged) samples in aqueous environment. In most cases the surfactants were deuterated in or close to the head group, which revealed headgroup dynamics and/or orientation, and in some cases conclusions on the chain dynamics were drawn, which in one study was investigated directly by deuteration in the chain.

Ionics. A double chain benzene sulfonate, deuterated on the benzene ring, and adsorbed to alumina showed a single resonance at low coverage, while at full coverage a discrete powder pattern was overlapping a single resonance, indicating the formation of a liquid crystal-like molecular environment in the layer [28, 31], see Fig. 9, lower right spectrum.

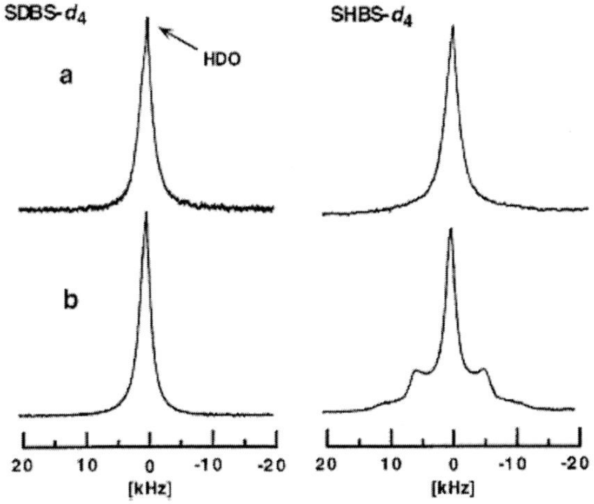

Fig. 9: 2H spectra of single chain (left) and double chain (right) benzene sulfonates on alumina. a) at low coverage, b) at high coverage. Figure adapted from [31], Fig. 2b), 2c) with permission.

This anisotropic molecular environment was attributed to the inner part of a bilayer, while the outer layer was motionally less restricted and thus isotropically averaged. It is interesting to note that the exchange between the layers appeared to be slow.

For the corresponding single chain compound, investigated in dependence of surface coverage [32], in all cases an isotropic line was detected, with a linewidth decreasing with increasing surface coverage, implying relatively restricted molecular motion at low surface coverage and increased motion at higher coverage. The 2H results were related to the known regions of the adsorption isotherm, and in a model assuming fast exchange between an inner and outer bilayer with different molecular motion, models of surfactant adsorption could be probed. It was shown that neither an approach involving the formation of a full reversed inner layer, nor the formation of bilayer patches was consistent with the data, thus an intermediate model was suggested [32]. Additionally, also single chain [33] and double chain [34] methylammonium surfactants on silica had been studied by 2H NMR, and their motional modes discussed.

The dependence of the 2H lineshape on the correlation time of slow motions was demonstrated in an investigation of SDS on alumina [35]: An increase of the particle size led to a transition from an isotropic to an anisotropic spectral shape, which hinted at lateral diffusion taking part in a slowly averaging motional mode. The data were treated by modelling different aggregate geometries (such as porous double layers, prolate or oblate micelles etc.) by making use of the sensitivity of the quadrupolar interaction to the curvature of the aggregate. The number of possible topologies could be narrowed down by the simulation [36].

Nonionics. With selective deuteration in the alkyl chain, the internal gradient of mobility in adsorption layers could be monitored in a 2H-NMR-study of the non-ionic surfactant $C_{12}E_5$ adsorbed to colloidal silica [37, 38]. The surfactant was selectively deuterated in the α, β, or γ-position of the alkyl chain, respectively. Surface aggregate spectra were isotropically averaged with a Lorentzian lineshape, see Fig. 10, consequently the existence of an isotropic motional mode present in surface aggregates was concluded.

In this case the lineshape is determined by homogeneous broadening only, so that relaxation rates R_2 can be derived from the linewidth, however, since R_2 depends on both τ_S and the order parameter S (see equ. 19), it is not possible to calculate absolute values of S from the relaxation rates.

Therefore the concept of a *relative* order parameter profile S_{rel} of the C-2H bond was introduced, defined as $S_{rel} = S/S_\alpha$, assuming that the motional correlation time τ_S is the same in samples with identical composition but different label position. Adsorption layers could thus be directly compared to micelles, lamellar and hexagonal phases with respect to their internal mobility along the chain. An increase of S_{rel} with distance of the 2H label position from the headgroup was found for all bulk aggregate types (see Fig. 11), which was interpreted in terms of packing constraints and chain interaction [38]. Order was induced rather by the hydrophobic aggregation, than by the separation at the hydrophobic/hydrophilic interface, which is the driving force for ordering in ionic or lipid aggregates.

The slope of S_{rel} was found to reflect the aggregate curvature in bulk aggregates of known shape, with the largest increase obtained for small micelles, and the smallest increase for the lamellar phase. These results were explained in terms of the aggregation of the alkyl chain: For $C_{12}E_5$, due to strong hydration of the ethyleneoxide, the cross sectional area of the head group is much larger than that of the alkyl chain, leading to increased mobility of the C atoms close to the head group.

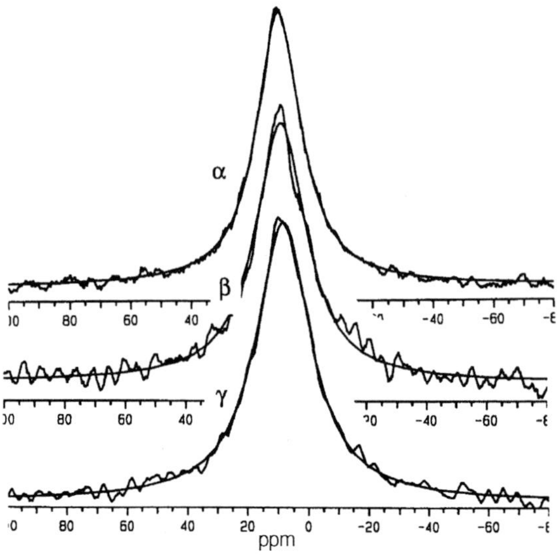

Fig. 10: 2H spectra of $C_{12}E_5$, selectively deuterated in the α, β or γ position, respectively, and adsorbed to colloidal silica, displayed together with Lorentzian fits, demonstrating a homogeneous line broadening.

Relative order parameters of surface aggregates (fig. 11, right) were also increasing with label position, with a slope of S_{rel} similar to that of the lamellar or hexagonal phase, indicating surface structures in the curvature range between long rodlike aggregates and bilayers. Employing the concept of S_{rel} profiles, it thus becomes feasible to obtain information about aggregate shapes from relaxation data. The interpretation of flat bilayerlike surface aggregates was further supported by discussing the experimentally determined motional correlation time in the range of $\tau_s \approx 1$ μs, which corresponded to lateral diffusion along the silica surface in case of closed bilayer aggregates [38].

Zwitterionics. Investigations on zwitterionic n-alkylphosphocholine surfactants, which were deuterated in the methyl groups of the choline head, and adsorbed to latex particles were performed by the group of Macdonald et al. [29, 30, 39, 40]. At a particle size of several hundred nm, diffusion does not average the quadrupolar splitting, and Pake pattern spectra were detected. The quadrupolar splitting increases with surface coverage and thus yields information about the local packing and order in the head group [29].

Using mixtures with ionic surfactants, the sign of their charge was found to influence the quadrupolar splitting: Fig. 12 gives the 2H spectrum of the zwitterionic phosphatidylcholine lipid HDPC-γ-d6 as trace C. With the addition of cationic surfactants (B, A) the quadrupolar splitting is increasing, while it is decreasing on addition of anionic surfactants (D, E). This effect is based on a change of the tilt angle of the choline head group, i.e. its dipole moment, with respect to the surface. Since the tilt angle of the dipolar head group is dependent on the local charge, and 2H NMR can monitor such orientations, this was proposed as a new method to detect the surface electrostatic charge of colloidal particles [30].

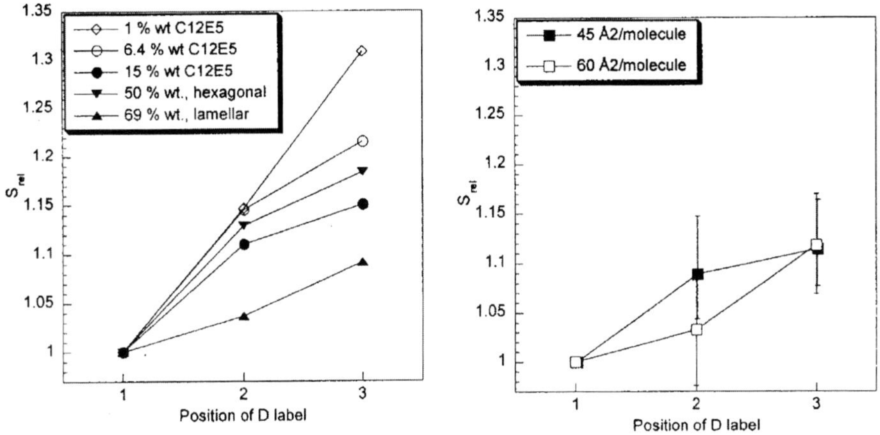

Fig. 11 Left: Relative order parameters in a series of volume aggregates, in which the aggregate curvature is decreasing with increasing surfactant concentration (micelles, hexagonal, lamellar phase). Right: Relative order parameters in adsorption layers at different surface coverage. Figures taken from [38] with permission.

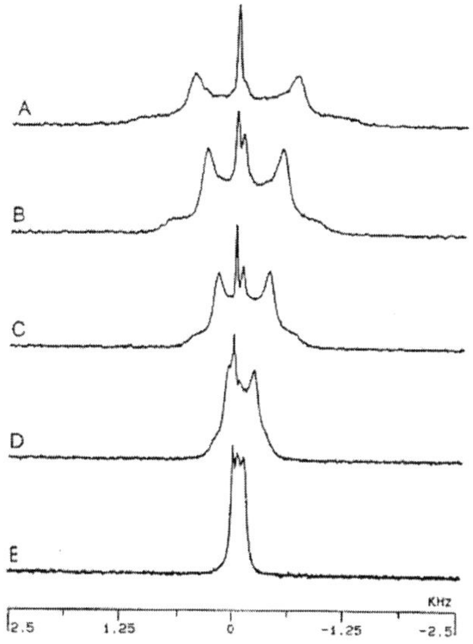

Fig. 12: 2H spectra of HDPC-γ-d6 on latex particles. The outer two maxima of each spectrum form the Pake pattern, while the central two narrow lines arise from free surfactant and HDO, respectively. The quadrupolar splitting is decreasing from a cationic towards an anionic environment (A to E). Figure taken from [41] with permission.

Fast motions. As shown in the above examples, 2H spectra are powerful in the investigation of adsorbed surfactants, giving combined information about structure and slow motional modes. Additionally, T_1 relaxation was studied in some cases to investigate the head group dynamics, i.e. the corresponding fast motions, which were found to be restricted in the adsorbed layer as compared to free surfactants. With increasing temperature an Arrhenius behaviour demonstrating the activation of local head group motions was found [39, 34].

3.2. Covalently attached Alkyl Chains

Surfactant monolayers at the solid-water interface, as described in the previous chapter can exhibit a broad range of dynamic features, and for example show motionally restricted or liquid-like dynamics in dependence of the local packing and the aggregate geometry. In the case of a covalent bond between the surfactant head group and the particle surface groups, the structure and mobility in monolayers is significantly changed. In contrast to physisorbed chains, the local packing in the head group region can be very high, leading to solid structures, often with ordered all-trans chains. At the solid-air interface due to the lack of hydration, alkyl chains are even less mobile and form ordered states corresponding to a solid phase. Such dry colloidal particle samples can be packed densely, and are mechanically stable, so that magic angle spinning is easily performed. Not so much dynamic, as rather conformational aspects are of interest in solid systems, and have been studied predominantly by 13C NMR. In this chapter, 'solid' alkyl chain layers are described, which will, due to the similarity of experimental techniques applied, include chemisorbed chains in air or swollen in solvent, or physisorbed chains in air.

Already in the early stages of NMR of adsorption layers, grafted alkyl chains were studied by various solid state methods, motivated by chromatographic applications (see references in [42, 1]). For example in 2H NMR studies of alkylsilanes on silica, deuterated close to the surface bond, the mobility was investigated by the 2H spectral shape. 2H spectra were anisotropic in the dry state at low temperatures, and mobility could be induced by a temperature increase or swelling with solvent, leading to isotropic spectra. The relevant motions determining the linewidths were interpreted and simulated as two-site jumps [42]. Other approaches involved 13C CP MAS studies, including measurements of different relaxation rates [43, 44]. Alternatively to MAS, 1H CRAMPS was employed as a method to decouple 1H- 1H dipolar interactions, and to study surface protons in a study of SDS physisorbed to porous alumina. 1H signals of different origin were detected: In dried samples the 1H resonances of Al-OH groups, residual water and aliphatic chain protons could be distinguished, while 1H MAS at modest spinning rates (3 kHz) was less efficient to average strong 1H- 1H dipolar interactions [45].

In the following, some examples are given from the large number of recent investigations of grafted alkyl chains, employing a range of different head groups and particles, which was performed by the group of Reven et al. Particles of diameters as small as some nm are used and investigated in densely packed samples, mainly by 13C CP MAS and 2D WISE experiments.

From 13C chemical shifts in the solid state the same information about conformations can be obtained as in the liquid state, but the configurations are far better defined: 13C CP MAS spectra were for example taken for alkanethiols on gold, where a resonance from the interior methylene carbons of the alkyl chain appeared at 33 ppm, corresponding to an all trans configuration, and a second resonance appears at 30 ppm, which is the average of the equilibrium population of trans and gauche chains and thus indicates the presence of disordered domains [46]. The spectra in fig. 13 show the increase of the fraction of disordered chains with temperature.

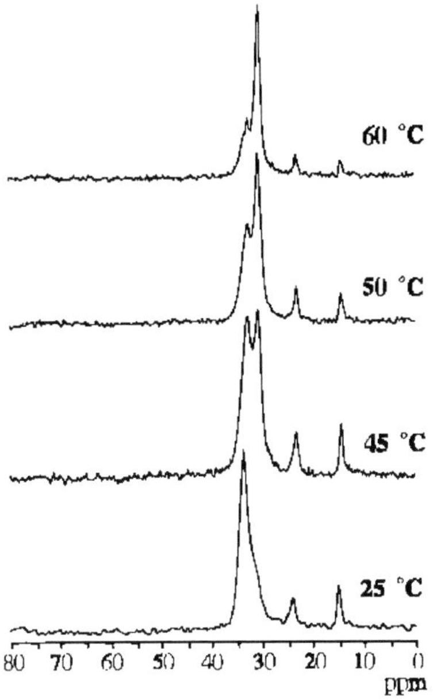

Fig. 13: 13C CP MAS spectra of $CH_3(CH_2)_{17}SH$ on colloidal gold. Figure taken from [46] with permission.

In a similar 13C study, this was also found for octadecylsiloxane (OTS) chemisorbed to silica [47]. For long chain thiols on gold particles the introduction of a terminal COOH group was found to induce conformational order [48]. Thiols on gold were further studied by relaxation and dipolar dephasing experiments, in the latter it was analysed to what extent the heteronuclear dipolar interaction is averaged by molecular motion [46].

While 13C spectra revealed conformations, dynamic information from the attached protons could be obtained simultaneously by WISE spectra. Fig 14 gives an example of a surface spectrum of an alkylphosphonic acid on alumina. It is clearly seen by the broad 1H slice, taken at 33 ppm 13C shift, that the protons attached to segments in the trans conformation had a low mobility, while the slice from disordered chains at a carbon shift of 30 ppm was further averaged due to segmental motions. Thus conformation and 1H mobility could be related.

In a similar way carboxylic acids adsorbed on ZrO_2 were studied, revealing similar properties for the alkyl chains at the solid/air interface as compared to the above mentioned covalently bound compounds [50]. For SDS physisorbed to porous alumina however, no changes of the 13C shifts in dependence of coverage were detected [45].

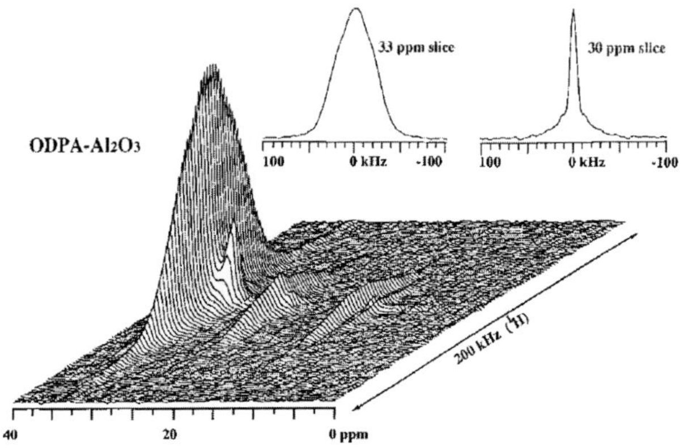

Fig. 14: WISE spectrum of ODPA-Al_2O_3 correlating the 1H lineshapes to the 13C resonance frequencies. The slices are taken from the 13C resonance at 33 ppm and 30 ppm, respectively, and display spectra of protons with different mobility. Figure taken from [49] with permission.

Altogether, the variety and combination of solid-state techniques led to a detailed picture of the conformation and dynamics of various self-assembled monolayers at the solid-air interface, as has recently been reviewed [51]. Due to a high degree of order of covalently attached chains, and the availability of conformational information in the solid state, a detailed picture of chemisorbed surfactants in monolayers was obtained.

3.3. Lipid Bilayers

In contrast to surfactants, lipids adsorbed on hydrophilic surfaces can be expected to form planar bilayers, due to their large spontaneous radius of curvature. A double chain amphiphile forming a bilayer on silica was already discussed in chapter 3.1.2 in the context of 2H NMR investigations of water soluble amphiphiles. Bilayers from water insoluble lipid amphiphiles have been adsorbed to large spherical silica particles by condensation of unilamellar vesicles from aqueous solution, and a series of studies explored different NMR methods suitable for the measurement of lateral diffusion coefficients in such 'supported bilayers'.

One possibility is to determine the time scale for diffusion around a spherical particle by relaxation experiments, since transverse relaxation rates are sensitive to slow motional time scales. The 2H relaxation time T_2 was measured in a quadrupolar CPMG experiment, in which the delay between pulses, τ, in the pulse sequence was varied. With increasing τ, the experiment increasingly allows for diffusion as a slow motional mode to contribute to the relaxation rate, and to decrease T_2. With this principle, from the dependence of $R_2 = T_2^{-1}$ against τ^2 the diffusion coefficient D is extracted, as shown in fig. 15, where the slope is proportional to D [52]. Diffusion coefficients of the phospholipid POPC were determined at different temperatures [52], the diffusion of the inner and outer monolayer of DPPC on silica could be separated [53], and recently a diffusion coefficient was determined for lipid bilayers on polymeric support [54].

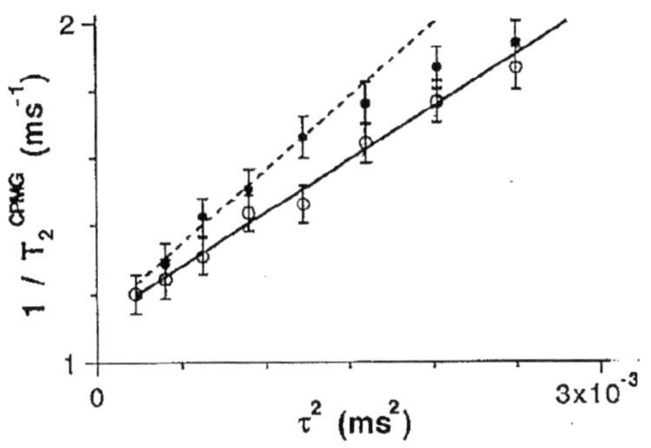

Fig. 15: Relaxation rate $R_2=T_2^{-1}$ in dependence of τ^2. The slope gives the diffusion coefficient. Figure taken from [52] with permission.

Another approach consisted of 2H 2-dimensional exchange spectra, where each spectral position and thus angle with respect to B_0 was correlated to the respective angle after a mixing time t_m in a second dimension, so that different positions of a molecule on a sphere at time *0* and t_m were correlated. 2H 2D exchange spectra were measured for POPC on silica, and additional random walk simulations were performed to model the spectral shapes at different ratios of the mixing time to the correlation time of diffusion around the sphere. By comparison of simulations and experiments, the lipid diffusion coefficient is obtained [55]. Extensive simulations and data treatment were necessary in this method, however, its advantage is to allow the variation of the experimental length scale by changing t_m. This was applied to investigate a lipid mixture in the coexistence region, where the diffusion in the fluid phase was restricted by the presence of domains of a higher ordered phase. By varying t_m, diffusion coefficients corresponding to different length scales were determined and the connectivity of the fluid phase was probed [56].

3.4. Polymer Adsorption and Grafted Layers

According to established adsorption models, the segments of adsorbed polymer chains at solid/liquid interfaces can be divided into two sites: Rigidly bound segments ('trains') are either in direct contact with the interface, or due to covalent binding to directly surface attached segments they are similarly reduced in their mobility. Other segments can extend far into the liquid phase and exhibit dynamic properties resembling polymer chains in solution, these form the more mobile 'loops' and 'tails' (see Fig. 16).

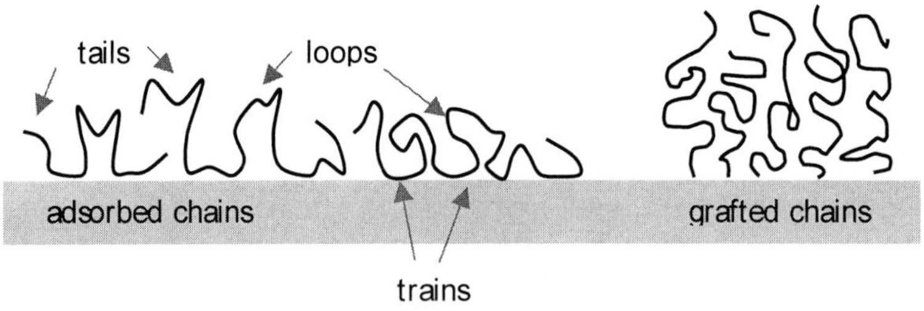

Fig. 16: Physisorbed polymers with trains, loops and tails, and end-grafted chains at the surface.

Due to slow segmental motions being present, in NMR experiments loops and tails provide isotropically averaged signals with a narrow liquid spectrum, whereas the train signals are subject to static interactions (dipole-dipole, or quadrupolar, rsp.) and can thus be broadened by orders of magnitude to a wide solid state spectrum. A detection of both components in one spectrum is theoretically possible, but practically hindered by signal-to-noise and dynamic range problems, since several orders of magnitude difference in linewidth cause similar contrasts in intensity. However, NMR offers other methods to separate trains from mobile segments, as described further below.

It has to be noted, that the assumption of two states of mobility is largely simplified, since in reality a continuous gradient of mobility can be expected with increasing distance from the surface.

Grafted chains exhibit one terminal point of bonding only per chain, and are therefore on average less immobilised than chains in the adsorption case. However, they can be described by a motional gradient as well, and the experimental methods applied to them are the same, mainly aiming at the characterisation of segmental mobility. Therefore this chapter covers both systems, organised by the technique of investigation.

3.4.1. 1H AND 13C NMR

1H and 13C spectra had been applied to polymers at interfaces already as early as in the 1980s, using liquid state spectra and relaxation to separate the segments into trains vs. tails and loops by their segmental mobility. Several of the early investigations involved complementary studies of ESR and NMR, and generally a two-component behaviour of mobility was confirmed. A series of investigations had been performed on grafted PEO by the group of Hommel, Legrand et al. and involved spectral analysis as well as relaxation measurements. A beautiful early example was given by a 1H and 13C T_2 and T_1 relaxation study of Faccini and Legrand [10], where from the relaxation information the spectral density function was constructed at different frequencies. A summary of the initial studies is given by Blum [1]. Most 1H techniques applied to polymers adsorbed from solutions have focussed on the solid-liquid contrast, where the first attempts were to establish the train fraction in different ways: A combined pulse sequence monitoring solid and liquid spins simultaneously was introduced by Cosgrove and Barnett [57], and from relaxation data, in combination with neutron scattering, segment density profiles in layers were obtained [58]. A review of these early 1H experiments was given by Cosgrove [2].

In a number of more recent 1H investigations the solid- liquid contrast was monitored by the simple detection of liquid state spectra, since under liquid NMR experimental conditions the detection of the FID is not fast enough to observe solid spins. This simplified approach leads to conclusions about the 'disappearing' (solid) components from the liquid spectrum. This was employed to investigate a number of specific problems in adsorption layers, such as stereospecificity, phase transitions, and mobility of grafted chains:

Stereospecificity. The adsorption of isotactic, stereotactic and atactic PMMA to silica from $CDCl_3$ was studied by 1H liquid spectra, where the methyl peaks spectrally separate into isotactic, heterotactic and syndiotactic sequences. The intensity ratios of the liquid 1H signals of polymers adsorbed to silica were compared to solution spectra. In Fig. 17 the comparison of traces A and D shows that the isotactic component (mm) has significantly decreased in intensity relative to the other two components in spectrum D, and a selective adsorption of i-PMMA was concluded. The stereospecificity decreased with adsorbed amount, and was also reduced for adsorption above the conformational transition temperature. Comparison to adsorption to a hydrophobised surface, which did not show stereospecific immobilisation, proved that the silica hydroxyl groups were essential to attract the isotactic component specifically [59].

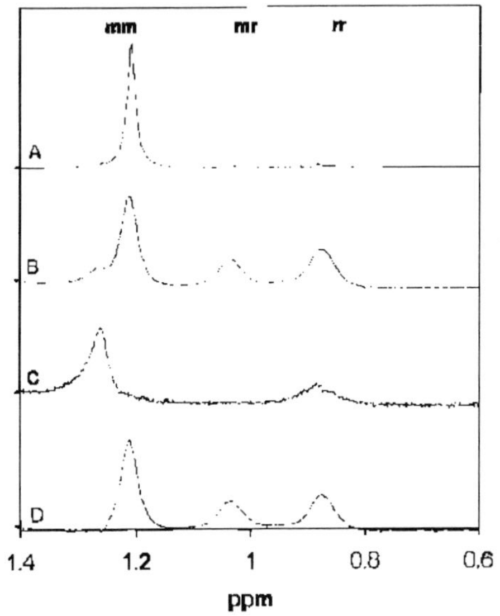

Fig. 17: 1H spectra of A) isotactic PMMA in $CDCl_3$, B) after silica addition and D) after subtraction of the silica spectrum C). Spectra taken from [59] with permission.

Phase transition. 1H liquid signals have further been applied to investigate the phase transitions of polymers in adsorption layers by monitoring the change in mobility of polymer segments. The thermoreversible Poly-N-isopropylacrylamide (PNIPAM) shows a coil to globule transition in solution at the lower critical solution temperature (LCST), where the water solvent quality is drastically decreasing, and thus solubilized polymer coils with liquid type mobility transform into in solid particles consisting of immobile segments [60].

This phase transition was investigated in the restricted geometry of adsorption layers of PNIPAM and a negatively charged PNIPAM-copolymer on colloidal silica. The methyl protons dominated the polymer signal (see Fig. 17a), and the spectrum was substantially broadened as compared to the polymer in solution. This indicates a reduced mobility of the segments observed, which were identified with loops and tails. Solid echo relaxation measurements confirmed the existence of an additional, rigidly adsorbed solid fraction of segments[61].

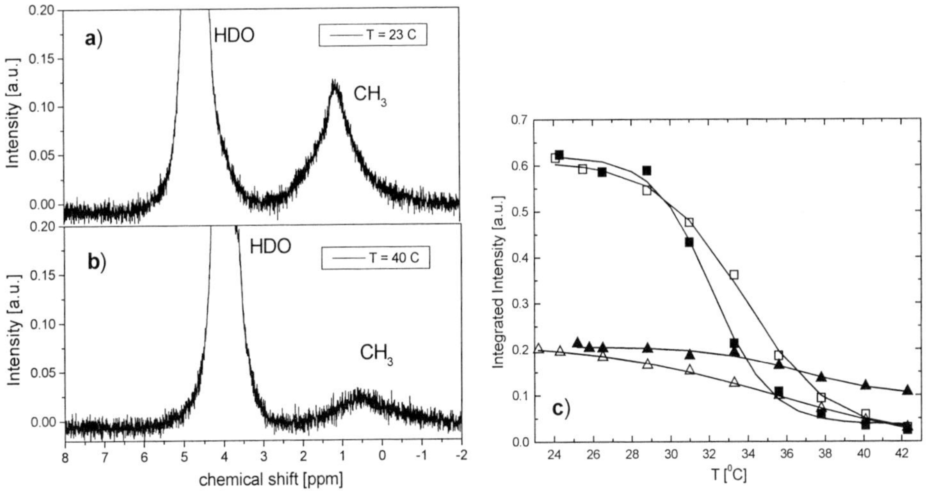

Fig 18: Spectra of PNIPAM on silica a) below and b) above the phase transition. The chemical shifts are slightly varying with T. c): Integrated intensities of the 1H liquid signal in dependence of temperature. Open symbols: PNIPAM, filled symbols: charged copolymer. The polymer/silica wt ratio is 0.85 (squares) and 0.25 (triangles), respectively.

By monitoring the liquid signal in dependence of temperature, the phase transition of the tails and loops was followed (see Fig. 18c). The transition in the adsorption layer was found to be substantially broader than in solution. This was the case especially at low surface coverage, thus resulting in a picture of the width of the phase transition depending on the distance of the segments from the surface. For a charged copolymer the transition was not complete, as above the LCST a liquid signal from loops and tails remained. Electrostatic repulsion from the surface led to an incomplete phase transition, since the repulsion seems to prevent aggregation. This occurs predominantly at low surface coverage, i.e. below a critical density of segments. At higher coverage, a sufficient number of charged segments was present to enable aggregation to the solid form [61].

A similar experiment was performed on PNIPAM *grafted* to latex particles using 13C NMR [62], where above the transition temperature a similar decrease of the NMR intensity due to a solidification of liquid spins was observed. Additionally, in grafted layers a signal increase occurred below the transition temperature, which was not found in 1H spectra of adsorption samples. It should be noted that the grafted and the physisorbed systems differ significantly in the number of contact points on the surface, and consequently also in mobility and layer thickness.

Residual linewidth. As an alternative way to investigate mobility of polymers at surfaces, the degree of line narrowing in 1H MAS experiments was employed. For PEO grafted to silica fast motions partially average the dipolar interaction, and with increasing rotation speed the linewidth was decreased due to the successive narrowing of slow segmental motions. The dependence of the residual linewidth on the frequency of rotation was evaluated using a model of the spectral density functions to obtain correlation times of the fast and slow motions [63]. This technique works in systems where linewidths are small enough (several kHz) to be further narrowed by conventional rotation frequencies.

The fraction of immobilised segments was further used to investigate the adsorption from a mixture of two immiscible polymers from solution [64]. A 1H decoupled 13C spectrum was employed to identify the block of a block-copolymer, which was adsorbed in trains, by detection of the mobile segments [65].

All these approaches assessed the segmental dynamics qualitatively in systems with partially liquid chains. Applying 1H NMR, the sensitivity is good and no labelling is required. In polymer systems, which exhibit dynamics in the intermediate range between liquid and solid, as it is the case for adsorption from solution, chemical shifts contained no further information.

3.4.2. 2H NMR Studies

In contrast to 1H or 13C, 2H studies of interfacial chains provide the possibility of quantifying information on mobility or ordering. 2H experiments have been performed on a series of hydrophobic polymers adsorbed from organic solvents to small inorganic particles, and on PDMS grafted chains on silica. Studies exist for the solid/air interface, and for layers swollen with solvents. In swollen polymer layers, relaxation rates of tails and loops provided a quantitative analysis of their chain dynamics, while for polymers at the solid/air interface wide line 2H NMR investigations were performed to assess the degree of order in a solid-like environment, and the influence of immobilisation at a surface on the glass transition.

Adsorbed chains at solid/liquid interfaces. Poly(methyl methacrylate)-d (PMA) was adsorbed to silica from toluene solutions and investigated by T_1 and T_2 liquid state relaxation experiments, probing tails and loops [14]. Relaxation times T_1 and T_2 were compared for PMA in toluene solution, and in swollen adsorption layers, see Fig. 19a.

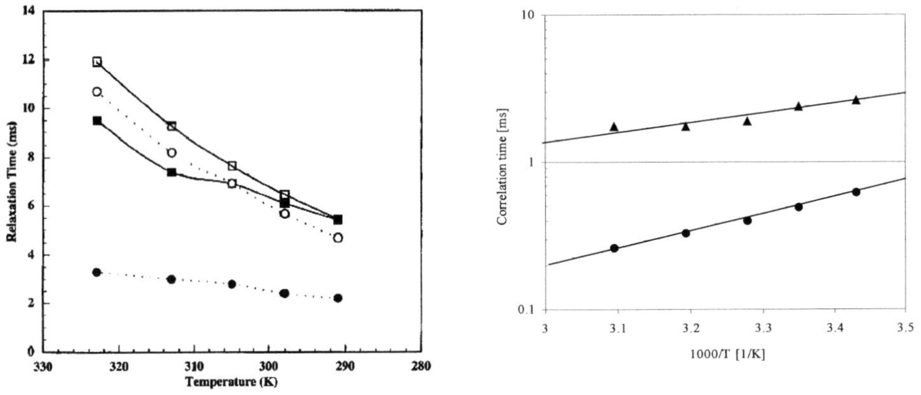

Fig. 19: Left: 2H relaxation times for PMA-d in toluene solution (open symbols) and in swollen adsorption layers (filled symbols), squares: T1, circles T2. Right: Mean correlation times for PMA-d in toluene solution (circles) and in swollen adsorption layers (squares) as evaluated from a log-normal distribution of τ_c. Both figures taken from [14] with permission.

In comparison to solution (open symbols), T_1 and T_2 values were both decreased, i.e. motions were hindered in the layer. T_1 is dominated by fast (local) motions, and its marginal decrease showed that these are not significantly affected in adsorbed chains. T_2 values were much more decreased, which was attributed to stronger limitations of the slow, long range segmental motions at the interface.

These results were furthermore interpreted by a quantitative analysis involving a log- normal distribution of correlation times. Fig. 19b gives τ_c data, which show an activation of the slow motions with temperature, and demonstrates that the tails and loops, if swollen in a good solvent, were only slowed down by one order of magnitude in τ_c. Alternatively the data were analysed by a Hall-Helfland distribution. Though resulting in different rate values, both models were found to be in qualitative agreement concerning the reduction of the slow motions due to the presence of the surface [14].

Other studies dealt with block-copolymers, where polystyrene-vinylpyridine (PS-*b*-VP), deuterated in the backbone of the styrene block, was adsorbed to colloidal silica. The vinylpyridine blocks being attached to the surface, the styrene segments are solvated and exhibit liquid type mobility. Detecting the signal of their liquid 2H resonance, relaxation experiments were performed and compared to the polymers in solution. In solution both T_1 and T_2 and thus fast and slow motions strongly depend on the concentration. To compare the segmental mobility at corresponding concentrations in the layer, the ratio T_1/T_2 is taken at the same value of T_1. T_1/T_2 thus serves as a parameter characterising the slow motions, where a large value $T_1/T_2>1$ signifies slower or more anisotropic slow motions [66, 67]. Additionally, employing a model (Hall-Helfland), rate constants of the fast local and the slow segmental motions were extracted from the relaxation rates, and were again compared to corresponding values in solution [65]. It was found that in a good solvent like toluene the vinylpyridine blocks can even exhibit an *enhanced* rate of segmental motions as compared to the dynamics in solution at the corresponding concentration [66]. These effects did not occur for a shorter M_w PS-*b*-VP [67], or for adsorption layers swollen with poorer solvents [65], as similar investigations showed. Thus the surface could either increase or reduce the mobility, probably in dependence of the local packing at the interface.

Adsorbed chains at solid/air interfaces. Some 2H NMR-investigations were performed on hydrophobic polymers adsorbed to the solid/air interface. If T_g is not too low, the dynamics in dry systems are more solid-like, and wide line solid spectra are resulting. Motional averaging depends mainly on the glass transition temperature, so that 2H wide line NMR could be applied to study glass transitions in adsorption layers.

Polymethylacrylate (PMA), which has a bulk T_g of about 50°C, was investigated at the silica/air interface by 2H spectra in dependence of temperature and surface coverage [69]. Fig. 20 gives spectra at 52°C at varying surface coverage, from which components with a different degree of mobility could be identified: An immobile Pake component dominated the spectrum at low coverage and was not observed in the bulk spectrum at this temperature. This component was attributed to segments close to the surface, which were immobilised (trains), while the narrower component, which was increasing with surface coverage, was attributed to segments close to the polymer/air interface. This component is even narrowed as compared to the bulk spectrum, showing an increase in mobility, probably induced by a larger motional freedom of chains at the polymer/air interface [69]. Detecting such spectra at variable temperature, the shift of the glass transition of both fractions was monitored.

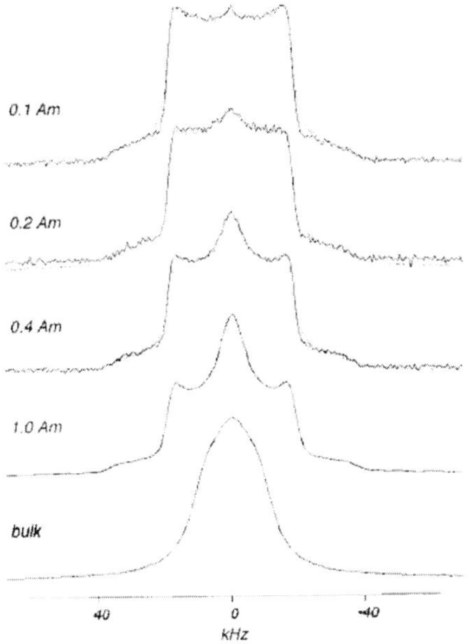

Fig. 20: 2H spectra of surface adsorbed PMA-d_3 at 52°C with immobile and mobile components compared to bulk. Figure taken from [69] with permission.

In comparison to this study, a lower molecular weight PMA was more mobile and of larger motional heterogeneity in bulk than the high M_w compound. At the solid-air interface the low M_w compound was less mobile, which was attributed to a more flat configuration as compared to the higher M_w chains [70].

A related study was concerned with poly(vinyl acetate) (PVA) on silica, where in a similar way a solid component, induced by the surface, was persistent even at $T > T_g^{bulk}$ [68].

Furthermore, the block-copolymer PS-*b*-VP described in the last section, was investigated at the solid/air interface and showed wide-line Pake spectra and reduced mobility as compared to bulk [71].

Grafted chains. Zeghal et al. have performed a series of 2H NMR investigations of PDMS grafted to silica, partly on planar stacked silica slides, and on porous silica with large pore sizes of 400 nm in air [7, 72] and in different solvents [73].

Planar stacked layers offered the possibility to investigate the quadrupolar interaction at a fixed angle Ω of the surface relative to B_0. By detecting the splitting $\Delta\nu_Q(\Omega)$ in dependence of Ω it was confirmed that the segmental motions are uniaxial around the surface normal [7].

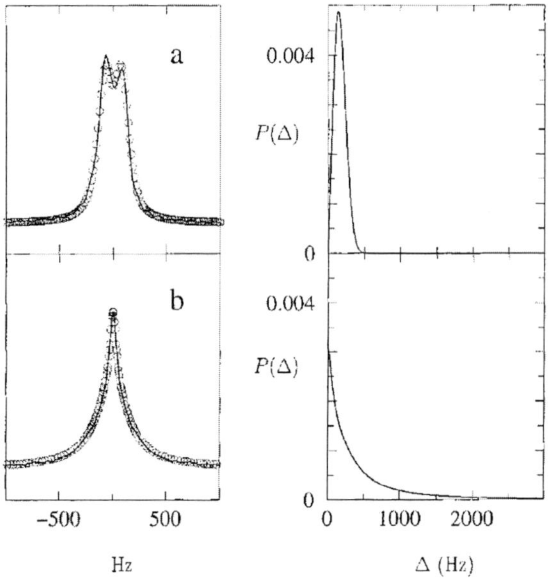

Fig. 21: Left side: 2H NMR spectra (solid line) of grafted PDMS chains on planar substrate at $\Omega = 90°$ (a) in air and (b) in CCl_4, together with fits (open circles) using the distributions of residual interactions $P(\Delta)$ shown on the right. Figure taken from [73] with permission.

For different solvents, different spectral shapes of the 2H polymer signal resulted, which allowed conclusions on the orientational distribution of the polymer segments. The spectra in Fig. 21 were simulated as a superposition of quadrupolar splittings Δ with a distribution function $P(\Delta)$. In air (a) the spectral shape was explained by a distribution function $P(\Delta)$ with a maximum at Δ =200 Hz. The spectral shape in CCl_4 (Fig. 21b) resulted from a distribution with a maximum at zero splitting.

The result for poor solvent cases (a) was interpreted as strongly squeezed chains, and since the segments are oriented preferentially parallel to the surface, it is $\langle S \rangle < 0$. The order parameter, and thus the anisotropy is quite uniform throughout the layer, leading to the maximum of $P(\Delta)$. In the good solvent case (b), the anisotropy was increasing with the solvent quality, and interpreted as a preferentially perpendicular alignment of segments relative to the surface, so that it is $\langle S \rangle > 0$. Here, the distribution of order throughout the layer was broad and thus probably dependent on the distance from the surface. Distributions of quadrupolar splittings were obtained for a range of solvents and discussed in terms of chain conformations [73].

3.4.3. Solvent studies

Solvent investigations provided indirect information, mainly for strongly binding polymers, where it can be difficult to directly observe strongly broadened and fast relaxing segments. The first solvent relaxation experiments in dispersions of polymer coated colloidal particles had been introduced with the aim to determine the bound fraction, i.e. fraction of trains, in a monolayer. Specific relaxation rates of water in silica dispersions increased in the presence of adsorbed uncharged polymers (PEO or PVP). The increase was almost independent of molecular weight, and thus the relaxation rate enhancement was attributed to solvent immobilisation by surface-bound segments (trains), while it was rather insensitive to the amount of tails and loops [17, 74]. Assuming flat train adsorption at low coverage the relaxation rate enhancement due to adsorption is proportional to the number of trains, Γ_{tr}. Further assuming that at low coverage this equals the total number of adsorbed segments, $\Gamma_{tr} = \Gamma_{total}$, the initial slope k of R_2 curves vs. adsorbed amount Γ_{total} was evaluated (see Fig. 22). It is then $\Gamma_{tr} = k^{-1} R_{2sp}$, which led to a separation of the effect of trains vs. loops or tails on the water immobilisation, and thus a calculation of Γ_{tr} in dependence of the absorbed amount [17].

In another study, a reduction of the adsorbed amount in a PEO layer due to the addition of ionic surfactant was concluded from a substantial decrease of the solvent relaxation with increasing surfactant concentration [75].

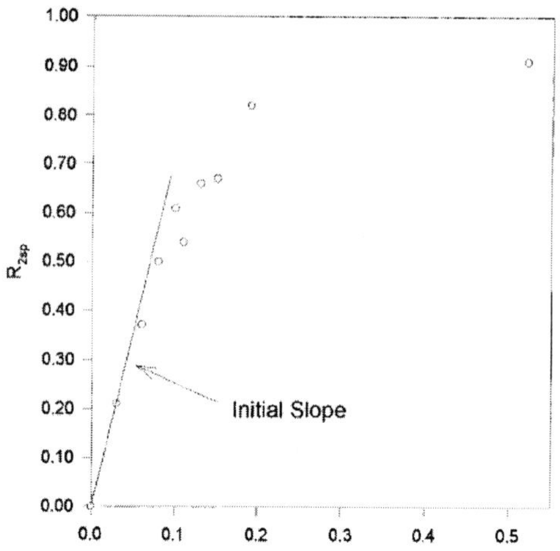

Fig. 22: Determination of the train fraction, Γ_{tr} from the initial slope of R_{2sp}. Figure taken from [75] with permission

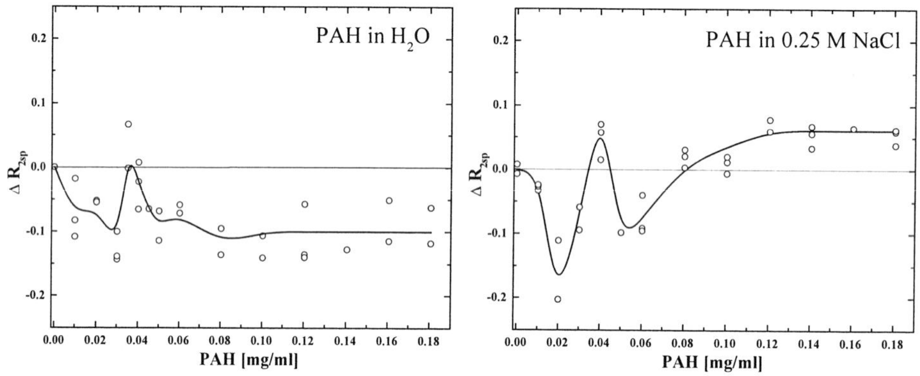

Fig. 23: ΔR_{2sp}, i.e. difference in specific water relaxation induced by adsorption of PAH to latex particles. Left: in aqueous solution. Right: in 0.25 M NaCl solution.

For the adsorption of polyelectrolytes a completely different behaviour of the relaxation rate was observed, since instead of a monotonous increase several contributions caused minima or maxima in R_{2sp} in dependence of surface coverage [76, 77]: For Na-PSS, adsorbed to positively charged latex, the relaxation rate R_{2sp} was first increasing to form a maximum with increasing polymer coverage, which was attributed to bridging flocculation [76]. At higher polymer concentrations, a decrease of R_{2sp} below the value of the pure particle solution was observed. This was in contrast to the findings for uncharged polymers described above, and was explained by a release of water from the particle surface due to the adsorption of polymer segments replacing surface bound water. It was further argued that in this case, without strong solvent- surface interactions present, but strong hydration of the chain, the relaxation rate at high coverage reflects the immobilisation of water in loops and tails [76]. The influence of different water fractions on relaxation rates could be further analysed in a study of polyelectrolyte adsorption to particles at concentrations below bridging flocculation. Positively charged PAH was adsorbed to negatively charged latex particles, and ΔR_{2sp} was found to have positive and negative regions, see Fig. 23. A typical shape of the ΔR_{2sp}-curve was identified for different polyelectrolytes and consisted of at least one minimum, followed by a maximum and a plateau region [77]. This structure was attributed to the contributions arising from different water populations: At low polymer concentration the replacement of particle bound water by adsorbed polymer trains led to negative values of ΔR_{2sp}, since the water mobility was increasing. With increasing concentration, this contribution became compensated by the positive contribution from hydration water of the tails and loops, which is immobilised on adsorption and led to a maximum. The latter contribution also led to larger R_{2sp} plateau values at saturation in the case of adsorption from salt solution (see Fig. 23, right), which was consistent with a higher polymer coverage in the presence of salt, observed by other methods.

In another solvent relaxation study, the adsorption of the milk protein β-casein, a charged polyelectrolyte consisting of a hydrophobic and a negatively charged block, to silica was investigated. Adsorbing with the hydrophobic block to the surface, this polymer exhibited a behaviour similar to the uncharged polymers, and the train fraction was extracted in the same way [78].

As an alternative to relaxation rates, the solvent diffusion coefficient was another parameter employed to distinguish between adsorbed and free solvent: In concentrated dispersions of PEO on silica the decrease of the water diffusion due to binding in the layer was analysed in terms of the average water concentration in the layer [79].

Relaxation studies yield information about the overall solvent dynamics, which in some cases can be difficult to interpret, as contributions to the solvent relaxation rate arise from various origins, such as changes in surface hydration, polymer hydration, coagulation, conformational changes which might lead to water release on adsorption. Additional contributions to the relaxation rate can arise from changes of the degree of dissociation, from cross-relaxation in the layer, or from susceptibility variations. Care has to be taken to interpret the origins of changes of R_{2sp} correctly. In spite of these drawbacks of being an indirect method, solvent relaxation is still powerful in aqueous systems where the organic adsorbed component is extremely rigid and difficult to observe directly.

3.5. *Polymer Self-Assembly of Composite Layers*

In recent years self-assembly processes involving electrostatic interactions have been used in order to build up multilayered materials. Polyelectrolyte multilayers can be formed by alternating exposure of a charged substrate to solutions of positive or negative polyelectrolytes [80]. This principle of layer formation has not only been achieved by adsorption to planar substrates, but even to colloidal particles [81], which offers the possibility for investigations by NMR methods.

However, the direct observation by 1H MAS spectra was not successful in these materials, since the interactions exceed the standard rotation speed; and 2H NMR wide-line approaches provided no sufficient signal. Both problems can be attributed to the strong electrostatic interaction between layers, which makes these multilayers extremely rigid materials, where dipolar or quadrupolar interactions are large. As a further complication, the preparation of multilayers requires comparatively large (> 50 nm) particles, so that sensitivity issues become critical in combination with very broad inhomogeneous lines.

Very recently, two different approaches were made to circumvent these problems: For a solid state approach samples were dried and treated by ultrafast MAS spinning of 30 kHz in a double quantum filter experiment. In another approach aqueous dispersions of coated particles are investigated, and the water relaxation is employed to indirectly monitor multilayer properties.

3.5.1. Solid State Techniques

Dried samples of up to 4 layers of self-assembled polyelectrolytes on silica particles have been investigated by solid state NMR techniques [82]: 1H MAS spectra at 30 kHz rotation showed the signals of residual water molecules associated with the polyelectrolyte, while no signal of the polymer was detected. By comparison to spectra of dried polymer bulk complexes water sites associated with the silica surface and the multilayers, respectively, were identified. To suppress the signal from such comparatively mobile protons, which dominate the spectrum even after vacuum treatment of the sample, the recently developed double quantum filter (DQF) experiment was performed (see refs 7-10 in [82]). The experiment allows the detection of dipolar coupled 1H pairs under fast MAS selectively for very small 1H-1H distances (0.35 nm). Due to suppression of the water signal, the DQF 1H spectra taken with 30 kHz MAS could indeed detect the polymer protons, appearing at 7.5 ppm (aromatic, PSS) and 3 ppm (methyl, PDADMAC), see Fig. 24. With increasing layer number the built-up can be seen by the increase of the aromatic resonance due to PSS adsorption, and the methyl resonance due to PDADMAC adsorption.

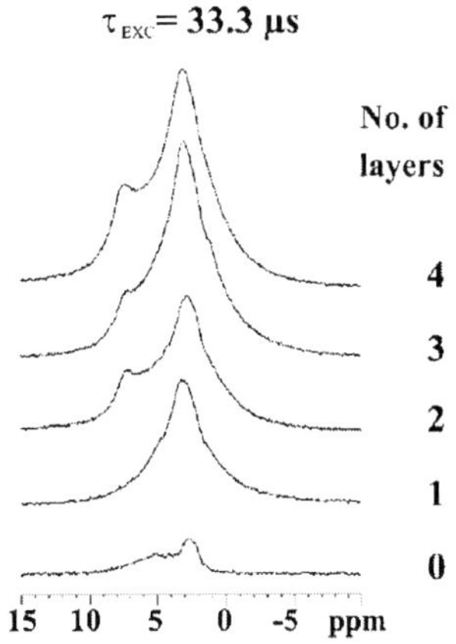

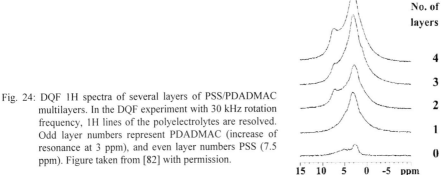

Fig. 24: DQF 1H spectra of several layers of PSS/PDADMAC multilayers. In the DQF experiment with 30 kHz rotation frequency, 1H lines of the polyelectrolytes are resolved. Odd layer numbers represent PDADMAC (increase of resonance at 3 ppm), and even layer numbers PSS (7.5 ppm). Figure taken from [82] with permission.

Furthermore, in a 2D DQF 1H spectrum, a cross peak indicated a very close proximity of the PSS aromatic protons and the PDADMAC methyl protons, which is a direct proof of the complexation between the two polymers in multilayers [82].

3.5.2. Solvent Relaxation

For adsorption of a PAH layer to particles precoated with multilayers, the adsorption isotherm monitored by the relaxation rate is dominated by an immobilisation contribution, while no release of water as in the monolayer case is detected. Fig. 25 shows the monotonous increase of ΔR_{2sp} (compare to Fig. 23 for monolayers). The effect of salt on the adsorption process was again an increase of the water immobilisation, similar to the findings for the adsorption of the first monolayer. Conclusions on the structure (rod or coil shape) and hydration of an adsorption layer on precoated particles as compared to the adsorption of the first layer and in dependence of the salt concentration were drawn [77].

Multilayer assemblies were furthermore investigated in dependence of the number of layers, *n*, where an interesting influence of the outer layer potential on the multilayer system was found (see Fig. 26). Here, R_{2sp} is displayed, which is a measure for the total 1H immobilisation in the multilayer arrangement. In PDADMAC/PSS multilayers (squares) a monotonous increase of R_{2sp} with layer number was observed and attributed to the additional water immobilised in each adsorbed layer.

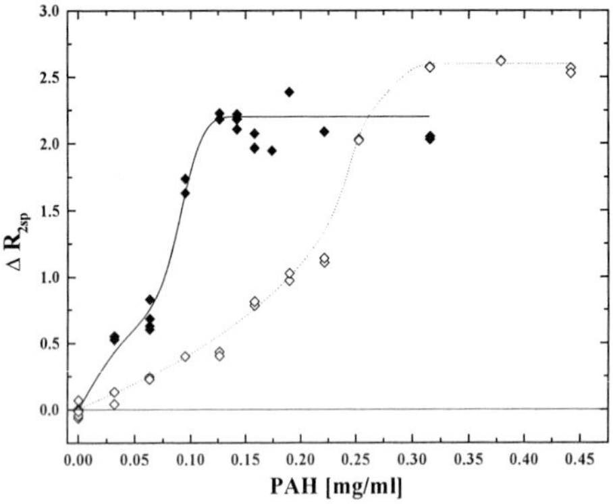

Fig 25: ΔR_{2sp} for the adsorption of PAH to latex particles precoated with two double layers of (PAH/PSS). Solid symbols: PAH adsorbed in H_2O, open symbols: in 0.25M NaCl. The precoated particles serve as reference, and ΔR_{2sp} represents the water immobilisation due to adsorption of the fifth layer.

In PSS/PAH layers, however, the 1H immobilisation was alternating (triangles in Fig. 26): An *increase* of water immobilisation occurred on adsorption of a positive layer, and a *decrease* on adsorption of a negative layer, indicating that the negative layer is leading to 1H *release* from the multilayers. By adsorbing PDADMAC as the last layer to a PAH/PSS assembly, it could be shown that the alternating behaviour is not dependent on the electrolyte strength of the last layer, but on the internal layers. The increase of the changes in R_{2sp} with *n* supports their interpretation as an integrated effect of the internal layers.

The alternating behaviour was thus attributed to an influence of the electric potential of the outer layer on the inner multilayer system. As potential origins of this odd/even effect not compensated charges within the multilayer assembly were discussed, which can lead to a reversible swelling and de-swelling, controlled by the surface potential. An alternative interpretation were changes of the dissociation equilibrium or ion content within the layer, since in the case of a weak polyelectrolyte R_{2sp} is sensitive to changes of the dissociation equilibrium via the fast relaxation rate of covalently bound and fast exchanging protons. From the increase of the curve in Fig. 26 a decay length of the potential of the outer layer of at least 6 layers was deduced [83].

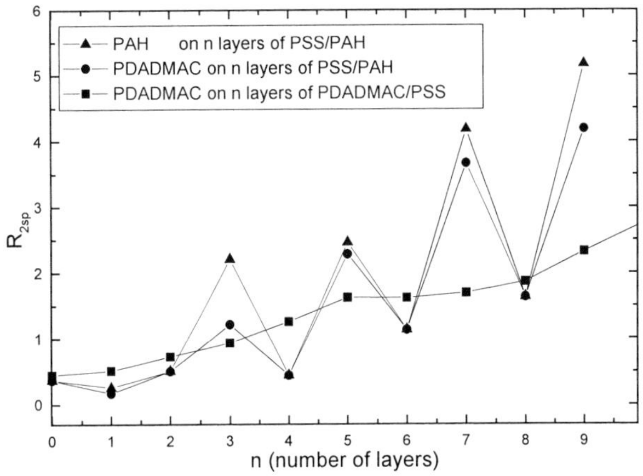

Fig. 26: Specific relaxation rates of multilayer systems for different polyelectrolyte pairs and for different polymers in the last layer. The odd/even effect occurs only if the multilayers contain a weak polyelectrolyte (PAH), and is controlled by the charge, not the electrolyte strength, of the outer layer.

4. CONCLUSIONS

NMR methods are powerful and versatile tools to investigate organic materials. For a large variety of different organic adsorption systems described in this review unique and new information could be obtained by applying a suitable NMR technique. In the past ten years a number of applicable principles and suitable conditions for performing the experiments in dependence of structural and motional aspects, have been established. Due to the broad range of structures and mobilities of organic material at interfaces, the range of employed NMR techniques is large as well, including various liquid state NMR methods applied to adsorption layers in solution, and solid state techniques applied to grafted solid layers.

For most material classes the studies still have exploratory character in the sense that most methods have only been applied to a few compounds. For example studies of grafted chains are limited to PEO and PDMS, surfactant 2H studies are limited to hydrophilic surfaces, and solvent relaxation was applied to water in aqueous systems only. In most cases this is not due to experimental limitations, so that in the near future, with the basic principles of applicability being established, a more systematic coverage of different compounds can be expected.

The majority of the results presently available was obtained for rather simple adsorption systems such as homopolymer monolayers, and amphiphilic mono- or bilayers. It can be expected that the field will expand further towards more complex systems like the examples of the last chapter or other coupled or composite layers. Though such systems seem fairly complex for experimental techniques just being established in organic adsorption layer research, magnetic resonance has a particular advantage, especially in complex systems: Due to the local origin of the NMR signal, selected molecular functionalities can be probed, while the majority of surface techniques rather detect overall layer parameters such as thickness, density etc. In this context examples of practical interest could be the coupling between lipids and polymeric layers, which is relevant for biological model systems, or the probing of functional components like temperature- or pH- sensitive materials in adsorption layers.

In general, due to the tremendous range of dynamic time scales and varying molecular structures which organic molecules can assume at solid interfaces, the task for NMR investigations will remain to probe and find different methods which resolve the issues of interest. Since an enormous variety of NMR techniques is available and established in bulk materials, a huge potential of locally resolved structural and dynamic investigations is arising.

ACKNOWLEDGEMENT

The author would like to thank H. Möhwald and O. Söderman for critically reading the manuscript and making helpful suggestions.

5. REFERENCES

1. F. D. Blum, *Ann. Rep. NMR Spectr.* 28(1994)277.
2. T. Cosgrove and P. C. Griffiths, *Adv. Coloid. Interface Sci.* 42(1992)175.

3. F. Grinberg and R. Kimmich, *J. Chem. Phys.* 105(1996)3301.
4. S. Stapf and R. Kimmich, *Macromolecules* 29(1996)1638.
5. P. A. Mirau, S. A. Heffner and M. Schilling, *Chem. Phys. Lett.* 313(1999)139.
6. L.-Q. Wang, J. Liu, G. J. Exarhos and B. C. Bunker, *Langmuir* 12(1996)2669.
7. M. Zeghal, P. Auroy and B. Deloche, *Phys. Rev. Lett.* 75(1995)2140.
8. O. Söderman and P. Stilbs, *Prog. NMR Spectr.* 26(1994)445.
9. G. Lindblom and G. Orädd, *Prog. NMR Spectr.* 26(1994)483.
10. L. Facchini and A.P. Legrand, *Macromolecules* 17(1984)2405.
11. H. Hommel, *Adv. Colloid Interface Sci.,* 54(1995)209.
12. J. H. Davis, *Biochim. Biophys. Acta* 737(1983)117.
13. H. Wennerström, B. Lindman, O. Söderman, T. Drakenberg and J.B. Rosenholm, *J. Am. Chem. Soc.* 101(1979)6860.
14. M. Liang and F.D. Blum, *Macromolecules* 29(1996)7374.
15. C.K. Hall and E.J. Helfland, *J. Chem. Phys.* 77(1982)3275.
16. R. Kimmich, 'NMR Tomography, Diffusometry, Relaxometry', Springer, Berlin 1997.
17. G.P. van der Beek and M.A. Cohen Stuart, T. Cosgrove, *Langmuir* 7(1991)327.
18. J. Kärger, H. Pfeifer and W. Heink, *Adv. in Magn. Reson.* 12(1989)1.
19. W. S. Price, *Ann Rep on NMR Spectr* 32(1996)51.
20. J.P.H. Zwetsloot and J.C. Leyte, *J. Colloid Interface Sci.* 181(1996)351.
21. J. Grandjean and J.-L. Robert, *J. Magn. Reson.* 138(1999)43.
22. J.F. Haggerty and J.E. Roberts, *J. Appl. Polym. Sci.* 58(1995)271.
23. M. Schönhoff and O. Söderman, *J. Phys. Chem. B* 101(1997)8237.
24. E. Söderlind and P. Stilbs, *Langmuir* 9(1993) 1678.
25. T. Yonezawa, N. Toshima, C. Wakai, M. Nakahara, M. Nishinaka, T. Tominaga and H. Nomura, *Coll. Surf. A* 169(2000)35.
26. M. Schönhoff and O. Söderman, *Magn. Reson. Imaging* 16(1998)683.
27. E. Söderlind and P. Stilbs, *J. Colloid Interface Sci.* 143(1991)586.
28. E. Söderlind and F.D. Blum, *J. Colloid Interface Sci.* 157(1993)172.
29. P.M. Macdonald, Y. Yue and J.R. Rydall, *Langmuir* 8(1992)164.
30. Y. Yue, J.R. Rydall and P.M. Macdonald, *Langmuir* 8(1992)390.

31. K. Nagashima and F.D. Blum, *Colloid Surfaces A* 176(2001)17.
32. K. Nagashima and F.D. Blum, *J. Colloid Interface Sci.* 214(1999)8.
33. E. Söderlind and P. Stilbs, *Langmuir* 9(1993) 2024.
34. E. Söderlind, M. Björling and P. Stilbs, *Langmuir* 10(1994) 890.
35. E. Söderlind, *Langmuir* 10(1994) 1122.
36. P.-O. Quist and E. Söderlind, *J. Colloid Interface Sci* 172(1995) 510.
37. M. Schönhoff, O. Söderman, Z.X. Li and R.K. Thomas, *Bull. Magn. Reson.* 20(1999)25.
38. M. Schönhoff, O. Söderman, Z.X. Li and R.K. Thomas, *Langmuir* 16(2000)3971.
39. S.C. Kuebler and P.M. Macdonald, *Langmuir* 8(1992) 397.
40. P.M. Macdonald and Y. Yue, *Langmuir* 9(1993)1206.
41. P.M. Macdonald, *Colloid Surfaces A* 147(1999)115.
42. R.C. Zeigler and G.E. Maciel, *J. Am. Chem. Soc.* 113(1991)6349.
43. A. Tuel, H. Hommel, A.P. Legrand, H. Balard, M. Sidqi and E. Papirer, *Colloids Surfaces* 58(1991)17.
44. R.C. Zeigler and G.E. Maciel, *J. Phys. Chem.* 95(1991)7345.
45. G. Piedra, J.J. Fitzgerald, C.F. Ridenour and G.E. Maciel, *Langmuir* 12(1996)1958.
46. A. Badia, W. Gao, S. Singh, L. Demers, L. Cuccia and L. Reven, *Langmuir* 12(1996)1262.
47. W. Gao and L. Reven, *Langmuir* 11(1995)1860.
48. H. Schmitt, A. Badia, L. Dickinson, L. Reven and R.B. Lennox, *Adv. Mat.* 10(1998)475.
49. W. Gao, L. Dickinson, C. Grozinger, F.G. Morin and L. Reven, *Langmuir* 12(1996)6429.
50. S. Pawsey, K. Yach, J. Halla and L. Reven, *Langmuir* 16(2000)3294.
51. A. Badia, R.B. Lennox and L. Reven, *Acc. Chem. Res.* 33(2000)475.
52. T. Köchy and T.M. Bayerl, *Phys. Rev. E* 47(1993)2109.
53. M. Hetzer, S. Heinz, S. Grage and T.M. Bayerl, *Langmuir* 14(1998)982.
54. J. Schmitt, B. Danner and T.M. Bayerl, *Langmuir* 17(2001)244.
55. C. Dolainsky, M. Unger, M. Bloom and T.M. Bayerl, *Phys. Rev. E* 51(1995)4743.

56. C. Dolainsky, P. Karakatsanis and T.M. Bayerl, *Phys. Rev. E*, 55(1997)4512.
57. T. Cosgrove and K.G. Barnett, *J. Magn. Reson.* 43(1981)15.
58. T. Cosgrove and K. Ryan, *J. Chem. Soc., Chem. Commun.*, 21(1988)1424.
59. P. Carriere, Y. Grohens, J. Spevacek and J. Schultz, *Langmuir*, 16(2000)5051.
60. A. Larsson, D. Kuckling and M. Schönhoff, *Colloids Surfaces A* 2001, *in press.*
61. M. Schönhoff, A. Larsson, D. Kuckling and P. Welzel, 2001, *in preparation.*
62. P.W. Zhu and D.H. Napper, *Colloids Surfaces A*, 113(1996)145.
63. S. Azizi, T. Tajouri and H. Bouchriha, *Polymer* 41(2000)5921.
64. Y. Lipatov, T. Todosijchuk and V. Chornaya, *J. Colloid Interface Sci.* 184(1996)123.
65. B.R. Sinha, F.D. Blum and F.C. Schwab, *Macromolecules* 26(1993)7053.
66. F.D. Blum, B.R. Sinha and F.C. Schwab, *Macromolecules* 23(1990)3592.
67. M. Xie and F.D. Blum, *Langmuir* 12(1996)5669.
68. F.D. Blum, G. Xu, M.H. Liang and C.G. Wade, *Macromolecules* 29(1996)8740.
69. W.Y. Lin and F.D. Blum, *Macromolecules* 30(1997)5331.
70. W.Y. Lin and F.D. Blum, *Macromolecules* 31(1998)4135.
71. M. Xie and F.D. Blum, *J. Polym. Sci. Polym. Phys.* 36(1998)1609.
72. M. Zeghal, B. Deloche, P.-A. Albouy and P. Auroy, *Phys. Rev. E* 56(1997)5603.
73. M. Zeghal, B. Deloche and P. Auroy, *Macromolecules* 32(1999)4947.
74. T. Cosgrove, P.C. Griffiths and P.M. Lloyd, *Langmuir* 11(1995)1457.
75. S.J. Mears, T. Cosgrove, L. Thompson and I. Howell, *Langmuir*, 14(1998)997.
76. T. Cosgrove, T. M. Obey and K. Ryan, *Colloids Surfaces* 65(1992)1.
77. B. Schwarz and M. Schönhoff, *Colloids Surfaces A*, 2001, *accepted.*
78. T. Cosgrove, S.J. Mears and P.C. Griffiths, *Colloids Surfaces A*, 86(1994)193.
79. T. Cosgrove and P.C. Griffiths, *Colloids Surfaces A* 84(1994)249.
80. G. Decher, *Science*, 277(1997)1232.
81. G.B. Sukhorukov, E. Donath, H. Lichtenfeld, E. Knippel, M. Knippel, A. Budde and H. Möhwald, *Colloids Surfaces A*, 137(1998)253.
82. L.N.J. Rodriguez, S.M. De Paul, C.J. Barrett, L. Reven and H.W. Spiess, *Adv. Mater.* 12(2000)1934.
83. B. Schwarz and M. Schönhoff, 2001, *in preparation.*

Novel Methods to Study Interfacial Layers
D. Möbius and R. Miller (Editors)

THE FABRICATION OF A SELF-ASSEMBLED MULTILAYER SYSTEM CONTAINING AN ELECTRON-TRANSPORTING CHANNEL

Dongho Kim[a], Hyunjin Chae[a], Haeseong Lee[a], Jaegeun Noh[b], Masahiko Hara[b], Wolfgang Knoll[c] and Haiwon Lee[a,*]

[a] Department of Chemistry, Hanyang University, Seoul 133-791, Korea, e-mail: haiwon@email.hanyang.ac.kr

[b] Frontier Research Program, The Institute of Physical and Chemical Research (RIKEN), Hirosawa, Wako, Saitama 351-0198, Japan, e-mail: masahara@postman.riken.go.jp

[c] Max-Plank-Institute for Polymer Research Ackermannweg 10, 55128 Mainz, Germany, e-mail: knoll@mpip-mainz.mpg.de

Contents

* corresponding author

Keywords: SPR; Kinetics; self-assembly; metal phosphonate film; peryleneimide

The formation of self-assembled multilayer system on various substrates was studied using an N,N'-bis(ethyldihydrogenphosphate)-3,4,9,10-perylene (dicarbox imide) (EPPI), a well-known electron donor. The substrates used in this study were gold, silicon, quartz, and glass. In order to attach the EPPI to the substrates, a layer of an anchoring agent was initially formed on the substrates. Also zirconium was introduced between the adhesive layer and the EPPI layer in order to stabilize this multiplayer system. The characterization and self-assembling process of EPPI on a Zr-MUDP layer were elucidated using *in-situ/ex-situ* surface plasmon resonance (SPR), UV-Vis, and ellipsometry.

1. INTRODUCTION

Nanotechnology becomes an essential field in the 21st century industry due to a high demand of tera-level or higher integrity in the fabrication of electronic devices [1, 2]. In order to achieve such a high integrity, the possibility to use molecules as a component has been investigated. Not only monolayer formation but also multilayer formation plays an important role in fabricating molecular level devices. Especially, the latter is very difficult in controlling uniformity and stability of each layer. In this article the methodology to accomplish self-assembled multiplayer systems will be presented.

Among self-assembled films, the preparation of zirconium phosphonate multilayer films on silicon and gold has been elucidated using a variation of self-assembling techniques [3, 4]. The multilayer systems produced by the self-assembling technique have many attractive advantages such as easy preparation, uniformity, and physical stability. Also, the multilayer films of zirconium phosphonate have another strength that a variety of physical and photochemical properties can be modified by introduction of a new functional group or a molecule to the phosphonate layer [5]. With the consideration of the thermal and photochemical stability, the N,N'-bis(ethyldihydrogenphosphate)-3,4,9,10-perylene(dicarboximide) (EPPI) was chosen to be added to the phosphonate layer since it shows unique photoactive and photoconductive properties [6 - 8]. The photoconductivity enables the EPPI to provide an electron-transporting channel for the multilayer system. Also, a two-dimensional network produced by Zr^{4+} ions provides an extra stability for our system. The molecular structure of EPPI is shown in Fig. 1.

The thickness and the formation of the self-assembled Zr-EPPI multilayer were observed using ellipsometry and UV, respectively. The growing process was investigated using *in-situ* and *ex-situ* surface plasmon resonance (SPR). Also SPR was used to confirming the former factors. A surface plasmon is a quantised oscillation of a free electron which is propagated along a metallic surface or semiconducting surface under specific conditions. When light is introduced to a conducting substrate at a specific angle, the incident light is resonated with surface plasmon. The output light contains detail information on physical and electrical properties. Accordingly, SPR is a useful technique in the field of chemical and biochemical sensing and surface characterization [9 - 13]. Among many optical techniques such as ellipsometry, multiple internal reflection spectroscopy, and differential reflectivity, SPR is one of the most sensitive techniques in investigating surface and interface properties. In this study, the growth of Zr-EPPI layer is monitored by *ex-situ* SPR and the average thickness of one layer of Zr-EPPI is calculated from *in-situ/ex-situ* resonance curves measured by SPR.

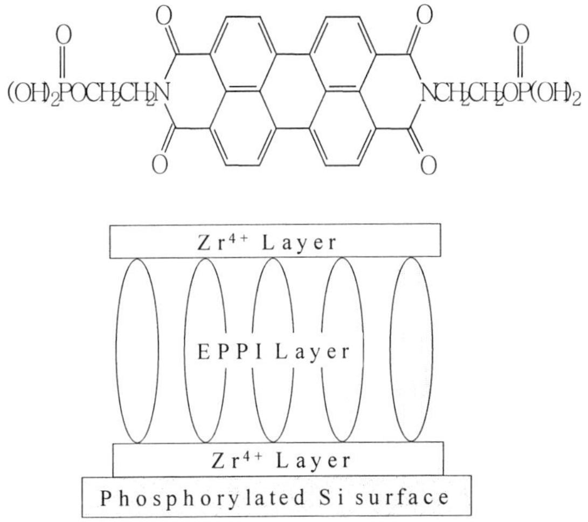

Fig. 1. The molecular structure of EPPI and model of Zr-EPPI layer .

Typical metal substrates such as gold and silicon are selected for the SPR measurements. Also, gold surface is modified by another organic molecules such as (11-mercaptoundecyl) dihydrogen phosphate (MUDP) for the construction of a self-assembled multilayer.

In this monograph we will report the trends in the adsorption process of EPPI on various surfaces at different concentrations using *in-situ* SPR method.

2. EXPERIMENTAL

2.1. Preparation of gold substrate

Gold (99.99% purity or higher) was evaporated onto glass slides (BK-270) cleaned by ultrasonication in ethanol using a vapour deposition apparatus. After the complete deposition of gold, the substrate was cleaned with a detergent solution (Hellmanex, Hellma, Germany) and excess amounts of Milli-Q water. For the SPR measurements chromium (20 Å thickness) and gold films (550 Å) were consecutively formed on a glass slide in a vacuum chamber at 5×10^{-6} mbar.

2.2. Preparation of Zr-EPPI film

Si wafers (Siltron Co., Korea) were mechanically polished to obtain (100) surface. Quartz plates (Wilmad Co., Germany) were used as a substrate in UV measurements. Deionized (D.I.) water was purified by a water purification system (Milli-Q reagent grade: Millipore Co., USA) until its resistivity was reached to 18 MΩcm and was used in all experiments. All solutions were filtered through 0.2 μm filter before an experiment. Zirconyl chloride octahydrate ($ZrOCl_2$), phosphorous oxychloride ($POCl_3$) and 2,4,6-collidine were obtained from Aldrich Chemical Co. Acetonitrile with HPLC grade (a solvent for $POCl_3$) was purchased from Fisher Scientific. Nitrogen gas with 99.99% purity was used for purging all solutions.

The silicon wafers and quartz plates were cleaned prior to use as substrates in piranha solution (30 % H_2O_2: 98 % H_2SO_4, 1 : 3 by volume). Subsequently silicon and quartz substrates were rinsed with acetonitrile. Then these surfaces were phosphorylated with 10 mM solutions of $POCl_3$ and 2,4,6-collidine in acetonitrile for 12 hours at 70 °C. To remove a remaining anchoring agent, the surfaces modified by the phosphonic acid were further washed with acetonitrile and D.I. water. The treated substrates were dipped into 5 mM $ZrOCl_2$ solution and 1 mM EPPI in KOH at room temperature to combine with Zr^{4+} ion and EPPI, respectively. This process may be repeated to obtain more than a two-layered film. To avoid the formation of bulk Zr-EPPI, a substrate was always rinsed with D.I. water thoroughly after it was dipped into the solution. Gold substrates were prepared by thermal evaporation of gold on mica and glass.

The formation of a Zr-EPPI layer on gold surface was much enhanced in the presence of an anchoring agent such as MUDP having -SH and phophonic acid at both ends. Substrates were dipped into 0.01 mM MUDP solution in ethanol. After 24 hrs, the surface was completely covered with phosphorylate. The modified substrate was dipped in 5 mM solution of $ZrOCl_2$ for 30 minutes to add the layer on phosphorylated substrates. Then an EPPI layer was formed on the Zr^{4+}/Phosphorylate/MUDP/Au in 1 mM aqueous solution of EPPI whose pH was adjusted at 4.6 with KOH and HCl at room temperature for 30 minutes. The time period of 30 minutes was enough to react because the chemical reaction between Zr^{4+} ion and POO^- ion was preceded very fast. The pH condition is very important because the Zr-EPPI system is decomposed at pH = 10 or higher and EPPI molecule is precipitated at pH = 3 or lower. The suitable pH condition (pH = 4.6) was found in growing crystalline Zr-EPPI. One layer of Zr-EPPI is defined as the layer between Zr^{4+} layer and next Zr^{4+} layer.

2.3 Ellipsometry

Ellipsometric measurements were carried out using a Rudolph auto-EL II ellipsometer using 632.8 nm radiation from halogen lamp, at an angle of incidence $\phi = 70°$ and always taken after zirconium deposition.

2.4. Surface plasmon resonance [14 - 16]

Surface plasma wave represents the oscillations of surface charge (free electrons in metal) which are stimulated by an external electric field. The amplitude of the wave shows a maximum intensity on the metallic surface and an exponential decay when it is propagated inside the sample. Surface plasmons are quantized oscillations of the wave. The dispersion pattern of a plasmon contains characteristic properties of a thin film and its interfacial properties. The dispersion relation for a nonradiative plasmon is given in the following expression for a plasmon wavevector.

$$k_{sp} = \frac{2\pi}{\lambda}\sqrt{\frac{\varepsilon_m \varepsilon_d}{\varepsilon_m + \varepsilon_d}} \tag{1}$$

λ is the wavelength of the excitation light, e_m is the real part of the dielectric constant of the metal, and e_d is the dielectric constant for the medium outside the metal. The frequencies that are of interest are those at which e_m is negative. If the excitation light is incident of the surface with an angle of theta, then the wave vector of the light incident on the surface is given by the following equation:

$$k_x = \frac{2\pi}{\lambda} n_p \sin\theta \tag{2}$$

n_p is the refractive index of the prism. It must be noted that the dielectric constant is the square root of the refractive index. A thin dielectric layer, e.g., an organic thin film, deposited on top of the metal film causes an increase in the plasmon wave vector and consequently shifts the resonance to an angle $\theta_1 > \theta_0$. Using Fresnel's equations, the optical thickness, the refractive index, and geometrical thickness of the thin film can be calculated subsequently.

In this study SPR measurements were performed in air with p-polarized He-Ne laser (wavelength, 632.8 nm) as a light source. An attenuated total internal reflection setup using the Kretschmann geometry was adopted for the excitation of the surface plasmon. Winspall 2, which is data analysis software (developed at the Max-Planck-Institute for Polymer Research), was applied to simulate the reflectivity curves.

3. RESULTS AND DISCUSSION

3.1. Characterization of Zr-EPPI multilayer films

The characterization of self-assembled Zr-EPPI multilayer films was accomplished using UV-Vis, ellipsometry, SPR, and XRR. The UV-Vis spectra of the EPPI solution and the self-assembled Zr-EPPI film revealed the adsorption of EPPI molecules on a modified substrate. The π- π^* absorption bands of EPPI monomers, EPPI n-mers in the EPPI solution, and self-assembled Zr-EPPI film appeared at 472 nm, 497 ± 2, 543 ± 5 nm, respectively. The π- π^* absorption band of EPPI monomers at 472 nm was shown not in Zr-EPPI film but in EPPI solution.

The absorption bands of EPPI n-mers and absorption bands by intermolecular interactions between EPPI n-mers were appeared at the longer wavelength, 496-498 nm and 538-548 nm. The absorption band of Zr-EPPI film was red shift about each 2 nm and 10 nm comparing to that of EPPI solution. These results mean that the distance between EPPI monomers is shorter because the free molecules of EPPI solution were packed into solid film. Figure 3 shows the

UV-Vis spectra of Zr-EPPI multilayer films on quartz as the number of layers and the plot of absorbance of Zr-EPPI multilayer films as a function of the number of layers increase at maximum wavelength, 498 nm. The absorbance of the film increases linearly as the number of layers increases at 498 nm. The linear increase in absorbance indicates that the same amount of EPPI molecules is being deposited in each treatment but does not verify uniform layer growth.

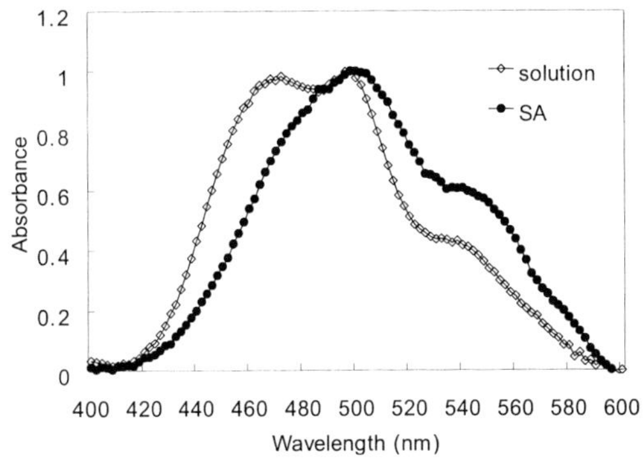

Fig. 2. The UV-Vis spectra of EPPI solution and self-assembled Zr-EPPI film.

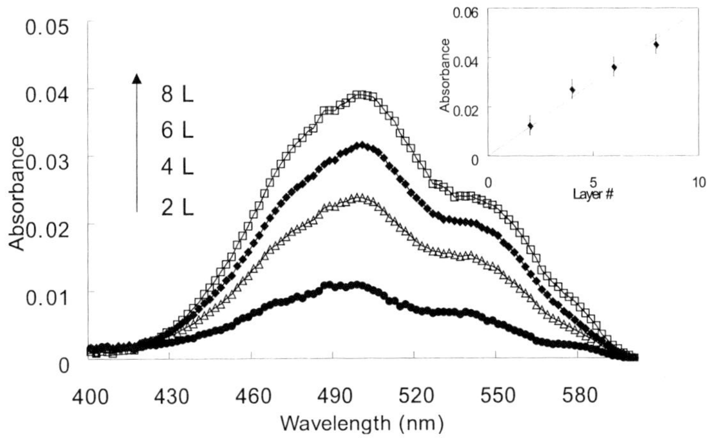

Fig. 3. The UV-Vis spectra of self-assembled Zr-EPPI multilayer films on quartz with a plot of the absorbance of Zr-EPPI multilayer films as a function of the number of layers increase.

Ellipsometric data are a better indication of uniformity and complete coverage since it measures an average thickness; however, it can also be misleading since the measurement is an average over a macroscopic area. If the layers are relatively tightly packed and uniform, the measured layer thickness should be close to the value seen in the bulk solid, and the deviation in the measurements at various points on the substrate should be small. As the number of layers increases, the thickness of the film increases linearly from the plot of the film thickness of Zr-EPPI multilayer on Si-wafer surface vs the number of layers by ellipsometry (as not shown here). The slope of the line gives a layer thickness of 13.1 Å/layer, which means the average thickness unit Zr-EPPI layer. The thickness measured with ellipsometry at various points on the substrate surface show very little deviation ($\pm$5 Å), as expected for a lamellar thin films. SPR measurement was done to calculate accurate thickness of self-assembled Zr-EPPI multilayer film, starting with a gold substrate prepared. SPR are available for the detection of the effective refractive index change of thin dielectric films. Experimentally the resonant excitation can be observed using an attenuated total reflection. The coupling angle of the plasmon is indicated where the reflectivity curve has a narrow dip. The resonant excitation of the surface plasmon is significantly influenced by the existence of a thin dielectric coating on the thin metal layer. The coating leads to a shift in the resonance angle toward a higher angle. The optical thickness of the coating can be calculated from the shift. Surface plasmons excited at the self-assembled Zr-EPPI film/Au interface were detected by angular scans of the reflectivity. For the purpose of the analysis of the data it was assumed that the dielectric constant (ε) of the EPPI is 2.25 ($n = \varepsilon^{1/2} = 1.54$). Resonant excitation of the surface plasmon was observed even in the presence of a Zr-EPPI film on a thin gold layer. Fig. 4 displays the SPR reflectivity curves obtained as a function of incident angle and the simulated thickness for a series of self-assembled Zr-EPPI film.

The reflectivity curve obtained for the film was shifted to higher angle as the number of layer increases around 42-45°. The thickness of the films was calculated from the resonance angle shift by a modified Fresnel equation. The calculated average thickness of self-assembled Zr EPPI film was 13.1 Å and this value was similar to ellipsometric results.

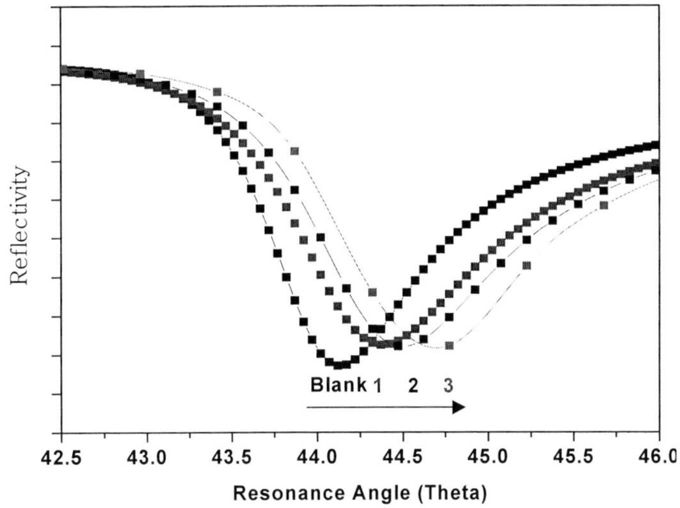

Fig. 4. The SPR reflectivity curves of Zr-EPPI multilayer.

Table 1. The resonance angle and average film thickness of Zr-EPPI multilayer obtained from SPR measurement.

Number of Layer	Resonance Angle(°)	Simmulated Thickness(Å)
Blank (Bare Au)	42.75 °	Cr : 22.4 Å Au : 544.9 Å
Zr-EPPI 1L	43.4 °	12.9 Å
Zr-EPPI 2L	43.7 °	24.9 Å
Zr-EPPI 3L	44.1 °	37.9 Å

3.2. Adsorption kinetics of Zr-EPPI layer

The SPR curve due to plasmon generation is extremely sensitive to variations in the refractive index and thickness of the layer at the metal-dielectric interface. During self-assembly of the EPPI molecule, the thickness and complex refractive index or dielectric for the SAM film changes the reflectivity of the combined films. From these reflectivity curves we can determine EPPI adsorption process on Zr-MUDP monolayer and its thickness using *in-situ* SPR measurements. The self-assembling process of EPPI on Zr-MUDP monolayer was monitored on various concentrations, from 10 mM to 0.01 mM, using an *in-situ* SPR method as shown in Fig. 5. In SPR study, we varied the concentration of EPPI in pH-controlled solution with HCl and KOH to define the adsorption process of EPPI and correlation of the formation rate with the concentration of the solution. SPR data for the EPPI monolayer formation shows that an increase in the overall film growth rate as concentration increases.

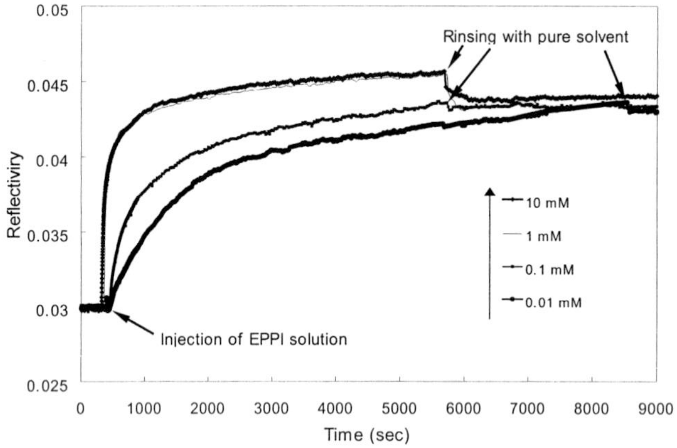

Fig. 5. The reflectivity change as a function of deposition time measured in various pH-controlled EPPI solutions (from the top, 10 mM, 1 mM, 0.1mM, and 0.01 mM).

The initial formation of the monolayer is rapidly occurred within 30 minutes. If we assume that coverage is proportional to total amount of film material and all surface sites are equivalent, we can use Langmuir adsorption model. We compared experimental EPPI adsorption trend with theoretical Langmuir isotherm determined from *ex-situ* quartz crystal microbalance data that was indicated that the hypothetical Langmuir isotherm matches with the experimental isotherm in aspects of deposition time and trend of adsorption. The Langmuir isotherm dictates that fractional surface coverage is given by [17 - 19]

$$\frac{d\theta}{dt} = \left(\frac{\mathrm{k_a}}{N_0}\right)c(1-\theta) - \left(\frac{k_d}{N_0}\right)\theta \tag{3}$$

where θ is the fractional surface coverage, t is the adsorption time, k_a and k_d are the adsorption and desorption rate constants, N_0 is the surface adsorbate concentration at full coverage, c is the solution concentration of adsorbate. If we assume that the gold film is microscopically smooth and the area of the EPPI molecule projected into the surface with a perpendicular orientation is 50.2 $Å^2$ (based on the usual bond lengths and the van der Waals radii) [19 - 21], the N_0 should be $3.307 \cdot 10^{-10}$ mol/cm^2. We can obtain the values of k_a and k_d from Langmuir isotherm equation (Eq. (3)) using N_0 value. For small c ($\theta \rightarrow 0$), initial monolayer formation process, adsorption rate becomes very close to the initial rate of monolayer formation ($k_a c/N_0$). A plot the initial rates of adsorption versus the concentration of EPPI are shown in Fig. 6 and initial rates on various concentrations are tabulated in Table 2.

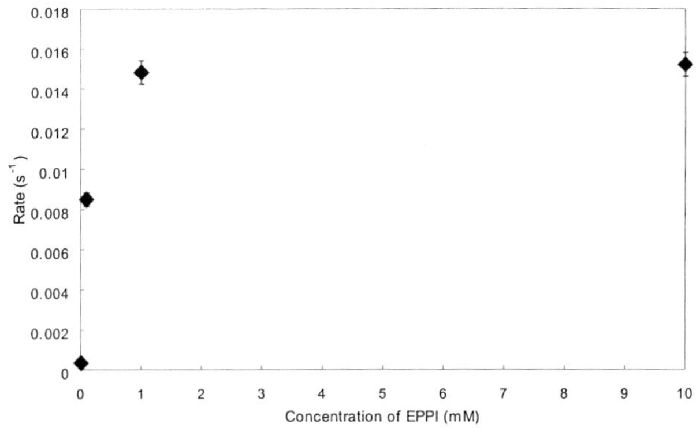

Fig. 6. The initial rate of Zr-EPPI film formation at various concentrations measured by SPR.

Table 2. Concentration dependence for initial rate of self-assembled EPPI film on Zr-MUDP modified gold surface using SPR.

Concentration of EPPI (mM)	Initial rate (s^{-1})
0.01	3.40×10^{-4}
0.1	8.50×10^{-3}
1	1.48×10^{-2}
10	1.52×10^{-2}

The initial rate increased sharply with an increase in the EPPI concentration until the concentration became 1mM. At sufficiently high concentrations of EPPI the adsorption rate is independent of the concentration, which means that the initial adsorption rates are same over 1 mM solution of EPPI. The thickness of EPPI adsorbed on Zr-MUDP modified gold surface is proportional to the amount of material in the film and can be calculated from the reflectivity change. The measured thickness of EPPI monolayer after full adsorption is about 13 Å. In second adsorption process, the adsorption rate is nearly zeroth order (Fig. 5) and the beginning of this stage appears to require a critical surface coverage. The final thickness (reflectivity) change of EPPI films on various concentrations are all same after rising with pure solvent, indicating the existence of a limiting coverage in the formation of higher concentration EPPI solution. For the higher concentrations, self-assembled monolayers are formed more quickly than lower one but rinsing with pure solvent removed physisorbed molecules from the EPPI films and reduced the thickness to that of a single monolayer.

4. Conclusions

A self-assembled multilayer system of Zr^{+4}-EPPI-Zr^{+4}-anchoring agent was successfully fabricated on the various substrates. The characterization of Zr-EPPI multilayer with UV-Vis, ellipsometry, and SPR proved the formation of well-ordered Zr-EPPI layer. Ellipsometry and SPR measurements provided quantitative information on the thickness of the Zr-EPPI multilayer films. Each average thickness of one layer of Zr-EPPI was 13.1 ± 0.5 Å. From the *in-situ* SPR measurements at various concentrations of EPPI solution, the adsorption process of EPPI on Zr-MUDP-gold surface was confirmed as a two-step model. In the first step, a self-assembled Zr-EPPI layer was formed very quickly but molecular orientation was not well ordered. In the second step, the disordered layer was rearranged and consolidated due to the π-π^* interaction between the neighbouring EPPI molecules. The initial rate for Zr-EPPI layer formation increased the concentration of EPPI solution increase up to 1 mM and the initial adsorption rates were same over 1 mM solution of EPPI.

5. Acknowledgments

The authors thank Ms. Gretl Dworschak for the help of SPR measurements. This work was supported by a program of National Research Laboratory, the Ministry of Science and Technology (Grant Number: 99-N-NL-01-C-103).

6. References

1. H.S. Nalwa, Hanbook of Nanostructured Materials and Nanotechnology, Academic Press, 2000.
2. G. Timp, Nanotechnology, 1999.
3. H. Lee, L. J. Kepley, H.–G. Hong and T.E. Mallouk, J. Am. Chem. Soc., 110 (1988) 618.
4. H. Lee, L.J. Kepley, H.–G. Hong, S. Akhter and T. E. Mallouk, J. Phys. Chem., 92 (1988) 2597.
5. S. Akhter, H. Lee, H.–G. Hong, T.E. Mallouk and J. M. White, J. Vac. Sci. Technol. 7(1989)1608
6. H. Langhals, Chem. Ber., 118 (1985) 4641.
7. H. J. Chae, Y. I. Kim and H. Lee, Bull. Kor. Chem. Soc., 19 (1998) 27.
8. H. J. Chae, E. R. Kim and H. Lee, Mol. Cryst. Liq. Cryst., 337 (1999) 149.
9. W. Knoll, MRS Bull., 16 (1991) 29.
10. L. S. Williams, S. D. Evans, T. M. Flynn and N. Boden, Langmuir, 13 (1997) 751.
11. F. Caruso, K. Niikura and Y. Okahata, Langmuir, 13 (1997) 3422.
12. J. Rao, L. Yan and G. M. Whitesides, J. Am. Chem. Soc., 121 (1999) 1629.
13. S. Boussaad, J. Pean and N. J. Tao, Anal. Chem., 72 (2000) 222.
14. O. Prucker, S. Christian and W. Knoll, Macromol. Chem. Phys., 199 (1998) 1435.
15. A. Badia, S. Arnold, V. Scheumann and W. Knoll, Sensors and Actuators B, 54 (1999) 145.

16. R. Advincula and W. Knoll, Colloids Surfaces A, 123 (1997) 435.

17. K. J. Laidler, Chemical Kinetics, 3rd ed.; New York, 1897; Chapter 7.

18. D. T. Grow and J. A. Shaeiwitz, J. Colloid Interface Sci., 86 (1982) 239.

19. A. Ulman, An introduction to ultrathin organic films; Academic Press: New York, 1991.

20. C. W. Block and P. J. George, Mol. Struct., 122 (1985) 155.

21. N. Tillman and A. Ulman, Am. Chem. Soc., 110 (1988) 6136.

Novel Methods to Study Interfacial Layers
D. Möbius and R. Miller (Editors)

COMPOSITE POLYELECTROLYTE SELF-ASSEMBLED FILMS FOR CHEMICAL AND BIO - SENSING

A.V. Nabok[1], A.K. Ray[1], A.K. Hassan[1], N.F. Starodub[2]

[1] Sheffield Hallam University, School of Engineering, Physical Electronics and Fibre Optics Research Laboratories, City Campus, Pond Street, Sheffield, S1 1WB

[2] Palladin Institute of Bio-Chemistry, National Academy of Sciences of Ukraine, 9 Leontovich Street, 252030, Kiev, Ukraine

Contents

Keywords: polyelectrolyte self-assembly, organic dyes, enzyme/indicator optrode, immunoglobuline, UV-visible spectroscopy, SPR, interferometry

The method of polyelectrolyte self-assembly (PESA) was employed for the development of composite multifunctional sensing membranes. Different organic dye molecules, namely copper phthalocyanine (CuPc) and cyclo-tetra-chromotropylene (CTCT), both tetra-sulfonic sodium salts, were incorporated into the polymer matrix by alternation with polyallylamine (PPA). The response of CuPc containing films to low concentration (5ppm) of NO_2 was observed from the changes in their UV-vis spectra. Films containing CTCT have been exploited for the developing enzyme sensors because of their potential for registration of ammonia. Molecules of the enzyme Urease deposited on top of CTCT/PAA films form an optrode for registration of the reaction of urea decomposition. Monolayers of immunoglobuline (anti-immunoglobuiline) were successfully immobilised on top of PAA layers and studied with both surface plasmon resonance (SPR) and planar polarisation interferometric (PPI) techniques. The registration of the immune components of the concentrations less then 1ng/ml was shown to be possible.

1. Introduction

Development of chemical- and bio-sensors depends upon the formation of composite sensitive membranes consisting of several chemical components and preferably combining several functions, e.g. sensing (recognition) and transducing. A novel technique of polyelectrolyte self-assembly (PESA) [1, 2] can provide such opportunities. Many other electrically charged objects, such as macro-molecules, particularly, organic dyes [2-4], proteins [5, 6], inorganic nano-particles [7] can be introduced in the film by simple alternation with adequate polyions. Composite films containing different components in required sequences can be produced by this method. The precision of film deposition and homogeneity of the films produced is very high and comparable with those for Langmuir-Blodgett (LB) method. Additionally, PESA films demonstrate much better thermal and mechanical stability, as compared to LB films. All factors mentioned above make the method of PESA very much suitable for development of optical sensors.

Several optical methods, namely UV-visible optical spectroscopy, surface plasmon resonance (SPR) and planar polarisation interferometry (PPI), were exploited in this work for different sensor applications. SPR has become a routine method in chemical and bio-sensing, and several commercial SPR instruments are available nowadays [8-11]. However, the interpretation of SPR results is not straightforward sometimes and requires deep understanding of physical processes behind. Method of interferometry provides a highest sensitivity to small changes in optical parameters of thin films and thus extremely promising for sensor development, especially considering recent progress with planar Mach-Zhender [12, 13] and polarisation [14, 15] interferometers. Well established method of optical spectroscopy, the most powerful tool in analytical chemistry, can provide much more information about chemical processes then simple registration of changes in refractive index and thickness with SPR and interferometry methods. Recent achievements of integrated optics technology, which allow to produce miniaturised optical spectrometers [16] , gave a second life to this method in sensing applications.

As has been shown previously [17, 18], polymer films containing enzyme and respective organic indicator molecules can be used for development of the optrode, the element combining recognition and optical transducing functions in the same membrane. In our earlier publication [19], the method of PESA was adopted for the same reason. The films containing cyclo-tetra-chromotropylene (CTCT) were produced, and their UV-visible absorption spectra were found to be very sensitive to ammonia due to the multiple deprotonation of the organic chromophore [19]. Furthermore, composite PESA films containing both CTCT and enzyme Urease were shown similar spectral response on ammonia molecules released as a result of urea decomposition [19, 20].

The present work is a continuation of our research activity towards the development of chemical and bio-sensors of this type. Two types of composite films were produced with PESA method and studied throughout this work: PAA/CuPc, and CTCT/PAA/Urease, Attempts to incorporate some other organic indicators, such as methyl red and thymol blue into PESA films have also been made.

The method of PESA has been successfully implicated for immobilisation of different proteins onto the solid substrates of different types [2-4, 21, 22]. Protein molecules incorporated into the hydrophilic polymer environment retain their structure and functions for a long time [22]. In the present work, the method of PESA was used for immobilisation of immunoglobuline.

Along with surface plasmon resonance (SPR) measurements, the traditional method of studying the adsorption of proteins, a novel, very sensitive method of planar polarisation interferometry [14, 15] was adopted here. Adsorption of both antigen and antibody as well as their specific interaction were studied with above methods.

2. Experimental Details

2.1. Deposition of composite PESA films

Two basic polyelectrolytes were chosen for film deposition, i.e. poly(alylamine) hydrochloride (PAA) and poly(sterenesulfonate) sodium salt (PSS) both purchased from Aldrich. Organic indicators, namely, cyclo-tetrachromotropylene tetra-sulfonic sodium salt (CTCT) (synthesised at Sheffield University [19]), Copper Phthalocyanine tetra-sulfonic sodium salt (CuPc), Thymol Blue (TB) sodium salts (purchased from Aldrich) were embedded into the polymer matrix. Chemical formula of all compounds used are shown in Fig. 1. Aqueous solutions of 1 mg/ml concentration of the above polyions and organic dyes were used for film deposition.

An in-house built, computer controlled experimental set-up for polyelectrolyte deposition, was used in this work [23]. The set-up schematically shown in Fig. 2 comprises a plastic dish (1) rotated by a stepper motor (2). Four glass beakers are placed on the dish: two of them (3, 4) containing polyion solutions and the other two (5, 6) filled with water. The sample (7) can be fixed in a holder of a dipping mechanism (8). Both stepper motor and dipping mechanism are interfaced to PC (10) via IO card (9). The deposition routine consisted of sequential dipping the sample in the solutions of PAA and organic dye for 10 min, respectively, with an intermediate rinsing in Millipore water. In order to charge glass substrates negatively, they were sonificated at $60^{O}C$ in 1% NaOH solution in 60% ethanol [23]. Freshly deposited Au layers, usually slightly negatively charged, were used as they are. For those substrates, PAA layer was deposited first. In the case of Si_3N_4, having positively charged surfaces due to the presence of NH_2 groups, the deposition was started with PSS layer.

Several types of samples of composite films were produced. Since both CuPc and CTCT molecules contain four SO_3^- groups (see Fig.1), they were deposited by simple alternation with PAA layers. The structure of the films produced is shown schematically in Fig. 3a. Situation with TB molecules, having only one SO_3^- group, is much more complicated. Several attempts of their deposition by PESA method were not successful. Surprisingly, reasonably good results were achieved by simple sequential dipping into TB and PAA solutions, although the films produced were still patchy. Most likely, TB molecules form aggregates with SO_3^- groups oriented randomly. Such aggregates can interact with PAA layers in a standard manner, and form multilayer PAA/TB structure.

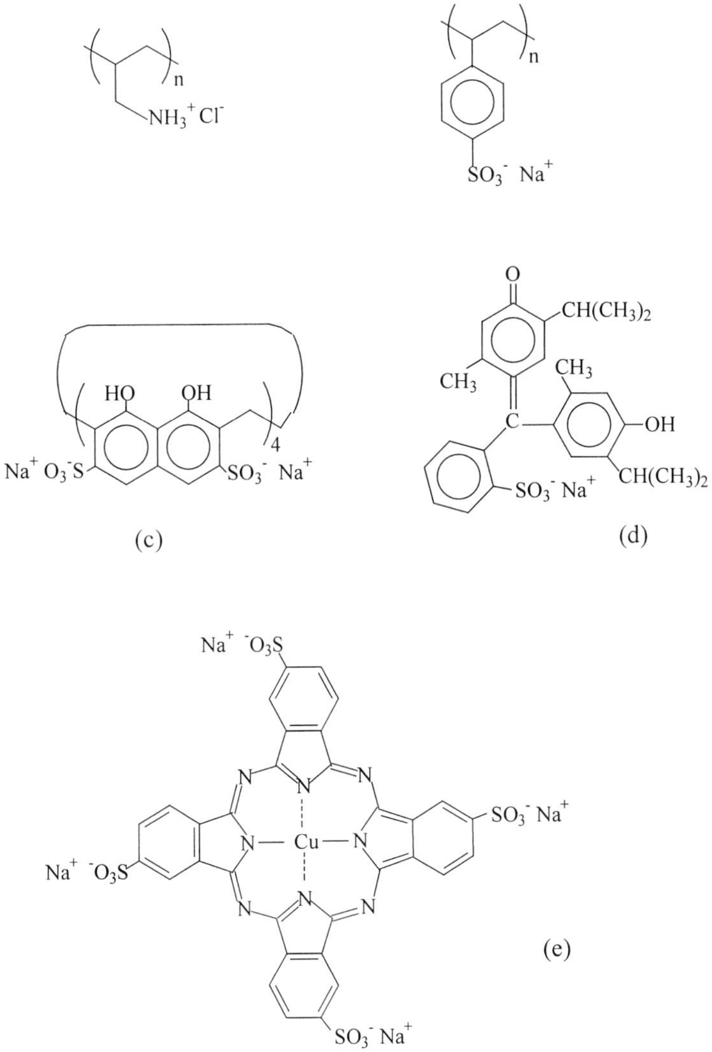

Fig. 1 Chemical formula of the compounds used: (a) PAA, (b) PSS, (c) CTCT, (d) TB, (e) CuPc.

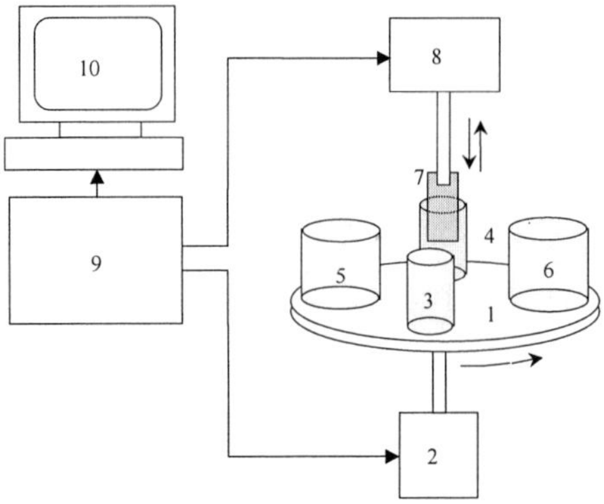

Fig. 2 Experimental set-up for polyelectrolyte self-assembly.

The enzyme Urease (SIGMA), which was negatively charged in 1mg/ml solution in the Trisma base/HCl buffer having pH 8.0-8.3, was deposited by alternation with PAA. A few layers of Urease/ PAA were deposited on top of CTCT/PAA films. Composite membranes with the typical structure of glass/PAA/(CTCT /PAA)$_n$/(Urease/PAA)$_m$ with n=10-40 and m=1-5 were finally produced (see the scheme in Fig. 3b).

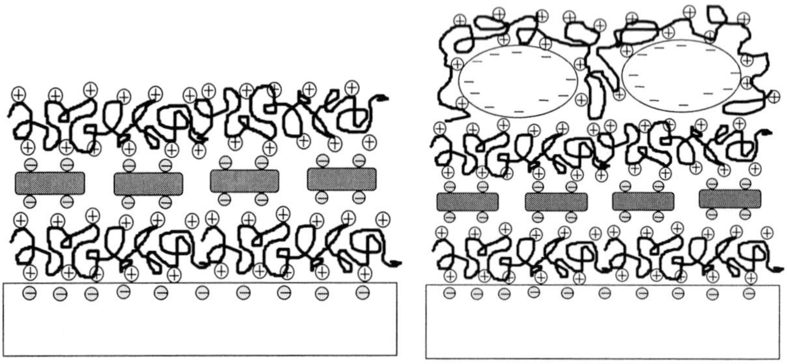

Fig. 3 Schematic layer-by-layer structure of the composite PESA films: (a) PAA/(Indicator/PAA)$_n$, (b) PAA/(Indicator/PAA)$_n$ (Enzyme/PAA)$_m$.

Human Immunoglobuline (IgG) and goat-on human polyclonal antibodies to it (anti-IgG) were produced in the Institute of Biochemistry (Kiev, Ukraine). Both immune components being negatively charged in the above buffer solution were immobilised on top of PAA layer. In order to improve orientation of the immune components, a layer of protein-A (Staphylococcus

Aureus from Miles Laboratories, USA) was used as described in [24]. Bovine Serum Albumin (BSA) (from SIGMA) was used to exclude non-specific binding of the immune components.

The produced composite films were characterised with UV-vis spectroscopy and SPR. UV-visible absorption spectroscopy was also used to study the response of composite PESA films to NO_2 gas, and urea.

2.2. *Film characterisation with SPR and UV-vis spectroscopy.*

Adsorption of the immune components as well as the immune reactions were investigated using an in-house made SPR experimental set-up [25], which is shown schematically in Fig. 4.

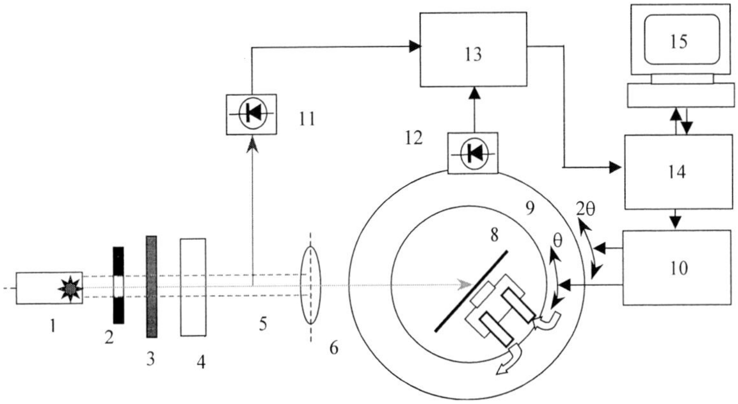

Fig. 4. Experimental set-up for SPR measurements.

A 632.8 nm light beam from HeNe laser (1) after passing through the aperture (2), chopper modulator (3) and polariser (4) was focused with the lens (6) on the back surface of the semi-cylindrical prism (7). The sample (8), e.g. standard microscopic glass slide with about 45 nm thermally evaporated gold overlayer, was brought into optical contact with the prism using index matching fluid (Ethyl salysitate from Aldrich). The prism was placed onto θ-2θ rotated stage (9) driven by a stepper motor (10). The reflected beam is collected by photodetector (12), which is also mounted onto the stage. In this Kretschmann type SPR set-up [26], turning the prism on the angle θ accompanied with the rotation of the photodetector on 2θ. Signal processing was carried out by lock-in amplifier (13). The beam reflected from the semitransparent mirror (5) and collected by photodetector (11) serves as a reference. The rotated stage operates by the control unit (14), and the whole system controlled by PC (15) through the IEEE card. A PTFE cell of about 2 cm^3 in volume was sealed by the sample through the rubber O-ring. The cell (16), having an inlet and outlet pipes, allows us to perform SPR measurements in different media (both gaseous and liquid).

Two types of SPR experiments can be performed with the above set-up: (i) measurements the whole SPR curve by sweeping the angle of incidence, and (ii) kinetic measurements by monitoring the reflected light intensity at a fixed angle of incidence, which was normally chosen on the linear part of the SPR curve (on its left side) close to the minimum.

In order to evaluate the changes in the effective thickness of the film as a result of adsorption of bio-molecules, the experimental SPR curves were fitted to the Fresnel's theory by using a least square technique, as was described previously [27, 28].

UV-visible optical spectroscopy was employed in this work in order to study the adsorption spectra of composite PESA films containing organic chromophores. The measurements were performed using a double beam UV4 UNICAM spectrophotometer.

2.3. *Method of planar polarisation interferometry (PPI)*

A novel method of planar polarisation interferometry (PPI) [14, 15] was adopted here for monitoring the immune reaction with enhanced sensitivity. Planar waveguides, schematically shown in Fig. 5a, were fabricated by planar technology on silicon wafers and consists of 1.3 μm thick layer of SiO_2, 190 nm thick Si_3N_4 layer and 1μm thick layer of phosphorosilicate glass.

A sensing window was etched in the upper layer exposing the surface Si_3N_3 to the environment. In this method, a *p*-component of the light is extremely sensitive to changes in refractive index and thus to adsorption onto the nitride surface, while *s*-component is not sensitive to adsorption, and can be used as a reference. Outcoming light intensity depends on the phase shift between *s*- and *p*- components of the polarised light and thus forms a multiperiodic response [15]:

$$I_{out} \sim I_0 \cos[(k_p - k_s)L] = I_0 \cos[2\pi n(\sin\varphi_p - \sin\varphi_s)L/\lambda] \quad (1)$$

Here *L* and λ are the cuvette length (6mm) and light wavelength (632.8 nm), respectively; k_s, k_p are the wave vectors for; φ_s and φ_p are the propagation angles for *s*- and *p*-components, respectively; n is the refractive index of Si_3N_4 core layer. Due to a high L/λ ratio of about 10^4, even a slight changes in the propagation angles lead to a considerable number of interference fringes appeared at the output. Therefore, the sensitivity of PPI devices to changes in the refractive index of the environment within the sensing window is similar to that for Mach-Zhender interferometer and can be estimated as 10^{-5} [29].

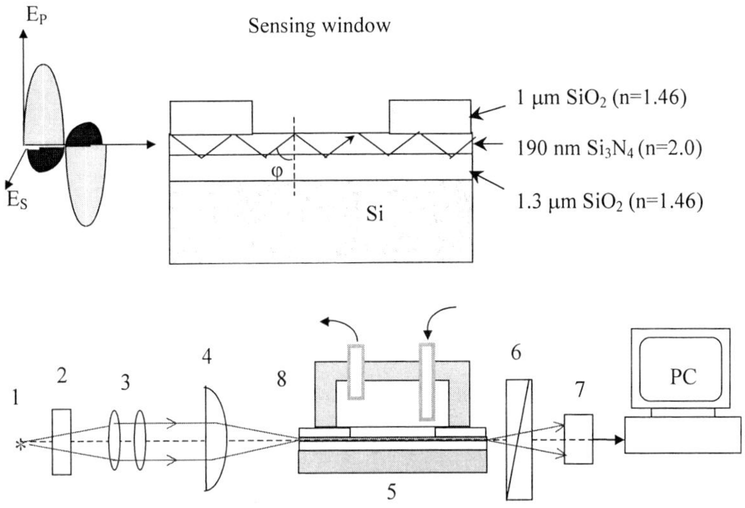

Fig. 5 Schematic diagram of the PPI chip (a) and PPI experimental set-up (b)

In the PPI experimental set-up (see Fig. 5b), a circular polarised beam from a HeNe laser (1) was focused by cylindrical lens (2) in a narrow stripe on the edge of PPI waveguide (3). The outcoming beam after passing the polariser (4) was collected by photodetector (5) connected to PC via IEEE card. A PTFE cell (6) having an inlet and outlet was sealed by the sample and allowed us to study the adsorption of different molecules in both gaseous and liquid media.

In both SPR and PPI methods, adsorption of the immune components, the protein-A and BSA, as well as a specific immune IgG/anti-IgG reaction were performed by injection of the respective solutions into the cell having a volume of 2 cm^3. The amount of solution injected was normally exceeded the cell volume in 3 times. Washing the samples between adsorption steps was carried out by pumping the excessive amount of buffer solution (10-20 times greater than the cell volume).

3. Results and Discussion

3.1 PESA films containing CuPc

Absorption spectra of PAA/CuPc composite films, shown in Fig. 6, demonstrate linear dependence of the absorbance on the number of PAA/CuPc layers. The spectra of PAA/ CuPc films are blue-shifted in respect to the spectrum of CuPc aqueous solution. It is believed to be caused by formation of H-aggregates of CuPc molecules in the composite PESA films. The value of the energy shift (ΔE) of the exciton band for the molecular stack of N molecules (ΔE_N) in respect to a monomer band (E_m) can be described as [30]:

$$\Delta E = E_N - E_m = \frac{N}{N-1}\frac{|M|^2}{r^3}(1 - 3Cos^2\theta) \tag{2}$$

where M is the transition dipole moment, r is the distance between centres of CuPc molecules in the stack, θ is the angle between CuPc moieties and stacking axes (see the scheme on the inset in Fig. 7). Analysis of this expression yields that at $\theta > 54^O$, $\Delta E > 0$, i.e. a blue shift of the exciton band is observed.

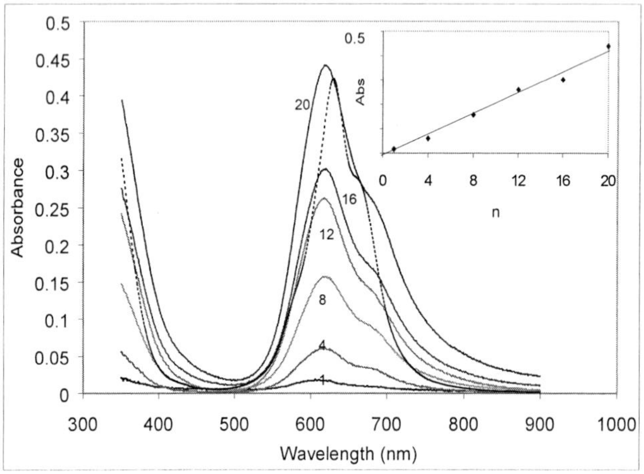

Fig. 6 Spectra of $(PAA/CuPc)_n$ films (number of layers n is shown near respective curves) and of CuPc aqueous solution (dotted line). The inset shows the dependence of the maximum absorbance on the number of layers.

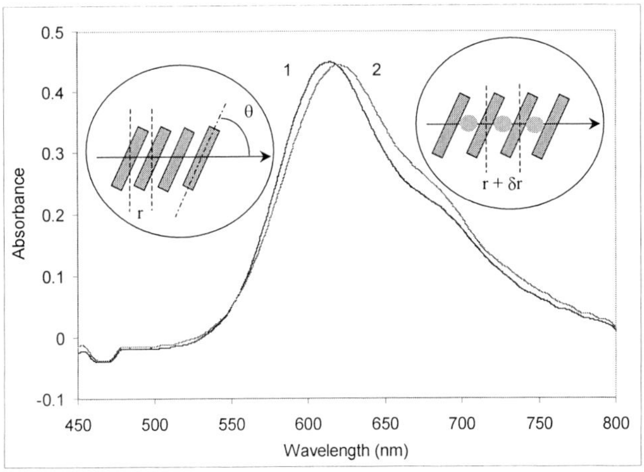

Fig. 7 Spectra of PESA films containing CuPc: (1) before and (2) after exposure to 5 ppm of NO_2 gas for 10 min. Insets shows schematically the stack of CuPc molecules and the effect NO_2 molecules on it.

The effect of exposure of CuPc-PESA films to NO_2 gas depends on its concentration. At low concentration (5 ppm) of NO_2, both the small red shift and reducing of the maximal absorbance of the Q-band were observed, as shown in Fig. 7. It can be interpreted by small increase in the intermolecular distance within CuPc H-aggregates due to incorporation of NO_2 molecules. Changes in the intermolecular distance can be calculated by differentiation of Eq. (2):

$$\delta r = 3r \frac{\delta(\Delta E)}{\Delta E} \tag{3}$$

For this estimation the values of r = 0.378 nm as an intermolecular distance in α-CuPc crystals [31] and ΔE = 0.048 eV from Fig. 6 were used. It was also assumed that the position of Q-band for CuPc aqueous solution corresponds to a monomer. The observed shift $\delta(\Delta E)$ = 0.016 eV, caused by exposure to NO_2, yields $\delta r \sim r$ = 0.378 nm, which corresponds approximately to the size of NO_2 molecule. The scheme on the inset in Fig. 7 illustrates the effect of inclusion of NO_2 molecules into CuPc molecular. It should be noted that such changes within CuPc aggregates are unlikely to occur in the crystalline forms but become possible in the flexible polymer matrix.

The changes in both Q-band position and absorption intensity caused by exposure to 5 ppm NO_2 and the following flushing with dry nitrogen are summarised on the bar-chart diagrams in Fig. 8.

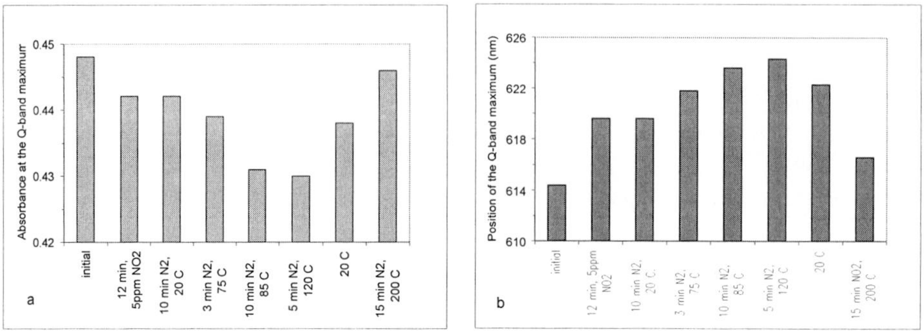

Fig. 8 Bar-chart diagrams of changes in the Q-band maximal absorption (a) and its position (b) due to exposure to 5 ppm NO_2 gas followed by recovery under the nitrogen flow.

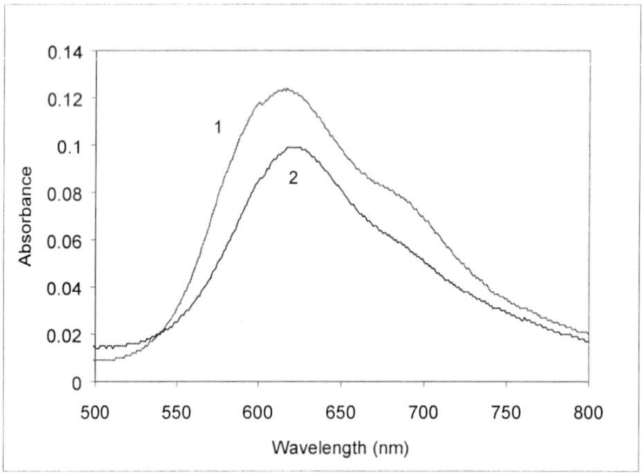

Fig. 9. Spectra of PESA films containing CuPc: (1) before and (2) after exposure to 400 ppm of NO_2 gas for 10 min.

Although the response was reasonably fast (a few minutes), the recovery was very poor. Increases in temperature up to $200^{0}C$ led to partial recovery of the spectrum. After exposure to NO_2 of a high concentration (400 ppm), there was no further shift of the Q-band observed, but the decrease in absorbance was much more pronounced, as shown in Fig. 9. The latter observation may be interpreted in terms of the interaction of the NO_2 molecules with the central Cu atom and π-electrons in its close vicinity [31]. The samples exposed to 400 ppm NO_2 did not show any recovery at room temperature. There was no significant difference in the relative response for the samples having different numbers of PAA/CuPc layers, which indicates the absence of the diffusion limitation of the reaction with NO_2. The observed changes of absorption spectra at low concentrations of NO_2 are believed to be sufficient for registration by sensitive optical techniques such as SPR and, particularly, interferometry.

3.2 Composite PESA films containing indicator/enzyme pairs

Optical spectral transformations of composite PESA films having a structure PAA(CTCT/PAA)$_n$(Urease/PAA)$_m$ were recorded by using UV-visible optical spectroscopy in order to study the reaction of urea decomposition. Typical UV/visible absorption spectra of these films are shown in Fig. 10. When the sample was soaked in urea solution, the band at 550 nm was found to be slightly shifted towards higher wavelengths with an additional band appeared at 440 nm.

These observations are similar to those previously reported for (CTCT/PAA)$_n$ composite films after exposure to ammonia and explained by deprotonation of CTCT molecules [19, 20]. The observed spectral changes are believed to be caused by ammonia released during urea decomposition in the presence of the enzyme Urease. Test samples without Urease layer did not show any spectral changes after soaking in urea solution. The observed spectral changes seem to be permanent since the spectra were not affected by washing in pure Trizma-base/HCl solution for few hours. The present results agree well with those reported earlier [19], where there was a little recovery observed after soaking the sample in aqueous ammonia solution. This can be explained by limited permeation of the liquid through the polymer matrix. The existence of the diffusion limit has also been confirmed by studying the effect of the thickness of (CTCT/PAA)$_n$ layer. In particular, the value of the relative changes of abrobance at 440 nm was found to be 43% for thin films having $n = 8$, while thicker films ($n = 34$) yield smaller changes of 33%.

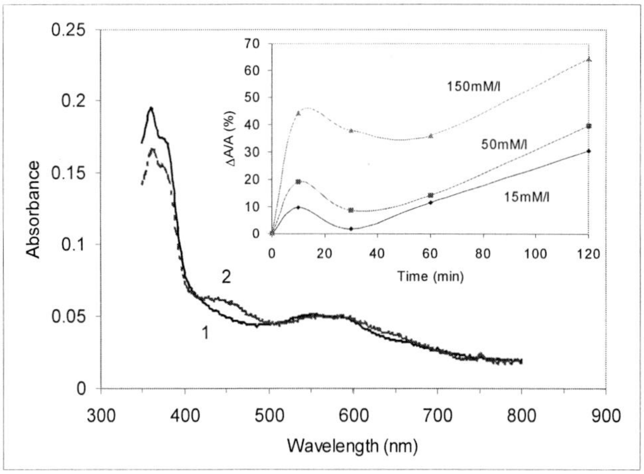

Fig. 10 Spectra of a (PAA/CTCT) (PAA/Urease) film: (1) initial, (2) after soaking in 150 mM/l solution of urea for 1 hour. The inset shows dependencies of the relative changes in absorbance at 440 nm on the urea concentration and exposure time.

The effect of urea concentration and exposure time on the relative absorption intensity of the 440 nm band was summarised in the inset of Fig.10. It can be seen that the kinetics of the spectral response is not monotonic. An initial increase in the absorption intensity after 10 min soaking in Urea solutions was followed by its slight decreasing after 20-30 min until further rising of the absorbance started. Two mechanisms of the response having different characteristic times are assumed in order to account for such a complex behaviour. Firstly, it can be the fast decomposition of the urea within few upper layers containing Urease. Ammonia molecules produced during this stage react with the top layers of CTCT. Secondly, increases in response may be caused by a long time diffusion of ammonia molecules through the film and simultaneous deprotonation of the CTCT-chromophores in the bulk. The observed response on 15 mM/l urea seems to be well above the detection limit, and urea concentrations in one order of magnitude less can be registered.

Several attempts to incorporate other organic indicators, such as methyl red (MR) and thymol blue (TB) sodium salts, into the PESA films has been made in order to form the optrode pairs with other enzymes. The organic indicators mentioned above have only one anionic group per molecule, which makes them difficult to alternate with cationic polymer. MR and TB were mixed with PSS, and then mixed layers of TB/PSS or MR/PSS were alternated with PAA layers. Another way was to form composite films of PAA/indicator/PSS. Here the layers of the indicator were deposited from its diluted aqueous solution on top of PAA layer with some NH_3^+ groups left available for further interaction with PSS. The transfer was very poor in both cases and resulted in inhomogeneously coloured films. However, the simple alternation of TB and PAA yields much better quality films although still patchy. Further work in this direction is underway. We hope to produce composite PESA films containing different pairs enzyme/indicator for the development of optical enzyme sensor-array in near future.

3.3 Immunoglobulines immobilised by polyelectrolyte self-assembly

Before studying the adsorption of the immune components and the immune reaction, optical parameters, i.e. refractive index n and thickness d of both the polymer film and layer of the protein-A were obtained. It should be noted that n and d for transparent thin films can not be evaluated simultaneously from a single SPR measurement [32] To overcome this difficulty, an approach of measurements of two SPR curves in two media having different refractive indices was proposed in [33] and successfully exploited in our previous publications [34, 35]. In this work we measured SPR curves for different protein coatings in air (n = 1) and in the buffer solution (n = 1.334). Two sets of pairs (n, d) were obtained by fitting of respective SPR curves and presented by two $n(d)$ dependencies in Fig. 11. Intercept of these two curves gives a true solution, e.g. values of n and d. In particular, for protein-A layer, calculations yield n = 1.441 and d = 1 nm. The obtained value of n is typical for proteins [36], and it was therefore implemented as a fixed parameter for SPR fitting of all bio-layers throughout this work. Therefore, in the current work, all changes in SPR curves due to adsorption of proteins and immune reactions were related to the changes in the effective film thickness.

Different sequences of adsorption were studied with SPR measurements. One of the typical set of SPR curves, corresponding to an adsorption sequence of PAA/protein-A/IgG/anti-IgG, is shown as an example in Fig. 12. Adsorption of relatively small molecules of the protein-A causes a little shift of the SPR curve, while adsorption of bulky IgG and anti-IgG molecules leads to a substantial shift. The values of the effective thickness of the adsorbed layers were evaluated by fitting of the experimental data to Fresnel's theory, and the results are presented on the inset to Fig. 12 as a bar-chart diagram.

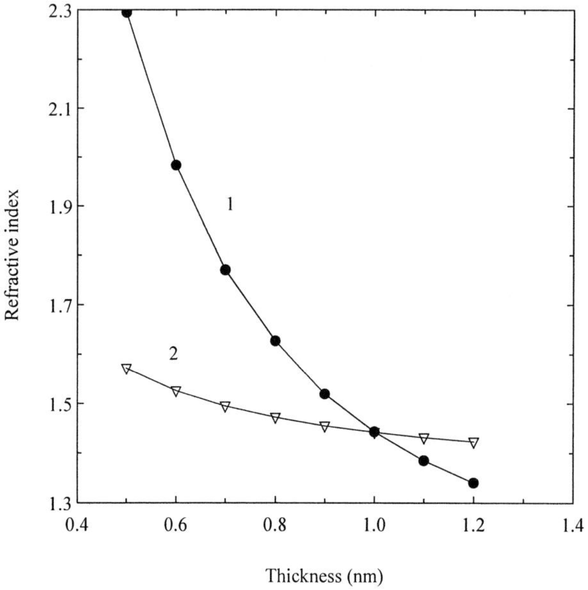

Fig. 11 n(d) dependencies obtained by fitting of SPR curves measured in two media: (1) air and (2) water

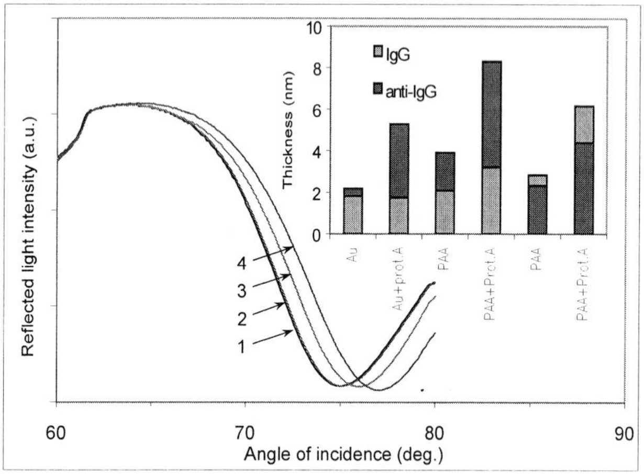

Fig. 12 Typical SPR curves for the adsorption sequence of (1) PAA layer on top of gold, (2) protein-A on top of PAA, (3) IgG on top of protein-A, (4) anti-IgG/IgG specific interaction. The inset shows the effective thicknesses of the adsorbed layers

These results can be summarised as follows: (i) the adsorption of proteins on PAA layer is more effective than one on bare gold; (ii) the presence of the protein-A improves orientation of the immune components and thus increases the efficiency of the immune reaction; (iii)

adsorption of anti-IgG on top of IgG is more efficient than that in reverse order, possibly, because of the difference in the number of binding cites available.

IgG/anti-IgG specific binding was registered in the range of concentration of 0.03 - 100μg/ml for both immune components with a linear response over the concentration range. The reaction was found to be permanent since washing with the twin-buffer removes only non-specifically bound components.

The sensitivity or registration of the immune reaction can be substantially improved by using the method of PPI. The adsorption of IgG (anti-IgG) as well as the immune reaction was studied with the PPI in the range of concentration of both IgG and anti-IgG of 1ng/ml - 100μg/ml. Typical results are shown in Fig. 13.

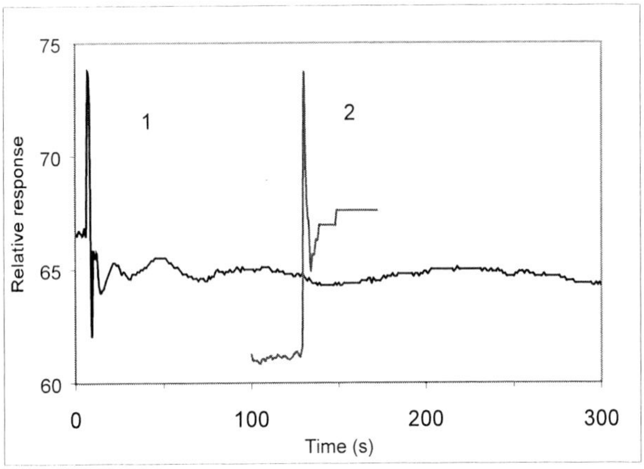

Fig. 13 Typical response of PPI on adsorption of IgG from its 0.1 mg/ml solution (1) and specific binding of anti-IgG of 1ng/ml (2)

The multiperiodic response reflects the changes in the phase between *p*- and *s*- components of the light due to adsorption. Adsorption of anti-IgG molecules from its 100 mM/ml solution onto the PAA layer has resulted in more then 6 periods phase change, while specific binding of 1ng/ml of IgG on top produced much smaller changes. The method is found to be very sensitive and thus promising for bio-sensing, since the concentration of IgG of 1ng/ml causes the phase change for nearly a whole period (see Fig. 13).

Further progress in PPI sensors development is going to be achieved by using an advanced experimental set-up based on lock-in-amplification principle.

4. Conclusions

Method of polyelectrolyte self-assembly was successfully implemented for designing and fabrication of composite sensitive membranes containing different macromolecules, such as organic dyes, enzymes, immunoglobulines.

PESA films containing organic dyes, such as phthalocyanines, were proved to be a good candidate for optical gas sensing elements. Porous polymer matrix provides good gas permeability and enough structural flexibility to sustain changes in Pc-aggregates.

Enzyme/indicator optrodes for registration of enzyme reactions and their inhibitors, such as heavy metal ions and pesticides, can be produced by PESA technique. Composite films containing the enzyme urease and cyclo-tetra-chromotropylene as indicator molecules show some characteristic spectral transformations caused by urea decomposition. The reaction of inhibition of urease by heavy metal ions can be also registered with this method. Further development of the enzyme sensors and sensor arrays lies in finding suitable pairs of enzyme/indicator, their deposition by PESA method and studying the enzyme reactions (including inhibition) with UV-vis spectroscopy.

Immunoglobulines can be successfully immobilised with PESA technique providing substantially higher surface coverage and stability as compared to their direct immobilisation on gold surface. Sensitive optical methods of SPR and, especially, PPI enabled us to register very low concentrations of the antibodies (or antigens) in the range of 30 ng/ml for SPR and 1 ng/ml for PPI. Further improvement in sensitivity is expected by using PPI method.

5. Acknowledgements

This work has been partly supported by INCO-Copernicus Research Project (Contract No IC15-CT96-08818). The authors wish to thank Prof. Yu.M. Shirshov, Dr. T. Richardson and Dr. F.Davis for fruitful collaborations.

6. References

1. Y. Lvov, G. Decher and H. Möhwald, Langmuir, 9 (1993) 481

2. Y.M. Lvov and G. Decher, Crystallography Reports, 39 (1994) 628

3. K. Ariga, Y. Lvov and T. Kunitake, J. Am. Chem. Soc., 119 (1997) 2224

4. M.R. Linford, M. Auch and H. Möhwald, J. Am. Chem. Soc., 120 (1998) 178

5. Y. Lvov, K. Ariga, I. Ichinose and T. Kunitake, J. Am. Chem. Soc., 117 (1995) 6117

6. Protein architecture. Interfacing molecular assemblies and immobilization technology, Y. Lvov and H. Möhwald (Eds.) Marcel Dekker, Inc, 1999, 125

7. J.H. Fendler, F.C. Meldrum, Adv. Mater., 7 (1995) 607

8. R.P.H. Kooyman, H.E. De Bruijn, R.G. Eenink and J. Greve, J. Mol. Structure 218 (1990) 345

9. E. Fontana, R.H. Pantell and S. Strober, Applied Optics, 29 (1990) 4694

10. S. Lofas, M. Malmqvist, I. Ronnberg and E. Stenberg, Sensors & Actuators B, 5 (1991) 79

11. B. Liedberg, I. Lundstrom and E. Stenberg, Sensors & Actuators B, 11 (1993) 63

12. T. Schubert, N. Haase, H. Kuck and R. Gottfried-Gottfried, Sensors & Actuators A, 60 (1997) 108

13. M. Weisser, G. Tovar, S. Mittlerneher, W. Knoll, F.Brosinger, H. Freimuth, M.Lacher and W. Ehrfeld, Biosensors & Bioelectronics, 14 (1999) 409

14. Yu.M. Shirshov, S.V. Svechnikov, A.P. Kiyanovskii, Yu.V. Ushenin, E.F. Venger, A.V. Samoylov and R. Merker, Sensors & Actuators A, 68 (1998) 384

15. Yu. M. Shirshov, B. A. Snopok, A.V. Samoylov, A.P. Kiyanovskij, E. F. Venger, A.V. Nabok and A.K.Ray, Bisensors & Bioelectronics, 2001, in press.

16. R. Wechsung, in Microstructure Components in Polymers. First industrial application, MST news 14/95, MicroParts

17. Y. Kavabata, H. Sugamoto and T. Imasaka, Anal. Chim. Acta, 283 (1993) 689

18. R. Koncki, G.J. Mohr and O.S. Wolfbeis, Bisoensors & Bioelectronics, 10 (1995) 653

19. A.V. Nabok, F. Davis, A.K. Hassan, A.K. Ray, R. Majeed and Z. Ghassemlooy, Materials Science & Engineering C, 8-9 (1999) 123

20. A.V. Nabok, A.K. Ray, A.K. Hassan, R. Yates and R. Majeed, Proceedings of SPIE's 6th Ann. Intern. Symposium on Smart Structures and Materials, 1-5 March, 1999, 3673

21. T. Cassier, K. Lowack and G. Decher, Supramol. Sci., 5(1998) 309

22. F. Caruso, K. Niikura, D.N. Furlong and Y.Okahata, Langmuir, 13(1997) 3427

23. A.V. Nabok, A.K. Hassan and A.K. Ray, Materials Science & Engineering C, 8-9 (1999) 505

24. K.P.S. Dancil, D.P. Greiner and M.J. Sailor, J. Am. Chem. Soc., 121 (1999) 7925

25. A.K. Ray, A.K. Hassan, M.R. Saatchi and M. Cook, Phil. Mag B, 76 (1997) 961

26. E. Kretschmann, Z. Phys, 241 (1971) 313

27. V.I. Chegel, Y.M. Shirshov, E.V. Piletskaya and S.A. Piletsky, Sensors & Actuators B, 48 (1998) 456

28. A.V. Nabok, A.K. Hassan. A.K. Ray, O. Omar and V.I. Kalchenko, Sensors & Actuators B, 45 (1997) 115

29. K. Fisher and J. Muller, Sensors & Actuators, B, 9 (1992) 209

30. M. Pope and C.E. Swenberg, Electronic Processes in Organic Crystals, 1982, Oxford, Clarendon, p.45

31. J. Simon and J.J. Andre, Molecular Semiconductors, 1984, Berlin, Springer, p.86

32. I. Pockrand, Surface Science, 1978, 72, 577

33. H.E. de Bruin, B.S.F. Altenburg, R.P.H. Kooyman and J. Greve, Optical Communications, 1991, 82 425

34. A.K. Ray, A.V. Nabok, A.K. Hassan, J. Silver, P. Marsh and T. Richardson, Phil. Mag. B, 79 (1999) 1005

35. A.K.Hassan, A.K.Ray, A.V.Nabok and S.Panigrahi, IEE, Proc. Meas. Technol. 147 (2000) 137

36. M. Weisser, G. Tovar, S. Mittlerneher, W. Knoll, F.Brosinger, H. Freimuth, M.Lacher and W. Ehrfeld, Biosensors & Bioelectronics, 14 (1999) 409

Novel Methods to Study Interfacial Layers
D. Möbius and R. Miller (Editors)

A CONDUCTIMETRIC pH SENSOR BASED ON A POLYPYRROLE LB FILM

L. Rossi*, R. Casalini[#], and M.C. Petty

Centre for Molecular and Nanoscale Electronics and School of Engineering, University of Durham, South Road, Durham DH1 3LE, UK

Contents

Keywords: Langmuir-Blodgett film, pH sensors, conductimetric sensors, polypyrrole.

* Centro 'Enrico Piaggio', School of Engineering, University of Pisa, Via Diotisalvi 2, 56100 Pisa, Italy.
# INFM e Dipartimento di Fisica, University of Pisa, Via F. Buonarroti 2, 56127 Pisa, Italy.

A pH sensor, based on a.c. conductivity measurements of a thin polymer film, has been developed. The sensor consists of a planar interdigitated electrode array coated with a polypyrrole multilayer, built-up using the Langmuir-Blodgett technique. Impedance spectroscopy has been used to investigate the complex admittance of the device when exposed to aqueous solutions of different pH. The experimental data have been fitted to the theoretical response of an equivalent electrical network of capacitors and resistors. A response over the pH range 3.5 to 8 has been measured.

1. INTRODUCTION

Conductimetric transduction is a promising modality for chemical sensing because of the relative simplicity of the measurement technique and the ease with which miniature sensors can be fabricated. A device of this type is usually fabricated by coating a planar interdigitated electrode array with a thin film of a material designed to respond to the analyte under observation. An important factor in the design is the choice of the sensing layer. Conductimetric gas sensors, often referred to as chemiresistors, make use of inorganic or organic thin films; typical materials are tin dioxide or conductive polymers [1]. The sensing layer must respond selectively to a particular analyte and produce a measurable change in its electrical impedance.

Polypyrrole is one the most widely studied conductive polymers and has been exploited for chemical sensing. The material may readily be synthesised, is stable in the conducting form and exhibits interesting electrochemical behaviour [2-5]. In particular, polypyrrole chloride (i.e. chlorine is the counter ion) shows changes in its physical and chemical characteristics in solutions with different pH values [6,7].

Here, we describe the construction and evaluation of a conductimetric pH sensor, based on a multilayer polypyrrole film [8]. The sensor is an interdigitated electrode array coated with the polymer by using the Langmuir-Blodgett (LB) technique [9]. Impedance spectroscopy has been used to investigate the complex admittance of the device when exposed to aqueous solutions of different pH.

2. Experimental

2.1 Electrode fabrication

Pairs of interdigitated electrode arrays, each 0.5 cm × 1 cm, were fabricated by patterning a film of Au deposited on glass microscope slides. Both the width of the individual digits and the spacing between them were 240 μm. The substrate was first coated with a few nanometres of chromium by thermal evaporation in an Edwards 306A vacuum coating system (pressure ~ 10^{-6} mbar); this was to improve adhesion of the gold to the glass. Gold (50 nm in thickness) was then deposited without breaking the vacuum. Photoresist solution (Shipley Microposit S1813) was distributed across the surface of the metallised substrate using spin-coating. The resulting film was baked on a hotplate at 80°C for 30 min. to evaporate the solvent. An image of the electrodes was patterned in the photoresist by exposure through a mask to U.V. radiation and developed in Shipley Microposit MF312 developer. A solution consisting of 4 g of potassium iodide and 1 g of iodine in 40 ml in ultrapure water was used to etch away the unwanted gold areas. The patterned resist was dissolved with Aristar grade acetone. Finally, the unwanted chromium film was etched away using solutions of potassium hexacyanoferrate (1 g per 3 ml of water) and sodium hydroxide (1 g per 2 ml of water) mixed in the volume ratio 3:1.

2.2 Polypyrrole multilayer preparation and characterisation

One of the two interdigitated electrodes was coated with a multilayer polypyrrole film. The preparation has been documented elsewhere [8, 10, 11]. A monolayer of ferric palmitate was formed on the water surface of an LB trough by spreading a solution of palmitic acid in chloroform (1 g l^{-1}) onto a subphase of ferric chloride ($FeCl_3$) dissolved in ultrapure water (concentration 2 mg l^{-1}). Using the vertical LB technique, 12 monolayers were transferred to the glass substrate; the deposition was Y-type, with a transfer ratio of 1.0 ± 0.1 The film was briefly exposed (5 min.) to hydrochloric acid vapour. This solid state reaction converted the multilayer to palmitic acid and hydrated ferric chloride. The acid film was then exposed to pyrrole vapour in a partially evacuated desiccator containing a few millilitres of pyrrole monomer, and left for one day. During this stage, the oxidising ferric chloride polymerised the pyrrole monomer and doped the resultant polypyrrole polymer. The thickness of the film was estimated using an Alfa-Step profilometer and a value of 6.0 ± 0.5 nm per layer was obtained.

2.3 Sensing measurements

All measurements were performed with the test cell at room temperature. A Philips 9420 pH meter was used to monitor pH of the test solution. Solutions having pH values ranging from 2.4 to 10 were prepared in ultrapure water (produced from reverse osmosis, deionisation, filtration and UV sterilisation) using potassium chloride (or NaCl), HCl, H_2SO_4 and NH_4OH. A Hewlett-Packard 4192A Impedance Analyser was used to measure the admittance of the devices over the frequency range 5 Hz to 2 MHz. The amplitude of the applied potential was 10 mV r.m.s. To compensate for the effects of parasitic elements in the measurement circuit, the complex admittance was recorded without the sample connected. This value was subtracted from subsequent measurements.

Following immersion in water, the conductivity of the film decreased an order of magnitude over a period of a few hours. The measurements reported below were taken on samples that had been immersed previously in ultrapure water for about 12 hours, after which the electrical behaviour was stable. The time between consecutive measurements in solutions of different pH values was 20 minutes. This was found to be sufficient to allow for stabilisation of the polymer film.

3. Results and Discussion

To understand the sensor response, two devices were studied: the uncoated interdigitated gold electrode and the electrode structure covered with the polypyrrole multilayer.

3.1 Uncoated gold electrode

The conductance and the capacitance were first measured for an uncoated gold electrode immersed in solutions of various HCl concentrations. For all pH values, the conductance increased monotonically with frequency and approached a value that increased approximately linearly with HCl concentration. At frequencies greater than 5 kHz, the conductance was found to saturate for every solution investigated. The saturation values represented the electrolytes' conductance. In contrast, the capacitance decreased monotonically with frequency following a $\omega^{-1.6}$ relationship, and saturated at high frequencies. The high frequency value represented the capacitance of the electrolyte, i.e. the capacitance of a parallel plate configuration formed by the interdigitated electrodes separated by the HCl electrolyte.

The results for the pH 4.15 solution are shown in Figure 1. Here, the full lines represent the fitting of the experimental data to the equivalent circuit shown in the figure.

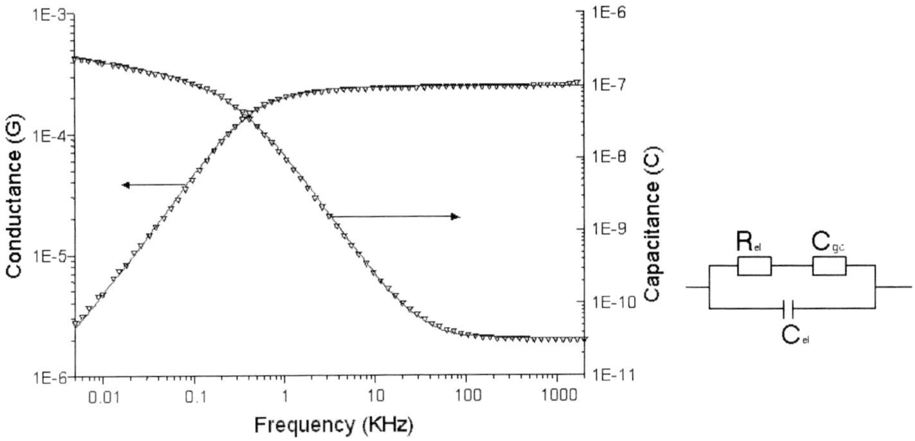

Fig. 1 Admittance spectra for an uncoated gold electrode in a solution of pH 4.15. Points are experimental; solid lines represent best theoretical fit. The equivalent electrical network used for the fitting is also shown. The values of the best-fit parameters are: R_{el} = 4040 Ω, C_{el} = 32 pF, C_{gc} : C_0 = 487 nF and n = 0.819.

The elements of the equivalent circuit can be interpreted physically as follows: R_{el} represents the resistance of the electrolyte, which from the fitting has a value of 4040 Ω; C_{el} is the capacitance of the electrolyte, equal to 32 pF; and C_{gc} represents the Gouy-Chapman capacitance that describes the diffuse layer of charge at the gold-electrolyte interface [12-13]. This is modelled by a universal capacitor, characterised by two frequency-independent parameters, C_0 and n. The complex capacitance $C_{gc}(\omega)$ is given by

$$C_{gc}(\omega) = C_0 (i\omega)^{n-1} = C_0 [\sin(n\pi/2) - i\cos(n\pi/2)]\, \omega^{n-1} \qquad (1)$$

where C_0 and n are constants and ω is the angular frequency. The exponent n is in the range $0 \leq n \leq 1$. The admittance $Y_{CPE} = C_0 (i\omega)^n$ is referred to as a constant-phase element (CPE) because its phase angle is independent of frequency. For n = 1, this circuit element is an ideal capacitor, while it represents an ideal resistance in the case n = 0. From the curve fitting, values of 487 nF and 0.819 were obtained for C_0 and n, respectively. Similar behaviour has been observed for uncoated interdigitated gold electrodes immersed in solutions of KCl and NaCl of various pH values [14].

3.2 *Polypyrrole-coated electrode*

3.2.1 Equivalent circuit

The Figs. 2 and 3 show, respectively, the conductance and the capacitance for the polypyrrole-coated electrode immersed in solutions of various pH and containing $3x10^{-3}$ M KCl . At low frequencies, the conductance has a value considerably higher than for the uncoated electrode, indicating that the LB film is acting as a layer of high specific conductance and that conduction at low frequencies is taking place mainly in the in-plane direction, through the multilayer film.

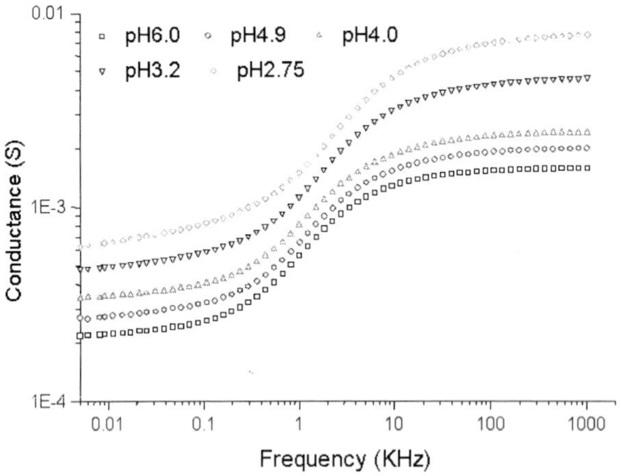

Fig. 2 Conductance versus frequency as a function of pH for the polypyrrole-coated electrode. Polymer film thickness = 72 nm. The pH of the solution was adjusted by the addition of HCl to 10^{-3} M KCl in deionised water.

The data for the coated electrode immersed in the pH 4.0 solution are shown again in Fig. 4. The full lines represent the fitting obtained with the equivalent circuit shown. As before, C_{el} is the capacitance of the electrolyte formed from the interdigitated electrodes.

From the fitting, this has a value of 36 pF (slightly greater than obtained with the uncoated electrode). R_{ppy} represents the in-plane resistance of the polypyrrole-palmitic acid film, which from the curve fitting is found to be 3030 Ω. The capacitance of the polypyrrole multilayer coated interdigitated electrodes measured in vacuum was only a few pF has been ignored in the model.

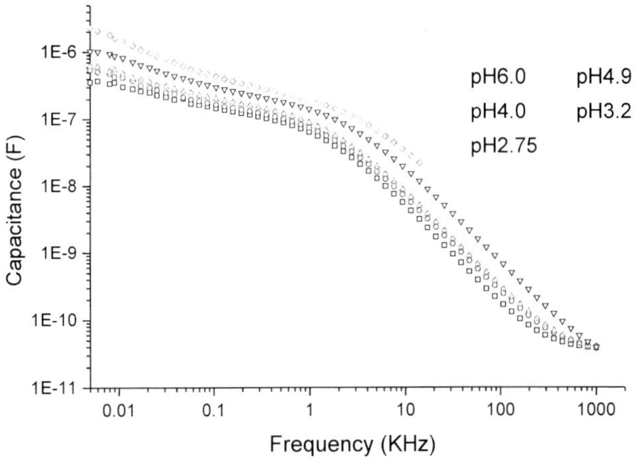

Fig. 3 Capacitance versus frequency as a function of pH for the polypyrrole-coated electrode. Polymer film thickness = 72 nm. The pH of the solution was adjusted by the addition of HCl to 10^{-3} M KCl in deionised water.

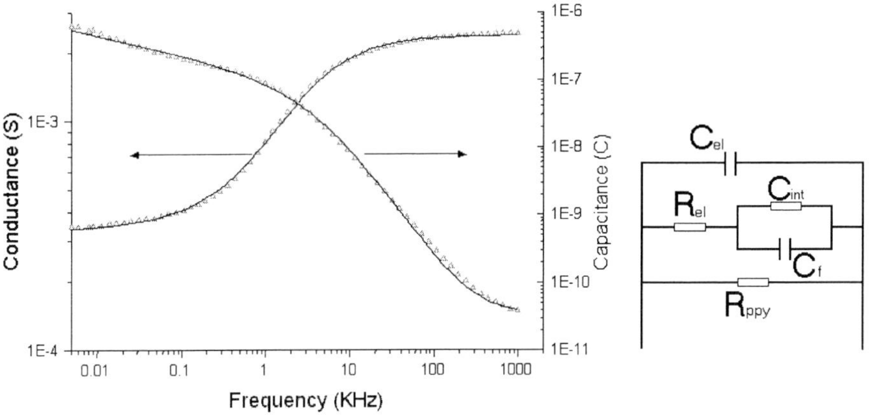

Fig. 4 Admittance spectra for a polypyrrole-coated electrode in a solution of pH 4.0. Polymer film thickness = 72 nm. Points are experimental; solid lines represent the best theoretical fit. The equivalent electrical network used for the fitting is also shown. The values of the best–fit parameters are: R_{el} = 488 Ω, R_{ppy} = 3030 Ω, C_{el} = 36 pF, C_f = 6.43 nF, C_{int} : C_0 =1.60 μF and n = 0.714.

At frequencies greater than 3×10^5 Hz, the conductance saturates at a value representing the sum of the conductance of the electrolyte (R_{el}^{-1}) and the conductance of the polypyrrole film in the in-plane direction (R_{ppy}^{-1}). Thus the polypyrrole film supports the in-plane conduction, while the electrolyte is responsible of the out-of-plane ionic conduction. This is due to the small film thickness (70 nm) and to the presence of defects (generated during the exposure of the ferric palmitate film to the HCl vapour). The Alfa-Step measurements on the polymer films revealed a rough surface (compared to measurements on simple fatty acid LB films). In the proximity of a defect, the film is short-circuited by the solution and the electrolyte comes into direct contact with the gold surface.

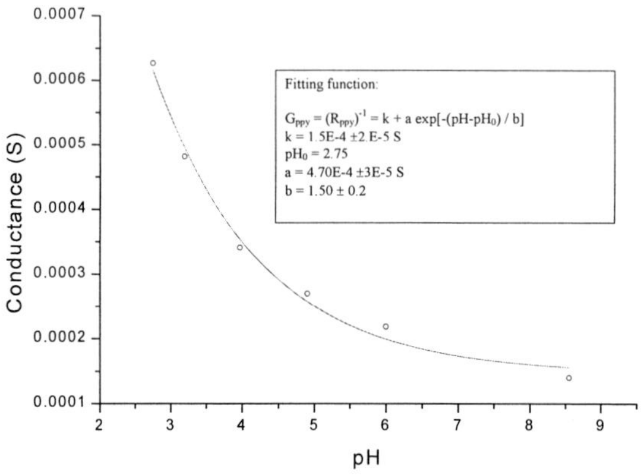

Fig. 5 Conductance at 5 Hz versus pH for polypyrrole-coated electrode. Polymer film thickness = 72 nm. Points are experimental (from Figure 2). The inset gives the fitting function used to obtain the full line.

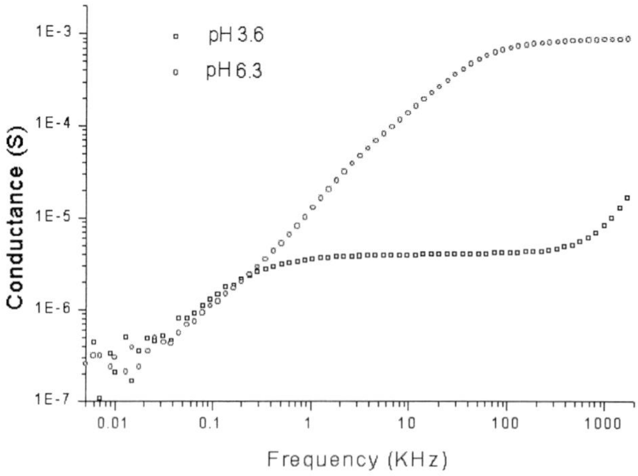

Fig. 6 Conductance versus frequency for an interdigitated electrode coated with 12 LB layers of tricosanoic acid at pH values of 6.3 and 3.6.

The behaviour of the system at low frequencies (less than 3×10^5 Hz) is best explained by a universal capacitor C_{int} representing the film/solution interface; C_0 and n were found to be 1.60 μF and 0.714, respectively. Finally, in Fig. 4, C_f represents the capacitance of the film between the electrodes and the solution, 6.43 nF from the fit. At 5 Hz the impedance of the branch of the circuit given by C_{int}, R_{el} and C_f is more than two orders of magnitude greater than R_{ppy} and the conduction can be considered to be due to the film.

3.2.2 pH sensing

Figure 5 shows the pH response of the device. These data are those shown at 5 Hz in Figure 2, here plotted as a function of pH. The curve in Figure 5 represents an exponential fitting of the data. The film conductivity G_{ppy}, as a function of pH, follows, to a good approximation, the relationship

$$G_{ppy} = (R_{ppy})^{-1} = k + a \exp [- (pH - pH_0) / b] \qquad (2)$$

pH_0, k, a and b have the values 2.75, 150 ± 20 μS, 470 ± 30 μS, and 1.5± 0.2, respectively. The pH response of the coated interdigitated electrode structure is due entirely to the polypyrrole layer. For comparison, the two sets of data in Fig. 6 show the conductivity of a gold interdigitated electrode coated with 12 layers of tricosanoic acid, and immersed in solutions of pH 6.3 and 3.6. At frequencies lower than 200 Hz, the two curves are coincident (within experimental error), indicating that the highly ordered fatty acid structure is behaving as a film of high specific resistance with no dependence on pH. The response of the sensor was reproducible over the pH range 3.5-8. Beyond these limits the reproducibility of the device deteriorated, probably due to a disruption of the film.

3.2.3 Cross sensitivity

To monitor the response of the sensor to the Cl^- counter ion, the following experiment was undertaken. First, the conductance of a polymer-coated electrode was measured at pH 6.2 (pure water) over the range of measurement frequencies. Potassium chloride (7×10^{-4} M) was then added to the water and the conductance re-measured. The measurements were finally repeated with the addition of ammonium hydroxide.

The results of the experiment are revealed in Fig. 7. The data for the pure water (i.e no added KCl) reveal a conductance that is invariant with frequency. The fixed value of about 400 μS represents the conductivity of the polypyrrole (the high frequency conductivity of deionised water at pH 6.3 is approximately 6 μS – much less than that of the polymer).

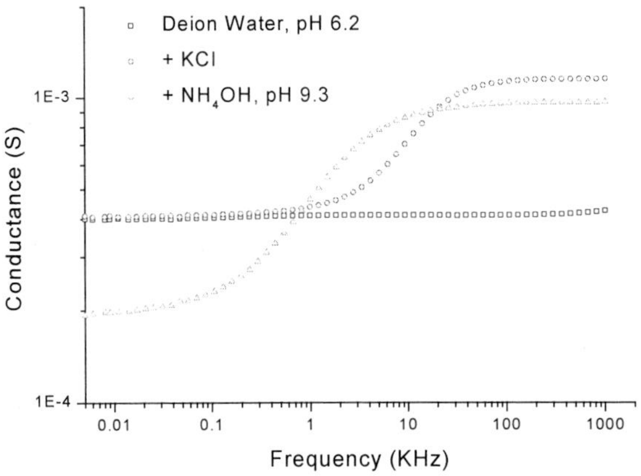

Fig. 7 Conductance versus frequency for polypyrrole-coated electrode. Measurements taken in pure water, KCl and NH_4OH.

The addition of KCl to ultrapure water produces an increase in the conductance (due to the solution) at high frequencies while the film response observed at low frequencies is unaffected. The subsequent addition of NH_4OH resulted in a change of pH from 6.2 to 9.3. This produced a decrease in the low frequency conductance of about 50% (consistent with the data in Figure 5), while the change in solution conductivity was small at high frequencies. This confirms that the dependence of conductivity in the polypyrrole film is determined mainly by pH.

4. Conclusions

A conductimetric pH sensor based on a polypyrrole chloride thin film has been developed. Impedance spectroscopy has been used to investigate the complex admittance of the device when exposed to aqueous solutions of different pH. The experimental data have been fitted to the theoretical response of an equivalent electrical network of capacitors and resistors. It has been found that the polypyrrole conductivity shows a good selective response to pH. This study has provided an understanding of the electrical behaviour and of the complex admittance spectra of a simple thin-film electrode system and could be a useful aid for the development of aqueous sensors based on conductive polymer thin films. Further work will focus on the reliability and lifetime of these devices.

5. Acknowledgements

L.R. would like to thank Professor De Rossi at the 'Centro Piaggio' of the School of Engineering of the University of Pisa, for providing the opportunity to study in the Centre for Molecular Electronics at Durham.

6. References

1. J. Janata, Anal. Chem., 64 (1992) 196R.
2. A.F. Diaz and J. Bargon, in T.A. Skotheim (ed.), Handbook of Conducting Polymers, Vol. 1, Marcel Dekker, New York (1986).
3. J.W. Gardner, Microsensors, Wiley, Chichester (1994).
4. E. Milella, F. Musio and M.B. Alba, Thin Solid Films, 284-285 (1996) 908.
5. M. Penza, E. Milella, M. B. Alba, A. Quirini and L. Vasanelli, Sensors and Actuators B, 40 (1997) 205.
6. Q. Pei and R. Qian, Electrochim. Acta, 37 (1992) 1075.
7. D. Tsamouras, E. Dalas, S. Sakkopoulos and E. Vitoratos, Physica Scripta, 48 (1993) 521.
8. R. Casalini, L.M. Goldenberg, C. Pearson, B.K. Tanner and M.C. Petty, J. Phys. D: Appl. Phys., 31 (1998) 1504.
9. M.C. Petty, Langmuir-Blodgett Films, Cambridge University Press, Cambridge (1996).
10. R.B. Rosner and M.F. Rubner, J. Chem. Soc. Chem. Commun., (1991) 1449.
11. A. Paul, D. Sarkar and T.N. Misra, J. Phys. D: Appl. Phys., 28 (1995) 899.
12. A.J. Bard and L.R. Faulkner, Electrochemical Methods, Wiley, New York (1980).
13. A.K. Jonscher, Dielectric Relaxation in Solids, Chelsea Dielectrics, London (1983).
14. V.A. Howarth and M.C. Petty, J. Phys. D: Appl. Phys., 29 (1996) 179.

Novel Methods to Study Interfacial Layers
D. Möbius and R. Miller (Editors)

DESIGNED NANO-ENGINEERED POLYMER FILMS ON COLLOIDAL PARTICLES AND CAPSULES

Gleb B. Sukhorukov

MAX PLANCK INSTITUTE OF COLLOID AND INTERFACES, 14424, POTSDAM/GOLM, GERMANY

Phone: +49-331-567-9429, Fax: +49-331-567-9202, E-mail: gleb@mpikg-golm.mpg.de

Contents

A novel approach to fabricate nano-engineered films on colloidal particles is based on layer-by-layer adsorption of oppositely charged macromolecules. Different templates with size ranging from 50 nm to tens of microns, such as organic and inorganic colloid particles, protein aggregates, biological cells and drug nanocrystals can be coated with multilayer films. Various materials, e.g. synthetic polyelectrolytes, biopolymers (proteins, DNA, polysacharides), lipids, multivalent dyes and magnetic nanoparticles, have been used as layer constituents to fabricate the designed shell to adjust required stability, biocompatibility and affinity properties of the capsules. Some colloidal templates can be decomposed at conditions where the polymer shell is stable, what leads to the formation of hollow capsules with defined size, shape and shell thickness. The permeability through the capsule wall can be regulated afterwards by pH. Several approaches on macromolecule encapsulation into these capsules are elaborated. The molecular weight selective permeability provides capturing of enzymes while the small substrates and products of enzymatic reactions can penetrate the capsule wall. These polyelectrolyte capsules can be used as carriers for biological species, for the controlled release and targeting of drugs and as microcontainers to perform chemical reactions in restricted volumes.

1. INTRODUCTION.

Nano-engineering of colloidal surfaces and design of functional colloid particles is currently an interesting topic of applied chemistry and biochemistry in the field of developing new materials with tailored properties. Research on composite colloidal particles (core-shell structures) has faced sufficient interest due to various applications expected in the areas of coatings, electronics, photonics, catalysis, biotechnology, sensorics, medicine, ecology and others [1]. In general, the research on core-shell structure and encapsulation implies the formation of a colloidal core of defined content and size, the preparation of a shell providing the required stability, permeability, compatibility, release of core material, and catalytic or affinity properties. Tailoring of the different components of one particle becomes important in order to develop these functionalised colloids, i.e. to combine several properties in one core-shell structure. The desired properties may be adjusted to facilitate the interaction of the core with the solvent or to add certain desired chemical properties. The shell may also have magnetic, optical, conductive, or targeting properties for directing and manipulating the core containing bioactive material.

This review is devoted to recently introduced novel pathway to fabricate nano-engineered core-shell structures, which can employ a great variety of substances as shell constituents as well as incorporated into hollow spheres. The development of this method emanated from formation ultrathin polymer film on macroscopic support. The alternated adsorption for assembling films was proposed by Iler [1] in 1966. and later developed by Mallouk et al. [2]. In 1991, Decher et al. proposed a method of forming polyelectrolyte films by using the alternated adsorption of polycations and polyanions [3]. A crucial factor for polyionic assembly is the change f the sign of the surface charge upon polyelectrolyte adsorption. A general scheme of assembling multilayer is given on Fig. 1.1.

A solid charged (let say positively) support is immersed into a containing an anionic polyelectrolyte. The polyion adsorbs, and under certain condition due to formation loops and tails the surface is re-charged. After the support is washed in clean water to remove polyanion solution. Then it is put in polycation solution. Upon adsorption of polycation the surface charge becomes positive. By repeating the cycle, a multilayer polyelectrolyte film can be obtained. Naturally, pH of the solution should provide for high degree of ionisation of the polyelectrolyte. The universal character of the method does not impose any restriction on the type of polyelectrolyte. To date, this method has been used for more than 50 various charged macromolecules including synthetic polyelectrolytes, conductive polymers and biopolymers (proteins and nucleic acids) [4-9]. The universal procedure makes possible to operate with charged nanoparticles such as ceramics, viruses and lipid vesicles. The latter usually spread over the oppositely charged surface forming bilayers. The potential of the method has been tested in several laboratories [10-14]. It has been established that films can contain over 100 layers. Thus, the ultrathin ordered films from 1 to 1000 nm thick can be formed with precision of one adsorbed layer thickness, which is about 1nm.

The incorporation of proteins and nucleic acids in multilayer films may lead to the application in biosensors and biotechnology [15-17]. The latter may even provide the base for new developments in ultrathin multistep chemical catalysis. The applications can be also envisaged in optical device fabrication [18, 19] and gas separation membrane [20]. One of the most important peculiarities of these multilayers is the selective permeability for different species. The film permeability depends on layer thickness, porosity, structure, chemical composition of the layers and the size of permeable compounds. Several investigations in this direction [21-23] showed that these films exclude macromolecules with large molecular weight whilst small polar molecules can readily penetrate through these films.

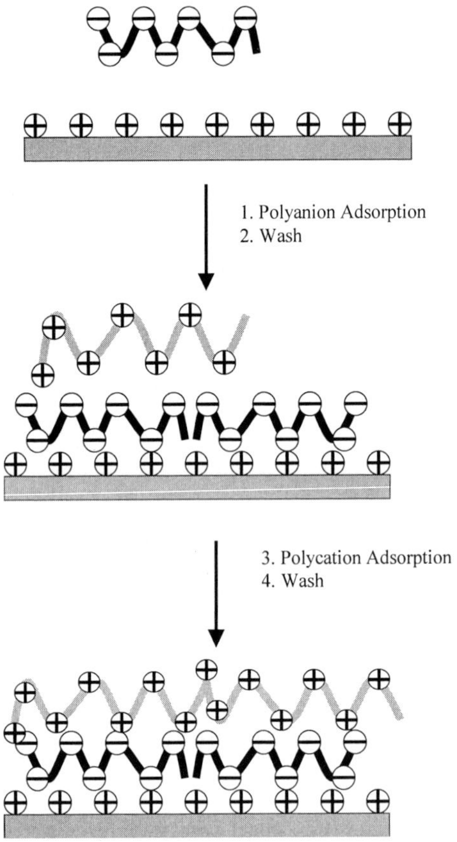

Fig. 1.1. Scheme of alternative adsorption of positively and negatively charged species onto solid support.

Few years ago this concert of LbL assembling of charged species was transferred to coat micron and sub-micron sized colloidal particles [24-29]. The idea is to employ the nano-engineered properties of multilayers as shell structures formed on colloidal particles. This paper outlines the recent works on step-wise shell formation on various colloidal cores, fabrication and properties of hollow capsules, regulation of capsule wall permeability and approaches to encapsulate different materials into these capsules.

2. COATING OF MICRON- AND SUBMICRON SIZED COLLOIDAL PARTICLES.

Here, we consider the two general approaches to coat colloidal particles with ultrathin film tunable in nanometer range. These methods differ by class of species employed for shell build-up, type of used colloidal particles and conditions for shell fabrication. First one is layer-by-layer (LbL) assembly of oppositely charged macromolecules or nanoparticles, which emanates from LbL assembly on macroscopic flat films developed in the early nineties by Decher and co-workers. Another approach is controlled precipitation of macromolecules or nanoparticles from the solution on surface of micron and submicron sized colloidal particles. The surfaces of these particles serves as collector for precipitating materials.

2.1. Layer-by-Layer assembly on colloidal particles.

The main problem in transferring the layer-by-layer technology from macroscopic flat to surface of colloidal particles (Fig. 2.1) is how to separate the remaining free polyelectrolytes from the particles prior to next deposition circle. Fig. 2.2 illustrates the behaviour of ζ–potential of particles and amount of oppositely charged polyelectrolyte non-bounded on the particle surface as a function of total concentration of added polyelectrolyte [29]. The amount of free fluorescence labelled polyelectrolyte molecules was measured in supernatant after particles have been centrifuged. The course of the ζ–potential shows that with increasing bulk concentration of the adsorbing polyelectrolyte recharging of the surface takes place.

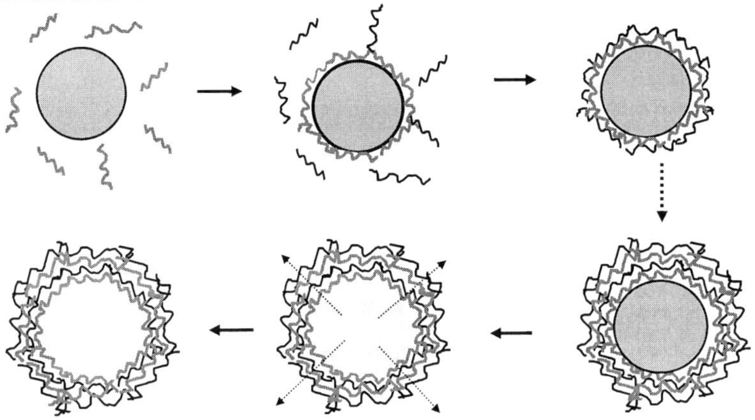

Fig. 2.1. Consecutive adsorption of positively (gray) and negatively (black) charged polyelectrolytes onto negatively charged colloid particles. After dissolution of colloidal core (e) a suspension of polyelectrolyte capsules is obtained (f).

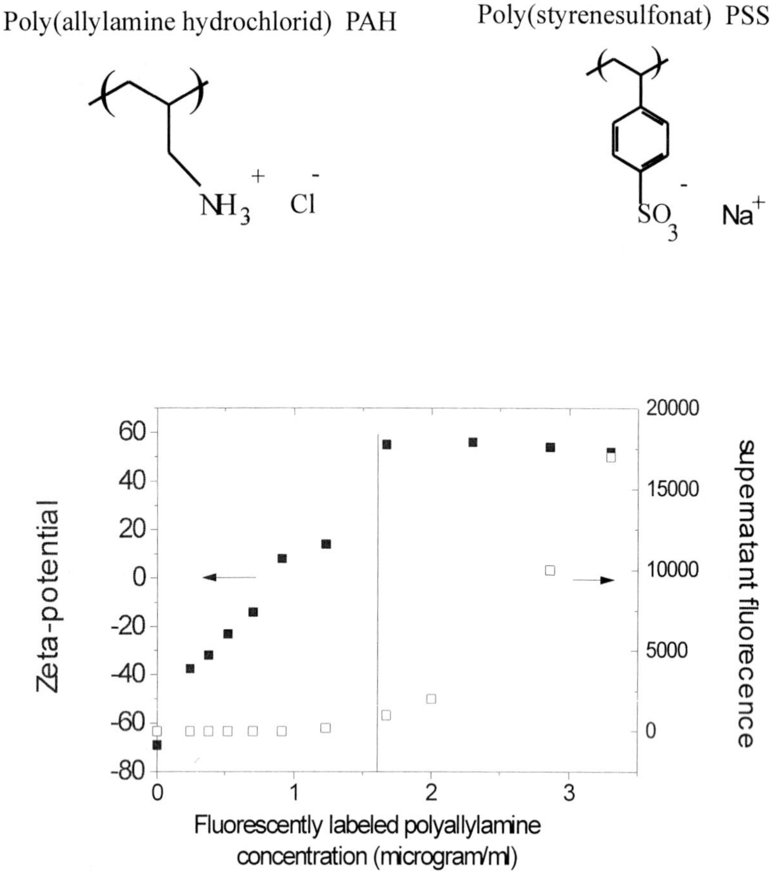

Fig. 2.2. Chemical formulas for PAH and PSS. ζ-potential change (■) of polystyrene sulphate latex particles (diameter 640nm) and fluorescence intensity (□) due to presence of FITC-labelled PAH as a function of total concentration of added FITC-PAH. Particle concentration was $2.2*10^8$ per ml. Adsorption in 0.5M NaCl

At the certain concentration the ζ–potential levels off and the polyelectrolyte is mostly adsorbed on the particles depleting the bulk from the majority of polyelectrolyte molecules. The further increasing of bulk polyelectrolyte concentration does not influence on ζ–potential and amount of polyelectrolyte on the particles. Thus, the layer-by-layer assembling of the polyelectrolytes onto colloidal particles can be performed in two different ways [28-30]. Either the concentration of polyelectrolyte added at each step is just sufficient to form a saturated layer or adsorption is curried on at excess of polyelectrolyte concentration. (Fig. 2.3).

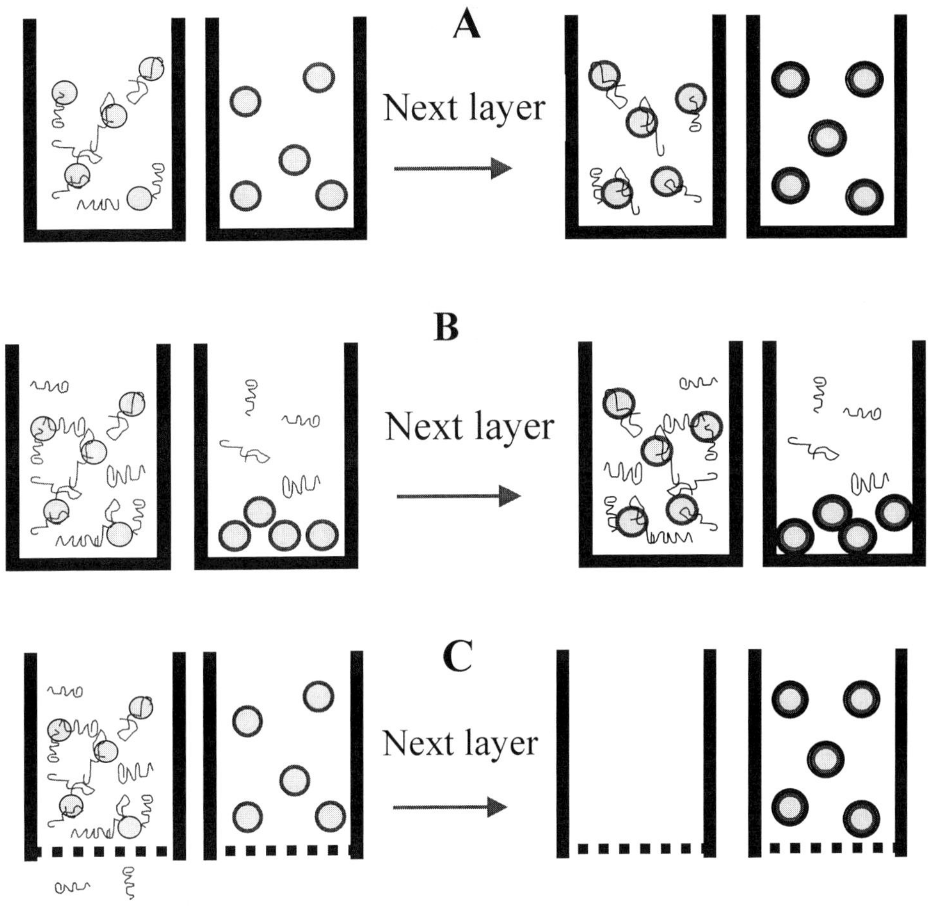

Fig. 2.3. The approaches to assemble polyelectrolyte films onto colloidal particles. A- sequential adding at matched concentration, B- centrifugation, C- filtration.

At latter case the non-bounded rest of polyelectrolyte molecules must be removed before adding of next polyelectrolyte in order to avoid formation of polyelectrolyte complexes out of colloidal particles. The removal of free polyelectrolyte molecules can be done by several times repeated centrifugation washing in pure water followed by re-suspension or by using filtration set-up where the colloidal particles are kept above the filter, but polyelectrolytes are small enough to pass through pores. In these methods after adsorption of each polyelectrolyte layer the suspension is washed in pure water to perform complete removal of polyelectrolytes before adding oppositely charged polyelectrolyte to the particles suspension. As was explicitly described in [28, 30] these different approaches provide sequential layer growth of multilayers on colloidal particles with some discrepancy in adsorbed amount, number of formed particle aggregates, easiness to vary the substances, material loss during the preparation and possibilities to scale-up production of coated particles. It should be stressed main advantages and drawbacks of these three approaches. The main problem is to suppress the tendency of the particles to form more or less strong aggregates at different stage of preparation. It is rather difficult to re-suspend particles of strong aggregates without incalculable modification or even damage of prepared surface structure. The centrifugation makes problematic the following re-suspension. Another drawbacks of centrifugation are difficulties to settle down the smaller particles and significant loss of particles during preparation. For instance, at the assembling 20 layers of polyelectrolytes on 640 nm particles the accumulated total loss of the particles was 80% of originally introduced. The washing of particle suspension by means of centrifugation is rather time-consumable process.

In the case of sequential adding of polyelectrolyte at matched concentration (Fig. 2.3a) one could prevent the direct formation of aggregates. In the contrast with centrifugation and filtration protocols here there is no loss of polyelectrolytes used as layer constituents. Here, at once determined conditions the multilayer coating might be performed into short time-scale because adsorption of one polyelectrolyte layer in used concentration takes not longer than few minutes. Whereas, at this approach the probability to form polyelectrolyte complexes out of particles and particles aggregates as revealed by SPLS are rather high [28]. Keeping rather diluted particle suspension is favourable to reduce the particle aggregation. Also one to control the concentration of all suspension components pedantically during whole process. Working with excess amount of polyelectrolytes facilitates the variation in shell composition. Indeed, to introduce new component one has just to estimate that the added concentration is enough to complete coverage of all colloidal particles.

The filtration method (Fig. 2.3c) is the only practicable to scale-up the coating large variety of colloidal particles [30]. The particles are not stressed with exception of permanent, but adjustable stirring. The liquid levels above the filter can be exactly controlled by the interplay between inflow of the ingredients including the suspension medium and pressure applied for pass through filter. The filtrate can be recovered easily. Thus, the shell preparation could be automated, in principle. The membrane filtration has adaptation capacity to meet all requirements of great variety of chemical and colloid chemical systems. But dealing with smaller particles slows down the process obliges use of filters with small pore size, what leads to necessarily for frequent substitution of filters due to corking up of the pores by adsorbing of polyelectrolytes.

The monitoring of the process of film formation, i.e. charge reversal and continuously layer growth, was followed at each step by electrophoresis, dynamic light scattering, SPLS and fluorescent intensity measurements [28-31]. Fig. 2.4 summarises the ζ–potential changes recorded upon layer deposition for polycation / polyanion pairs PDADMAC/PSS, PAH/PSS, BSA/PAH and PDADMAC/DNA. The ζ–potential alternates between positive and negative values indicating the successful recharging of the particle coated with the adsorbed polyelectrolyte multilayer upon each layer deposition.

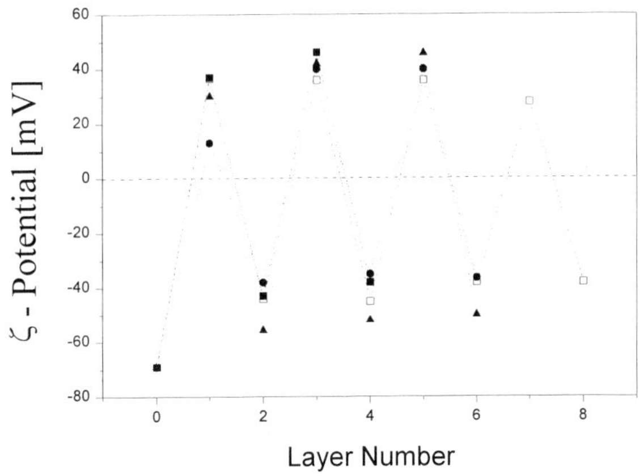

Fig. 2.4. ζ-potential as function of layer number for coated polystyrene latex particles, diameter 640nm, □ PSS/PDADMAC, ■ BSA/PDADMAC, ● PSS/PAH, σ DNA/PDADMAC.

The evidence for sequential layer growth has been obtained by means of Single Particle Light Scattering (SPLS) [28-30]. This method is capable to unambiguously identify coating particles and discriminate singlets, doublets and triplets of particles. The lower limit of layer thickness determination is about 1 nm. Currently, monodisperse polystyrene particles as small as 200 nm can be investigated. In Fig. 2.5 the intensity distribution of the control naked particles is compared with particles coated with 8 layers assembled by filtration technique [30]. From the shift of the peak of the intensity distribution the adsorbed mass can be derived. Given refractive index of the polyelectrolyte complexes [28] these data can be converted into a layer thickness of the adsorbed polyion multilayers. The increase of polyelectrolyte film thickness is proportional to the layer number (Fig. 2.5).

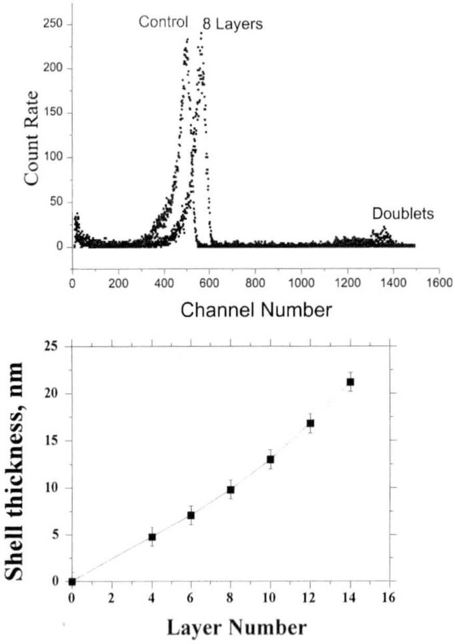

Fig. 2.5. Top - Normalised light scattering intensity distributions (SPLS) of PAH/PSS-coated polystyrene sulphate latex particles (diameter 640nm). Particles with 8 layers are compared with uncoated ones. Bottom – shell thickness as a function of layer number.

The mean polyelectrolyte layer thickness was found to be 1.5 nm for the case of PSS/PAH assembled from 0.5M NaCl. It should be noted that the average layer thickness of polyelectrolyte multilayers strongly depends of kind of polyelectrolytes and salt concentration used at polyelectrolyte assembling. More rigid polymers and increasing of salt concentration lead to thicker adsorption layer.

The mostly used methods to monitor LbL deposition on monodisperse PS-latex particles for various substances are SPLS method and microelectrophoresis. Inorganic (magnetite, silica, titania and fluorescent quantum dots) nanoparticles [32-34], lipids [35-37] and proteins (albumin, immunoglobulin and others) [29, 38, 39] were incorporated as building block for shell formation on colloidal particles. In paper [39] the construction of enzyme multilayer films on colloidal particles for biocatalysis was demonstrated. The enzyme multilayers were assembled on submicrometer-sized polystyrene spheres via the alternate adsorption of poly(ethyleneimine) and glucose oxidase. The high surface area bio-multilayer coated particles formed were subsequently utilized in enzymatic catalysis. The step-wise coating of different lipids alternated with polyelectrolytes was performed by adsorption of preformed vesicles onto the capsule surface. As was shown by flow cytometry, interlayer energy transfer measurements the structure of charged lipids layers in between polyelelctrolyte was presumably found as bilayers [36]. Fabrication of inorganic shells on colloidal particles envisages the application of such core-shell structures in catalysis and colloidal band-gap crystals [32, 34]. Introducing magnetic particles into the shell composition opens possibility to manipulate them by applying external field [33, 35].

When the sequential layer growth is proofed for certain multilayer composition one can apply these conditions to fabricate the multilayer film on surface of the colloidal particles with different size and shape discrepancy such as biological cells [40,41], dye nanocrystals [42,43] or protein aggregates [44].

2.2. Controlled precipitation on surface of colloid particles

In the light of shell formation on colloidal particles our consideration would not be complete if we do not mention about surface controlled precipitation. The described in [45] results demonstrate that slow heterocoagulation process of polymer at the suspension of colloidal particles is a way to controlled coating the particles by polymeric films (Fig. 2.6). Precipitation could be caused either by non-soluble complex formation between polyelectrolyte and multivalent ions (Fig. 2.6A) or by mixing the polymer solution with non-solvent (Fig. 2.6A).

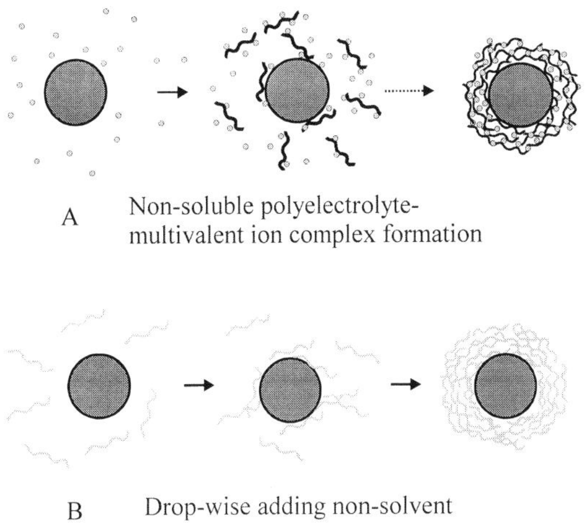

Fig. 2.6. Schematic representation of heterocoagulation process of polymers in suspension of colloidal particles. Polymers precipitation is caused either by complexation with multivalent ions (A) or by mixing with non-solvent (B)

The proper choice of concentration ration between polymers, particles and speed of coagulation provides coverage with defined amount of precipitated polymer on each particle. The particles statistically harvest the coagulated polymers and their complexes with multivalent ions on colloidal surface. The theoretical consideration of this heterocoagulation of polymers in the presence of collector is presented in [45]. The experimental observations of smooth film formation of surface of colloid particles were demonstrated in [45-47], there were illustrated the possibilities to build the shells composed of only one component (DNA), utilising non-charged polymers (dextran) complexes between ions and polyelectrolytes (Me^{3+}/PSS). DNA and dextran were precipitated on particle surface by drop-wise adding ethanol in aqueous solution. As revealed from confocal microscopy images the resulted coating of particles is homogeneous for all investigated systems. The thickness of polymeric film on colloids can be tuned in the range of few monomolecular layers. The controlled precipitation (CP) method explores many compounds, which cannot be used by means of LbL assembly. Certainly, the conditions to precipitate, such as solvent change or adding of complex-ions, might be found for wide class of polymers and nanoparticles as well. As demonstrated in [47] by TEM the homogeneous shell consisting on precipitated in ethanol 4 nm semiconductor CdTe nanocrystals can be formed with tunable thickness on submicron sized colloidal latex particles. Homogeneous and complete coverage of colloidal cores with a shell of luminescent nanocrystals with desirable thickness in the range of 15-40 nm was achieved as confirmed by transmission electron microscopy (Fig. 2.7).

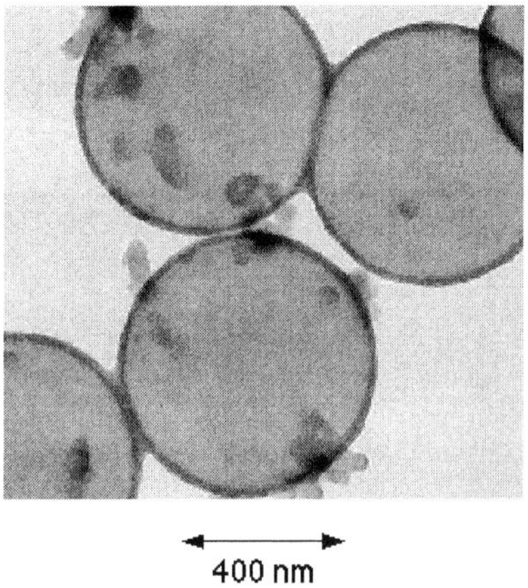

Fig. 2.7. TEM images of latex spheres covered by CdTe nanocrystals precipitated on surface of PS-latex particles by drop-wise adding of ethanol in aqueous solution.

The controlling precipitation on colloids implies the search for adequate window in range of concentration of polymers, particles and coagulation speed for each polymer one wishes to make shell composed of it. According to [45] the theoretical estimations of the concentration parameters in the system on the base of simple coagulation model give good agreement with experimentally observed optimal values. It should be mentioned that one more advantage of CP method to cover the colloidal particles comparably to LbL technique is significantly less time-consumption in order to build the film consisting of tens monomolecular layers. By CP methods the assembly is done in minute time-scale instead of hours for LbL method. Both of CP to LbL methods accomplish each other. Indeed, the assembling of two oppositely charged polyelectrolytes could not be done by controlled precipitation because of fast coagulation of polyelectrolyte complexes. In this case the film can be assembled only by LbL technique. Also the ordering of monomolecular layer in Z-direction to colloidal surface seems to be better arranged by LbL adsorption than for CP method. While the LbL technique usually results in formation of stable film, the CP assembled films might be easily decomposed by increasing again the polymer solubility or complex-ion extraction. The real benefit should be brought by combination of these both methods, what opens a lot of opportunities to assemble multicomposite film. A wide variety of materials despite on its charge and molecular weight can be incorporate as constituents in the film composition. The step-wise compilation of LbL- and CP-assembled building blocks promotes the solution of diverse tasks related to micro- and nanostructuring materials. It might be, for instance, inorganic reactions between the previously spatially located reagents in the shells or encapsulation of polymer in the polyelectrolyte capsules via capturing of CP-assembled polymers, like DNA or dextran, by sequentially LbL-assembled stable shell afterwards (see below in section 4).

2.3. Colloidal core decomposition and formation of hollow capsule.

Different colloidal cores can be decomposed after multilayers are assembled on their surface. If the products of core decomposition are small enough to expel out of polyelectrolyte multilayer the process of core dissolution leads to formation of hollow polyelectrolytes shells (Fig. 2.1, d-f). Up to now, various colloidal templates such as organic and inorganic cores, like MF-particles, organic crystals, carbonate particles and biological cells were used as templates for hollow capsule fabrication. Decomposition can be done by different means, such as low pH for MF- and carbonate particles [43], organic water miscible solvents for organic crystals [44] and the same MF-particles, strong oxidising agents (NaOCl) for biological cells [41]. The formation of hollow capsules was more intensively studied on MF-particles [31]. These particles dissolve onto oligomers at 0.1M NaCl and in some water-miscible solvents DMF or DMSO. Subjecting the coated with 8 PSS/PAH layers MF particles to low pH results in solubilisation of the core. The MF-oligomers, which have a characteristic cross-sectional extension of about 2-3 nm are expelled from the core and permeate through the polyelectrolyte layers forming the shells. This observation is consistent with the finding that polyelectrolyte-coated MF-particles are readily permeable to molecules of a few nanometers in size [31]. The MF-oligomers are finally separated from the hollow shells by centrifugation or filtration protocols.

The fabricated hollow polyelectrolyte capsules were characterized using scanning, transmission, atomic force and confocal fluorescent microscopy techniques [29-31, 35, 41, 48,49]. A scanning force microscopy (SEM) image of hollow polyelectrolyte capsules is shown in Fig. 2.8. The numerous folds and creases observed are attributed to collapsing of the hollow capsules under drying. The shells are flattened and some spreading is noticed. The diameter of the capsules shown in Fig.2.8 is larger than the diameter of the templated MF-particle (3.3 μm). This increase in diameter observed is ascribed to collapsing and adhesive forces attracting the polyelectrolyte shell to the surface. From the TEM image it was also deduced that the thickness of the polyelectrolyte film is of the order of 20 nm for the 9 layer polyelectrolyte film [29]. This value is consistent with the single particle light scattering data on layer thickness obtained for polyelectrolyte-coated polystyrene particles (Fig. 2.5).

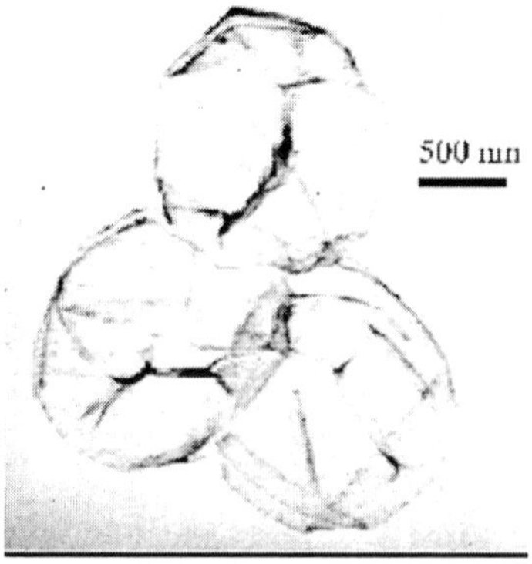

Fig. 2.8. Scanning electron microscopy. Dried collapsed polyelectrolyte capsules of 3μm. Scale bar is 0.5μm.

This suggests that the nature of the colloid does not significantly affect the thickness of the polyelectrolyte layers. It should be noted that the polyelectrolyte capsules completely repeat the shape the templating colloids as was shown on example of echinocyte cell, which have star-like shape [41]. The choice of the core is determined by concrete tasks. For instance, the carbonate cores are decomposable at pH<4 what makes them convenient to fabricate hollow capsules composed of biological polymers, what are not suffering very low pH or organic solvents for core decomposition. The capsule composed of biocompatible chitosan as polycation and chitosan sulfide as polyanion were fabricated in [30, 35, 48]. At present hollow polyelectrolyte capsules with diameters varying from 0.2 to 10 microns and wall thicknesses from few to tens of nanometers were obtained. Many compounds were used as layer constituents to build the hollow capsules. Uniform inorganic and hybrid inorganic-organic hollow microspheres have been produced by coating colloidal core templates with alternating layers of oppositely charged nanoparticles and polymer, and thereafter removing the core either by heating [50]. The multilayers were constructed by consecutively depositing the nanoparticles and polymer onto the colloidal templates. Hollow silica spheres were obtained by calcining polymer latex spheres coated with multilayers of silica nanoparticles (SiO2) bridged by polycations.

3. Permeability properties of macromolecule encapsulation in polyelectrolyte capsules.

In general, the polyelectrolyte capsule walls have semipermeable properties. High molecular weight compounds are excluded by the polyelectrolyte shell whilst small molecules, such as dyes and ions, can readily penetrate the capsule wall [35, 51]. Thus, the problem arises how to introduce macromolecules and how to switch and control capsule permeability for them. There are obviously a variety of materials, the encapsulation of which is desirable for application in different areas of technology, such as catalysis, cosmetics, medicine, biotechnology, nutrition and others.

The properties and structure of polyelectrolyte multilayers are sensitive to a variety of physical and chemical conditions of the surrounding media which might dramatically influence on structure of polyelectrolyte complexes and result on permeability of the capsules. In particular, the pH is one of the physico-chemical parameters, which influences the state of the inter-polyelectrolyte complex, especially in the case, if the charge of one polyelectrolyte in the complex depends on the pH.

The exclusion properties of hollow polyelectrolyte capsules templated on MF and $CdCO_3$ cores and consisting of 4 PSS-PAH bilayers have been investigated for FITC-labeled dextrans (MW 75.000 and 2.000.000) and FITC-labeled albumin as a function of pH in [52, 53]. Fig. 3.1 (left) provides a confocal image of capsules in the presence of FITC-dextran at pH 10. The interior of the capsules remains dark, while the background is fluorescent.

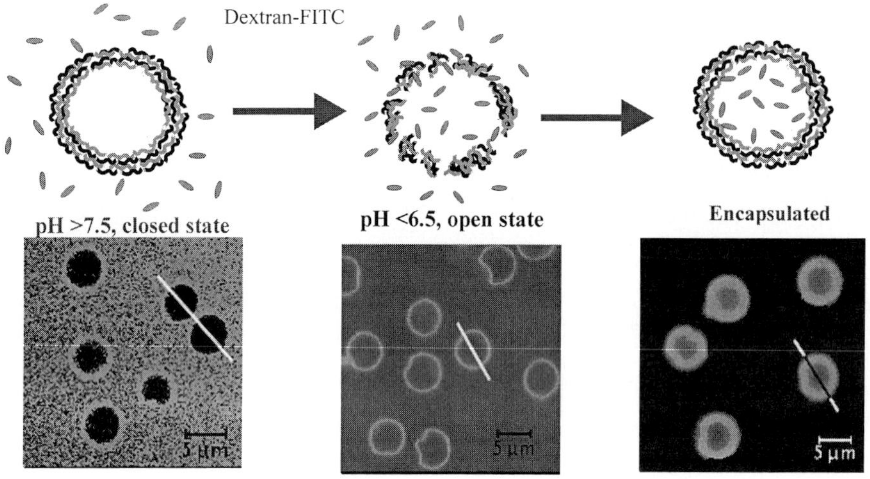

Fig. 3.1. Permeation and encapsulation of FITC-dextran (M.w. 75000) into polyelectrolyte multilayer capsules. Left - pH = 10, centre – pH = 3, right – pH increased to 10 after the capsules were loaded with FITC-dextran at pH = 3. The bulk FITC-dextran was removed by washings at pH = 10. Upper – scheme, bottom – confocal images of the capsules.

This proves that at this condition the capsule wall is not permeable for FITC-dextran. Even deformed capsules do not reveal any fluorescence inside. However, at pH 3 the capsule interior becomes as fluorescent as the bulk (Fig. 1, center). This can only be explained by opening of the capsules for FITC-dextran at this low pH value. It has to be mentioned that the "open" and "closed" states at relatively low pH and high pH, respectively, were observed for more than 90% of the capsules. The permeation was studied at different pH values in a range from 3.5 to 12. The open state for FITC-dextran was observed for pH values up to 6. From pH 8 onwards most of the capsules are closed. Accordingly to [53] there are no basic difference in permeability properties of the capsules templated on MF- or $CdCO_3$ particles illustration that core material does not essentially contribute on capsule properties afterwards.

At a pH value in between "open" and "closed" states, i.e. pH 7, open and closed capsules were simultaneously observed. At least at the measurable time, which is about hour, the fluorescence coming from capsule interior does not change. The opening of capsule wall presumably occurs as cooperative process and appears like defect formation. It might be that changes of the polyelectrolyte charge upon pH variation are able to induce pore formation of about 50-200nm as was observed on polyelectrolyte multilayers at flat surfaces [54] or to loosen the polyelectrolyte network, thus enabling polymer penetration. No "half-filled" capsules were found. From this behaviour one may conclude that the change of permeability can indeed be explained by a sharp transition regarding the wall properties as a function of pH.

The mechanism underlying the pH dependent capsule permeability is not yet fully understood. A possible explanation may be provided by considering the polyelectrolyte interactions in the shell wall. At the pH of capsule formation, pH 7, the charge densities on both polyelectrolytes determine their stoichiometric ratio during adsorption. Since the polymers are irreversibly adsorbed in the shell wall a pH decrease does not induce polymer desorption. However, charging of the PAH may occur, which would induce positive charge into the shell wall. This may alter the shell wall morphology by enhancing the mutual repulsion, which could lead to defects in the polymer network.

The possibility of loading is demonstrated in Fig. 3.1 (right) where the capsules were initially exposed to FITC-dextran solution at pH 3. Then the pH was shifted to 10 and the rest of FITC-dextran was removed from the bulk by centrifugation. It is remarkable that the capsules remain filled with fluorescent material as shown by the fluorescence profile through the confocal image. The interior of the capsule observes a bright and constant over time fluorescence while there is no fluorescence signal from solution. Similar experiments of changing the capsule wall permeability by pH changes and subsequent capsule loading were performed with FITC-dextran of a MW of 2.000.000 and FITC-labelled bovine serum albumin. The results are analogous to those with FITC-dextran of MW 75.000.

The structure of the capsules was studied by means of SFM as a function of pH. Samples were prepared by applying a drop of the buffered shell suspension onto a mica surface. After allowing the shells to settle the liquid phase was gently blown away and the sample was dried under a stream of nitrogen. SFM images of the shells incubated at pH 3.5 and pH 10 buffers are shown in Fig. 3.2.

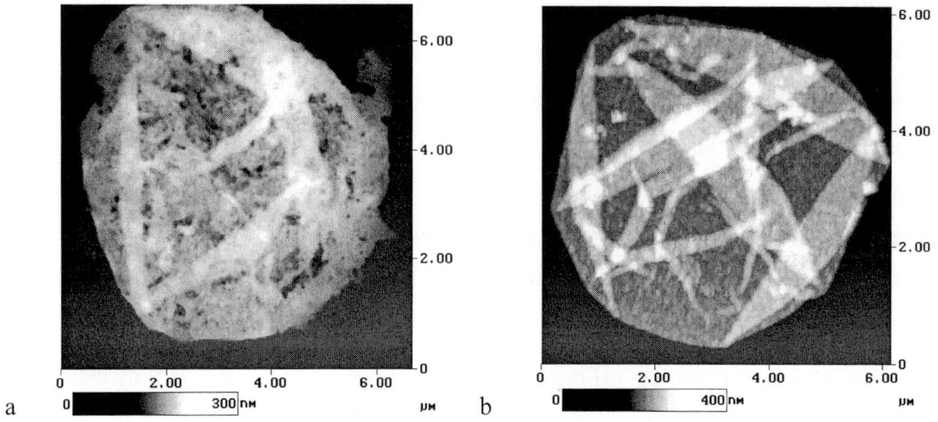

Fig. 3.2. SFM images of capsules treated with pH 3.5 (a) and pH 12 (b) buffers before drying. Capsules were prepared on MF particles followed by core dissolving. The porous structure of the shell wall treated with acidic solution is clearly visible.

Large holes up to 100 nm in diameter can be identified in the capsule walls, which had been exposed to acidic solution while no changes compared with the control occurred after incubation at pH 10. Mostly remarkable is the reversibility of these structure changes. Upon transfer of the capsules, exposed previously to pH 3.5, into an alkaline solution of pH 10 the holes could not be observed. Comparable observations were recently reported concerning the behaviour of PAH-PSS multilayers on flat surfaces [17] – the presence of holes up to hundred nanometers in diameter was established after the multilayers composed of strong and weak polyelectrolytes were incubated at certain pH.

The presence of holes in the dried and collapsed capsules does not necessarily imply that capsules in solution have them too, although it is highly likely that at low pH the shell wall becomes disturbed. This may result in hole formation upon drying as a result of layer shrinking. It indicates a weakening of the connectivity of the polyelectrolytes, which could be responsible for the enhanced permeability.

The mechanism underlying the pH dependent capsule permeability is not yet fully understood. A possible explanation may be provided by considering the polyelectrolyte interactions in the shell wall. At the pH of capsule formation, pH 7, the charge densities on both polyelectrolytes determine their stoichiometric ratio during adsorption. Since the polymers are irreversibly adsorbed in the shell wall a pH decrease does not induce polymer desorption. However, charging of the PAH may occur, which would induce positive charge into the shell wall. This may alter the shell wall morphology by enhancing the mutual repulsion, which could lead to defects in the polymer network.

This possibility of switching the capsule walls between an open and closed state provides a convenient and efficient tool to control the uptake and release of polymers, biopolymers and nanoparticles (cf. Fig. 3.3.). For instance, the capsules might be loaded at low pH and after increasing the pH the material is captured inside. The encapsulation of different substances can be done under mild conditions within several minutes. The demonstrated in [52] possibility of controlling loading and release of macromolecules into and from polyelectrolyte capsules may find widespread application. The described pH-induced permeability change of the polyelectrolyte network in the film may be a general mechanism for modifying the permeability of polyelectrolyte multilayers. Thus we expect a similar behaviour for other shell composed of weak and strong polyelectrolytes.

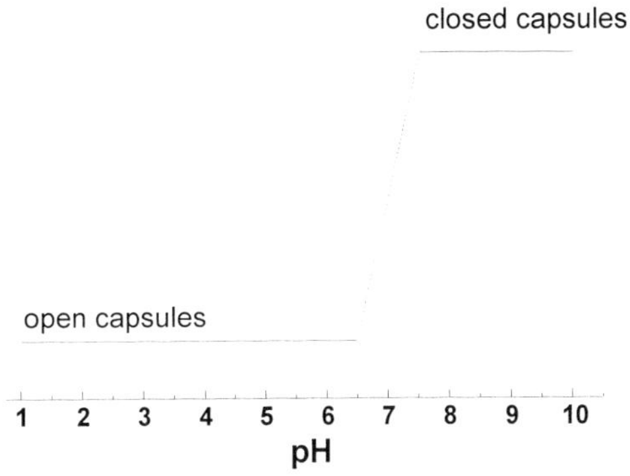

Fig. 3.3. Schematic representation of open and closed state of the capsules composed of 8 PSS/PAH layers assembled at pH 6.5

Summarizing permeability properties one can see a lot of opportunities to control uptake and release substances with different molecular weight. It opens new pathways for biotechnological, chemical, pharmaceutical technologies where controlled encapsulation and release of different substances is required. But at present stage it seems difficult to anticipate the permeability properties for every particular pair of oppositely charged polyelectrolytes used for capsule wall build-up. Let us stress main features observed so far. The capsule might undergo transition at pH close to pK of used polymers what reflects on possible segregation and formation of pores. Increasing of shell thickness more that 20 nm might cause barrier difficulties for low molecular weight compounds to pass through capsule wall. Further research on influence of preparation condition, i.e. salt, pH, at polyelectrolyte adsorption on permeability properties is envisaged. One also could introduce temperature and light sensitive polymers into the shell during capsule preparation to regulate the permeability by these physical factors.

4. ENCAPSULATION OF MACROMOLECULES BY MEANS OF PRE-PRECIPITATION ON COLLOIDAL PARTICLES.

Above we described the possibilities how to introduce the macromolecules into the hollow capsules by regulation of shell permeability. Pores in the polyelectrolyte shell can be opened and closed by variation of pH, salt and/or solvent exchange. The main drawback if this approach is low encapsulation efficiency for macromolecules. Indeed, one could capture in the hollow capsules only the macromolecules, which are floating in the capsule interior when pores are closed. In following we propose another approach to encapsulate macromolecules and nanoparticles which could provide rather high encapsulation efficiency. This approach consists in combination of layer-by-layer approach and controlled precipitation of polymers on surface of colloidal particles.

As was described in section 2 when the polymer molecules precipitate in the presence of colloidal particles the colloidal surfaces are harvesting the precipitating polymers. The proper choice of concentration of particles, polymers and speed of heterocoagulation allows to form smooth coverage of colloidal particles by precipitating polymers. The following idea might be to capture the precipitated polymers, which is now placed on surface of colloidal particles. The scheme is depicted on Fig. 4.1.

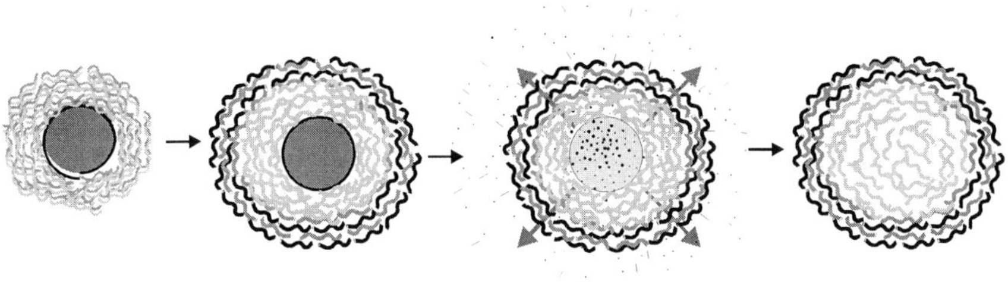

Fig. 4.1. Schematic representation of preparation of capsules loaded with polymers. The templates with precipitated polymers (a) are covered with stable shell by LbL approach (b). Then the templating cores are decomposed (c) and polymers dissolve into interior (d). The polymers might be precipitated either by complexation with multivalent ions or by adding of non-solvent.

The polyelectrolyte LbL shell is formed on particles already covered by precipitated polymers. After formation of stable outer shell, the colloidal template could be decomposed. The polymer molecules being underneath stable shell can be dissolved in shell interior. In fact, this approach compiles three stages: a) precipitation of polymers on colloidal surface; b) capturing of precipitated polymers on colloidal particles by stable shell formation; c) decomposition of the core, the products of decomposition expel through capsule wall, whilst the polymer dissolve and float in capsule interior. This approach has been tested for encapsulation of non-charged polysaccharides, i.e. dextran, and polyelectrolyte as PSS and PAH [46]. The more tricky thing of this procedure is to assemble the polyelectrolytes composing outer shell at the condition, where previously precipitated polymer does not dissolve from surface of colloidal particles. The polyelectrolytes, like PSS or PAH can be precipitated by complexation with multivalent ions, for instance, Me^{3+} and CO_3^{2-}, respectively. Then the polyelectrolyte alternation adsorption is performed in order to reach certain number of layers, i.e. shell thickness. After decomposition of colloidal core and removal of its products the multivalent ions are extracted from inner shell by applying salt solution and/or complex-ion, such as EDTA [46]. The polyelectrolytes at inner shell became free of complex and removes shell dissolving in inner volume (Fig. 4.2a).

It should be mentioned that the first layers of oppositely charged polyelectrolytes partially substitute multivalent ions in inner shell. This tells on entanglement of polyelectrolyte previously precipitated and then alternatively adsorbed. In fact, not all precipitated by complexation with multivalent ions polyelectrolytes are finally floating in inner volume. As seen on Fig.4.2 the fluorescence intensity from shell is higher then interior, that indicates some rest of precipitated polyelectrolyte entangled in polyelectrolyte network of alternatively adsorbed polyelectrolyte multilayers.

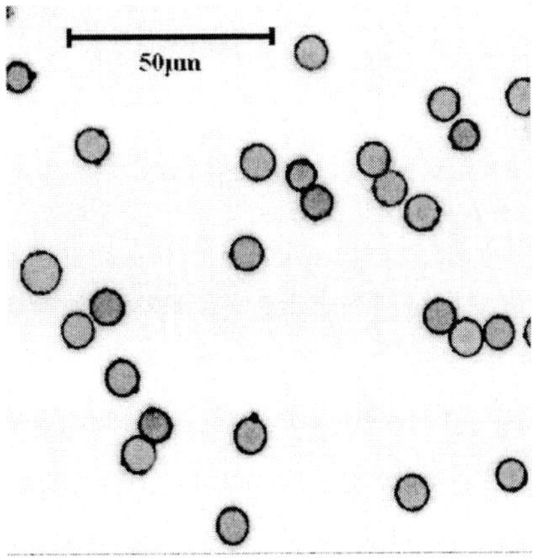

Fig. 4.2. Confocal microscopy image of the polyelectrolyte capsules loaded with labelled PSS via surface controlled precipitation followed by LbL assembling.

If the precipitation of polymer molecules on colloidal particles is caused by addition of another solvent the sequential LbL assembling should by done in the same solvent mixture [55]. Dextran has been encapsulated in the capsules by, at first, precipitating on colloidal particles by dropping of ethanol until volume % of ethanol reached 50%. Secondly, assembling LbL of PSS and PAH on colloidal particles covered with dextran from the same water/ethanol mixture which prevents dextran dissolution. Towards, our concept the cores (in our case it were MF-particles) were dissolved at the condition keeping dextran not soluble (acid/ethanol mixture). Then the capsules were transferred to aqueous solution. Dextran molecules became soluble and being captured by outer shell are displaced into the capsule interior [46]. As one can see the labelled dextran distributed all inner volume of the capsules. The calculated concentration of dextran in the capsule is about 0.2M on sugar residue. Because fluorescence intensity at each capsule looks the same one can deduce the dextran is equally distributed in every capsule. Some swelling of the capsule is pronounced due to osmotic pressure caused by dextran and coming from interior. Naturally, the amount of loaded polymer per capsule is amount of all added polymer divided by number of colloidal particles harvesting precipitating polymers. The possible loss of used polymers due to breakage of capsules did not exceed 10-15%. Such simplicity in procedure allows to dosage polymer contents in capsule in tiny, less than nanograms, amounts.

Summarizing approaches how to encapsulate the macromolecules by their precipitation and then capturing by LbL shell fabrication the positive and negative features should be stressed. In fact, the encapsulation via precipitation could reach encapsulation efficiency closed to 100%. Indeed, for encapsulation of PSS, PAH and dextran via harvesting of polymers on colloidal particles as was described above, we have achieved about 80-90% encapsulation efficiency [46]. The more essential peculiarities to use initial colloidal templates as polymer collectors are: i) The size distribution is determined by original templates; ii) amount of loaded polymers can be dosed and homogeneously distributed in each capsule; iii) there are possibilities to load the capsule with several different substances. The last point has not yet been demonstrated, but one would think that there are no significant difficulties to compose multicomponent shell precipitating on surface of colloidal particles several macromolecules one by another or at the same time if they could be precipitated at the same condition. It has just to be considered only compatibility of conditions for each precipitation step and further LbL fabrication of stable polyelectrolyte shell. The encapsulated macromolecules are kept in the capsule until the permeability is hold as non-permeable for macromolecules. It might be utilized for fabrication of enzyme microreactors, when the substrates and reaction products diffuse through capsule wall freely. The loaded macromolecules might release after certain pH treatment (cf. section 3).

5. Physico-chemical reactions in restricted volume of the capsules

Above one can see, that there are several methods to load the capsules with different polymer material. Selective capsule wall permeability allows to differ physico-chemical properties of the capsule interior from exterior. Incorporation of polymer into capsule excites the temptation to use such capsules with modified inner volume. Indeed, the presence of polymer on only one side of membrane might significantly change such physico-chemical characteristics of interior, as pH value. Let us consider the systems depicted on Fig. 5.1.

If polymeric acid is placed only outside capsule the inner volume must keep neutral pH value. Due to electroneutrality the amount of H+ and OH- ions should be compensated and hence neutral pH value in the interior must be kept. Thus the pH-difference across polyelectrolyte shell is established despite on penetration of H+ ions. Naturally, the presence of any other small ions would influence on proton distribution accordingly Donnan equilibrium. Increasing of salt concentration decreases pH-difference. The pH-value inside the capsules can be measured by introducing dye linked to polymer in capsule interior. The theoretical consideration of this effect, influence of salt and possible reflection of capsule on osmotic pressure produced by polymers is described in details [51]. As one can imagine there are several possibilities to establish pH-gradient across capsule wall in the dependence on where inside or outside polybase and polyacid is placed (see Table).

Placement of polyacid or polybase	Polyacid		Polybase	
	inside	outside	inside	outside
Inner pH	acidic	neutral	basic	neutral
Outer pH	neural	acidic	neutral	basic

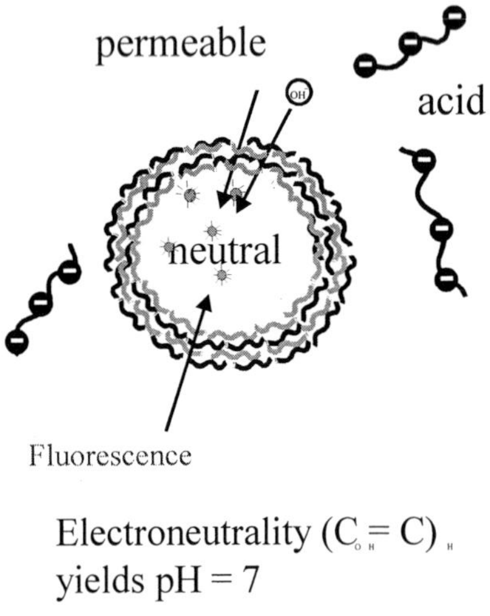

Fig. 5.1. Schematic representation of the Donnan equilibrium situation developing in the presence of bulk polymeric acid not capable to penetrate the wall of hollow polyelectrolyte capsule.

The idea is to set the conditions for penetrable capsule wall molecules where they could precipitate locally due to reduced solubility at shifted pH. Experimentally, this idea was performed with the capsules loaded with PSS, which have acidic pH. The pH value in the capsule as was measured using calibrated buffer solutions, which equalize pH inside and outside [51], revealed pH around 2.4 inside the capsule, whereas outside pH is 6.5 as measured by standard pH electrode. Two dyes 5-(and-6)-carboxytetramethyl-rhodamine (CR) and 6-carboxyfluorescein (6-CF) were chosen because of their low solubility at that low pH. (Fig. 5.2).

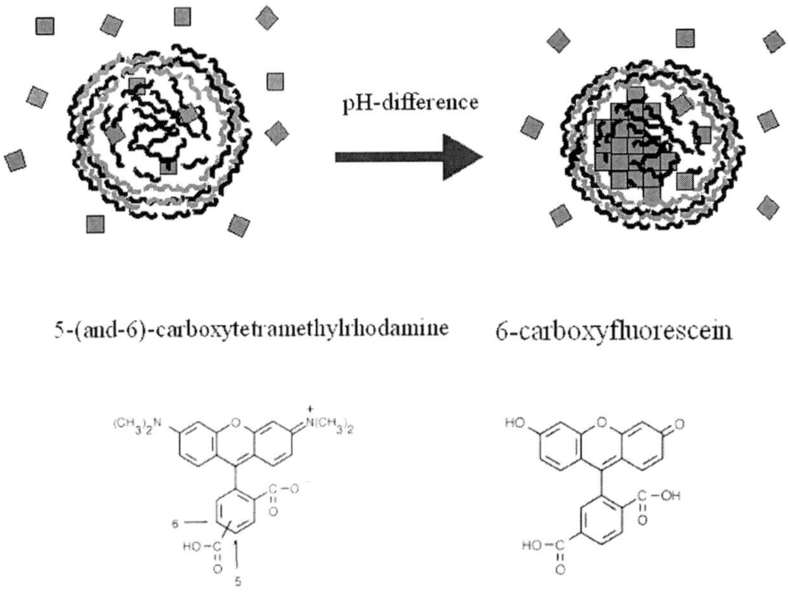

Fig. 5.2. Illustration for dye precipitation process in the capsules due to pH-difference caused by polyanion loading. Chemical formulas for 5-(and-6)-carboxytetramethyl-rhodamine (left) and 6-carboxyfluorescein right).

The precipitation was curried out from 10^{-4} M dye solution. The capsules containing encapsulated PSS were added to the dye solution. It should be noted that these dyes have no interaction with PSS due to the same charge. The typical fluorescence confocal image of the capsules is shown on Fig. 5.2 (left). The fluorescence background is naturally attributed to 6-CF solution. The interior of capsule looks dark and in transmission light the capsules are black. This is attributed to formation of 6-CF solid precipitates, which do not fluoresce due to self-quenching. It should be noted that capsules filled with PSS are swollen and 6-CF precipitates are formed only in filled swollen capsules. Occasionally one can find broken capsules, which are not swollen. The contents of such damaged capsules are about 10%. Remarkably, the interior of these capsules is filled with 6-CF concentration as outside, but no formation of solid precipitates. It confirms that only PSS filled capsules enable to cause 6-CF precipitation in capsule interior. The similar results have been obtained also for CR, which also forms solid precipitates at low pH. The capsules filled with PSS contain CR precipitates.

Fig. 5.2 (right) illustrates the typical SEM images of CR-precipitates formed in the capsules. The capsules look filled with solid materials. They are not flattened on surface due to collapse as empty capsules usually look like (Fig. 3.1). Dyes preferentially precipitates into the capsule interior. There were found almost no precipitates outside capsules. By the same manner the dye what precipitates in basic pH will be accumulated in the capsule loaded with polycation, whose pK is higher than pH value of precipitation of the dye.

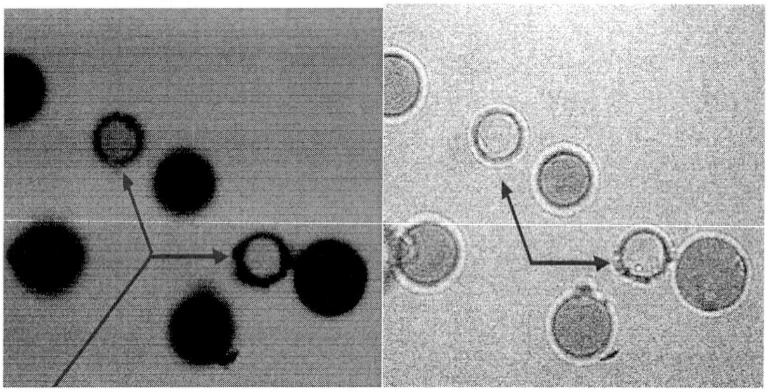

Fig. 5.3. Confocal fluorescence microscopy images of 6-CF precipitates (left) forming in the capsules loaded with PSS. Arrows show broken capsules without PSS. Scanning electron microscopy image (right) of CR precipitates formed [56].

Thus, the hollow and loaded polyelectrolyte capsules can be used as templates for selective precipitation or crystallisation of various organic and inorganic materials. The method yields monodisperse filled capsules. The ease with which one can manipulate the chemical composition of both the interior and exterior of the capsules, their size and permeability provides the possibility to tailor the 'microreactors' more subtlety for the required precipitation reaction. For example, the capsules can be loaded with macromolecules bearing functional groups that may enhance nucleus formation and growth. Loading of drug substances by means of controlled precipitation into polyelectrolyte capsules of a predetermined size is envisaged for elaboration of drug delivery systems and dye pigment preparation. Thus, these capsules have promising outlooks for applied studies on the development of drug formulation and crystal engineering.

6. CONCLUSIVE REMARKS.

The uniformity, simplicity and versatility of possible applications are significant points of the described above technology. The particular aspects coming from basic concept of the technology are outlined on Fig. 6.1.

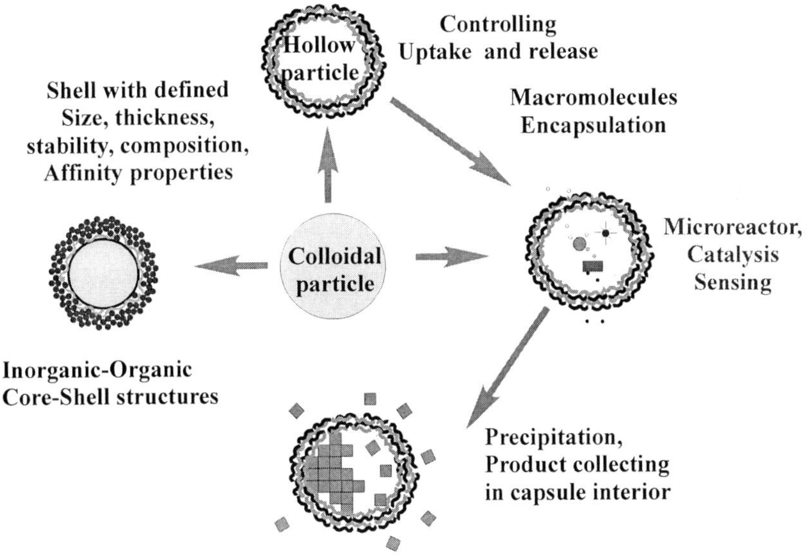

Fig. 6.1. Comprehensive illustration of applications of step-wise shell formation on colloidal particles.

As an initial matrix we have a colloidal template. Organic and inorganic particles, drug nanocrystals, biological cells, protein aggregates, in fact, any colloidal particles ranging in size from 50 nm till tens of microns are possible templates to fabricate shell. The shell can be composed of variety of materials. There are actually two ways to assemble nano-engineered shells on micron and submicron-sized colloidal particles. In first layer-by-layer approach, there must be a couple of oppositely charged species. Alternated adsorption of them results to form a shell with defined thickness, composition and, hence the properties, such as permeability, release, degradability, compatibility. Another approach of shell formation is surface controlled precipitation of polymers. In this case the colloidal particles are harvesting the precipitated polymers and can be covered with polymeric film of defined thickness as well. The choice of polymer to build the shell is determined by certain requests for application, in order to reach desired stability and comparability. Actually, one could choose the proper line to match certain tasks using scheme on Fig. 6.1 accordingly to concrete task taking into account desired characteristics of the elaborated core-shell structures or capsules.

Selective shell permeability makes possible keeping of macromolecules in the capsules. Macromolecules can be incorporate into the inner volume of the capsule by different means, either their aggregation in the solution or precipitation onto surface of collecting colloidal particles followed by their capturing by formation of stable polyelectrolyte shell afterwards or opening and closing pores in the capsule wall by pH, salt, temperature and solvent. These approaches to load macromolecules into the capsules are differed by size of the capsules, their monodispersity, low or high macromolecule concentration in the capsules, encapsulation efficiency and possibilities to incorporate several substances in one capsules. The macromolecules can be, of course, incorporated as layer constituents during shell build-up. LbL deposition on colloidal particles at the fabrication could supply LbL degradation of components afterwards. One could find the optimum time scale for this process as well. Generally speaking, one also has to choose more profitable way according to the aims these capsule should serve for. Talking about encapsulation of enzymes it should be stressed that selective permeability provides the elaboration of enzymatic microreactor where the proteins are placed into the capsule interior. The enzyme have higher stability and are prevented by outer shell against high molecular weight inhibitors and proteolytic agents. While encapsulated enzymes could do their catalytic reaction if substrates for the reactions are small molecules. The products of the one or multi-step enzymatic reactions in the capsules could be released out or collecting in the capsules interior. The capsules themselves can be easily withdrawn from one solution, washed and put to another by filtration, centrifugation or they can be driven by applying of magnetic field if the magnetite particles were used as layer constituents. We did not mention here any chemical treatments of the capsules after they have been assembled. Obviously, further modifications, like cross-linking, can significantly change the properties of the capsules, such as permeability or stability, and research on that is envisaged.

The fundamental aspects of research on these capsules should be mentioned. They represent an unique system to study chemical and physical phenomenon in micron and submicron sized volumes. In the comparison with liposomes the polyelectrolyte capsules, besides they have higher stability, give the possibilities to vary the contents of inner volume by controllable way. Defined inner composition allows to perform the chemical reactions in restricted volume, what have been demonstrated up to now only as precipitation reaction for small organic molecules. There are no doubts that such philosophy might be expanded further on. Selective permeability and related solubility and pH in interior and exterior are background parameters to modify the capsule interior. These capsules might be a device to study single molecule effects. Indeed, the concentration of substances 10^{-7}M means about one molecule of this substance in capsules of 300nm diameter. Defined input into the interior could register discrete chemical reactions occurring in the capsule interior.

Wide range of possible use, interplay of several approaches, solving different problems, focusing fundamental and application aspects in diverse areas of life and material sciences attract interest and stimulate further research on the development of decribed approaches for nano-engineering surfaces of colloidal particles and polymeric capsules.

Acknowledgment.

The author expresses his gratitude to Prof. Dr. H. Möhwald (MPI Colloids and Interfaces, Potsdam) for continuous strong support of the work on this topic and stimulating discussions. Dr. E. Donath is acknowledged for fruitful collaboration over the years. Members of group at MPI: Dr. A. Voigt, Dr. L. Dähne, Dr. S. Leporatti, S. Moya, I.L. Radtchenko and A.A. Antipov are thanked for providing a creative atmosphere, collaboration and help in the research on encapsulation.

7. References.

1. R.K. Iler, J. Colloid Interface Sci., 21 (1966) 569-572

2. H. Lee, L.J. Kepley, H.G. Hong, S. Akhter and T.E. Mallouk, J. Phys. Chem., 92 (1988) 2597-2601

3. G. Decher and J.-D. Hong, Macromol. Chem., Macromol. Symp., 46 (1991) 321-327

4. G. Decher, Science 1997, 277, 1232-1237

5. G. Decher, B. Lehr, K. Lowack, Y.M. Lvov and J. Schmitt, Biosensors & Bioelectronics, 9 (1994) 677-683

6. Y.M. Lvov and G.B. Sukhorukov, Membr. Cell Biol., 11 (1997) 277-303

7. G.B. Sukhorukov, M.M. Montrel, A.I. Petrov, L.I. Shabarchina and B.I. Sukhorukov, Biosensors & Bioelectronics, 11 (1996) 913-922

8. F. Caruso, K. Niikura, D.N. Furlong and Y. Okahata, Langmuir, 13 (1997) 3427-3433

9. Y.M. Lvov, G. Decher and H. Möhwald, Langmuir, 9 (1993) 481-486

10. N.A. Kotov, Nanostructured Materials, 12 (1999) 789-796

11. J.H. Cheung, A.F. Fou and M.F. Rubner, Thin Solid Films, 244 (1994) 985-989,

12. P. Berndt, K. Kurihara and T. Kunitake, Langmuir, 8 (1992) 2486-2490

13. P.T. Hammond, Current Opinion in Colloid & Interface Science, 4 (1999) 430-442

14. P. Bertrand, A. Jonas, A. Laschewsky and R. Legras, Macromol. Rapid Com., 21 (2000) 319-348

15. Y.M. Lvov, K. Ariga, I. Ichinose and T. Kunitake, J. Chem. Soc. Chem. Commun., 22 (1995) 2313-2314

16. M. Onda, Y.M. Lvov, K. Ariga and T. Kunitake, Biotechnology and Bioengineering, 51 (1996) 163-166

17. M. Onda, Y.M. Lvov, K. Ariga and T. Kunitake, J. Fermentation & Bioengineering, 82 (1996) 502-506

18. M. Gao, B. Richter, S. Kirstein and H. Möhwald, J. Phys.Chem., 102 (1998) 4096-4103

19. X.G. Wang, S. Balasubramanian, L. Li, X.L. Jiang, D.J. Sandman, M.F. Rubner, J. Kumar and S.K. Tripathy, Macromol. Rapid Com., 18 (1997) 451-459

20. P. Stroeve, V. Vasquez, M.A.N. Coelho and J.F. Rabolt, Thin Solid Films, 285 (1996) 708-712

21. R. Von Klitzing and Möhwald, H, Langmuir, 11 (1995) 3554-3559

22. L. Krasemann and B. Tieke, J. Memb. Sci., 150 (1998) 23-30

23. K. Pommerening, O. Ristau, H. Rein, H. Dautzenberg and F. Loth, Biomed. Biochim. Acta, 42 (1983) 813-823

24. R. Pommersheim, J. Schrezenmeir and W. Vogt, Macromol. Chem. Phys., 195 (1994) 1557-1567

25. S.W. Keller, S.A. Johnson, E.S. Brigham, E.H. Yonemoto and T.E. Mallouk, J. Am. Chem. Soc., 117 (1995) 12879-12880

26. Y.T. Chen and P. Somasundaran, J. Am. Chem. Soc., 81 (1998)140-144

27. P. Somasundaran, Y.T. Chen and D. Sarkar, Mater. Res. Innovations, 2 (1999) 325-327

28. G.B. Sukhorukov, E. Donath, H. Lichtenfeld, E. Knippel, M. Knippel, A. Budde and H. Möhwald, Colloids Surfaces A, 137 (1998) 253-266

29. G.B. Sukhorukov, E. Donath, S.A. Davis, H. Lichtenfeld, F. Caruso, V.I. Popov and H. Möhwald, Polymers for Advanced Technologies, 9 (1998) 759-767

30. A. Voigt, H. Lichtenfeld, H. Zastrow, G.B. Sukhorukov, E. Donath and H. Möhwald, Industrial & Engineering Chemistry Research, 38 (1999) 4037-4043

31. E. Donath, G.B. Sukhorukov, F. Caruso, S.A. Davis and H. Möhwald, Angewandte Chemie, International Edition, 37 (1998) 2201-2205.

32. F. Caruso, Chem. Eur. J., 6 (2000) 413-419

33. F. Caruso, A.S. Susha, M. Giersig and H. Möhwald, Advanced Materials, 11 (1999) 950-953

34. A. Rogach, A. Sucha, F. Caruso, G. B. Sukhorukov, A. Kornovski, S. Kershaw, H. Möhwald, A. Eychmüller and H. Weller, Advanced Materials, 12 (2000) 333-336.

35. G.B. Sukhorukov, E. Donath, S. Moya, A.S. Susha, A. Voigt, J. Hartmann and H. Möhwald, J. Microencapsulation, 17 (2000) 177-185

36. S. Moya, E. Donath, G.B. Sukhorukov., M. Auch, H. Bäumler, H. Lichtenfeld and H. Möhwald, Macromolecules, 33 (2000) 4538-4544

37. R. Georgieva, S. Moya, S. Leporatti, B. Neu, H. Bäumler, C. Reichle, E. Donath and H. Möhwald, Langmuir, 16 (2000) 7075-7081

38. F. Caruso, H. Fiedler and K. Haage, Colloids Surfaces A, 169 (2000) 287-293

39. C. Schuler and F. Caruso, Macromol. Rapid Com., 21 (2000) 750-753

40. E. Donath, G.B. Sukhorukov and H. Möhwald, Nachrichten aus Chemie, Technik und Laboratorium, 47 (1999) 400-405

41. B. Neu, A. Voigt, R. Mitlöhner, S. Leporatti, E. Donath, C.Y. Gao, H. Kiesewetter, H. Möhwald., H.J. Meiselman and H. Bäumler, J. Microencapsulation, 18 (2001) 385-395

42. F. Caruso, Y.W. Wenjun, D. Trau and R. Renneberg, Langmuir, 16 (2000) 8932-8936

43. A.A. Antipov, G.B. Sukhorukov, E. Donath and H. Möhwald, J. Phys. Chem. B, 105 (2001) 2281-2284

44. M.E. Bobreshova, G.B. Sukhorukov, E.A. Saburova, L.I. Elfimova, B.I. Sukhorukov and L.I. Sharabchina, Biophysics, 44 (1999) 813-820

45. V. Dudnik, G.B. Sukhorukov, I.L. Radtchenko and H. Möhwald, Macromolecules 34 (2001) 2329-2334

46. I.L. Radtchenko, G.B. Sukhorukov, H. Möhwald, Colloid Surfaces A, (2001), in press

47. I.L. Radtchenko, G.B. Sukhorukov, A.L. Rogach, A. Kornowski and H. Möhwald, Advan. Mat. (2001), in press.

58. S. Leporatti, A. Voigt, R. Mithöhner, G.B. Sukhorukov, E. Donath and H. Möhwald, Langmuir, 16 (2000) 4059-4063

49. H. Möhwald., H. Lichtenfeld., S. Moya, A. Voigt, H. Bäumler, G.B. Sukhorukov, F. Caruso and E. Donath, Macromol. Chem. Makromol. Symposia, 145 (1999) 75-81

50. F. Caruso, R.A. Caruso and H. Möhwald, Science, 282 (1998) 1111-1114

51. G.B. Sukhorukov, M. Brumen, E. Donath and H. Möhwald, J. Phys. Chem. B, 103 (1999) 6434-6440.

52. G.B. Sukhorukov, A.A. Antipov, A. Voigt, E. Donath and H. Möhwald, Macromol. Rapid Com., 22 (2001) 44-46

53. A.A. Antipov, G.B. Sukhorukov, S. Leporatti, I.L. Radtchenko, E. Donath and H. Möhwald, Colloid Surface A, (2001), in press

54. J.D. Mendelsohn, C.J. Barrett, V.V. Chan, A.J. Pal, A.M. Mayes and M.F. Rubner, Langmuir, 16 (2000) 5017-5023

55. S.T. Dubas and J.B. Schlenoff, Macromolecules, 32 (1999) 8153-8160

56. I.L.Radtchenko and G.B.Sukhorukov. Unpublished results

Novel Methods to Study Interfacial Layers
D. Möbius and R. Miller (Editors)

ULTRATHIN SELF-ASSEMBLED POLYVINYLAMINE/POLYVINYLSULFATE MEMBRANES FOR SEPARATION OF IONS

Ali Toutianoush and Bernd Tieke*

Institut für Physikalische Chemie der Universität zu Köln, Luxemburgerstraße 116, D-50939 Köln, Germany

Contents

Keywords. Polyelectrolyte membrane, self-assembly, ion transport, ion separation

* author to whom correspondence should be addressed

The ion transport across ultrathin membranes consisting of an alternating sequence of polyvinylamine (PVA) and polyvinylsulfate (PVS) is described. The membranes were prepared by means of electrostatic layer-by-layer adsorption and studied on their permeability for alkali and alkaline earth metal salts. It was found that the permeation rates P_R decline in the sequence $P_R(NaCl) > P_R(Na_2SO_4) \geq P_R(MgCl_2) > P_R(MgSO_4)$ because divalent ions of higher charge density receive stronger electrostatic repulsive forces from the equally charged parts of the membrane than the monovalent ions. Within the series of alkali and alkaline earth metal chlorides, the P_R values increase from LiCl to KCl, and from $MgCl_2$ to $BaCl_2$ due to different charge density of the cations originating from their different size. PVA/PVS membranes prepared from salt-containing polyelectrolyte solution show considerably higher separation factors α than corresponding membranes prepared from salt-free solutions. Typical α-values found for a PVA/PVS membrane of 60 layer pairs are $\alpha(NaCl/MgCl_2) = 44.3$, $\alpha(NaCl/Na_2SO_4) = 10.3$ and $\alpha(MgCl_2/MgSO_4) = 6.2$. Influences of the thickness and surface charge of the membrane and the concentration of the permeating salt solution on P_R are also reported.

1. Introduction

The alternating electrostatic layer-by-layer adsorption of polyionic compounds has been proven to be a simple, yet elegant and versatile technique for preparation of films with controlled structure and uniform thickness in the nanometer range [1-3]. The method is also useful for the preparation of a new type of composite membrane with ultrathin separation layer, if cationic and anionic polyelectrolytes are alternately adsorbed on a porous support [4-11]. First attempts to use these membranes for separation of gases [4-6], liquid mixtures [7-10] and ions [8, 11] have already been reported.

Probably one of the most interesting properties of the new composite membranes is their ability to reject di- and multivalent ions more strongly than monovalent ones. This property can be ascribed to the multi-bipolar architecture of the self-assembled polyelectrolyte membrane, which enables the rejection of ions by electrostatic repulsive forces [8, 11]. Some time ago, the effect has already been studied in solution-cast bipolar polyelectrolyte membranes of macroscopic thickness, and has been ascribed to the fact that divalent cations receive a much stronger repulsive force from a positively charged polyelectrolyte layer than the monovalent ones, thus being more strongly rejected [12, 13]. The same is true for the divalent anions which are rejected by the negatively charged polyelectrolyte layer. A corresponding rejection model for the polyelectrolyte multilayer membrane is depicted in Fig. 1.

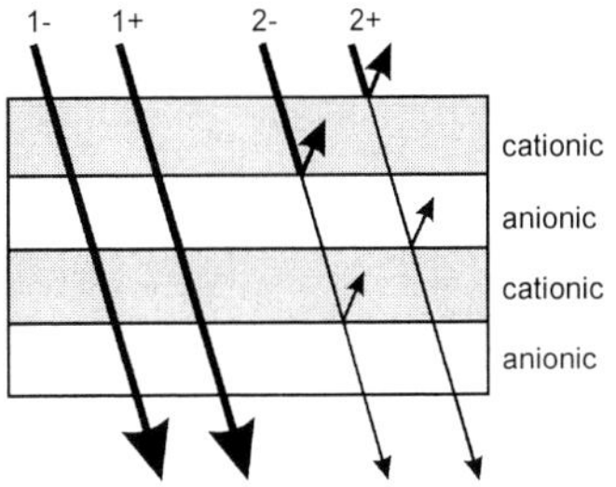

Fig. 1 Rejection model of multi-bipolar polyelectrolyte membrane

In a previous communication, we demonstrated the validity of this model [11]. For a membrane consisting of 60 layer pairs of polyallylamine/polystyrenesulfonate (PAH/PSS), for example, a separation factor α for NaCl/$MgCl_2$ of up to 112.5 and for NaCl/Na_2SO_4 an α-value up to 45.0 was found. In that communication we also showed that the rejection of divalent ions becomes progressively stronger, when the polyelectrolytes used for membrane preparation exhibit a higher charge density. Since most of the previous studies were carried out with the PAH/PSS membrane of only moderate charge density, it was of interest to also study the separation behaviour of a polyelectrolyte membrane with higher charge density in detail. We therefore have chosen the polyvinylamine/polyvinylsulfate (PVA/PVS) membrane, which is especially interesting as it has recently been proven to be useful in the pervaporation separation of alcohol/water mixtures [10]. Here we describe the ion transport behaviour of this membrane and report on the influence of various parameters on the ion separation.

2. Experimental part

2.1. Materials

Polyvinylamine (PVA) was kindly supplied by BASF, Ludwigshafen, while polyvinylsulfate potassium salt (PVS; molecular weight 350.000) was purchased from Acros.

PVA PVS

Both polymers were used without further purification. Milli-Q-water (resistance ≥18 MΩ cm^{-1}) was used as solvent. PAN/PET supporting membranes treated with oxygen plasma [6, 7] were kindly provided by Sulzer Chemtech GmbH, Neunkirchen.

2.2. Methods

Preparation of the separating membranes

For preparation of the separating membrane, aqueous solutions of PVA and PVS were used. The solutions contained the polymers in a concentration of 10^{-2} monomoles /l (monomol = mol of monomer units) and, if not especially stated, sodium chloride in a concentration of 0.1 mol/l. The solutions were acidified to pH 1.7 using aqueous HCl. For polyelectrolyte adsorption, the porous PAN/PET support was consecutively dipped into the solution of PVA, pure water, the solution of PVS, and pure water again. The four steps led to adsorption of a single PVA/PVS layer pair as schematically indicated in Fig. 2. In order to obtain the separation layer, the four steps were carried out up to 60 times. Immersion time for the individual steps was 30 min, the temperature was 20 °C. For the dipping procedure, a home-built, computerised apparatus was used. The size of the membranes was 12 x 12 cm^2.

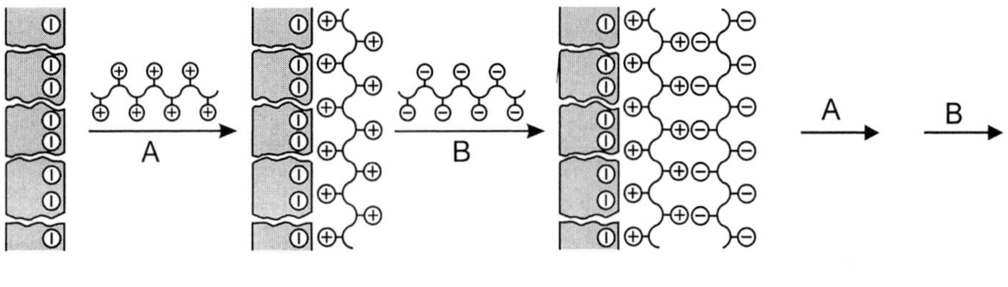

Fig. 2 Scheme of layer-by-layer adsorption of polyelectrolytes on plasma-treated porous supporting membrane. Multiple repetition of steps A and B leads to the ultrathin separation layer. Note that in reality the pore diameter is 20 - 200 nm, and the individual polymer layers are less ordered than indicated in the scheme.

Ion permeation

Measurements of ion permeation were carried out using a home-made apparatus. The membrane (area A = 4.52 cm^2) was mounted between two chambers with a volume V_0 of 60 ml each. Chamber one contained the electrolyte solution of concentration c = 0.1 mol/l (if not especially noted), chamber two pure water and the cell for measurement of the conductivity Λ. The permeation rate P_R of electrolytes across the membrane was determined by measuring the initial increase of conductivity $\Delta\Lambda$ per unit time Δt under constant stirring. The osmotic flow of water in the direction opposite to the ion transport was taken into account by measuring the volume increase ΔV in chamber one during the time period Δt of the conductivity measurement. P_R was calculated using the equation $P_R = (\Delta\Lambda/\Delta t)\,(V_0\text{-}\Delta V)\,\Lambda_m^{-1}\,(Ac)^{-1}$ with Λ_m being the molar conductivity of the corresponding salt solution and $(V_0 - \Delta V)$ being the volume in chamber two after correction of the osmotic flow. The change of the P_R value upon the correction was never larger than 2 percent. The theoretical separation factor α is given by the ratio of the permeation rates of the corresponding salt solutions, e.g. for NaCl and $MgCl_2$

$$\alpha\,(NaCl/MgCl_2) = \frac{P_R(NaCl)}{P_R(MgCl_2)}\,. \qquad (1)$$

3. Results and discussion

In a first set of experiments, the permeation rates of various alkali and alkaline earth metal chlorides were investigated. For these experiments, a composite membrane with 60 layer pairs of PVA/PVS as the separating layer was used. In Fig. 3, the permeation rates P_R of the various salts are compiled. It can be seen that the P_R values increase from LiCl to KCl and from $MgCl_2$ to $BaCl_2$, the values of the alkaline earth metal chlorides being significantly smaller than those of the alkali salts. Since the anions were identical in all cases, the different P_R values can only originate from different charge densities of the cations, a high charge density favouring a strong rejection by the membrane. Within the series of the alkali metal ions, the radius increases from Li to K so that the charge density of K is lowest and thus the P_R value is highest. Accordingly, the same holds for the alkaline earth metal chlorides except that the charge of the cations is always twice as large as for the alkali ions and thus the ions are generally more strongly rejected.

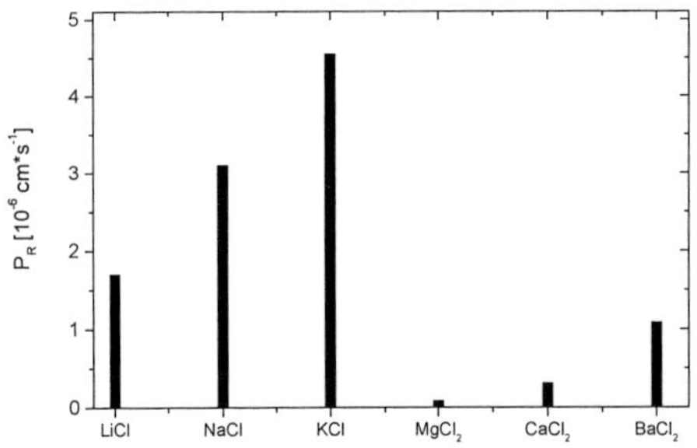

Fig. 3 Permeation rates P_R of various alkali and alkaline earth metal salts. Separating membrane: 60 layer pairs of PVA/PVS adsorbed from NaCl-containing polyelectrolyte solutions.

One may find the correlation between the naked ions and the transport behaviour surprising because ions in aqueous medium are generally believed to be strongly hydrated so that the transport behaviour might be rather determined by the hydrodynamic radius, which decreases from lithium to potassium, for example. However, our results clearly indicate that the hydration only plays a minor role. Obviously, the interactions of the water molecules within the membrane with the polyelectrolyte chains are so strong that the available energy is not sufficient to also cause a hydration of the metal ions. Similar arguments were already previously considered to explain the ion transport across membranes of macroscopic thickness [14].

In a second set of experiments, the effect of the concentration of electrolyte solutions on the ion transport was investigated. In the log-log-plot of Fig. 4, the dependence of the permeation rates of NaCl and $MgCl_2$ on their concentration in aqueous solution is shown. In a low concentration range up to about 0.3 mol/l, a proportionality between the two parameters is found, while at higher concentration, the permeation rate is strongly increased. This points to a change in the transport mechanism, possibly caused by structural rearrangements in the separation layer. The increase of P_R at high $MgCl_2$ concentration is stronger than at high NaCl concentration, so that the separation factor strongly decreases, if the salt concentration is high. Therefore, in order to get a good ion separation it is highly recommended to work at relatively low concentration, i.e. at concentrations below 0.3 mol/l.

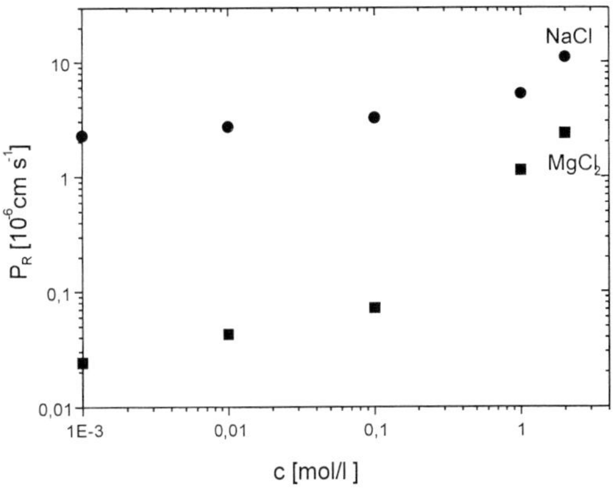

Fig. 4 Plot of permeation rates P_R of NaCl and $MgCl_2$ versus the concentration c of the aqueous salt solution. Separating membrane: 60 layer pairs of PVA/PVS adsorbed from NaCl-containing polyelectrolyte solutions.

In a further set of experiments, the influence of the membrane thickness on the permeation rates of NaCl and $MgCl_2$ was investigated. The experiments were carried out for membranes prepared from NaCl-containing polyelectrolyte solutions and from salt-free solutions. As indicated in Fig. 5a and b, the permeation rates decrease with increasing number of deposited layer pairs, as already suggested by the rejection model depicted in Fig. 1. If the separation layer is prepared from salt-free polyelectrolyte solution, the P_R value of $MgCl_2$ drops more strongly than for NaCl so that for a sample with the maximum number of deposited layer pairs being 60 a highest separation factor α (NaCl/$MgCl_2$) of 9 is found (Fig. 5a). This behaviour is analogous to PAH/PSS separation layers reported previously although the α-value of those membranes was higher [11].

If the separation layer is prepared from salt-containing polyelectrolyte solution, the permeation rates of NaCl and $MgCl_2$ are considerably lower (Fig. 5b). The decrease of P_R is especially pronounced for $MgCl_2$ so that after 10 dipping cycles the P_R value has dropped from 4.6×10^{-5} to 1.3×10^{-7} cm s^{-1}, and a separation factor α(NaCl/ $MgCl_2$) of 51 has been reached. Even higher α-values are only found, if substantially more layer pairs are adsorbed. For example, an α-value of 68 is only reached for a separation layer consisting of 90 PVA/PVS layer pairs. This value is comparable with PAH/PSS membranes. It is likely that the observed behaviour has its origin in different morphologies of the separation layer caused by the different preparation conditions. Presence of salt in the polyelectrolyte solution induces a reduction of the coil size so that the polyelectrolytes are rather adsorbed as coils than in the flat conformation obtained from a salt-free solution. Consequently, individual polyelectrolyte layers are much thicker [15] and thus less able to interpenetrate and neutralise their charges. As a result, more excess charges are present in the separation layer and the permeating ions are more effectively rejected, in agreement with the experimental observations.

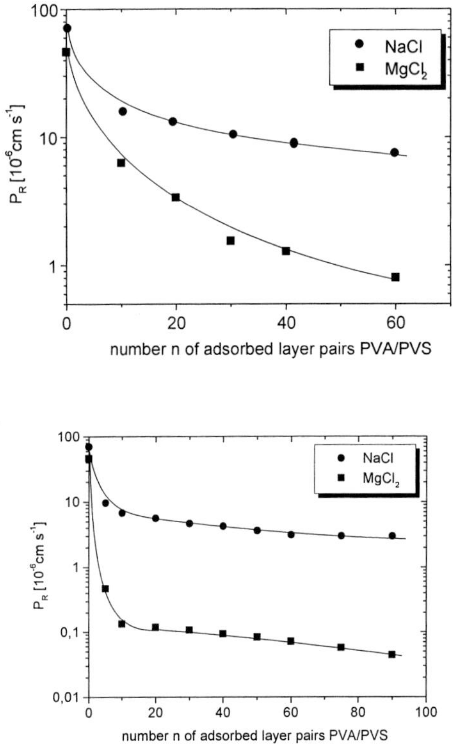

Fig. 5 Plot of permeation rates P_R of NaCl and $MgCl_2$ vs. the numbers of deposited layer pairs. Separating membrane: PVA/PVS prepared from NaCl-free (a) and from NaCl-containing polyelectrolyte solutions (b).

One may argue that the ion rejection is mainly caused by surface charges of the separation layer, because the concentration of excess charges in the membrane might be too low [16] to account for the observed rejection. In that case one should expect that the P_R values depend on the surface charge of the membrane, i.e. on whether a cationic or anionic polyelectrolyte is deposited in the uppermost layer. To study this effect, separating membranes consisting of 59 to 61 polyelectrolyte layer pairs were prepared, the uppermost layer either being PVA or PVS, so that the surface charge was alternately positive or negative. The permeation rates of NaCl, $MgCl_2$, Na_2SO_4 and $MgSO_4$ were investigated. The results are shown in the plot of Fig. 6. Clearly any influence of the surface charge is missing, i.e. the ion rejection is caused by the charges in the entire membrane as suggested by our model in Fig. 1.

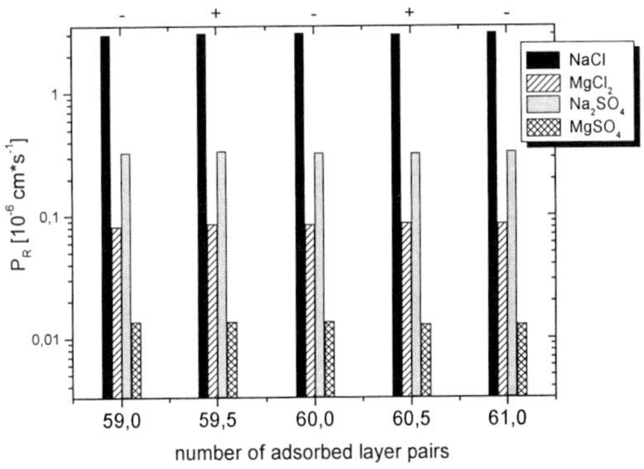

Fig. 6 Effect of the surface charge of the separating membrane on the permeation rates PR of NaCl, $MgCl_2$, Na_2SO_4 and $MgSO_4$. Separating membrane: 59 to 61 layer pairs of PVA/PVS adsorbed from NaCl-containing polyelectrolyte solution.

Fig. 6 also indicates that the permeation rates of the four electrolytes are significantly different. The permeation rate decreases in the sequence $P_R(NaCl) > P_R(Na_2SO_4) \geq P_R(MgCl_2) > P_R(MgSO_4)$ due to the increasing charge density of the cations as well as the anions. For example, the rejection of $MgSO_4$ is strongest because of the high charge density of the Mg^{2+} and the SO_4^{2-} ions. For a separating membrane consisting of 60 layer pairs of PVA/PVS the separation factors $\alpha(NaCl/MgCl_2)$, $\alpha(NaCl/Na_2SO_4)$ and α $(MgCl_2/MgSO_4)$ are 44.3, 10.3 and 6.2, respectively.

4. Summary and Conclusions

Our study shows that PVA/PVS membranes prepared upon alternate electrostatic adsorption are excellently suited for the separation of mono- and divalent ions. The separation factors are comparable with PAH/PSS membranes reported previously [11] although the number of layer pairs to be deposited for achieving a good separation is significantly lower. It is also shown that the differences in the permeation rates of the various salts originate from rejection of cations as well as anions, the rejection being most pronounced for the ions with the highest charge density. Therefore the permeation rates decline in the sequence $P_R(NaCl) > P_R(Na_2SO_4) \geq P_R(MgCl_2) > P_R(MgSO_4)$. The ion transport is very sensitive to the charge density of the ions so that pronounced differences of the P_R values are even found within the series of alkali and alkaline earth metal chlorides because of the differences in the charge density of the cations caused by their different size. The fact that the strongest rejection is found for the smallest cations with the highest charge density, i.e. for Li^+ and Mg^{2+}, is an indication that the ions permeate in the non-hydrated state across the membrane. This is reliable, especially if one assumes that the water within the membrane is strongly interacting with the polyelectrolyte chains so that the available energy is not sufficient to also built up a hydration sphere of the ions.

Also, it is demonstrated that the ion rejection is not merely a surface effect but a property of the bulk volume of the membrane, which supports our rejection model depicted in Fig. 1. Factors that lead to an increase of the concentration of excess charges within the membrane therefore also improve the ion separation. This is demonstrated for the membrane preparation from a salt-containing polyelectrolyte solution. In that case, the adsorbed polyelectrolyte layers are thicker [15] and thus less able to interpenetrate and neutralize their charges so that the concentration of excess charges is increased and the observed ion rejection improved. The larger thickness of the individual layers also causes that substantially less dipping cycles have to be applied in order to get a membrane with suitable ion separation.

Future work will be concerned with the study of the transport behaviour of transition metal salts and salts with organic ions of different size. Also, the ion transport under reverse osmosis conditions will be investigated.

ACKNOWLEDGEMENT.

The authors are grateful to Dr. H. Scholz, Sulzer Chemtech GmbH, Neunkirchen, for providing the plasma-treated PAN/PET supporting membranes, and to Dr. Mahr, BASF Ludwigshafen, for providing polyvinylamine. Financial support from the Deutsche Forschungsgemeinschaft (projects No. Ti 219/3-3 and 3-4) is also gratefully acknowledged.

5. REFERENCES

1. G. Decher, Science, 277, 1232 (1997)
2. P. Bertrand, A. Jonas, A. Laschewsky and R. Legras, Macromol. Rapid Commun., 21 (2000) 319
3. P.T. Hammond, Curr. Opin. Coll. Interf. Sci., 4 (2000) 430
4. P. Stroeve, V. Vasquez, M.A.N. Coelho and J.F. Rabolt, Thin Solid Films, 284-285 (1996) 708
5. J.-M. Leväsalmi and T. J. Mc Carthy, Macromolecules, 30 (1997) 1752
6. F.v. Ackern, L. Krasemann and B. Tieke, Thin Solid Films, 329 (1998) 762
7. L. Krasemann and B. Tieke, J. Membr. Sci., 150 (1998) 23
8. L. Krasemann and B. Tieke, Mater Sci. Eng. C, 8-9 (1999) 513
9. L. Krasemann and B. Tieke, Chem. Eng. Technol., 23 (2000) 211
10. L. Krasemann, A. Toutianoush and B. Tieke, J. Membr. Sci., 181 (2000) 221
11. L. Krasemann and B. Tieke, Langmuir, 16 (2000) 287
12. M. Urairi, T. Tsuru, S. Nakao and S. Kimura, J. Membr. Sci., 70 (1992) 153
13. T. Tsuru, S. Nakao and S. Kimura, J. Membr. Sci., 108 (1995) 269
14. W. Pusch and A. Walch, Angew. Chem. Int. Ed. Engl., 21 (1982) 660
15. G. Decher, J. Schmitt, Prog. Colloid Polym. Sci., 89 (1992) 160
16. J.B. Schlenoff, H. Ly and M. Li, J. Am. Chem. Soc., 120 (1998) 7626

Novel Methods to Study Interfacial Layers
D. Möbius and R. Miller (Editors)

THE DETECTION OF ORGANIC POLLUTANTS IN WATER WITH CALIXARENE COATED ELECTRODES

T. Wilkop[a], S. Krause[b], A. Nabok[a], A.K. Ray[a] and R. Yates[a]

[a] Sheffield Hallam University, School of Engineering, Physical Electronics and Fibre Optics Research Laboratories, City Campus, Pond Street, Sheffield, S1 1WB, UK

[b] University of Sheffield, Department of Chemistry, Dainton Building, Brook Hill, Sheffield S3 7HF, UK

Contents

Keywords: Calix[4]resorcinarenes; Langmuir-Blodgett films; permeability; cyclic voltammetry; impedance spectroscopy

A novel type of chemical sensor based on the changes in the membrane permeability, triggered by neutral organic analytes in water was investigated. The membrane consisted of several LB layers of calix[4]resorcinarene, deposited on a gold layer on top of a supporting glass substrate. The film permeability was evaluated with cyclic voltammetry, and a further characterisation of the membrane was carried out with impedance spectroscopy in the range of 0.1 - 10^5 Hz. Based on these results, an equivalent circuit model for the membrane was developed. It was shown that the presence of small amounts of organic analytes altered the permeability of the membrane by affecting its structural properties and hydrophobicity. The response to chloroform and acetone resulted in an increase of the permeability by several orders of magnitude. A decrease of the charge transfer resistance by a factor of 2.1 was observed in the case of chloroform.

1. INTRODUCTION

Since the 1980, calixarene derivatives have found many applications in the field of chemical sensing. They have advanced from a chemical curiosity to a rich source for molecular receptors [1]. The majority of the applications has so far concentrated on the detection of organic vapours in air [2, 3] and the detection or extraction of metal ions from aqueous solutions [4]. The calix[4]resorcinareneC_7H_{15} (C[4]RA) that was employed in our studies, has already shown its suitability for the detection of solvent vapours in air. Different transducing mechanisms such as QCM, SPR and ellipsometry were used, and an intensive characterisation of the membrane properties was carried out [5, 6]. The most probable mechanism of complexation includes the interaction of organic guest molecules with the alkyl chains, their condensation and accumulation within the bulk of the film followed by the swelling of the film [7]. The detection of organic analytes in water using calixarenes has only recently been investigated [8, 9].

In an attempt to mimic the highly sophisticated ion transport properties found in biological membranes for the development of an ion channel sensor, a comprehensive study on artificial systems was carried out [10]. It was shown that the permeability of orientated monolayers of various calixarene esters is strongly influenced by the complexation with metal ions [11]. It was reported that the membrane permeability of calix[6]arene esters for a neutral redox-active species increases on complexation with Na^+ ions [11]. The opposite was reported for strongly charged permeability markers such as $[Fe(CN)_6]^{4-}$. This behaviour was attributed to an increase in the electrostatic repulsion upon accumulation of likewise charged species in the membrane. Further examples for the control of the permeability through intermolecular voids upon analyte exposure were reported for electrodes coated with polypeptides [11]. First data confirming induced permeability changes in a Ta_2O_5- LB Calixarene - electrolyte structure was presented in [12].

The aim of this study was to investigate the changes in the membrane properties that occur when a neutral organic analyte is absorbed from the aqueous phase. An implementation of the investigated system into a dipstick like sensor for the real time detection of analyte in water would offer many benefits such as avoiding cumbersome sample collection and laboratory analysis.

2. EXPERIMENTAL DETAILS

2.1 Membrane preparation

A 0.5 mg ml^{-1} solution of C[4]RA in chloroform was spread on the surface of 18 $M\Omega cm^{-1}$ resistivity Millipore water in order to form a monolayer. The layers were then compressed to a surface pressure of 25 mN m^{-1}, and this pressure was held constant during the deposition. The dipping speed for the up- and down- stroke of the slide was set to 5 mm min^{-1}. The transfer ratios were close to unity. The substrates for the LB deposition were prepared by coating standard microscopic glass slides with a 3 nm seed layer of chromium and a 150 nm thick layer of gold using thermal evaporation in vacuum. The slides were left in saturated vapours of dimethyldisilazane for 24 hours in order to improve their hydrophobicity. A sensing membrane typically consisted of 12 monolayers of C[4]RA molecules.

2.2. Electrolyte

The background electrolyte was a pH 6.4 phosphate buffer containing 0.1M of NaCl and 1.5 mM phosphate. 10 mM of the redox-active species hydroquinone was added to the solution as a neutral permeability marker.

After adding the analyte to the electrolyte, the solution was vigorously stirred and sonicated. The absence of any light scattering confirmed a micelle free solution.

2.3 Measurement set-up

The experiments were conducted in an in-house built PTFE electrochemical cell shown in Fig. 1. The chamber was sealed onto the sample by using a rubber O-ring, and the spring pressure was carefully adjusted to avoid damaging the membrane. The effective area of the working electrode was 0.45 cm^2. The Ag/AgCl electrode was fabricated by pre-anodising a 0.25 mm diameter silver wire in a 1 M NaCl solution.

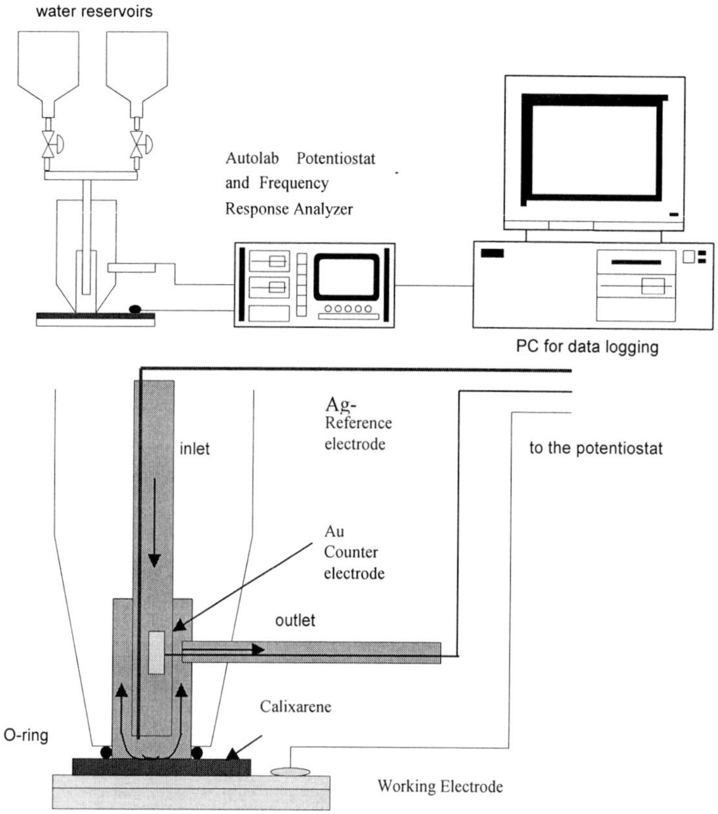

Fig. 1 Experimental set-up and the electrochemical cell

Cyclic voltammetry and impedance spectroscopy were performed using an Autolab potentiostat PG STAT 10 with an Autolab frequency response analyser FRA 2 both linked to a computer, running the software packages General Purpose Electrochemical System (GPES) Version 4.7 and Frequency Response Analysis (FRA) 4.7. For the cyclic voltammetry measurements the potential scan rate was set to 50 mV s^{-1} and each scan was started at 0 V sweeping first to the positive vertex potential and then back. Two scans were recorded for each run and the second one was evaluated. Recordings were taken after the system was allowed to stabilise for about 10 minutes, after 200 ml of solution were flushed through the chamber, which had a volume of about 3 ml.

3. Results and discussion

Fig. 2 shows the cyclic voltammogram of an uncoated gold electrode in the presence of hydroquinone (a) and its redox reaction (b). The separation of the two Faradaic peaks of 0.33 V is larger than theoretically predicted from the theoretical two electron transfer. This can be attributed to the specific cell properties and indicates a quasi-reversible system.

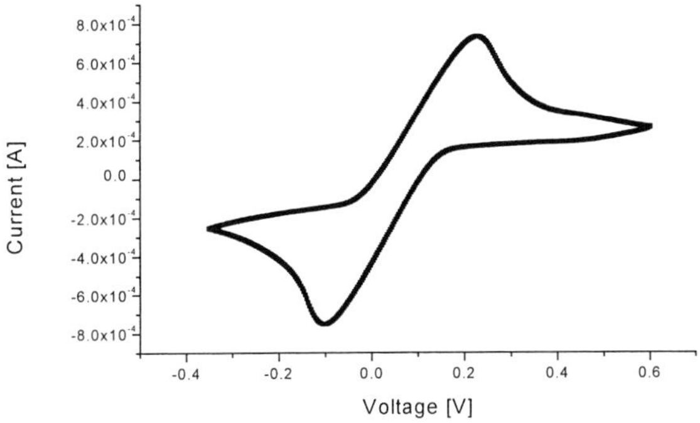

Fig. 2a Cyclic voltammogram of an uncoated gold slide in the presence of hydroquinone

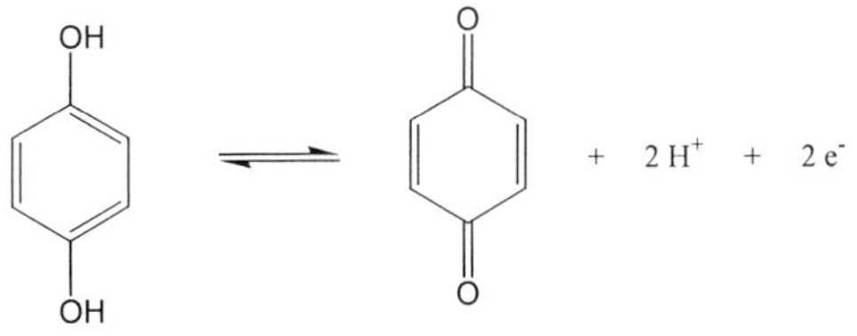

Fig. 2b Redox reaction of hydroquinone

In Fig. 3, typical cyclic voltammograms of the C[4]RA membrane in background electrolyte, without the marker present 3a, with hydroquinone present and under the addition of two different analytes are presented 3b + 3c. The oxidation peak of hydroquinone appeared at 0.2 V vs. Ag/AgCl electrode in all cases. The reduction of hydroquinone was observed at -0.27 V vs. Ag/AgCl electrode for the acetone exposure and -0.33 V vs. Ag/AgCl electrode for the chloroform exposure. Analysis of the peak heights, with respect to the baseline (see Fig. 3) for oxidation and reduction result in a averaged relative increase of the current and therefore the permeability of 265 % for a concentration of 13 mM of acetone and 1500 % for a concentration of 6.25 mM of chloroform.

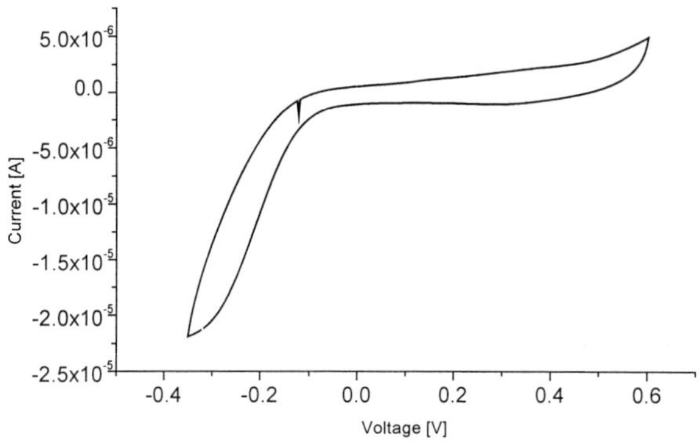

Fig. 3a Cyclic voltammograms of the C[4]RA membrane coated electrode in background electrolyte

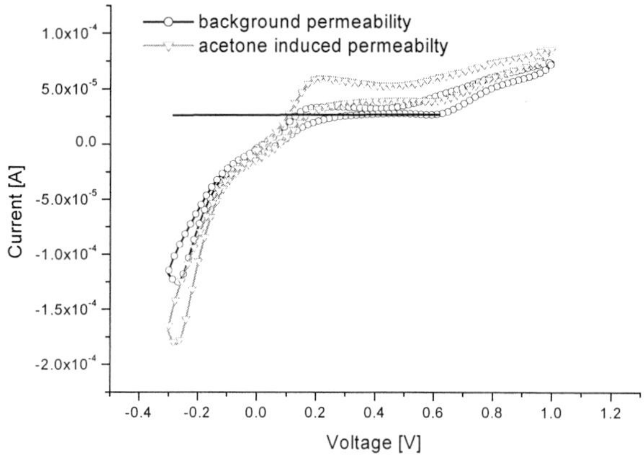

Fig. 3b Permeability changes induced by 13 mM of acetone

The oxidation and reduction peaks in the absence of analyte showed that there was substantial background permeability for the membranes. This can have two possible origins: (i) microscopic imperfections such as pinholes, and (ii) permeation of the hydroquinone into intermolecular voids and complexing with the C[4]RA itself. A further notable feature is the substantial shift of the reduction peak to more negative values, more pronounced in the case of the chloroform than in the case of the acetone. A shift of the redox peaks positions has also been reported in [5].

A possible reason for this increased irreversibility of the hydroquinone redox-reaction is an alteration of the charge transfer characteristic depending on the position of the hydroquinone in the membrane and a possible interaction of the analyte with the marker. In all cases there was imperfect recovery upon flushing with background electrolyte, possibly because of a substantial difference in solubility of the hydroquinone (70g/l) and the benzoquinone (insoluble).

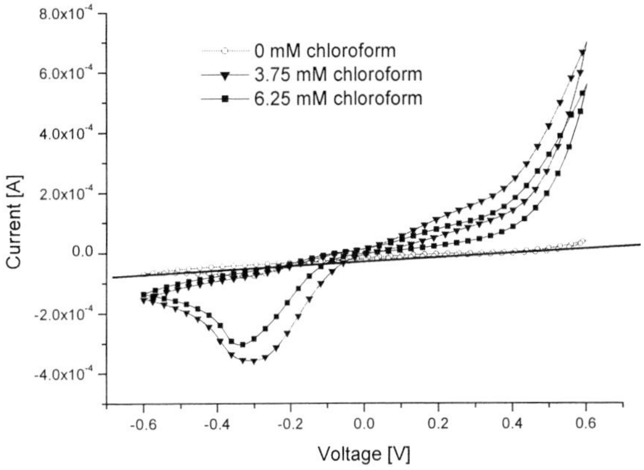

Fig. 3c Permeability changes induced by 3.75 mM chloroform and 6.25 mM chloroform

For a further analysis of the membrane properties, the system was characterised using impedance spectroscopy over a frequency range of 0.1-10^5 Hz. Fig. 4 shows the Bode plot of the spectra recorded for different concentrations of analyte.

At high frequencies, the impedance was nearly frequency independent, while at low frequencies, an increase of the impedance was observed. The system impedance shows increasing dispersion for frequencies below 100 Hz, reaching its maximum at the lower end of the spectrum.

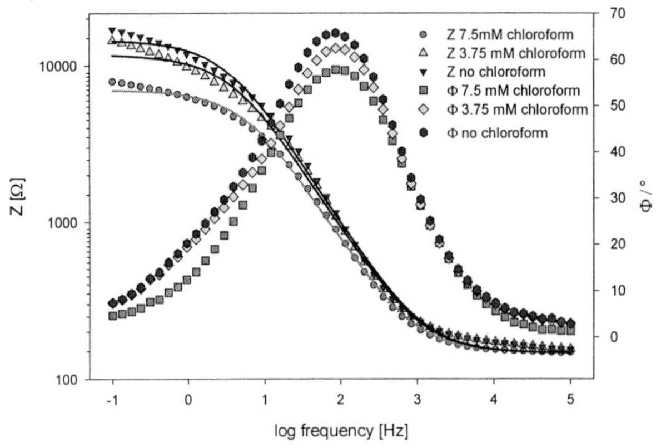

Fig. 4 Bode Plot of the membrane system for different concentrations of chloroform, lines represent mathematical fit of spectra based on equivalent circuit in Fig. 6a

A mathematical fit of the experimental data, using a least square fitting was carried out employing the equivalent circuit shown in Fig. 6a. and the fitted data for Z is shown in Fig. 4. The solution resistance determines the high-frequency behaviour, while the charge transfer resistance and the membrane capacitance dominate at low frequencies.

A constant phase element (CPE) was chosen to represent the capacitive structure consisting of the double layer on the calixarene membrane (Eq. 1),

$$Z_{CPE} = \frac{1}{(2\pi f Q)^n} \qquad (1)$$

where f is the frequency. For a rough surface, the fractal dimension (D) is between 2 and 3. This means that the surface fills between 2 (perfectly flat) and 3 dimensions (branching every-which-way through space resembling a porous cube). Mulder *et al.* [13] showed that for rough electrodes the interfacial impedance is modified by an exponent, n = 1/(D-1). For a smooth surface, the fractal dimension is D = 2.0 and n=1, but for a highly porous surface D = 3, and n = 0.5. A CPE describes real systems better then a purely capacitive structure, taking into account the frequency dispersion caused by the surface roughness and inhomogeneities of the membrane.

Table 1 lists the values of the components obtained from the fit. The solution resistance was constant. The charge transfer resistance was a measure of the permeability of the membrane since it described the Faradaic current caused by the redox-active marker. A decrease of the charge transfer resistance by a factor of 2.1 was observed in the presence of 7.5 mM chloroform, confirming the increased permeability of the membrane due to the calixarene-analyte interaction. However, the change of the charge transfer resistance was not as large as would have been expected from the results of the cyclic voltammetry (see above and Fig. 3c). This may be attributed to the increased irreversibility of the hydroquinone redox reaction under the influence of chloroform discussed above.

Table 1 The components of the equivalent circuit obtained by fitting the frequency response to the model 6a

Concentration of chloroform [mM]	$R_{solution}$ [Ω]	Q [F]	n	$R_{charge\ transfer}$ [kΩ]
0	145	$0.3829 \cdot 10^{-6}$	0.7936	14.49
3.75	145	$0.3103 \cdot 10^{-6}$	0.7945	11.57
7.5	145	$0.2852 \cdot 10^{-6}$	0.7954	6.86

The value of Q of the constant phase element decreased from $0.383 \cdot 10^{-6}$ F to $0.285 \cdot 10^{-6}$ F, whereas n is nearly identical in all cases. This also confirms a change in the structure of the calixarene membrane due to interaction with the analyte. The value of 0.79 for n can be regarded as reasonable for the given system with a substantial three-dimensional surface and a high degree of inhomogeneities.

The coated electrodes were also analysed in a system without the permeability marker. In this case the charge transfer resistance was infinity, and the equivalent circuit in Fig. 6b was used for fitting the experimental data. Fig. 5 shows the frequency response for two electrodes and the theoretical response from the derived equivalent circuit.

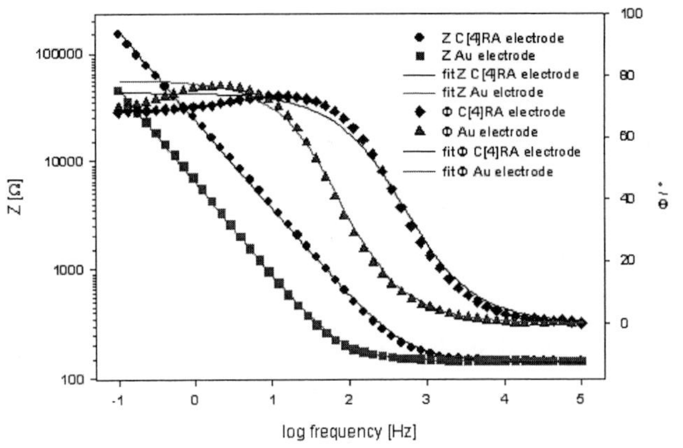

Fig. 5 Bode plot of the uncoated and coated electrode in the background electrolyte

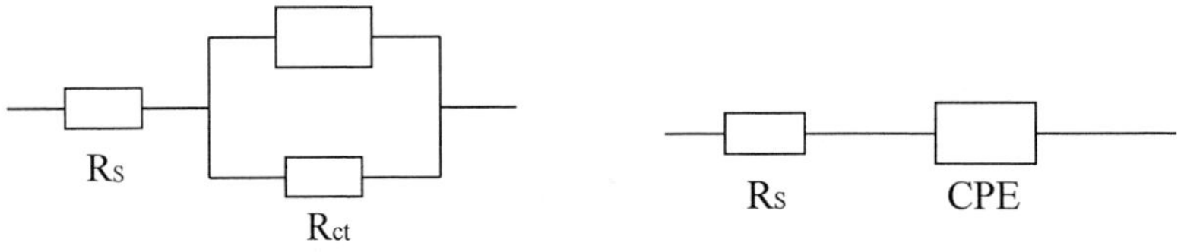

Fig. 6 Experimental data was fitted to equivalent circuits (a) with the permeability marker and (b) without the permeability marker

Values of $Q_{calix} = 0.8208 \times 10^{-6}$ F and $n_{calix} = 0.8208$ were obtained for the CPE in case of the coated electrode and $Q_{Au} = 0.7258 \times 10^{-5}$ and $n_{Au} = 0.8750$ in case of the uncoated electrode. This shows a substantial change in the three-dimensional character of the electrode surface upon coating with the C[4]RA. A comparison of these values with the ones obtained in the presence of the permeability marker shows that the interaction with hydroquinone decreases the capacitance of the calixarene membrane indicating an increase of the hydrophobicity and/or the thickness of the film.

4. Conclusions

In contrast to the well developed calixarene sensors of the potentiometric type, chemical sensors based on membrane permeability changes are still in its initial stage. In the present study the permeability of LB calix[4]resorcinarene was found to be controlled by the presence of chloroform and acetone. The changes in the membrane permeability are attributed to the changes in the intermolecular voids, restructuring the film and changes in the hydrophobicity of the film. A strong analyte induced alteration of the charge transfer resistance was found. The sensitivity of the system can be enhanced in several ways such as increasing the electrode area or optimising the thickness of the LB layers to minimise the background permeability and maximise its induced changes.

5. References

1. A. Casnati, Gazzetta Chimica Italiana, 127 (1997) 637
2. C.D. Gutsche, Calixarenes Revisited, The Royal Society of Chemistry, 1998, ISBN 0854045023
3. J. Auge, P. Hauptmann, J. Hartmann, S. Rösler and R. Lucklum, Sensors & Actuators B, 26 (1-3) (1995) 181
4. R. Mlika, H. BenOuada, N. JaffrezicRenault, I. Dumazet, R. Lamartine, M. Gamoudi and G. Guillaud, Sensors & Actuators B, 47 (1995) 181
5. A. Nabok, N. Lavrik and Z. Kazantseva, Thin Solid Films, 259 (1995) 259
6. A. Nabok, A.K. Hassan, A. K. Ray, O. Omar and V.I. Kalchenko, Sensors & Actuators B, Volume, 45 (1997) 115
7. A. Nabok, A. K. Hassan and A. K. Ray, J. Materials Chemistry,10 (2000) 189
8. S. Rosler, R. Lucklum, R. Borngraber, J. Hartman and P. Hauptmann, Sensors & Actuators B, 48 (1998), No.1-3, 415- 424
9. J.H. Buegler, J.F.J. Engbersen and D.N. Reinhoudl, Proceedings Eurosensor 2A4-1 (1997) 11
10. M. Maeda, Y. Fujita, K. Nakano and m. Takagi, J., Chem. Soc. Chem. Commun., (1990) 1529
11. K. Yagi, S.B. Khoo, M. Sugawara, T. Sakaki, S. Shinkai, K. Odashima and Y. Umezawa, J. Electroanalytical Chemistry, 401 (1996) 65-79
12. A.V. Nabok, N.V. Lavrik, Z.I. Kazantseva, B.A. Nesterenko, L.N. Markovski, V.I. Kalchenko and A.N. Shivaniuk, Conference Proceedings of LB 7, Ancona, Italy (1995)P1, 2
13. W.H. Mulder, J.H. Sluyters, T. Pajossi and L. Nyikos, J. Electroanal. Chem., 285 (1990) 103

Novel Methods to Study Interfacial Layers
D. Möbius and R. Miller (Editors)

DROP AND BUBBLE SHAPE ANALYSIS AS A TOOL FOR DILATIONAL RHEOLOGICAL STUDIES OF INTERFACIAL LAYERS

G. Loglio*[1], P. Pandolfini[2], R. Miller[3], A.V. Makievski[3,4], F. Ravera[5], M. Ferrari[5] and L. Liggieri[5]

[1] University of Florence, Department of Organic Chemistry, Via della Lastruccia, 6, I-50019 Sesto Fiorentino (Firenze), Italy
[2] Istituto di Ricerca sulle Onde Elettromagnetiche, CNR, Via Panciatichi 64, I-50127 Florence, Italy
[3] Max-Planck-Institut für Kolloid- und Grenzflächenforschung, Am Mühlenberg 1, D-14424 Potsdam/Golm, Germany
[4] SINTECH Surface and Interface Technology, Volmerstrasse 5-7, D-12489 Berlin, Germany
[5] Instituto di Chimica Fisica Applicata dei Materiali, CNR , Via De Marini 6, I-16149 Genova, Italy

Contents

Keywords: Drop and bubble shape tensiometer, interfacial tension, interfacial relaxation, dilational elasticity, adsorption of surfactants and proteins

* corresponding author

Drop and bubble shape tensiometry is a modern and very effective tool for measuring dynamic and static interfacial tensions. An automatic instrument with an accurate computer controlled dosing system is discussed in detail. Due to an active control loop experiments under various conditions can be performed: constant drop/bubble volume, surface area, or height, trapezoidal, ramp type, step type and sinusoidal area changes. The theoretical basis of the method, the fitting procedure to the Gauss-Laplace equation and the key procedures for calibration of the instrument are analysed and described.

The interfacial tension response to transient and harmonic area perturbations yields the dilational rheological parameters of the interfacial layer: dilational elasticity and exchange of matter function. The data interpretation with the diffusion-controlled adsorption mechanism based on various adsorption isotherms is demonstrated by a number of experiments, obtained for model surfactants and proteins and also technical surfactants. The application of the Fourier transformation is demonstrated for the analysis of harmonic area changes. The experiments shown are performed at the water/air and water/oil interface and underline the large capacity of the tensiometer.

1. Introduction

Among the methods for measuring the interfacial tension of liquid interfaces the drop and bubble shape tensiometry is known for a long time [1]. However, it became applicable with an acceptable accuracy only about 15 years ago with the availability of a video technique that can be directly linked to a high performance computer [2]. The first set-ups were run with expensive work stations, while today typically PCs are used as their performance is much higher now than the work stations of the first instruments. The development of this experimental technique was extremely fast and quite a number of commercial instruments are at present available on the market.

The drop and bubble shape technique is superior over other methods due to quite a number of advantages. Most of all it is an absolute method and is applicable to any liquid interface. In many practical cases it is the only method of choice when only a small amount of the sample is available. Moreover, temperature control is simple to organise and even cells for measurements at several hundreds degrees [3] or at very high pressures [4] are known. The drop and bubble shape technique also allows studies of wetting and contact angles [5], however, this type of application will not be discussed in the present chapter.

Most of the instruments allow only the measurement of surface and interfacial tensions, without a sufficient control of the drop/bubble size. Advanced models provide very accurate controlling procedures. The instrument described here in detail represents the state of the art of drop and bubble shape tensiometers. The possibility to study bubbles in addition to drops opens a number of features not available by other instruments: less loss of molecules caused by adsorption from extremely diluted solutions (small reservoir in the small single drop), long time experiments with very small amounts of a sample, easy application of a pressure sensor for additional measurement of the capillary pressure inside the bubble. Moreover, high quality sinusoidal relaxation studies can be performed by inserting a piezo system which can be driven such that very smooth changes of the bubble surface area are obtained.

In this chapter first the theoretical basis of the drop and bubble shape method is described, and then details of its practical use are given. The various functionalities of the instrument are demonstrated then in the form of examples, such as dynamic surface and interfacial tensions of surfactant and protein solutions, and the dilational rheology of some selected systems.

2. Theoretical Background of Meniscus Shape Methods

The shape of liquid menisci is controlled by the competition of the capillary effects, caused by the surface (or interfacial) tension, γ, and the forces acting on the liquid volume. The action of capillary forces results in the capillary pressure, ΔP, across a curved liquid interface

$$\Delta P = \gamma \kappa \tag{1}$$

where κ is the mean curvature of the interface at the point considered. Although this latter equation is rigorous only in thermodynamic equilibrium and for macroscopic droplets (i.e. moderate curvature) [6], its validity can be easily assumed also out of equilibrium, if the characteristic time for the variation of γ is large with respect to the variation of κ.

The forces acting on the liquid volume may either have external origin, such as gravity or electromagnetic fields, or have a non-inertial character, as is the case for rotating menisci. In all these cases, the knowledge on equilibrium shape and stability conditions of menisci has a great technological relevance. For example, the effect of electric fields on liquid menisci [7, 8], is important in ink-jet technology and atmospheric physics, while liquid bridges are intensively investigated [9, 10] because of their utilisation in crystal growth technology. In some cases, for example in the presence of electromagnetic fields, intrinsic effects on the surface tension can also be observed [11], however they are generally weak and negligible.

This section describes the general theory for axis-symmetric liquid menisci to measure interfacial tension under both equilibrium and dynamic gravitational conditions, for drops and bubbles formed at the tip of a capillary or sitting on a solid substrate. Also, it demonstrates how measured data can be compared with theoretical curves.

2.1. General theory

A liquid meniscus subjected to gravity takes a shape which corresponds to the minimum of the total energy of the system, i.e., the sum of the bulk plus interfacial energy. The latter depends on the interfacial tension, γ, thus the analysis of meniscus shapes is largely used as the basis for many interfacial tension measurement techniques. Accordingly, it is possible to predict the shape of a meniscus by using variation methods imposing extremum conditions with constraints on the liquid volume and the contact angles with solid and liquid surfaces. However, though very general, such formulation is often useless for practical purposes, leading to differential equations containing cumbersome elliptical functions, which are extremely difficult to solve.

In most cases a mechanical approach is more adequate. This is based on the concept that in mechanical equilibrium at each point on the interface, the curvature is adjusted such that the difference in the pressures between the two phases is balanced by the capillary pressure. This approach is particularly fruitful when applied to axisymmetric menisci, like drops and bubbles. In this case, assuming a Cartesian coordinate system with the origin at the drop apex, O, and the vertical axes, z, in the symmetry axis and directed towards the interior of the drop, at any point, S, of the interface we have

$$P^0 + \Delta\rho g z = \gamma\left(\frac{1}{R_1} + \frac{1}{R_2}\right) \tag{2}$$

Here P^0 is the capillary pressure at the drop apex, $\Delta\rho$ is the density difference between the internal and the external phases, g is the z-component of the gravitational acceleration, and z is the vertical coordinate of S. R_1 and R_2 are the principal radii of curvature of the interface at point S.

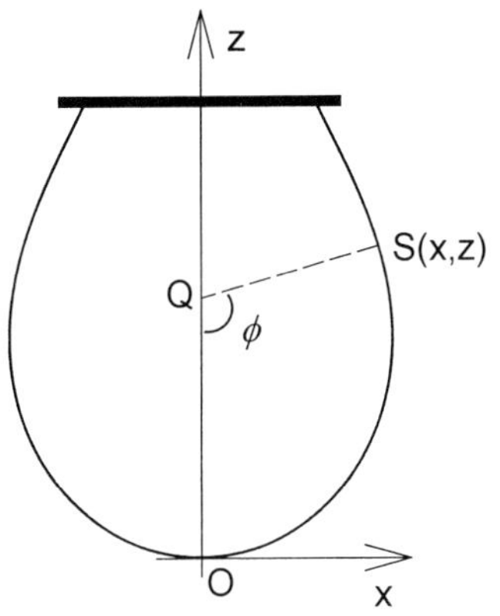

Fig. 1 Sketch of the meridian section of a pendant drop

Considering the meridian profile of the drop in the x-z plane (cf. Fig. 1), owing to the properties of revolution surfaces [12], the radii, R_1 and R_2 in the generic point S(x, z) coincide, respectively, with the radius of curvature, R, of the generating curve in S and with the distance between S and the symmetry axes along the normal to the curve, $x/\sin\phi$. Moreover, these radii are equal at the apex so that the pressure balance in S can be expressed as

$$\frac{2\gamma}{b} + \Delta\rho g z = \gamma\left(\frac{1}{R} + \frac{\sin\phi}{x}\right), \tag{3}$$

where b is the radius of curvature at the apex. By introducing the "shape factor" β, defined as

$$\beta = \frac{\Delta\rho g}{\gamma}, \tag{4}$$

this equation can be rewritten as

$$2 + \frac{\beta z/b}{b^2} = \left[\frac{1}{R/b} + \frac{\sin\phi}{x/b}\right]. \tag{5}$$

This equation was firstly used by Bashforth and Adams [13] and represents the starting point for most of the drop shape methods developed so far. By definition, β is positive for captive bubbles and sessile drops while is negative for emerging bubbles and pendant drops. Moreover, the larger the meniscus deformation, the larger is the absolute value of β. Under gravity, spherical drops can be obtained only under iso-dense conditions ($\Delta\rho=0$) which gives $\beta=0$.

On such a basis and by applying variational considerations, the equilibrium and stability properties of drops and bubbles have been widely investigated [13, 14, 15]. Moreover, since the pioneering work of Bashforth and Adams [13], solutions of Eq. (5) provided in tabular form [16, 17, 18, 19, 20], have been used to infer the interfacial tension of drops and bubbles by measuring the characteristic geometrical quantities of the profile, arranged in non-dimensional parameters. Often the goal of these methods was the determination of β and b, which, once the density difference is known yields the surface tension.

In the last 20 years, large improvements in numerical techniques were obtained by the availability of fast computers. Hence, drop shape methods became more popular due to the direct estimation of the surface tension by profile fitting techniques. Moreover, the utilisation of automatic imaging techniques also provided an easy way to perform measurements in almost real-time conditions. Several set-ups and minimisation procedures have been proposed [21 - 22, 23, 24, 25, 26, 27, 28, 29], in order to increase the measurement accuracy and the sampling rate. Typically, the output of these modern procedures are the best fit values of β and b.

A major advantage of the drop shape methods is their non invasive character, which makes them an ideal tool for measurements at high temperature of molten metals, alloys and polymers [21, 30] and for liquid-liquid interfaces with transparent external phases. In the latter case, limitations arise only for liquids with a very low density contrast. The negligible drop deformation dramatically decreases the accuracy of the estimated interfacial tension γ, which becomes undetermined when approaching $\beta=0$. Also very low surface tension systems give rise to large inaccuracies arising from the droplet flatness and consequently the vanishing apex curvature. A small drawback of this technique is the continuous variation of the drop area due to changes in the interfacial tension, which can be overcome by implementing automatic control loops.

2.2. *Profile fitting procedures*

The fitting software fits a Laplacian curve to the observed experimental drop/bubble profile. The fit model, that is the Laplacian curve, is the profile of the axisymmetric drop/bubble meridian section as sketched above in Fig. 1. The model profile is calculated by a fourth-order Runge-Kutta integration algorithm from the Laplace equation (5). This equation can also be represented in the form of a set of three first-order differential equations:

$$\frac{dx}{ds} = \cos(\phi) \tag{6}$$

$$\frac{dz}{ds} = \sin(\phi) \tag{7}$$

$$\frac{d\phi}{ds} = \frac{1}{R} = \pm\beta z + \frac{2}{b} - \frac{\sin(\phi)}{x} \tag{8}$$

where s is the arc length used here as an independent variable, and ϕ is the normal angle. The sign "plus" holds for sessile drops or captive bubbles and the sign "minus" holds for pendant drops or emerging bubbles.

As mentioned above, the Cartesian reference system has the origin at the drop/bubble apex, the z-axis along the vertical symmetry axis and the x-axis on the horizontal plane.

As can be seen by inspection of the set of three first-order differential equations, Eqs. (6) to (8), the model profile depends nonlinearly on the set of 2 unknown parameters, namely the apex radius b and the shape factor β. Moreover, as an additional third parameter, the apex correction error, ε , is also taken into account.

With respect to the number N of the observed points $S_{obs}(x_i, z_i)$, it is advantageous to calculate a (threefold to fivefold) redundant number, n, of model profile points, $S_{calc}(x_i, z_i)$, whose coordinates depend on the particular choice of the parameter values:

$$S_{calc}(x_i, z_i) = f(\beta, b, \varepsilon), \qquad i = 1, 2, \ldots\ldots n \tag{9}$$

Several algorithms are known for obtaining the best fit parameters, matching the calculated and the observed profile points of drops or bubbles [1].

Instead of using standard non-linear fitting routines (such as the steepest descendent or the Levenberg-Marquardt routine [31]), in the special case of the Laplacian curve it appears more convenient to adopt substantially the same approach as earlier used by Maze and Burnet [21, 32], which can be defined as an iterative linear method.

To measure the agreement between the measured (or experimental) profile and an assumed (or calculated) Laplacian curve, the objective function χ^2 is defined as follows:

$$\chi^2 = \sum_i \left[S_{obs}(x_i, z_i) - S_{calc}(x_i, z_i)\right]^2 \tag{10}$$

In other words, the experimental profile is compared with the calculated Laplacian curve by computing the normal distance between each observed point and the corresponding calculated point (properly obtained by interpolation from the Runge- Kutta points).

The best-fit parameters are determined by minimisation of the objective function χ^2. With non-linear dependence, however, the minimisation must proceed iteratively. Given trial values for the parameters, Maze and Burnet [21] developed a procedure that improves the trial solution, by computing parameter correction values. The procedure is then repeated until χ^2 stops (or effectively stops) decreasing.

Sufficiently close to the minimum of χ^2 , that is sufficiently close to the convergence of the parameters, the Laplacian curve, $x_i = x(z_i; \beta, b, \varepsilon)$ is expected to be well approximated by a Taylor series, truncated after the second term:

$$x_i = x(z_i; \beta_0, b_0, \varepsilon_0) + \Delta\beta \frac{dx}{d\beta} + \Delta b \frac{dx}{db} + \Delta\varepsilon \frac{dx}{d\varepsilon}, \tag{11}$$

where the partial derivatives are calculated at the points β_0, b_0, ε_0 .

Minimisation of χ^2 involves its derivatives, with respect to the parameters, to be equated to zero. The mathematics treatment leads to a linear form, consisting of a set of three linear equations with three unknowns. In matrix form, the linear least-squares problem, can be written as:

$$(\mathbf{F}^T \mathbf{F})\, \mathbf{A} = \mathbf{F}^T \mathbf{Z} \tag{12}$$

where $\mathbf{F}^T$ denotes the transpose of the matrix **F**, that is the Nx3 partial derivatives matrix, **Z** is the discrepancy vector, and **A** is the parameter correction vector.

If the approximation is good, within a single leap a jump occurs from the current trial of the parameters to the minimising ones. Actually, from the first estimation of the parameter values, just 3 - 4 successive loops are usually required. Note, the absolute accuracy of drop and bubble shape methods depends mainly on the main properties of the video technique and the experimental conditions needed for an error to be better than ± 0.1 mN/m are well defined in a very recent paper by Zhou and Gaydos [33].

2.3. *Estimation of parameters of the Laplacian curve*

Estimation of the curvature radius at the apex

The apex radius b of the drop/bubble is estimated by fitting an arc of circumference to the observed profile points, within a visual angle (from the circumference centre) of about 15 degrees. The obtained value (b_1) is corrected by an empirical factor, depending on the shape factor β:

$$b = b_1 + 0.0025\beta \exp[2.5b_1] \qquad (13)$$

Estimation of the shape factor

The shape factor β, as defined by Eq. (4), is related to the coordinates of the equatorial points of a drop or bubble, that is, the equatorial diameter and its distance from the apex. This relationship, however, is not known in an exact analytical form. Approximated (initial) values of β can be estimated from the drop/bubble geometry, either by using polynomials or by interpolating tabulated data. Several variants of these two methods are reported in the literature, involving measurements of diameters at particular distances from the apex or at particular angles between the z-axis and the geometrical tangent to the profile (cf. chapter 3 in [1]). The simplest case occurs when the profile coordinates at the vertical tangency (that is, of the equatorial point $S(x_m, z_m)$)) can be accurately measured. In this condition, for the sessile drop or the captive bubble, the following semi-empirical relationships holds [21]

$$\beta\ b^2 = \Sigma\ a_j\ (x_m / z_m)^j\ ,\ j = 0, 1, \ldots, 8 \qquad (14)$$

where the polynomial coefficients, a_j, are:

16.8621,	-53.395889,	26.5354,	95.097412,	-182.7348,
146.43645,	-60.568497,	12.668255,	-0.8999756	

Many software packages provide an initial estimation of the shape factor, i.e., the parameter β, by interpolating tabulated data (for example the software for PAT-1 from SINTECH). Numerical tables were created by the following procedure.

a) Computing Laplacian profiles up to a given height z (that is, up to the constant part with the x-coordinate, $z = 1.75\ x$), for each apex radius (in the interval $b = 0.2 - 4.5$ mm, at steps $b_i = b_{i-1} + 0.025$ mm) and for different values of the shape factor (in the interval $\beta = 0.02 - 0.65\ mm^{-2}$ and in the interval $\beta = -0.003 - 0.40\ mm^{-2}$, at steps $\beta_i = \beta_{i-1} * 1.015$, for a bubble and a drop respectively).

b) Calculating the volume, V_{sphere}, of the spherical segment whose meridian section passes through the drop/bubble apex (0, 0), and through the two symmetrical points (-x, z) , (x, z), which fulfil the condition $z = 1.75\ |x|$ (see Fig. 2) .

c) Defining an empirical deformation index, ID, i.e. the ratio between the drop/bubble volume V (bounded by the plane at the reference distance from the apex, $z = 1.75\ |x|$) and the relevant spherical-segment volume, that is $ID = (V/V_{sphere}) - 1$, for each pair of values, i.e., the apex radius b, and the shape factor β. Note that the sign of ID is the same as the sign of β (positive for a bubble and negative for a drop).

d) Finally, generating a numerical table for the function $\beta = f(z, ID)$ (see Figs. 3 and 4).

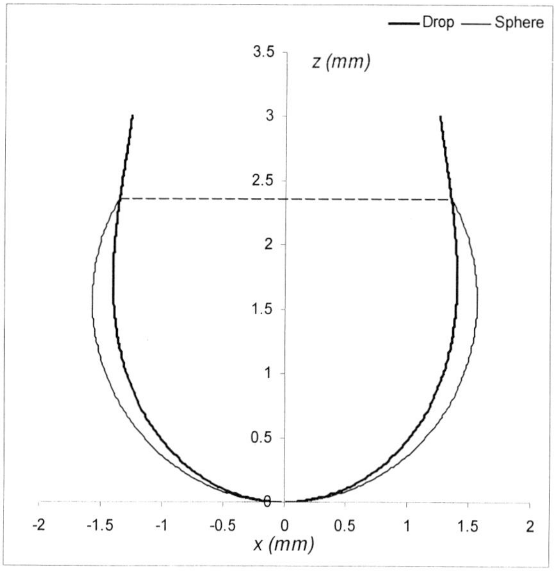

Fig. 2. Scheme of the meridian section of a drop and of the pertinent sphere which passes through the drop/bubble apex (0, 0), and through the two symmetrical points (-x, z) , (x, z), which fulfil the condition $z = 1.75\ |x|$.

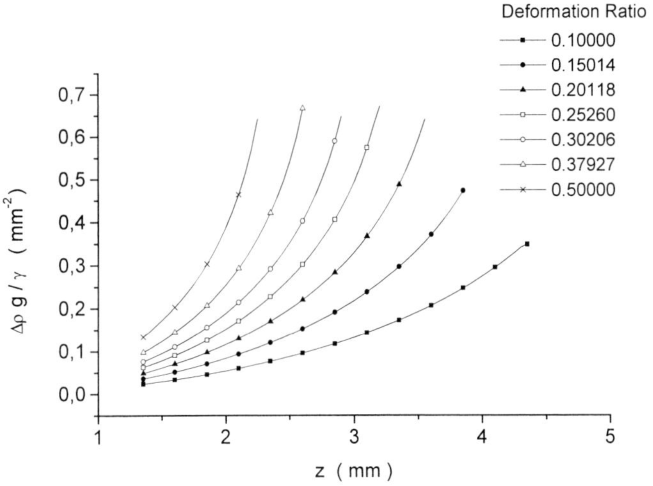

Fig. 3 Illustrative plot of some particular values of the numerical table for the function β = f(z, ID) , relevant to the bubble profiles.

When a drop/bubble profile is acquired from a real image, the pertinent values of z and ID are determined by using the same procedure as the above-described procedure, for the computed Laplacian profile. Then, the real β parameter is estimated from the two entries, z and ID, of the tabulated values by linear interpolation.

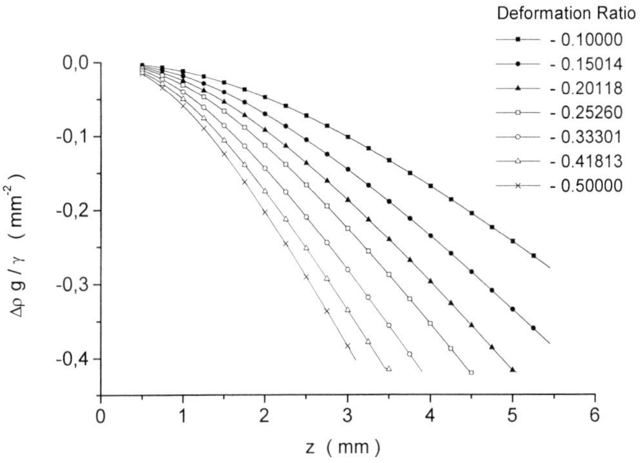

Fig. 4 Illustrative plot of some particular values of the numerical table for the function β = f(z, ID), relevant to the drop profiles.

Estimation of surface tension and of geometrical properties

Surface tension is determined from the shape parameter β, with the knowledge of the density difference of the two fluids and the local gravity constant.

The volume of the drop/bubble is estimated from the observed profile by summation (from the apex up to the contact line with the capillary tip) of the partial volumes of each segment bounded by the couple of diameters passing through ($-x_{i-1}$, x_{i-1}) and ($-x_i$, x_i), respectively:

$$V = \Sigma_i \, \pi \, (x_{i-1})^2 \, h_i, \tag{15}$$

where h_i is the vertical distance between two successive diameters.

Likewise, the drop/bubble surface area is estimated by summation of the partial zones:

$$A = \Sigma_i \, 2\pi \, (x_{i-1}) \, h_i. \tag{16}$$

3. EXPERIMENTAL SET-UPS FOR THE DROP AND BUBBLE SHAPE TECHNIQUE

First experiments on drop shapes were done a long time ago [34, 35, 36] using standard photo techniques. Although, at that time the efforts were huge, the results were much less accurate than those obtained from force measurements. Now, due to modern video technique there is quite a number of commercial instruments available on the market. The accuracy of most is comparable. However, the variety of measurement features is very different. Only the instruments which can control the size of the drop or bubble on-line during the measurement are suitable for more advanced studies. Transient or harmonic relaxation studies, as described in the monograph by Dukhin et al. [37], cannot be done properly without this important function. Hence, these devices are not able to provide interfacial rheological data of interfacial layers.

3.1. Main components of an experimental set-up

Hardware - Figure 5 shows the schematic diagram of an experimental set-up for the measurement of interfacial static and dynamic properties. It is a particular profile analysis tensiometer, which essentially consists of three main components: 1) the measurement assembly, 2) the optical system (illuminator - microscope objective - WV-BP310 B/W Panasonic CCD video camera) and 3) the computer (Intel-Pentium microprocessor, IBM-compatible PCI-Personal Computer, with a National Instruments IMAQ PCI-1408 Monochrome Image Acquisition Board).

The measurement assembly is composed of three elements, namely, 1) the sample vessel, 2) the capillary, 3) the Hamilton syringe with the relevant driver.

In the design of the circuit, connecting the syringe with the capillary, possible effects of fluid dynamic instability have been taken into account.

Software - Figure 6 illustrates the flow diagram of the instructions provided to the instrument- and experiment-parameter selection, instrument operation and drop/bubble dimension control. National Instruments LabVIEW graphical programming software ensures automatic instrument functioning as well as image acquisition, processing and storage (grabbing of drop/bubble image, sub-pixel resolution detection of the profile, saving into a file and presenting the results on the monitor takes about 200-250 ms). Sub-pixel resolution is of the order of 1.0 - 0.5 μm (depending on the magnification factor).

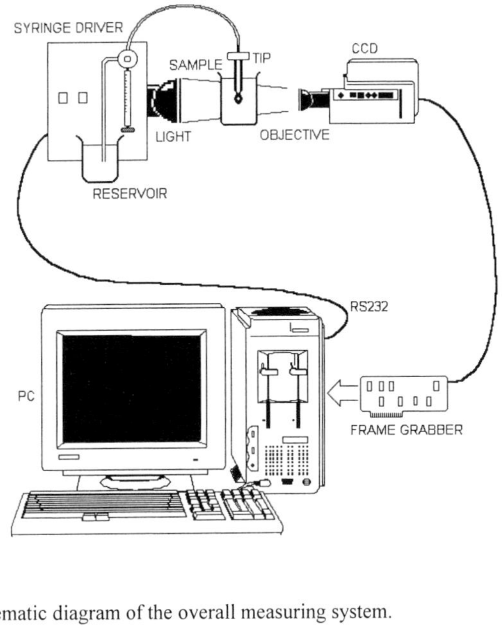

Fig.5 Schematic diagram of the overall measuring system.

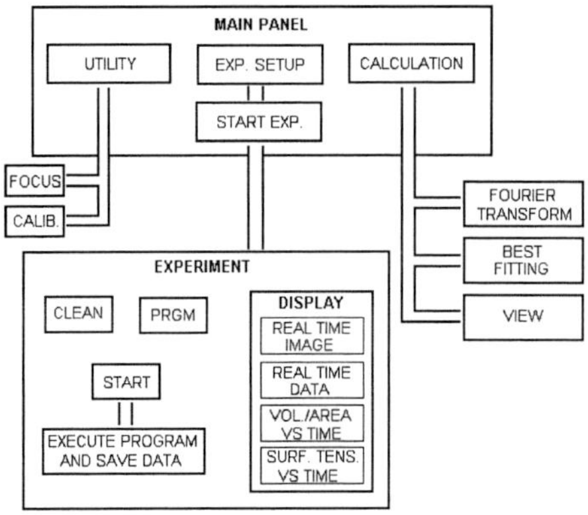

Fig. 6 LabVIEW graphical programming software for instrument setting and control: flow diagram of the instructions.

3.2. Necessary optical settings: magnification, contrast, focus, verticality

Proper settings of the PAT optical system is a key preliminary step, prior to any given series of experimental runs. The attainment of optimum quality in the image acquisition is mandatory for the subsequent data processing and, finally, for the reliability of the presented results.

The quality of the acquired drop/bubble profile depends on 4 optical settings, namely a) magnification, b) contrast, c) focus and d) verticality.

a) *Magnification.* After the observed object (drop or bubble) appears adequately centred inside the (640x480-pixel) screen, the magnification is adjusted in such a way that the drop/bubble image will occupy a good part of the screen area, during either a static or a dynamic experiment (allowing for all the programmed variations).

b) *Contrast.* The CCD-camera exposure is controlled via software. The light intensity can be partitioned into 256 grey levels, assigning 0-value to the darkest region of the screen and 255-value to the brightest one. An offset value is regulated in order to prevent saturation. If necessary, coarse control of the light source or of the optical-objective diaphragm is manually performed.

c) *Focusing.* The optimum focus for the observed object is first ascertained by visual inspection of the edge sharpness, all around the drop/bubble profile. Furthermore, PAT software provides a diagnostic tool for a quantitative evaluation of the edge characteristics. Actually, the software allows a (horizontal or vertical) bar to be moved up-down and left-right over the image screen and, in synchronism, the illumination is plotted for the ensuing sequence of each individual pixel. Also, the length of the bar can be increased or decreased in order to explore a greater or a smaller number of pixels. Thus, fine adjustments of the focus are finally accomplished by taking into account the steep change of the pixel illumination plot (Fig. 7).

d) *Verticality.* A drop/bubble, subjected to a gravitational field, possesses a geometrical axisymmetric shape provided that it is formed from the tip of a vertical capillary (or, in other words, from a circular contact line, belonging to a horizontal plane). As an additional requirement, the reference Cartesian system of the CCD-camera must be aligned with the vertical direction in such a way that the profile coordinates of the acquired image retain the axial symmetry. After proper adjustment, the overall mechanical orientation (of the capillary and of the CCD-camera) can be checked by the PAT software diagnostic method, by measurement of the deviation angle in respect to verticality (Fig. 8).

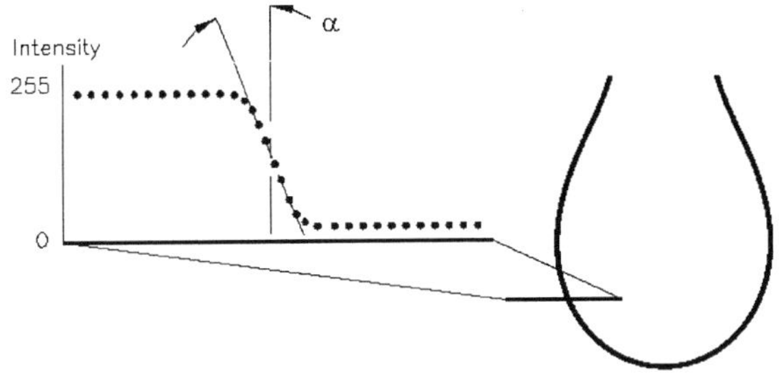

Fig. 7 Pictorial description of the diagnostic bar for the image exploration, measuring the light intensity impinging upon each individual pixel, belonging to the bar.

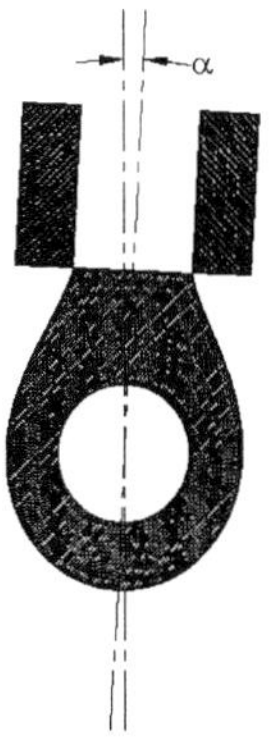

Fig. 8 Schematic representation of a possible misalignment of the CCD-camera, in respect to the vertical direction, that can be checked by PAT the diagnostic method through quantitative measurement of the deviation angle.

4. EXPERIMENTAL EXAMPLES

The fundamentals of drop and bubble shape analysis have been discussed in detail above. In the next section examples are given to demonstrate the various applications of the profile analysis tensiometer. Besides dynamic surface and interfacial tensions, results are shown for trapezoidal and sinusoidal relaxation experiments from which the dilational elasticity can be derived. The experiments selected are not only for model surfactants of high chemical purity but also for technical surfactants for which effective data can be deduced.

4.1. Dynamic surface tension

The characterisation of the dynamic surface tension in surfactant systems is very important, both from a technological point of view and to understand the adsorption mechanism.

The variation of surface/interfacial tension, due, for example, to the formation of a new interfacial area, is a consequence of adsorption dynamics. This transient phenomenon involves different time steps: exchange between the interface and the adjacent bulk layer, reorganisation in the interfacial layer and diffusion in the bulk. However, for most of the surfactants studied so far, the first two steps are much faster than the diffusion in the bulk, so they can be considered locally in equilibrium with respect to diffusion, which is, thus, controlling the adsorption process.

Typical experiments to study the adsorption kinetics are measurements of the surface tension during the ageing of a freshly formed surface (negligible initial amount of adsorbed molecules). In this case, for a diffusion controlled process, the time evolution of the surface concentration Γ is described by the Ward and Tordai equation [38]

$$\Gamma(t) = \sqrt{\frac{D}{\pi}}\left[2c_0\sqrt{t} - \int_0^t \frac{c_s(\tau)}{\sqrt{t-\tau}}\,d\tau\right], \tag{17}$$

where D is the diffusion coefficient and c_s is the concentration in the sublayer, i.e. the bulk layer adjacent to the interface. Since the exchange between the bulk and the interface and the processes internal to the interface are considered in equilibrium, the adsorption isotherm can be assumed as an additional relationship between c_s and $\Gamma(t)$. Thus, Eq. (17) becomes an integral equation in Γ which can be analytically or numerically solved. In this way the evolution of γ

with time is available via $\Gamma(t)$ and the respective surface equation of state. Thus, investigations of the adsorption dynamics offer a possibility for the validation of surface models and the parameters describing them. For this, however, methods with an accurate measurement of the dynamic surface tension are needed.

Today, thanks to the fast development of computer enhanced imaging techniques and numerical fitting procedures, the accuracy and the sampling rate of drop shape methods are substantially increased. Thus, this technique is an important tool for the investigation of adsorption dynamics, and it is particularly suitable for studying processes with characteristic times from a few seconds up to hours and even longer. In fact, there is a large number of experimental studies in which the drop shape technique is used to evaluate the adsorption equilibrium properties, like adsorption isotherms and the dynamic surface tension behaviour. The method is also extensively utilised in the study of surfactants and proteins both in liquid/liquid and liquid/air systems.

A typical experimental example for the characterisation of the adsorption properties of a surfactant, at the water/air interface is shown in Figs. 9 and 10, where the dynamic surface tension and the corresponding equilibrium data for the surfactant Triton X 100 are reported. The best fit for the surface tension isotherm is given by the two state model [39] which assumes that molecules can adsorb in two different states, 1 and 2.

The two states are characterised respectively by two different molar surface areas, ω_1 and ω_2, and two surface activities, related to the parameters b_1 and b_2. The distribution in the two adsorption states depends on the coverage, i.e. on the surface pressure Π,

$$\frac{\Gamma_1}{\Gamma_2} = \frac{b_1}{b_2}\exp\left[-\frac{\Pi(\omega_1 - \omega_2)}{RT}\right]. \qquad (18)$$

In practice, the model describes the situation where, according to the surface coverage, adsorbed molecules may rotate or change their conformation, varying their occupation area. The validation of this model has been given for some systems by using dynamic surface tension measurements acquired by the drop shape analysis [40, 41].

Another example of experimental data is reported in Figs. 11, where the same surfactant, $C_{10}EO_8$, is studied at the water/air and water/hexane interfaces. In Fig. 11b also the theoretical curves calculated from the diffusion controlled adsorption theory, coupled with the two state isotherm are shown, which fit the experimental data very well.

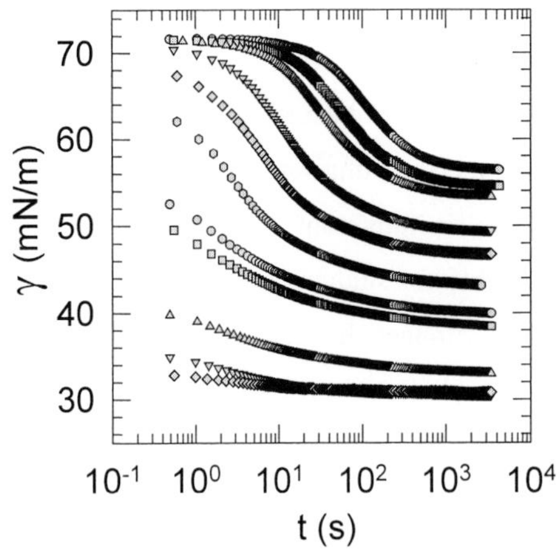

Fig. 9 Dynamic surface tension during the adsorption at a fresh interface of Triton X-100 aqueous solutions;. the bulk concentrations c_w are (from top): 5, 8, 10, 20, 30, 50, 80, 100, 200, 300, 500 * 10^{-9} mol/cm^3.

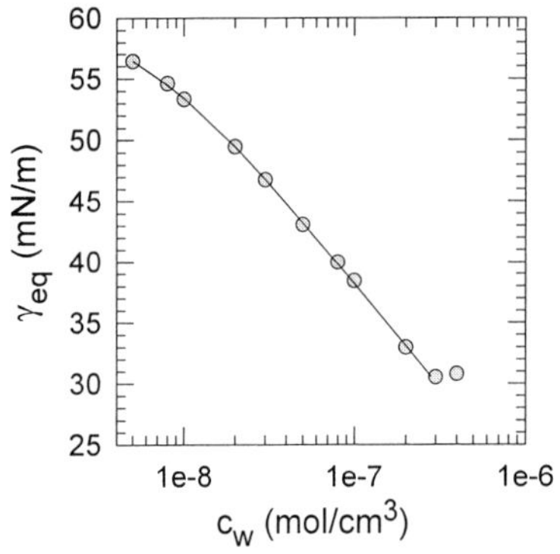

Fig. 10 Isotherm of Triton X100 at the water/air; equilibrium values are obtained from the dynamic surface tensions in Fig. 9 by extrapolation to infinity

It is clear that the given time is not the actual age of the surface but refers to the beginning of the profile acquisition. This is a typical problem of this method of measurement when it is applied under dynamic conditions. It is important especially for the comparison with the theoretical curves. Typically the ageing of a surface is started from an ideally fresh surface. However, the initial load of the drop/bubble surface can be found as a fitting parameter or via comparison of experimental data from other methods.

Theoretical and experimental studies [42, 43] have shown that the dynamic processes involved in adsorption kinetics have different characteristic times which makes it particularly interesting to investigate the dynamic surface tension of a system at different time scales.

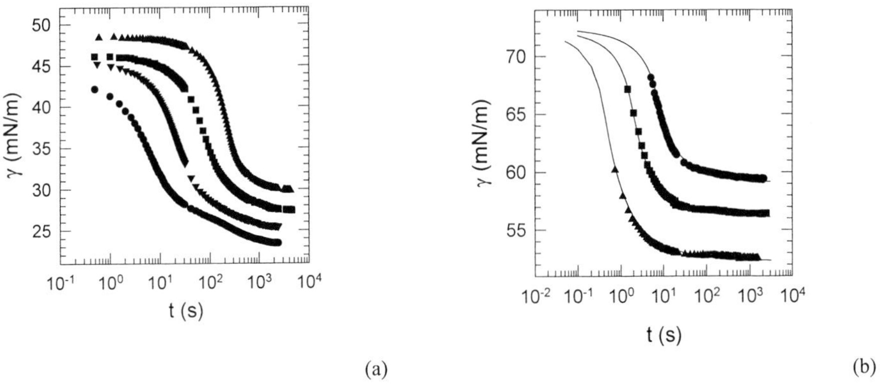

Fig. 11 Dynamic surface tension of aqueous $C_{10}E_8$ solutions for non pre-equilibrated conditions; From top, the initial bulk concentrations in water are 2, 4, 8, 20 * 10^{-9} mol/cm^3; a) water/hexane interface, b) water/air interface

For example, in some cases the reorientation of molecules in the adsorbed layer can play an important role in the adsorption dynamics. The characteristic time, according to the model reported in [43], is given by

$$\tau_{or} = \frac{1}{k_{12}\left[1+\frac{\Gamma_1}{\Gamma_2}\right]} \tag{19}$$

where k_{12} is the rate coefficient of the orientation process, which is different to the characteristic time of the diffusion. However, the need for such a multi-scale approach already exists for diffusion controlled adsorption. In this case, the characteristic time for the adsorption process, τ_D, can be estimated as

$$\tau_D = \frac{1}{D}\left(\frac{\Gamma^0}{c^0}\right)^2 . \tag{20}$$

Depending on the ratio Γ^0/c^0 between the equilibrium adsorption and bulk concentration, τ_D becomes typically smaller either when increasing the surface activity for a given surfactant or by increasing the surfactant bulk concentration. In practice, τ_D typically spans over several orders of magnitudes, from milliseconds to days which makes it necessary to study the process on different time scales. To this aim, the drop shape analysis has been fruitfully used together with other techniques, such as the capillary pressure techniques [44] and the maximum bubble pressure method [45]. An example of this approach is shown in Figs. 12, where the dynamic interfacial tensions of two $C_{10}EO_5$ solutions at the water/air interface are reported.

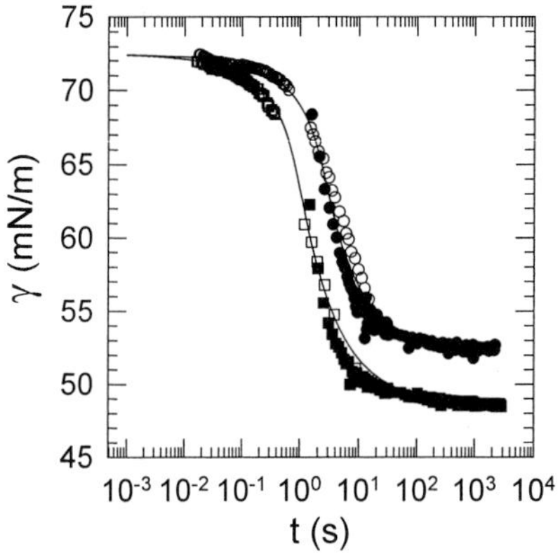

Fig. 12 Dynamic surface tension during the adsorption of $C_{10}E_5$ at water/air interface. From top, the bulk concentrations are 6 10^{-8}, and 10^{-7} mol/cm^3; the empty symbols refer to data acquired by the dynamic maximum bubble pressure method, while the filled ones to data acquired by the drop shape method; the solid lines are the theoretical prediction by the diffusion controlled adsorption with the two-state isotherm

In the interpretation of dynamic interfacial tension data received from the drop shape analysis, often valid equations for plane interfaces are used. This approximation can be adopted as long as the diffusion layer thickness δ is negligibly small with respect to the drop radius. Such depth is estimated by $\delta=(D\tau_D)^{1/2}$ [46]. The equations for a plane interface are no longer adequate for very diluted solutions or for slightly surface active systems. In these cases, a spherical symmetry can be adopted to the Ward and Tordai model [47]

$$\Gamma(t)=\left(\frac{D}{R}\right)\left[c_0 t-\int_0^t c_s(t-\tau)d\tau\right]+\sqrt{\frac{D}{\pi}}\left[2c_0\sqrt{t}-\int_0^t\frac{c_s(\tau)}{\sqrt{t-\tau}}d\tau\right] \tag{21}$$

which describes the diffusion controlled adsorption from the external volume at a spherical interface of radius R.

As mentioned above, in some particular cases the use of a bubble instead of a drop can lead to significantly different results. The data for two $C_{10}EO_8$ solutions are shown in Figs. 13, which were obtained using a bubble [48] and drop method [49], respectively.

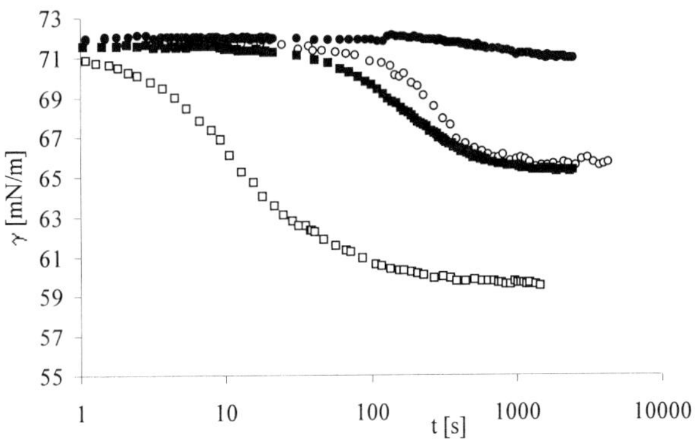

Fig. 13 Dynamic surface tension of two $C_{10}EO_8$ solutions at $c=10^{-9}$ (○●) and 10^{-8} mol/l (□■); (●■) – data of Ferrari et al. for a drop [49], (○□) – data of Chang et al. for a bubble [48].

One can see that, the surface tension change with time, i.e. the adsorption dynamics is much slower for the drop as compared with the bubble. This can be qualitatively described by the decrease in the $C_{10}EO_8$ bulk concentration within the drop in contrast to the measurements with a bubble, where loss of surfactant due to adsorption at the interface is negligible. Moreover, the diffusion from inside or outside a drop is quite different. For a diffusional adsorption mechanism the dynamic adsorption Γ(t) at a drop or bubble surface is approximately given by the relationship

$$\Gamma = \pm \frac{cDt}{r} + 2c\sqrt{\frac{Dt}{\pi}} \qquad (22)$$

where c is the bulk concentration, D is the diffusion coefficient, r is the drop radius, and t is the time. The sign "+" is for the case of adsorption from inside the drop, while "-" stands for adsorption from outside the drop or bubble [50].

The dynamic surface tension of β-casein solutions at three concentrations $5 \cdot 10^{-8}$, 10^{-7} and 10^{-6} mol/l are shown in Fig. 14. As one can see the results from the two methods differ significantly. For the bubble the surface tension decrease starts much earlier. The surface tensions at long times, and hence the equilibrium surface tension from the bubble experiment are lower than those from the drop. However, the establishment of a quasi-equilibrium for the drop method is more rapid at low β-casein concentrations while at higher β-casein concentrations this process is more rapid for the bubble method. This essential difference between solutions of proteins and surfactants was discussed in detail elsewhere [50]. In brief, it is caused by simultaneous effects of differences in the concentration loss, and the adsorption rate, which both lead to a strong difference in the conformational changes of the adsorbed protein molecules.

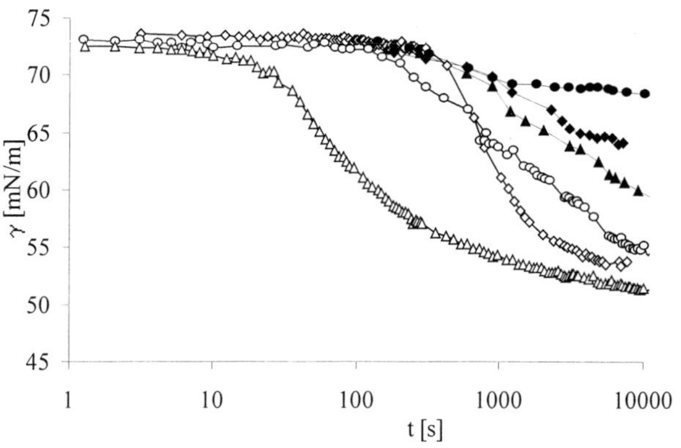

Fig. 14 Dynamic surface tension γ of β-casein solutions as a function of time, $5 \cdot 10^{-8}$ mol/l (○●), 10^{-7} mol/l (□■), 10^{-6} mol/l (△▲), (●■▲) – drop experiments, (○□△) - bubble experiment, according to [50]

Please note that at low initial concentrations and high adsorption activity, which is typical for proteins but also some surfactants, the final bulk concentration in the drop can become significantly lower than the initial value. By comparing the results obtained from drop and bubble experiments one can make use of this principle difference and estimate the value of the adsorption at the surface. Quantitative experiments are not known so far.

4.2. Dynamic interfacial tensions and transfer across the interface

Though liquid-liquid systems are very important in many technological applications they are not so extensively studied as compared to the liquid-air interface. In fact the adsorption processes in such systems deserve some specific considerations because, in most cases, their behaviour is more complex [51].

This is mainly caused by the solubility of the surfactant in both phases which is in practice never negligible. In fact, due to these characteristics the transfer of surfactant across the interface can become the main process controlling the adsorption, which in turn calls for an extension of the theoretical models and elaboration of suitable experimental techniques.

The drop shape method is possibly the most useful one for the investigation of the adsorptive transfer, i.e. the adsorption kinetics at the interface between two liquid phases containing the surfactant from the partition equilibrium. This phenomenon is particularly significant when situations far from the partition equilibrium are considered, in systems characterised by a high solubility of the surfactant in the recipient phase or by a large solubility of the surfactant in both phases. The latter case represents a typical situation for many types of ionic surfactants in water-oil and water-alkane systems, as demonstrated by the partition coefficients measured for various solvents [52, 53, 54, 55, 56].

In the adsorptive transfer the evolution of $\gamma(t)$, starting from the state of a fresh interface, can be divided into two stages. The first step is essentially controlled by the adsorption of surfactant from the rich phase and the second by the transfer of the surfactant into the second phase. Though the relative duration of these stages depends on the phase volumes and the distance of the initial conditions from the partition equilibrium, the first stage is typically faster than the second one. As far as large phases of comparable volume and initial conditions not far from the partition equilibrium are concerned, the first stage shows a relaxation-like evolution of $\gamma(t)$ as it is typical for adsorption processes, while in the transfer phase γ changes slowly and with a nearly linear slope. Such circumstance can make the evaluation of the attainment of equilibrium conditions difficult for these systems.
An example of such behaviour, studied by the drop shape method described here, is shown in Fig. 11b, where the dynamic surface tension during the adsorptive transfer of $C_{10}EO_8$ at a fresh water/hexane interface is shown. The diffusion controlled approach can be applied to model the

adsorption and transfer steps of the surfactant. Assuming semi-infinite volume phases 1 and 2 and a plane interface, the equivalent Ward-Tordai equation reads

$$\Gamma(t) = \frac{2c_{01}}{\sqrt{\pi}}\left[\sqrt{D_1} + \frac{c_{02}}{c_{01}}\sqrt{D_2}\right]\sqrt{t} - \frac{1}{\sqrt{\pi}}\int_0^t \left[\sqrt{D_1} + k_p\sqrt{D_2}\right]\frac{c_{1s}(\tau)}{\sqrt{t-\tau}}d\tau, \tag{23}$$

where c_{0i} are the initial concentrations and D_i are the diffusion coefficients in the two liquid phases, k_p is the partition coefficient, i.e. the ratio between the equilibrium concentrations in phases 2 and 1 [57].

This approach is adequate to describe the realistic situations because Eq (19) holds for any ratio of initial concentrations c_{01} and c_{02}, even out of the initial partition equilibrium. However, in the latter case, only a steady-state situation can be predicted, in which the amount of surfactant adsorbing from one phase balances the desorption into the other phase, and the transfer flux across the interface is constant.

Moreover, adsorptive transfer occurs often in limited size systems, for example at the interface of droplets in an emulsion or at the interface of bubbling droplets. The main characteristics in the study of such finite phase systems is the significant depletion of the bulk due to adsorption and transfer processes, which in some cases are the main factors controlling the adsorption dynamics. In fact, the depletion of the finite volume coupled to the transfer across the interface, can produce a non-monotonic behaviour of the dynamic adsorption $\Gamma(t)$, with minima in the dynamic interfacial tension $\gamma(t)$ [58, 59].

The problem of the adsorptive transfer in finite systems can be successfully studied with the drop shape method by arranging large volume ratios between the droplet and the surrounding liquid. Moreover, in these studies the contact between the two liquids must be avoided until the interface has formed. This can be conveniently arranged by leaving an air bridge between the two liquids into the capillary, that is blown out and eliminated by the buoyancy, when forming the drop.

Examples of such non-monotonic behaviour of $\gamma(t)$ are given in Fig 15b, corresponding to the adsorptive transfer of C_{13}DMPO at a freshly formed water/hexane interface [60]. In this case, drops of the aqueous surfactant solution were formed in a cell filled with pure hexane, arranging a ratio $Q = V_{hex}/V_{wat} = 10^3$ between the volumes of the two phases. The minimum interfacial tension value is well below the equilibrium value, which can be relevant for some technological processes, like the control of the droplet size in emulsification.

A similar experiment was performed in the opposite way by forming a drop of hexane inside the cell filled with the surfactant aqueous solution (volume ratio $Q=10^{-3}$), which yields a monotonic decrease of the interfacial tension (Fig. 15a).

Performing experiments with surfactants of smaller partition coefficients at the same volume ratio conditions [60, 61], yield a similar behaviour as observed providing Q>1. In fact, under these conditions, the supplying phase is quickly depleted by the adsorption at the interface and the transfer, reducing the adsorption flux. In the second phase, the desorbed molecules are rapidly diluted into the large volume, causing a large desorption flux. Eventually the net adsorption flux at the interface vanishes and then becomes negative, leading to the appearance of the minimum in the dynamic surface tension.

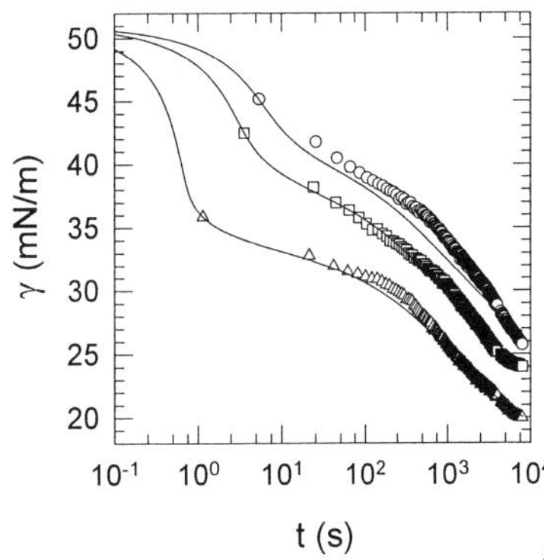

(a)

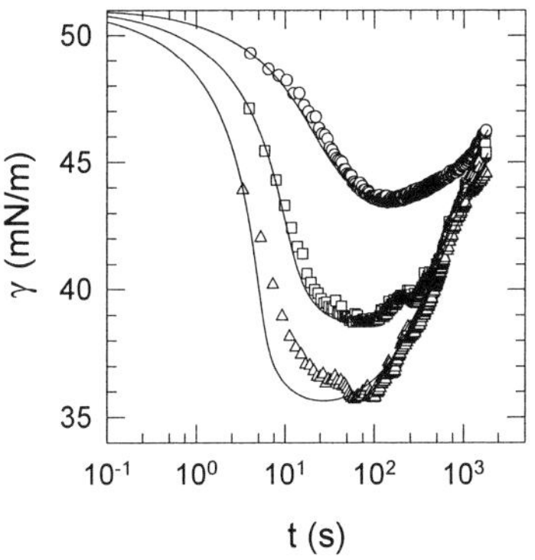

(b)

Fig. 15 Dynamic interfacial tension during the adsorptive transfer of C_{13}DMPO at a fresh water/hexane interface; a) a drop of pure hexane formed in the aqueous surfactant solution, $Q = 10^{-3}$, initial concentrations in water are: $c_{01} = 1.5,\ 2.3,\ 5.3*10^{-8}$ mol/cm³ (top to bottom); b) a drop of the aqueous surfactant solution is formed in pure hexane at a volume ratio of $Q = 10^{3}$, initial concentrations in water are $c_{01}=1, 2, 3*10^{-8}$ mol/cm³ (top to bottom); solid lines are obtained from model calculations

The observed features can be rigorously explained in the framework of a diffusion controlled model, for the adsorption at a spherical interface of a liquid surrounded by a shell of the second liquid [62] with the surfactant concentration being initially zero in one phase. Such a model requires a numerical solution of the diffusion problem in the bulk phases, coupled with the adsorption at the interface and, as shown in Figs 15, provides good agreement with the experimental data. The theoretical study reported in [62] has shown that the appearance of minima is preserved in a large span of diffusion coefficient ratios (D_1/D_2 from 0.1 to 10) and partition coefficients (k_p from 0.1 to 100).

4.3. Transient relaxations

Transient relaxation experiments are most suitable for diluted solutions as is generally the case for proteins [63]. First transient relaxations with a drop shape technique were performed by Miller et al. [64]. The adsorption and rheological behaviour of some model proteins at the water/air and water/oil interface were characterised in [65, 66].

The theoretical background of transient relaxations was recently summarised in [67]. For a trapezoidal interfacial area perturbation and a diffusion controlled exchange of matter mechanism the following set of equations results:

$$\Delta\gamma_1(t) = \frac{\Theta E_0}{2\omega_0}\exp(2\omega_0 t)\mathrm{erfc}(\sqrt{2\omega_0 t}) + \frac{2\Theta E_0\sqrt{t}}{\sqrt{2\pi\omega_0}} - \frac{\Theta E_0}{2\omega_0}, \quad 0 < t < t_1, \tag{24}$$

$$\Delta\gamma_2(t) = \Delta\gamma_1(t) - \Delta\gamma_1(t - t_1), \qquad t_1 < t < t_1 + t_2, \tag{25}$$

$$\Delta\gamma_3(t) = \Delta\gamma_2(t) - \Delta\gamma_1(t - t_1 - t_2), \qquad t_1 + t_2 < t < 2t_1 + t_2, \tag{26}$$

$$\Delta\gamma_4(t) = \Delta\gamma_3(t) - \Delta\gamma_1(t - 2t_1 - t_2), \qquad t > 2t_1 + t_2. \tag{27}$$

Here, the relative area change is denoted by $\Theta = \frac{d\ln A}{dt} = \frac{1}{t_1}\ln(1 - \frac{\Delta A}{A_0})$. t_1 and t_2 are the characteristic times of the trapezoidal perturbation, and ω_0 is the characteristic frequency defined by $\omega_0 = (dc/d\Gamma)^2(D/2)$. The parameter E_0 is the dilational elasticity module defined by the following relationship

$$E_0 = -\left(\frac{d\gamma}{d\ln\Gamma}\right)_A = \left(\frac{d\gamma}{d\ln A}\right)_\Gamma. \tag{28}$$

An experimental example is given below in paragraph 4.5.

4.4. *Harmonic relaxations*

The drop and bubble shape tensiometer allows one to perform oscillation experiments. However, in contrast to similar studies with small spherical drops it is limited to slow oscillations. Depending on the hydrodynamic conditions, mainly the rheological behaviour of the bulk phases, oscillation from a certain frequency onwards do not provide drops or bubbles with a Laplacian shape, so that the data analysis will fail and yield unrealistic data. Even for liquids of high grade of purity interfacial tension changes are simulated and hence misinterpretations can be the consequence. Thus, the use of high speed video technique is not really relevant for shape analysis tensiometry (although provided by several companies), as the hydrodynamic relaxation can take much time to yield Laplacian menisci.

The oscillation behaviour of the interfacial tension can be described by the complex elasticity E(iω) defined by the following equation

$$E(i\omega) = E'(\omega) + iE''(\omega) = \mathbf{F}\{\Delta\gamma(t)\}/\mathbf{F}\{\Delta \ln A(t)\}. \tag{29}$$

For a diffusion controlled exchange of matter we obtain

$$E(i\omega) = E_0 \frac{\sqrt{i\omega}}{\sqrt{i\omega} + \sqrt{2\omega_0}}, \tag{30}$$

where the real and imaginary parts are given by

$$E'(\omega) = E_0 \frac{1 + \sqrt{\omega_0 / \omega}}{1 + 2\sqrt{\omega_0 / \omega} + 2\omega_0 / \omega} \tag{31}$$

and

$$E''(\omega) = E_0 \frac{\sqrt{\omega_0 / \omega}}{1 + 2\sqrt{\omega_0 / \omega} + 2\omega_0 / \omega}. \tag{32}$$

Oscillation experiments with the elastic ring method in the frequency range from about 0.5 to 0.001 Hz for dodecyl dimethyl phosphine oxide solution were first conducted by Loglio [68].

Here, an example of experiments performed with the pendant drop apparatus for the surfactant Triton X-100 is given. This surfactant is a highly surface active substance and frequently used for many technical applications. At the studied concentration of 10^{-5} mol/l a surface tension value of about 58.8 mN/m is reached after about one hour adsorption. This surfactant is a technical product and therefore the surface tension is expected to continuously decrease even over 24 hours. Despite the fact that equilibrium is not reached after one hour from the start of the harmonic oscillations (Fig. 16a), the sinusoidal area changes were very accurate as demonstrated in Fig. 16b.

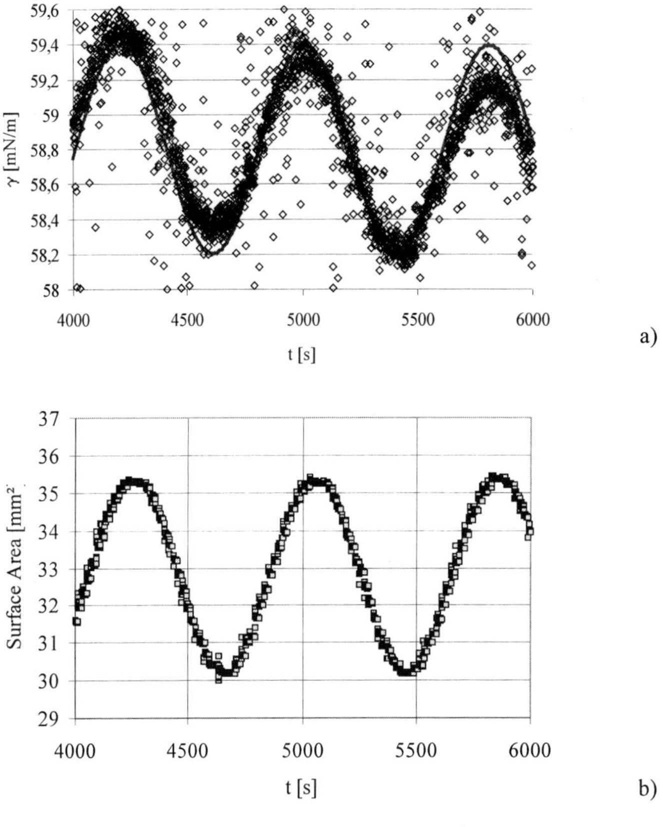

Fig. 16 Harmonic oscillation with a Triton X-100 solution at 10^{-5}mol/l; a) Surface tension response: symbols – experiment, solid curve –response function; b) Surface area change for the experiment shown in (a)

The surface tension response shows a slight drift downwards after each oscillation cycle. Both graphs contain the same number of experimental points. While the surface tension response shows a scattering around the sinusoidal change with a standard deviation of about

± 0.2 mN/m, the area is changing very smoothly. A sinusoidal function is also shown which essentially describes the change in surface tension with time.

In Fig. 17 oscillation experiments with the same solution are shown, however, instead of 800 s now 400 s is used per oscillation period and a smaller amplitude of the area changes. Also the sampling rate was reduced by one order of magnitude so that a much smaller number of points is received.

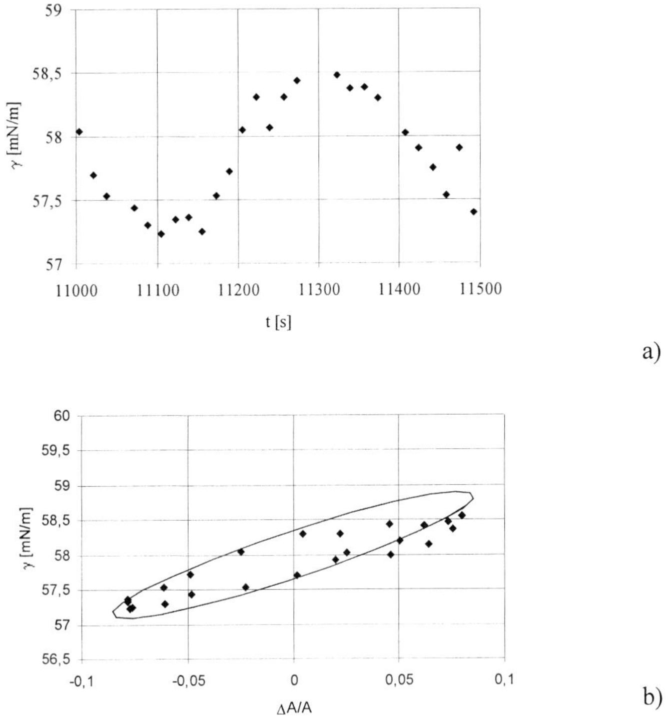

Fig. 17 Surface tension response for a Triton X-100 solution at $1 \cdot 10^{-5}$mol/l; a) change of surface tension with time; b) same as in (a) but plotted over the area change A(t)/A during oscillation

Another way of data interpretation is to plot the change in surface tension over the corresponding area change. As one can see an ellipse results from which the tilt angle and the thickness contain the rheological information. While the tilt is a measure of the dilational elasticity, the thickness is proportional to the exchange of matter rate, sometimes called dilational viscosity. Essentially this parameter corresponds to the phase shift between the generated area oscillation and the surface response (cf. Eq. 32). With increasing frequencies the thickness decreases while the tilt angle increases up to a final value which is identical to the dilational elasticity modulus.

The data of Fig. 16 (at 800 s oscillation period) yields a surface elasticity of E = 3.8 mN/m while a value of E = 7.2 mN/m is obtained from the data in Fig. 17, according to the definition of the dilational elasticity given by Eq. (28).

4.5. Practical applications: medical studies

The general application of tensiometry was shown in [69] and it was impressively demonstrated how large the capacity of interfacial studies for medical research is. For example, selected dynamic surface tension values of serum or urine correlate with the health state of patients suffering from various diseases. In the course of a medical treatment these values then change from a pathological level back to the normal values determined as standard for a certain group of people (age and sex).

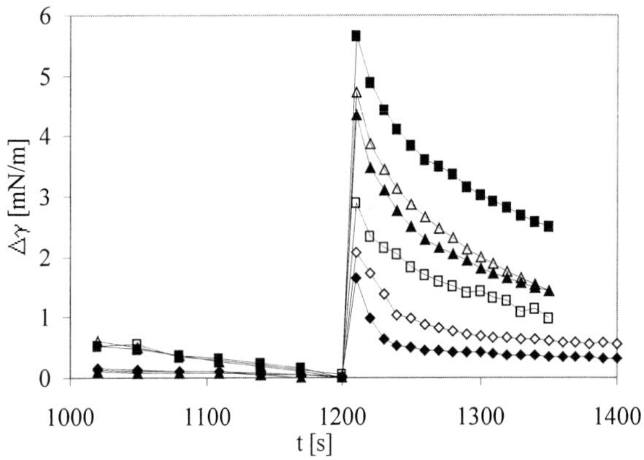

Fig. 18 Surface tension response of serum from a 46 years old patient suffering from an acute kidney insufficiency; admission to hospital (◇), therapy (◆, □), after haemodialysis (■), polyuria (△), leaving the hospital (▲)

Very recently examples of rheological studies on blood were published elsewhere [70]. As an example some results of these investigations will be discussed here. Fig. 18 shows the surface tension response after a step-type area change of a pendant drop area by about 10% for 6 serum samples from the same patient at different stages of his acute kidney insufficiency.

The dilational elasticities calculated from the measured jump via Eq. (28) are summarised in the following table together with the relaxation time τ calculated from the exponential function

$$\Delta\gamma(t) = \Delta\gamma_0 \exp[-\Delta t / \tau], \tag{33}$$

where Δt is the time interval after the area stress has been applied. An interpretation using the relaxation model given by Eqs. (24) to (27) is impossible as serum is a mixture of quite a number of surface active compounds of unknown concentrations. After admission to the hospital the elasticity is very low and even decreases a bit in the beginning of the therapy, while τ is maximum. After haemodialysis the elasticity goes through a maximum and levels off then at values close to the normal, determined as a standard for persons of this age and sex. The relaxation time decreases to less than half of the initial value.

Table Dilational elasticity ε and relaxation time τ determined from the data given in Fig. 18

Health state or treatment	E [mN/m]	τ [s]
admission to hospital	12.0	273
therapy	7.7	198
therapy	26.7	180
after haemodialysis	49.1	181
polyuria	35.0	129
leaving the hospital	29.5	139

The results allow one to conclude that rheological studies provide an interesting tool for medical practice and research, in particular due to the fact that very small quantities of a sample are needed.

4.6. Practical applications: surface tension of clouds and fogs

Surface active organic compounds, often present in atmospheric wet aerosols (i.e., clouds and fogs), alter the surface tension of the tiny liquid droplets. A large decrease of the surface tension value may change the processes of droplet nucleation and growth. As a consequence, the changes in droplet population significantly affect the cloud albedo as well as the formation of atmospheric precipitation.

The above-mentioned chain of events has been recently investigated by Facchini *et al.*, who first estimated the effects of interfacial properties in atmospheric microphysics [71, 72]. These authors established a semi-empirical relationship between the decrease of surface tension, γ , and the population, N, of cloud condensation nuclei (a decrease of 30% in γ, in respect to pure water, results in an increase of ~20% in N). An increase of droplet number is connected to smaller droplet size and thereby to a higher level of scattering of solar radiation (albedo). On the base of actual surface tension measurements for water samples collected from aerosols present in the real atmosphere, Facchini *et al.* estimated an approximate magnitude of the global climate forcing due to the effect of surface tension changes (the resulting figure for stratus clouds is ~ -1 W/m^2). In a subsequent paper [73], Facchini *et al.* further provided evidence about the presence of water soluble organic carbon (WSOC) in cloud/fog water samples and about the relevant influence to the state of the gas-liquid interface.

In practice, fine aerosol samples ($d < 1.5$ μm diameter) were collected on pre-fired 47-mm quartz-fibre filters, in the Po Valley (Northern Italy). Filters were cut into small pieces and then extracted with HPLC-grade water by sonication. Also, cloud/fog water samples were collected using an active string collector, entirely made of stainless steel. Some abundant water samples were partitioned into three fractions, separating three main classes of organic compounds (namely, neutral compounds; mono- and dicarboxylic acids; and polyacidic compounds). Surface tension was measured by the axisymmetric profile analysis technique on the original samples, as well as on each of three fractions, and a correlation was shown between γ-values and WSOC. The data obtained in [73] support the assertion that surface tension should be accounted for in cloud models, as an additional variable controlling atmospheric droplet formation and properties [71].

Figures 19 and 20 illustrate the performance of the profile analysis instrument for atmospheric studies, by displaying the relaxation of dynamic surface tension (at constant surface area) for a cloud water sample and, respectively, the dynamic surface tension response to harmonic surface area disturbances for a marine water sample [74].

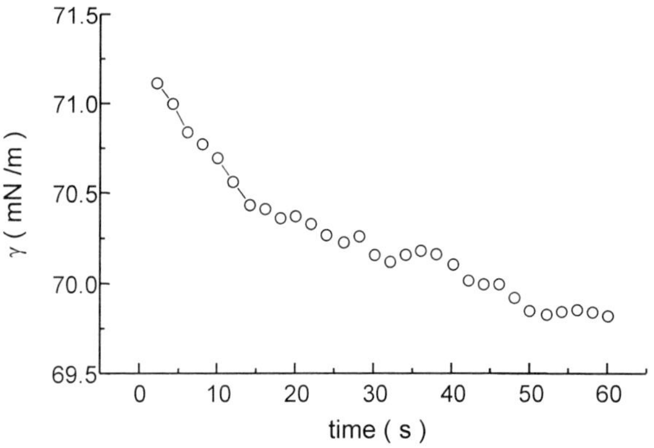

Fig. 19. Dynamic surface tension of a cloud-water sample, at constant air/water interfacial area, as observed a few seconds after formation of a bubble inside the liquid phase. Temperature T = 20 °C .

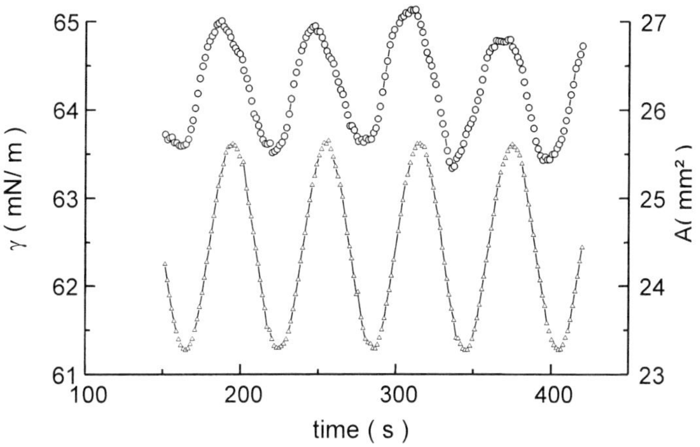

Fig. 20. Example of steady-state dynamic surface tension response, γ (t) (upper symbols, left axis scale), excited by harmonic changes of relative surface area, $\Delta \ln A$(t) (lower symbols, right axis scale), for a marine aerosol sample, extracted in water. Temperature T = 20.0 °C.

5. FOURIER TRANSFORMATION AS A TOOL TO ANALYSE HARMONIC RELAXATION STUDIES

In drop/bubble experiments, either working with transient (aperiodic) procedure or with harmonic (periodic) procedure, there never is a continuous function $g(t)$ to be analysed. There is instead a list of measurements of $g(t_i)$ for a discrete set of N time values t_i, where $g(t)$ represents the time-evolution of the inherent interfacial physical and geometrical properties (surface tension, differential pressure, interfacial area, *et cetera*).

5.1. Discrete Fourier Transform

Considering first the general mathematical topic, in the most common situations, the function $g(t)$ is sampled (i.e., its value is recorded) at evenly spaced intervals in time. Let Δ denote the time interval between consecutive samples, so that the sequence of sampled values is

$$g_n = g(n\,\Delta),\ n = 0;1;2;3; \tag{34}$$

The reciprocal of the time interval Δ is called the sampling rate. If Δ is measured in seconds, for example, then the sampling rate is the number of samples recorded per second [73]. A sequence of sampled values in a relevant time domain, starting from a given time origin, are considered, and the one-sided Fourier Transform is taken into account. When a discrete set of experimental points is available, the usual procedure is to approximate the integral, appearing in the analytical expression, by a discrete sum:

$$G(f_n) = \int_0^{\infty} g(t)\exp[2\pi i f_n t]dt \approx \sum_{k=0}^{N-1} g_k \exp[2\pi i f_n t_k]\Delta = \Delta \sum_{k=0}^{N-1} g_k \exp[2\pi i f_n t_k] \tag{35}$$

where k= 0;1;2; . . .; N - 1 ; and

$$g_k = g(t_k) \text{ and } t_k = k\Delta\,. \tag{36}$$

The final summation in Eq. (31) is called the *discrete Fourier transform* (DFT) of the N points [73]. In the general case, with N numbers of input in the time domain, we can produce N/2 independent numbers of output, in the frequency domain. In particular, the possible values of frequency are:

$$f_n = \frac{n}{N\Delta},\ n = 0,\ 1,\ 2,\ \ldots.,\ N/2. \tag{37}$$

The upper value of n corresponds exactly to the critical sampling frequency of two sample points per cycle (i.e., the Nyquist critical frequency). Thus, in general, the discrete Fourier transform maps N complex numbers into N/2 complex numbers [75].

5.2. Fourier analysis of data obtained for oscillating drops / bubbles.

Speaking in principle, a single cycle of a periodic phenomenon contains all the necessary physical information. In practice, in the harmonic oscillation experiments, the acquisition of data during a time interval longer than the single-oscillation period appears advantageous for a convenient experimental redundancy as well as for a verification of the transient or steady-state regime. Thus, in general, we obtain a given selected rectangular window comprising a temporal sequence of experimental points, which might not constitute a whole number of complete cycles. Moreover, usually, raw data of surface tension and of drop volume (and of other geometrical properties) are not obtained at exactly evenly-spaced intervals of time.

For the analysis and interpretation of the experimental results, in the first instance the software should perform a pre-processing step of raw data, extracting the required number of complete cycles and, optionally, the sequence of evenly-spaced interpolated points. Successively, the software gives forth the relevant Fourier analysis. The adopted numerical procedure is based on the above-mentioned discrete Fourier transformation (DFT) algorithm (i.e., the summation over all points in Eq. (31)). As shown in the previous paragraph, the algorithm allows any number n and any particular values of periods $N\Delta / n$ (frequencies, f_n) to be selected, among all the possible discrete values, for the analysis of observed data.

Several programming packages provide built-in DFT and FFT (fast Fourier transform) routines (or virtual instruments). However, such standard algorithms perform the data analysis at a great number of frequencies, $f_n = n / N\Delta$, but not necessarily at the exact frequency of the observed data. In our case, we know exactly the period (or we can estimate the frequency by the Buneman formula) of the imposed perturbation of bubble/drop volume (area). Rather than the frequencies, the relevant quantities to our experiments are the amplitude of a) the input disturbance and b) of the issuing response and the inherent difference of phase-angle.

In the case of an infinite number of oscillations, the Fourier transform amplitude appears as a spike at the oscillation frequency (that is, a Dirac delta). Actually, experimentally a (small) finite number of oscillations are obtained. In other words, there is a rectangular impulse of oscillations. As known from the Fourier transform theory, in the case of an impulse of oscillations there is a spectrum constituted of sinc-shaped amplitudes, rather than of spike-shaped amplitudes. Unless the sampling rate is an exact multiple of the frequency of the imposed perturbation, no frequency value of our discrete spectrum coincides with the actual experimental frequency. Thus a standard DFT (or FFT), routine may result in unreliable values of amplitude and of phase angle for the observed oscillating quantities. The adoption of a different window method (e.g., such as the triangular or the Hamming window), instead of the rectangular window, does not seem useful for our case.

Actually, as we are interested in the peaking value of the sinc-shaped spectrum, the most reliable procedure consists in executing the above-mentioned summation of Eq. (35), after assigning a (single) pre-selected known value (or estimated value) to the frequency, f_n. Finally, to obtain the correct amplitude value, the summation is divided by the number of cycles, Z_{cycl}, and by half the period, T/2.

We can obtain the real and the imaginary part of each quantity FT transformed, as follows:

$$\mathrm{Re}\{G(f_n)\} = \frac{2\Delta}{Z_{cycl}T}\sum_{k=0}^{N-1} g_k \cos(2\pi f_n t_k) \tag{38}$$

$$\mathrm{Im}\{G(f_n)\} = \frac{2\Delta}{Z_{cycl}T}\sum_{k=0}^{N-1} g_k \sin(2\pi f_n t_k) \tag{39}$$

When the g(t) values are not sampled at evenly-spaced time intervals, instead of multiplying the total summation by the collected Δ, each term of the summation in Eqs. (38) and (39) must be multiplied by the proper time interval.

Considering a periodic time-function, with period T, on inspection of Eqs. (38) and (39) we note that the amplitude values, per unit period, of the samples (taken at the frequencies n/T) of its Fourier transform are coincident with the coefficients of the Fourier series, representing the periodic function itself.

The Fourier analysis, in case of bubble/drop experiments, is mainly used to obtain the value of the surface dilational modulus. In such a case, from the values of the surface tension oscillation amplitude, of relative interfacial area amplitude and phase-angle difference, we determine the value of the above-mentioned modulus at the fundamental frequency of the oscillation, $f_1 = 1/T$.

As a differential dilational modulus is not defined, just the fundamental frequency is significant for the determination of the surface dilational modulus. Higher-frequency analysis may carry information about possible an-harmonicity of the input perturbation or possible non-linearity of the output response.

The results up-to-now obtained with the DFT algorithm, in the reduced form for the oscillating drop/bubble experiments, appears more reliable than the results obtained by DFT / FFT standard routines. In Fig. 21 an example plot is reported, as obtained from raw experimental data, showing the time evolution of surface tension and of drop volume, relevant to a dilute aqueous surfactant solution.

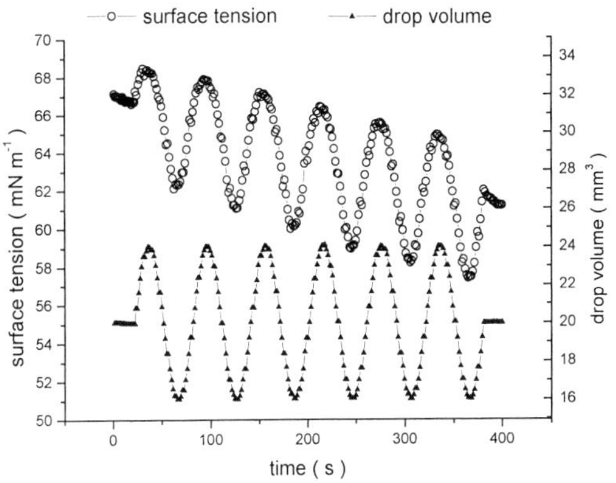

Fig. 21 Plot of (not smoothed) data of surface tension and of drop volume oscillations, relevant to a dilute aqueous surfactant solution, as observed by PAT-1.

Figure 22 displays a (favourable) comparison between the amplitudes of DFT (column graph) for the surface tension data, as reported in Fig.1, and the corresponding theoretical sinc-funtion (that is, the Fourier transform of the sinusoidal signal, within a rectangular window). Considering the satisfactory agreement between experimental and theoretical behaviour, we can assert the good harmonicity of the imposed perturbation and linearity of the issuing response.

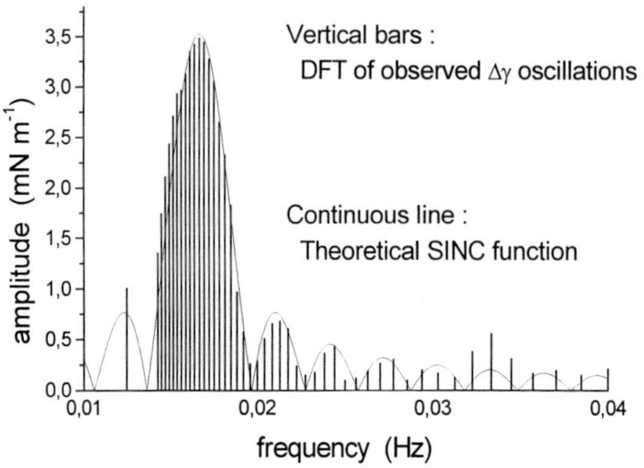

Fig. 22 Column graph of amplitudes of DFT of surface tension data, as reported in Fig. 21

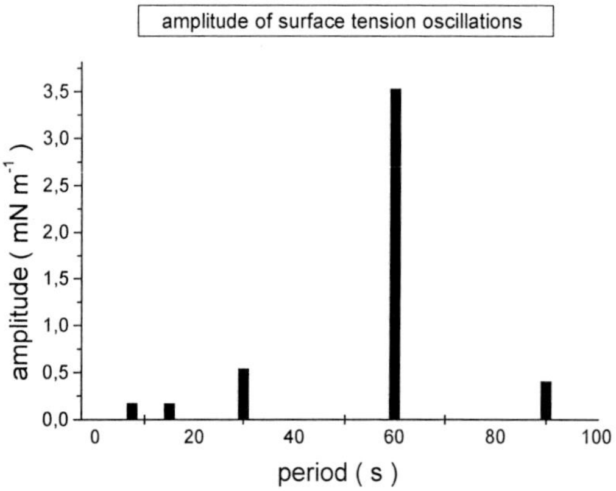

Fig. 23.- Column graph of the oscillation amplitude of surface tension, as reported in Fig. 21.

Fig. 23 illustrates the performance of the adopted numerical algorithm (based on Eqs. (38) and (39)). The amplitude value, $\Delta\gamma$, of the surface tension oscillations

$$\Delta\gamma = \sqrt{\operatorname{Re} G^2 + \operatorname{Im} G^2} \qquad (40)$$

has a physical significance just for the cycle-period of 60 seconds, while such a value is a meaningless mathematical phantom for the cycle-periods of 90, 30, 15 and 7.5 seconds, due to the sinc-function effect.

6. SUMMARY AND CONCLUSIONS

The examples have shown that the drop and bubble shape tensiometer PAT1 from SINTECH Berlin, Germany, is an accurate instrument with a large number of measurement procedures. It is shown that the mode of keeping the surface area constant is a vital prerequisite for experiments with bubbles. Only experiments with constant surface area give the opportunity for an easy data analysis of adsorption processes. It is demonstrated that experiments at the liquid/liquid interface provide extra information about distribution of the studied surfactant between the two adjacent phases, which is a measure for the HLB of this surfactant [76].

The relaxation studies with transient and harmonic area perturbations allow the determination of the dilational elasticity of interfacial layers. For protein the properties of adsorption layers at the water/air interface are very much different to those obtained at the water/oil interface which can be explained by the structure of the molecules in the interfacial layer. Consequently, the dilational elasticity and relaxation behaviour is very different at these two interfaces.

Contact angles can be measured with this tensiometer as well, where drops can be produced and increased/decreased on a solid surface. This growing and shrinking of the drop gives access to the so-called advancing and receding contact angles, which are needed to characterise the energetic behaviour of a solid surface [77]. Also studies at elevated temperatures and pressures can be performed, which however is not shown here as this chapter is dedicated only to the surface rheological characterisation of liquid interfaces.

Although no direct comparison with other commercial products are given we can state that the instrument PAT1 discussed here has the best features in respect to interfacial rheology studies. It provides a comfortable function generator for any type of transient and harmonic relaxation studies and also the theoretical tools to analyse trapezoidal and sinusoidal relaxation experiments. The on-line control of the interfacial area changes is very accurate and the oscillations performed in the range between 0.01 and 0.2 Hz are ideally smooth sinusoidal functions in contrast to experiments performed with other instruments.

ACKNOWLEDGEMENTS

The work was financially supported by the ESA MAP Project FASES.

7. LIST OF ABBREVIATIONS

A	surface area of a drop/bubble
b	radius of curvature at the drop apex
c_0	concentration in the bulk
c_s	concentration in the sublayer
c_w	concentration the aqueous phase
D	diffusion coefficient
E	surface elasticity
$E(i\omega)$	complex elasticity
E_0	dilational elasticity module
f	frequency of oscillation
g	gravitational acceleration
k_p	partition coefficient
P	pressure
Q	volume ratio
r	drop radius
R_1, R_2	principal radii of curvature
s	arc length
t	time
V	volume of a drop/bubble
x	horizontal coordinate
z	normal coordinate
β	shape factor
χ^2	objective function
δ	diffusion depth
ΔP	capillary pressure
$\Delta\rho$	density difference

Γ	surface concentration
γ	surface or interfacial tension
κ	mean curvature of the interface
Π	surface pressure
Θ	relative area change
τ	relaxation time
τ_D	characteristic time of diffusion
τ_{or}	characteristic time of orientation
ω_1	molar surface area in state 1
ω	circular frequency ($2\pi f$)

8. References

1. A.I. Rusanov and V.A. Prokhorov, *Interfacial Tensiometry*, in "Studies in Interface Science", Vol. 3, D. Möbius and R. Miller (Editors), Elsevier, Amsterdam, 1996
2. P. Chen, D.Y. Kwok, R.M. Prokop, O.I. del Rio, S.S. Susnar and A.W. Neumann, in *Drops and Bubbles in Interfacial Research*, in „Studies in Interface Science", Vol. 6, D. Möbius and R. Miller (Editors), Elsevier, Amsterdam, 1998, pp. 61
3. A. Passerone and E. Ricci, in *Drops and Bubbles in Interfacial Research*, in „Studies in Interface Science", Vol. 6, D. Möbius and R. Miller (Editors), Elsevier, Amsterdam, 1998, pp. 475
4. S.S. Susnar, H.A. Hamza and A.W. Neumann, Colloids Surfaces A, 89 (1994) 169
5. F.K. Skinner, Y. Rotenberg and A.W. Neumann, J. Colloid Interface Sci., 130(1989)25
6. J. Gaydos, in *Drops and Bubbles in Interfacial Research*, in „Studies in Interface Science", Vol. 6, D. Möbius and R. Miller (Editors), Elsevier, Amsterdam, 1998, pp. 1
7. G.I. Taylor, Proc Roy Soc, A280 (1964), 383
8. F.K. Wohlhuter, O.A. Basaran, J. Fluid Mech., 235 (1992) 481.
9. J. Meseguer and A. Sanz, J. Fluid Mech., 153 (1985) 83.
10. S.R. Coriell, S.C. Hardy, M.R. Corfes, J. Colloid Interface Sci, 60 (1977) 126.
11. L. Liggieri, A. Sanfeld, A. Steinchen, Physica A, 206 (1994), 299.

12. M.M. Lipshutz, Differential Geometry, McGraw-Hill, New York 1969.
13. F. Bashforth and C. Adams, An attempt to test the theories of capillary action, Cambridge Univ. Press, 1883.
14. J.F. Padday, Phil. Trans. Royal Soc., A269 (1971) 265.
15. J.F. Padday, Phil. Trans. Royal Soc., A275 (1973) 489.
16. S. Hartland and R. Hartley, Axisymmetric fluid-liquid interfaces, Elsevier, Amsterdam, 1976.
17. S. Fordham, Proc.Royal Soc.(London), 194A (1948) 1.
18. J. F. Padday, in E. Matijevic (Ed.), Surface and Colloid Science series vol. 1, Wiley-Interscience, New York 1969. p.101.
19. N.E. Dorsey, J. Washington Acad. Sci., 18 (1928) 505.
20. D.W.G. White, A supplement to the tables of Bashforth and Adams, Mines Branch, Phys. Metall. Div., Rep.PM-I-67-4, May 1967
21. C. Maze and G. Burnet, Surface Sci., 13 (1969) 451.
22. L. Liggieri and A. Passerone , High Temperature Tech.,7 (1989) 80.
23. Y. Rotenberg, L. Boruvka and A.W. Neumann , J.Colloid Interface Sci. , 93 (1983) 169.
24. N.R. Pallas and Y. Harrison, Colloids Surfaces, 43 (1990) 169.
25. C. Huh and R.L. Reed, J. Colloid Interface Sci., 91 (1983) 472.
26. S.-Y. Lin, K. McKeigue and C. Maldarelli, AIChE J., 36 (1990) 1785.
27. L.L. Schramm, D.B Fisher, S. Schürch and A. Cameron, Colloids and Surfaces A, 94 (1995) 145.
28. J. Benjamins, A. Cagna and E.H. Lucassen-Reynders, Colloids and Surfaces A, 114 (1996) 245.
29. A. Semmler, R. Ferstl and H.-H., Kohler, Langmuir, 12 (1996) 4165.
30. R. Sangiorgi, G. Caracciolo and A. Passerone, J. Material Sci., 17 (1982) 2895.
31. J.J. Moré and S.J. Wright, Optimization Software Guide, SIAM, Philadelphia, 1993
32. C. Maze and G. Burnet, Surface Sci., 24 (1971) 335
33. Y.Z. Zhou and J. Gaydos, Colloids Surfaces A ,in press
34. A.M. Worthington, Proc. Roy. Soc. A., 32 (1881) 362

35. A. Ferguson, Phil. Mag. Ser. 6, 23 (1912)417
36. J.M. Andreas, E.A. Hauser and W.B. Tucker, J. Phys. Chem., 42 (1938) 1001
37. S.S. Dukhin, G. Kretzschmar and R. Miller, Dynamics of Adsorption at Liquid Interfaces: Theory, Experiment, Application, in Studies in Interface Science, Vol. 1, D. Möbius and R. Miller (Editors), Elsevier, Amsterdam, 1995
38. A.F.H. Ward and L. Tordai, J. Phys. Chem. 14 (1946) 453
39. V.B. Fainerman, R. Miller, R. Wüstneck and A.V. Makievski, J. Physical Chemistry, 100 (1996) 7669.
40. M. Ferrari, L. Liggieri, F. Ravera, J. Phys. Chem. B, 102 (1998), 10521.
41. L.Liggieri, M. Ferrari, A. Massa and F. Ravera, Colloids Surfaces A, 156 (1999), 455.
42. J. K. Ferri, K. J. Stebe, Adv. in Colloid and Interface Sci. 85 (2000), 61.
43. Ravera, F.; Liggieri, L.; Miller, R. Colloids Surfaces A, 175 (2000) 51.
44. L. Liggieri and F. Ravera, in *Drops and Bubbles in Interfacial Research*, in „Studies in Interface Science", Vol. 6, D. Möbius and R. Miller (Editors), Elsevier, Amsterdam 1998., p. 239.
45. V.B. Fainerman, R. Miller, in *Drops and Bubbles in Interfacial Research*, in „Studies in Interface Science", Vol. 6, D. Möbius and R. Miller (Editors), Elsevier, Amsterdam 1998., p. 279.
46. W. Jost, Diffusion in Solid, Liquid, Gases, Academic Press, New York 1952, chapter 1.
47. K.J. Mysels, J. Physical Chemistry, 86 (1982) 4648.
48. H.-C. Chang, C.-T. Hsu, and S.Y. Lin, Langmuir, **1998**, 14, 2476.
49. M. Ferrari, L. Liggieri and F. Ravera, J. Phys. Chem. **1998**, 102, 10521
50. A.V. Makievski, G. Loglio, J. Krägel, R. Miller, V.B. Fainerman and A.W. Neumann, J. Phys. Chem. 103(1999)9557
51. F. Ravera, M. Ferrari and L. Liggieri, Adv. Coll. Int. Sci., 88 (2000) 129.
52. F.A. Villalonga, R.J. Koftan and J.P. O'Connell, J. Colloid Interface Sci., 90(1982)539.
53. R. Aveyard, B.P. Binks, S. Clark and P.D.I. Fletcher, J. Chem. Soc. Faraday Trans., 86 (1990) 3111.
54. E.H. Crook, D.B. Fordyce and G.F. Trebbi, J. Colloid Science, 20 (1965) 191.

55. F. Ravera, M. Ferrari, L. Liggieri, R. Miller and A. Passerone, Langmuir, 13 (1997) 4817.

56. J.G. Göbel and J.R. Joppien, J. Colloid Interface Sci., 191 (1997) 30.

57. R. Miller, V.B. Fainerman, A.V. Makievski, J. Krägel, D.O. Grigoriev, F. Ravera, L. Liggieri, D.Y. Kwok and A.W. Neumann, Characterisation of water/oil interfaces, in Encyclopaedic Handbook of Emulsion Technology, J. Sjöblom (Ed.), Marcel Dekker, New York 2001, pp. 1

58. W.W. Mansfield, Aust. J. Sci Res., A5 (1952) 331.

59. E. Rubin and C.J. Radke, Chemical Eng. Sci, 35 (1980) 1129.

60. M. Ferrari, L. Liggieri, F. Ravera, C. Amodio and R. Miller, J. Colloid Interface Sci., 186 (1997) 40.

61. F. Ravera, M. Ferrari. L. Liggieri and R. Miller, Progr. Colloid Polymer Sci., 105 (1997) 346.

62. L. Liggieri, F. Ravera, M. Ferrari, A. Passerone and R. Miller, J. Colloid Interface Sci., 186 (1997) 46.

63. *Proteins at Liquid Interfaces*, in "Studies in Interface Science", Vol. 7, D. Möbius and R. Miller (Editors), Elsevier, Amsterdam, 1998

64. R. Miller, R. Sedev, K.-H. Schano, Ch. Ng and A.W. Neumann, Colloids & Surfaces, 69(1993) 209

65. R. Miller, J. Krägel, A.V. Makievski, R. Wüstneck, J.B. Li, V.B. Fainerman and A.W. Neumann, Proceedings of the *2nd World Emulsion Congress*, Bordeaux, 1997, Vol. 4, 153

66. A.V. Makievski, V.B. Fainerman and R. Miller, *Proceedings of the 2nd World Emulsion Congress*, Bordeaux, 1997, Vol. 2, 2-2-288

67. R. Miller, G. Loglio, U. Tesei and K.-H. Schano, Adv. Colloid Interface Sci., 37(1991)73

68. G. Loglio, R. Miller, A.M. Stortini, U. Tesei, N. Degli Innocenti and R. Cini, Colloids Surfaces A, 95(1995)63

69. V.N. Kazakov, O.V. Sinyachenko, V.B. Fainerman, U. Pison and R. Miller, *Dynamic Surface Tensiometry in Medicine*, in "Studies in Interface Science", Vol. 8, D. Möbius and R. Miller (Editors), Elsevier, Amsterdam, 2000

70. A.F. Vozianov, V.N. Kazakov, O.V. Sinyachenko, V.B. Fainerman and R. Miller, *Interfacial tensiometry and rheometry of biological liquids in nephrology*, Izd. Donetsk. Med. Univ., Donetsk, 1999 (in Russian)

71. M.C. Facchini, M. Mircea, S. Fuzzi and R.J. Charlson, Nature, 401 (1999) 257

72. H. Rodhe, Nature, 401 (1999) 223

73. M.C. Facchini, S. Decesari, M. Mircea, S. Fuzzi and G. Loglio, Atmospheric Environment, 34 (2000) 4853

74. G. Loglio, P. Pandolfni, U Tesei and B. Noskov, Colloids Surfaces A, 143 (1998) 301

75. W. H. Press, S. A. Teukolsky, W. T. Vetterling, and B. P. Flannery, *Numerical Recipes in Fortran 77, The Art of Scientific Computing*, 2nd Edition, Volume 1 of Fortran Numerical Recipes, Cambridge University Press, New York, 1997

76. P.M. Krugljakov, *Hydrophilic Lipophilic Balance: Physicochemical Aspects and Applications*, in "Studies in Interface Science" Vol. 9, D. Möbius and R. Miller (Editors), Elsevier, Amsterdam, 2000

77. Users Manual of the instrument PAT1, SINTECH Berlin, Germany

Novel Methods to Study Interfacial Layers
D. Möbius and R. Miller (Editors)

Oscillating bubble and drop techniques

V.I. Kovalchuk[1], J. Krägel[2], E.V. Aksenenko[3], G. Loglio[4] and L. Liggieri[5]

[1] Institute of Biocolloid Chemistry, Ukrainian National Academy of Sciences, 42 Vernadsky Avenue, Kiev 03142, Ukraine

[2] MPI für Kolloid- und Grenzflächenforschung, Am Mühlenberg 2, D-14476 Golm, Germany

[3] Institute of Colloid Chemistry and Chemistry of Water, Ukrainian National Academy of Sciences, 42 Vernadsky Avenue, Kiev 03142, Ukraine

[4] University of Florence, Institute of Organic Chemistry, Via G. Capponi 9, Florence, Italy

[5] CNR - Istituto di Chimica Fisica Applicata dei Materiali, Via De Marini 6, I-16149 Genova, Italy

Contents

Keywords: Oscillating drop method, surfactant solutions, dilational elasticity, microgravity experiments, hydrodynamic theory

The analysis of the dynamic behaviour of an oscillating bubble or drop provides the amplitude- and phase-frequency characteristics which take into account the dependency on the viscoelastic properties of the fluid-fluid interface. Corresponding experiments can be organized in different ways depending on the type of the measuring cell (open or closed), generation of bubble and drop oscillations (either by external pressure variations or volume variation in the cell), and registration of the measured signal (meniscus volume or pressure changes in the cell). All experimental schemes can be considered in the framework of a common theory based on the application of Fourier transforms. The method demonstrates a high sensitivity of the measured signal to changes in the surface elasticity and viscosity, which depends on the presence of surfactants in the solution and on the mechanisms of surface relaxations. The good qualitative agreement with experiments underlines that the theory is adequate to describe the main properties.

1. Introduction

Many technologies and natural phenomena involve processes of fast expansion or compression of fluid interfaces covered with surfactant adsorption layers. The dynamic system properties depend on the mechanisms and rate of equilibrium restoration after a deformation. At small magnitudes of deformation the mechanical relaxation of an interface can be described by the complex dilational viscoelastic modulus [1, 2]. For sinusoidal deformations it is defined as the ratio of complex amplitudes of interfacial tension variation and the relative surface area variation $\varepsilon(i\omega) = d\gamma / d \ln A$ being a function of frequency. This modulus may include contributions of many relaxation processes at the interface, such as diffusion transfer and adsorption-desorption of surfactant molecules [1-4], re-orientation or conformational changes of adsorbed molecules [5-7] or micellisation [8]. Under dynamic conditions the interfacial tension depends not only on the current state of the interface but also on the previous changes of the surface area. This suggests that the interfacial tension can be expressed by the convolution integral with the kernel $\varepsilon(t)$ the Fourier image of which is $\varepsilon(i\omega)$ [9-11]. With the aid of this integral the time dependency of the interfacial tension can be obtained for any given relative surface area change provided the complex viscoelastic modulus is known (direct problem). In turn this modulus can be obtained by using the Fourier transformations procedure when the interfacial tension and the relative surface area change are known as functions of time (inverse problem) [9, 10, 12].

The goal of many experimental studies is to obtain the complex dilational viscoelasticity as a function of frequency $\varepsilon(i\omega)$ from the measured interfacial tension response because this function contains important information about the relaxation processes in the interface. Different methods are utilised for this purpose which show a variety of types of surface deformation and can be used in different frequency ranges. The oscillating barrier [1, 2] and elastic ring [9, 10, 12] methods are used for relatively small frequencies. The capillary wave technique is most suitable for high frequencies [7, 8]. The oscillating bubble and drop techniques [3, 4, 13, 14] work in the intermediate frequency range of 0.1 Hz to 400 Hz. Many relaxation processes belong to this frequency range, and what is very important, is the characteristic times of diffusion for usual surfactant solutions corresponds also to this frequency range.

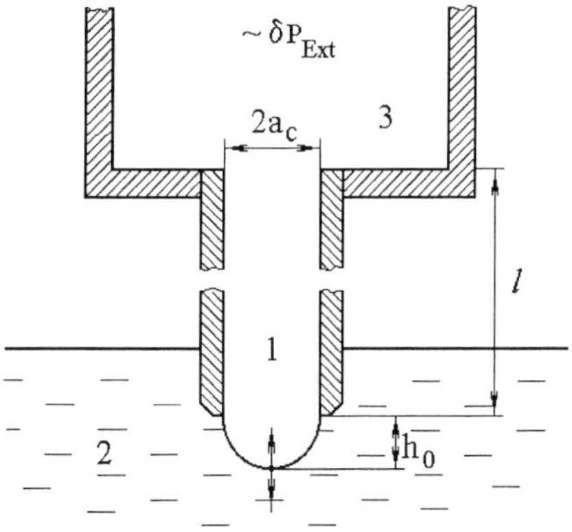

Fig. 1. The oscillating bubble (drop) system with an open measuring cell; 1 – capillary, 2 – phase A (solution), 3 – phase B (gas or liquid).

The most important types of oscillating bubble instruments have been reviewed recently in [4]. The main difference between the instruments follows from two points: the way to generate forced oscillations of the meniscus attached to the capillary tip, and what is the system response, i.e. what is the measured value. The meniscus oscillations can be generated either by external pressure variation applied though the capillary or a pressure variation in the cell filled with the solution into which the capillary is immersed produced via variation of the cell volume. Very often the measuring cell is open and the meniscus oscillations are generated by pressure changes at the opposite end of the capillary (Fig. 1) [13, 15]. In this case the measured value is the meniscus volume variation which can be determined either photometrically or by a CCD camera. In other cases the measuring cell is closed and allows the measurement of the pressure variation in the cell by a pressure transducer (instead of meniscus volume variation, Figs. 2 and 3).

The applied signal can be either the pressure change at the opposite end of the capillary (Fig. 2) [16, 17] as in the previous case, or the volume variation in the cell produced by a pulsating rod or a piezodriver at constant pressure near the opposite capillary end (Fig. 3) [4, 14, 18-20]. In all cases from the comparison of the applied and the measured signals the complex dilational viscoelasticity $\varepsilon(i\omega)$ can be obtained as a function of frequency after elimination of all contributions caused by the bulk phase behaviour. It is possible also to measure both the pressure in the cell and the meniscus volume and to compare them [21-23].

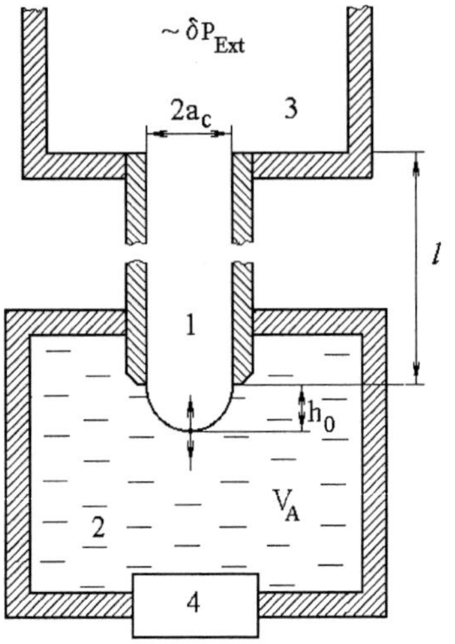

Fig. 2. The oscillating bubble (drop) system with a closed measuring cell and excitation by the pressure change at the external end of the capillary; 1 – capillary, 2 – phase A (solution), 3 – phase B (gas or liquid), 4 – pressure transducer.

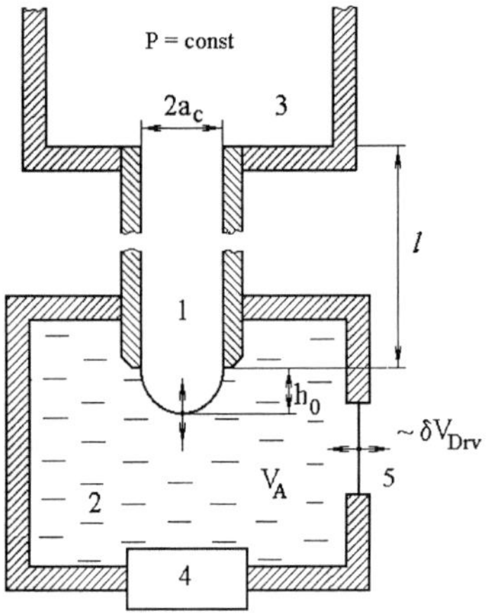

Fig. 3. The oscillating bubble (drop) system with a closed measuring cell and excitation by volume variation in the cell; 1 – capillary, 2 – phase A (solution), 3 – phase B (gas or liquid), 4 – pressure transducer, 5 – piezodriver.

Different hydrodynamic effects in the bulk phases contribute to the measured signal. The most important of them are inertia, viscosity and bulk compressibility of the mediums moving together with the meniscus [3, 11, 13, 14, 24].

The system response depends not only on the relaxation processes in the interfacial layer but also on the hydrodynamic relaxation in the fluid bulk phases. These hydrodynamic contributions vary with the system geometry and physical properties of the contacting media. Therefore, the system behaviour can be very different depending on the experimental conditions. Nevertheless, there are common regularities, characteristic for all types of oscillating bubble or drop systems and they can be considered in the framework of a general theory as discussed below.

2. Measuring principle of oscillating bubble and drop methods

Let us consider a closed measuring cell filled with a surfactant solution and the bubble or drop formed on the tip of a capillary submerged into this solution (Fig. 4). In equilibrium the pressure difference between the solution (phase A) and the external gaseous or liquid phase (phase B) is compensated by the Laplace pressure of the meniscus of the surface which is covered by the equilibrium surfactant layer.

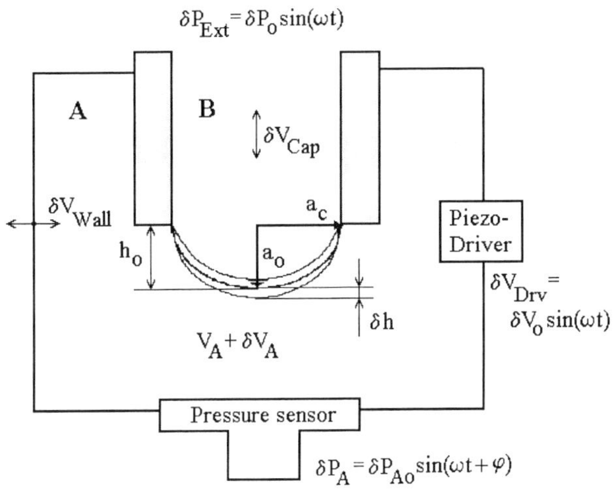

Fig. 4. Measuring principle of oscillating bubble and drop methods.

The meniscus oscillations can be excited either by pressure variations in the external phase δP_{Ext} or by volume variations produced by the piezo-driver inside the closed cell δV_{Drv}. The meniscus oscillations are accompanied by a volume flow through the capillary δV_{Cap}, the variations of the volume of liquid A due to its compressibility δV_A, and the inner cell volume variations due to walls deformations δV_{Wall}. The gas volume variations should also be considered δV_{Gas} when unwanted small gas bubbles are included inside phase A. The volume balance can be expressed as

$$\delta V_{Drv} = \delta V_{Cap} - \delta V_{Wall} + \delta V_A + \delta V_{Gas} \tag{1}$$

It is accepted here that δV_{Cap}, δV_{Wall}, δV_A, and δV_{Gas} are positive when the flow through the capillary is directed to the cell and the cell volume, the volume of the solution and the gas volume are increasing relative to their equilibrium values; δV_{Drv} is positive when the piezo-driver pulls the solution outwards from the cell. The solution volume and the gas volume depend on the pressure inside the cell

$$\delta V_A = -\frac{V_A}{B_A}\delta P_A \tag{2}$$

$$\delta V_{Gas} = -\frac{V_{Gas}}{B_{Gas}}\delta P_A \tag{3}$$

where V_A and V_{Gas}, B_A and B_{Gas} are the volumes and the bulk elasticity modules of the solution and the gas, respectively, and δP_A is the pressure variation in the cell. The deformation of the cell walls can be expressed as

$$\delta V_{Wall} = \frac{dV_{Cell}}{dP}\delta P_A \tag{4}$$

The coefficient $\frac{dV_{Cell}}{dP}$ describes the cell volume dependency on the pressure, which is determined by the elastic properties of the walls. Combining Eqs. (2)-(4) one obtains

$$\delta V_{Wall} - \delta V_A - \delta V_{Gas} = \frac{V_A}{B_{ef}}\delta P_A \tag{5}$$

where the effective cell elasticity modulus is defined by

$$B_{ef} = \frac{B_A}{1 + \frac{B_A}{V_A}\frac{dV_{Cell}}{dP} + \frac{V_{Gas}B_A}{V_A B_{Gas}}} \tag{6}$$

This modulus is always smaller than the bulk elasticity modulus of the phase A: $B_{ef} < B_A$. When the gas is absent in the solution and the cell is rigid (small coefficient dV_{Cell}/dP) then the effective modulus is determined only by the solution bulk elasticity: $B_{ef} \approx B_A$. When the cell is not rigid (deformational) or the volume of the solution V_A is very small then the effective modulus is determined only by the cell deformation $B_{ef} \approx V_A \left(\frac{dV_{Cell}}{dP}\right)^{-1}$.

The adiabatic and isothermic bulk elasticity modules for water only slightly differ. The adiabatic modulus for water is $2.2 \cdot 10^9$ Pa. The bulk elasticity modulus for gas can be obtained from the equation of state. For ideal gas the isothermic modulus is approximately equivalent to P_A, and the adiabatic modulus is equivalent to γP_A, where P_A is the pressure inside the cell, $\gamma = 1.4$ the adiabatic constant. The isothermic modulus can be used in the case of infinitely slow processes. In the case of pressure oscillations at sound frequencies the adiabatic modulus should be used. Gas is much more compressible than liquids. The bulk elasticity modulus for gas is four orders of magnitude smaller than for water. Therefore, even small amounts of gas much smaller than the volume of the solution ($V_{Gas} \ll V_A$), can mimic small values of the effective cell elasticity modulus Eq. (6), i.e. a strong decrease of the cell resistance to pressure variations. It is extremely important to avoid the presence of any small amounts of gas in the solution because it can lead to uncontrolled changes of the effective cell elasticity modulus.

Substituting Eq. (5) in Eq. (1) one obtains

$$\delta V_{Drv} = \delta V_{Cap} - \frac{V_A}{B_{ef}} \delta P_A \tag{7}$$

One can apply the Laplace-Fourier transformation to this equation and obtains

$$\delta V_{Drv}(i\hat{\omega}) = \delta V_{Cap}(i\hat{\omega}) - \frac{V_A}{B_{ef}} \delta P_A(i\hat{\omega}) \tag{8}$$

where $\delta V_{Drv}(i\hat{\omega})$, $\delta V_{Cap}(i\hat{\omega})$, and $\delta P_A(i\hat{\omega})$ are the Fourier images of the corresponding time functions (here and below, the argument $i\hat{\omega}$ indicates the Fourier images). The volume flow through the capillary is determined by the difference of the pressures in the cell and in the external phase B outside the capillary, and by the resistance of the capillary with the attached meniscus

$$\delta V_{Cap}(i\hat{\omega}) = \frac{\delta P_{Ext}(i\hat{\omega}) - \delta P_A(i\hat{\omega})}{R_C(i\hat{\omega})} \tag{9}$$

The resistance of the capillary with the attached meniscus ($R_C(i\hat{\omega})$) is a complex value. That means the volume flow through the capillary does not change simultaneously with the pressure change. Transitional processes take place in the system because of the relaxation of the surface tension and the hydrodynamic relaxation in the bulk phases. As usual in such cases the volume flow through the capillary is determined not only by the pressure difference at the given time moment but also by the prehistory of the process, and should be expressed with the help of a convolution integral [25, 26]. In the general case, the cell wall deformation (Eq. (4)) should be also described by a complex coefficient [24]. When the cell dimensions are much smaller than the wave length and both the sound emission from the cell and the energy dissipation in the walls are negligibly small then the response of the cell on the deformation will be purely elastic. In this case the coefficient dV_{Cell}/dP is a real value and can be considered as a constant in the framework of a linear approximation. The thermal relaxation in the cell can also be usually ignored.

Accounting for Eqs. (8) and (9) one obtains

$$\delta V_{Cap}(i\hat{\omega}) = \frac{\frac{B_{ef}}{V_A}\delta V_{Drv}(i\hat{\omega}) + \delta P_{Ext}(i\hat{\omega})}{\frac{B_{ef}}{V_A} + R_C(i\hat{\omega})} \tag{10}$$

and

$$\delta P_A(i\hat{\omega}) = \frac{B_{ef}}{V_A} \cdot \frac{\delta P_{Ext}(i\hat{\omega}) - R_C(i\hat{\omega}) \cdot \delta V_{Drv}(i\hat{\omega})}{\frac{B_{ef}}{V_A} + R_C(i\hat{\omega})} \tag{11}$$

These two equations allow the understanding of the measuring principle of the oscillating bubble or drop method. Oscillation can be generated either by external pressure variation δP_{Ext} or by volume variation in the cell δV_{Drv}, while the measured signal is the meniscus volume δV_m (which is equivalent to the volume passing through the capillary $\delta V_m = \delta V_{Cap}$) or the pressure in the cell δP_A. When comparing the externally applied signal with the measured signal, and using Eq. (10) or (11) we obtain the resistance of the capillary with the attached meniscus $R_C(i\hat{\omega})$. This resistance contains the information about the relaxation processes in the interface and in the bulk phases. Separating the influence of the bulk phases, information about the relaxation of the interfacial tension is obtained. When the response of the cell walls to deformations is not purely elastic its contribution has also to be separated from the measured signal.

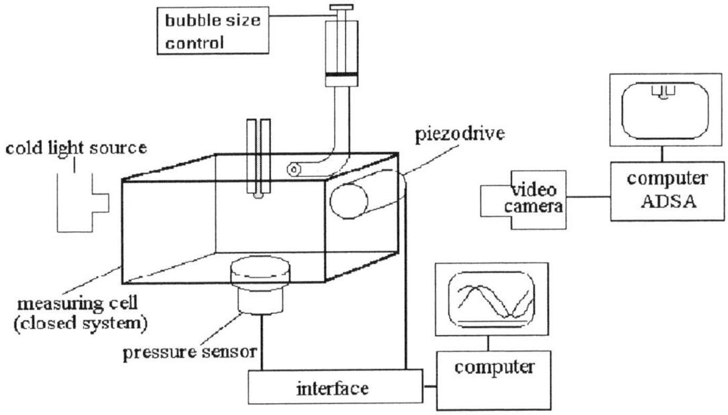

Fig. 5 Scheme of the experimental set-up of oscillating bubble and drop methods.

When the solution volume is large (or the cell is open to the atmosphere) then $\frac{B_{ef}}{V_A} \rightarrow 0$ and one obtains a system shown in Fig. 1 with the excitation by an external pressure variation δP_{Ext} and the meniscus volume δV_m as the measured signal (Eq. (10)). For a closed cell it is more convenient to measure the pressure in the cell (Eq. (11)) instead of the meniscus volume, while the excitation can be either via external pressure variation (the second system, Fig. 2) or volume variation in the cell (the third system, Fig. 3). In principle, both the variation of the meniscus volume δV_m and the pressure in the cell δP_A can be measured simultaneously and compared with each other. Experiments with a drop formed at the capillary tip outside the cell can be organised in the same way (cf. [21-23]). The viscoelastic bulk properties of non-Newtonian liquids can be studied with the same equipment [27].

The externally applied signal can be sinusoidal or of any other form (triangular, rectangular, trapezoidal, etc.) provided it is small enough that linear approximations are valid. Any kind of transitional experiments can be performed with the same system [9, 10, 21-23]. Devices can be easily connected to a computer which allows an automatic data acquisition, comparison of signals, separation of bulk phase effects, and a complete control of the experiment (Fig. 5).

3. Hydrodynamic resistance of the capillary with attached meniscus

The resistance of a capillary with an attached meniscus $R_C(i\hat{\omega})$ can be determined via the corresponding hydrodynamic problem which relates the volume flow through the capillary with the pressure in the fluid phases at its two ends. The pressure drop between the phases A and B includes the pressure drop over the capillary and over the meniscus. Under dynamic conditions the pressure drop at the meniscus can be obtained from the viscous stress balance at the interface [3, 11]. The contribution of added liquid mass moving with the meniscus [11, 26] has also to be included. The Fourier images of the pressure drop can be expressed by [26]

$$\delta P_{In}(i\hat{\omega}) - \delta P_A(i\hat{\omega}) = \frac{2}{a_0}\delta\gamma(i\hat{\omega}) - \frac{2\gamma_0}{a_0^2}\delta a(i\hat{\omega}) + \left(i\hat{\omega}\frac{\chi'_{1A}\eta_A + \chi'_{1B}\eta_B}{a_C} - \rho_A a_C \chi_2 \hat{\omega}^2 \right)\delta h(i\hat{\omega}) \qquad (12)$$

where $\delta P_{In}(i\hat{\omega})$ is the pressure variation inside the bubble or drop attached to the capillary; $\delta\gamma(i\hat{\omega})$, $\delta a(i\hat{\omega})$, and $\delta h(i\hat{\omega})$ are the surface tension variation, the meniscus radius of curvature variation and the meniscus height variation; γ_0, a_0, and a_C are the equilibrium surface tension, equilibrium radius of curvature and capillary radius; χ'_{1A}, χ'_{1B} and χ_2 are dimensionless coefficients depending only on the ratio h/a_C of the meniscus height to the capillary radius; η_A, η_B, and ρ_A are the dynamic viscosities and the density. The first two terms on the right hand side describe the Laplace pressure variation due to the surface expansion or compression and the radius of curvature variation. The last term describes the viscous stress contribution at the interface and the inertia of the liquid A moving together with the meniscus. χ_2 is the added mass coefficient. Note that it is supposed here that the bubble or drop dimensions are much smaller than the cell dimensions [24]. The pressure inside the bubble or drop $\delta P_{In}(i\hat{\omega})$ is considered here as uniform. The inertia of the phase B inside the bubble or drop can be neglected as compared to that inside the capillary because the capillaries typically used are much longer than their radius. The viscous contribution in the bulk phase A is usually small when the viscosity η_A is not extremely large [24]. The correction coefficients χ'_{IA}, χ'_{IB} and χ_2 are necessary because of the deviations from an ideal spherical meniscus deformation [11, 24].

The surface tension variation in Eq. (12) can be expressed with the help of the complex dilatational modulus $\varepsilon(i\hat{\omega})$ [9-11]

$$\delta\gamma(i\hat{\omega}) = \varepsilon(i\hat{\omega})\delta[\ln S(i\hat{\omega})] \qquad (13)$$

where $\delta[\mathit{ln}\, S(i\hat{\omega})]$ is the variation of the relative interfacial area.

The volume flow through the capillary is characterised by a hydrodynamic relaxation time $t_h = a_C^2/\nu$, where aC is the capillary radius and $\nu = \eta B/\rho B$ is the kinematic viscosity of gas or liquid in the capillary. When the characteristic time of the pressure variation is smaller than t_h then the mobility of the flow inside the capillary should be taken into account. In this case the velocity distribution over the capillary cross-section is not parabolic and the resistance of the capillary is not described by the Poiseuille law [25-28]. Hydrodynamic relaxation influences the volume flow through the capillary and as a consequence the meniscus volume variation. If phase B is a gas and the capillary is long enough then the gas compressibility can also influence the flow through the capillary. Because of compressibility the flow through the entrance of the capillary can differ from that through the opposite end. The pressure and velocity distributions along the capillary can be described in terms of the direct waves and those reflected from the meniscus [26]. The volume flow at the capillary outlet (i.e. the inflow to the bubble or drop) can be obtained as

$$\delta V_{Cap}(i\hat{\omega}) = \delta V_m(i\hat{\omega}) = -\frac{\pi a_C^2 \cosh\beta l}{B_B \beta \sinh\beta l}\left[\delta P_{In}(i\hat{\omega}) - \frac{\delta P_{Ext}(i\hat{\omega})}{\cosh\beta l}\right] \qquad (14)$$

where

$$\beta^2 = -\frac{\rho_B \hat{\omega}^2}{B_B} \frac{I_0\left(a_C\sqrt{\frac{i\hat{\omega}}{\nu}}\right)}{I_2\left(a_C\sqrt{\frac{i\hat{\omega}}{\nu}}\right)} \tag{15}$$

$I_0(x)$ and $I_2(x)$ are modified Bessel functions of zero and second order, B_B and ρ_B are the bulk elasticity modulus and the density of phase B, l is the capillary length ($l \gg a_C$). Substituting Eqs. (12) and (14) into (8) one can obtain the general expressions for the bubble (drop) volume variation $\delta V_m(i\hat{\omega})$, the pressure variations inside the bubble (drop) $\delta P_{In}(i\hat{\omega})$, and inside the cell $\delta P_A(i\hat{\omega})$ at the given δP_{Ext} and δV_{Drv}. The effect of both the flow mobility and the compressibility of the medium in the capillary on the oscillating bubble or drop measurements is analysed in details in [26]. Here only the simpler case of quasi-stationary flow of incompressible media in the capillary is described.

When the bulk elasticity modulus B_B is large and the capillary is not very long and narrow, the incompressible medium approximation can be used. It is valid when the pressure variations are not very fast, and the corresponding frequency limit is obtained [26]

$$f \ll \frac{B_B a_C^2}{16\pi\nu\rho_B l^2} \tag{16}$$

This yields the frequencies $f \ll 10^4\,\text{Hz}$ for air and $f \ll 4\cdot 10^6\,\text{Hz}$ for water inside a capillary of typical radius $a_C = 0.01$ cm and length $l = 1$ cm. Usually pressure variations are not so fast and this approximation can be used with good accuracy. In this case the condition $|\beta l| \ll 1$ is valid and Eq. (14) can be simplified to

$$\delta P_{In}(i\hat{\omega}) \approx \delta P_{Ext}(i\hat{\omega}) - \frac{B_B l \beta^2}{\pi a_C^2} \delta V_m(i\hat{\omega}) \tag{17}$$

Combining this equation with Eqs. (9), (12) and (13) one obtains the resistance of the capillary with the attached meniscus

$$R_C(i\hat{\omega}) = \frac{\delta P_{Ext}(i\hat{\omega}) - \delta P_A(i\hat{\omega})}{\delta V_m(i\hat{\omega})} =$$
$$\frac{B_B l \beta^2}{\pi a_C^2} + \frac{2\varepsilon(i\hat{\omega})}{a_0}\frac{d\ln S_0}{dV_m} - \frac{2\gamma_0}{a_0^2}\frac{da_0}{dV_m} + \left(i\hat{\omega}\frac{\chi'_{1A}\eta_A + \chi'_{1B}\eta_B}{a_C} - \rho_A a_C \chi_2 \hat{\omega}^2 \right)\frac{dh_0}{dV_m} \qquad (18)$$

where S_0 and h_0 are the interfacial area, and meniscus height at equilibrium. The coefficients $\frac{d\,ln S_0}{dV_m}$, $\frac{da_0}{dV_m}$, and $\frac{dh_0}{dV_m}$ can be obtained from the meniscus geometry. In the limits of a linear approximation they are constant.

When the characteristic time of the pressure change is larger than the hydrodynamic relaxation time (or the oscillation frequency is small $f \ll (t_h)^{-1}$) the quasi-stationary approximation holds

$$\beta^2 \approx \frac{8\nu\rho_B}{B_B a_C^2} i\hat{\omega} - \frac{4\rho_B \hat{\omega}^2}{3B_B}. \qquad (19)$$

This approximation is valid for flow regimes where the velocity distribution over the capillary cross-section is parabolic. It corresponds to frequencies $f \ll 1500$ Hz for air and $f \ll 100$ Hz for water inside the capillary of radius $a_C = 0.01$ cm. With this approximation the expression for the resistance of the capillary with attached meniscus can be simplified to

$$R_C(i\hat{\omega}) = \frac{8\eta_B l}{\pi a_C^4} i\hat{\omega} - \frac{4\rho_B l \hat{\omega}^2}{3\pi a_C^2} + \frac{2\varepsilon(i\hat{\omega})}{a_0}\frac{d\ln S_0}{dV_m} - \frac{2\gamma_0}{a_0^2}\frac{da_0}{dV_m} + \left(i\hat{\omega}\frac{\chi'_{1A}\eta_A + \chi'_{1B}\eta_B}{a_C} - \rho_A a_C \chi_2 \hat{\omega}^2 \right)\frac{dh_0}{dV_m}. \qquad (20)$$

The first term on the right-hand side is the Poiseuille resistance of the capillary and the second term is the inertial term which includes the first correction to the Poiseuille resistance.

4. TRANSITIONAL OSCILLATIONS

With account for Eqs. (10), (11) and (20) one obtains the equations for the meniscus volume and pressure variation in the cell which are valid for the case of quasi-stationary flow of an incompressible medium in the capillary

$$\delta V_m(i\hat{\omega}) = \frac{\frac{B_{ef}}{V_A}\delta V_{Drv}(i\hat{\omega}) + \delta P_{Ext}(i\hat{\omega})}{G_2\left[\omega_0^2 - \hat{\omega}^2 + 2\lambda \cdot i\hat{\omega} + \kappa\varepsilon(i\hat{\omega})\right]} \qquad (21)$$

$$\delta P_A(i\hat{\omega}) = \frac{B_{ef}}{V_A} \cdot \frac{\delta P_{Ext}(i\hat{\omega}) - G_2\left[\omega_m^2 - \hat{\omega}^2 + 2\lambda \cdot i\hat{\omega} + \kappa\varepsilon(i\hat{\omega})\right] \cdot \delta V_{Drv}(i\hat{\omega})}{G_2\left[\omega_0^2 - \hat{\omega}^2 + 2\lambda \cdot i\hat{\omega} + \kappa\varepsilon(i\hat{\omega})\right]} \tag{22}$$

where $G_2 = \frac{4\rho_B l}{3\pi a_C^2} + \rho_A a_C \chi_2 \frac{dh_0}{dV_m}$ is the inertial coefficient, ω_0 is the characteristic frequency,

$$\omega_0^2 = \frac{1}{G_2}\left(\frac{B_{ef}}{V_A} - \frac{2\gamma_0}{a_0^2}\frac{da_0}{dV_m}\right) \tag{23}$$

ω_m is the meniscus frequency in the open cell,

$$\omega_m^2 = -\frac{2\gamma_0}{G_2 a_0^2}\frac{da_0}{dV_m} \tag{24}$$

λ is the damping coefficient,

$$\lambda = \frac{1}{2G_2}\left(\frac{8\eta_B l}{\pi a_C^4} + \frac{\chi'_{1A}\eta_A + \chi'_{1B}\eta_B}{a_C}\frac{dh_0}{dV_m}\right) \tag{25}$$

and $\kappa = \frac{2}{G_2 a_0}\frac{d\ln S_0}{dV_m}$.

When the externally applied signals $\delta V_{Drv}(t)$ and $\delta P_{Ext}(t)$, and the function $\varepsilon(i\hat{\omega})$ characterizing the surface properties, are given, then the time dependencies for the meniscus volume $\delta V_m(t)$ and the pressure in the cell $\delta P_A(t)$ can be obtained via the inverse transformation of Eqs. (21) and (22). This is the so-called direct problem [11]. In oscillating bubble or drop experiments we have usually to solve the reverse problem: the function $\varepsilon(i\hat{\omega})$ characterizing the surface properties is unknown and has to be found from experimental data. To solve this problem one can use the approach proposed in [12]. By applying the Laplace-Fourier transform to the measured time dependencies $\delta V_m(t)$ and $\delta P_A(t)$ and to the applied signals $\delta V_{Drv}(t)$ and $\delta P_{Ext}(t)$, the corresponding functions $\delta V_m(i\hat{\omega})$, $\delta P_A(i\hat{\omega})$, $\delta V_{Drv}(i\hat{\omega})$ and $\delta P_{Ext}(i\hat{\omega})$ are found in the frequency domain. Substituting these functions into Eqs. (21) or (22) the unknown function $\varepsilon(i\hat{\omega})$ is obtained. However, to understand the system behavior at any condition the corresponding direct problems have to be solved, which requires a preliminary idea about the function $\varepsilon(i\hat{\omega})$. The simplest case is associated with a clean surface in contact with a fluid free of surfactant, which corresponds to $\varepsilon(i\hat{\omega}) = 0$ in Eq. (21)

$$\delta V_m(i\hat{\omega}) = \frac{\frac{B_{ef}}{V_A}\delta V_{Drv}(i\hat{\omega}) + \delta P_{Ext}(i\hat{\omega})}{G_2\left[\omega_0^2 - \hat{\omega}^2 + 2\lambda \cdot i\hat{\omega}\right]}. \qquad (26)$$

When for $t > 0$ the harmonic pressure oscillation $\delta P_{Ext}(t) = \delta P_0 \sin(\omega t)$ is applied at $\delta V_{Drv}(t) = 0$ (Fig. 2) the inverse transform of Eq. (26) for condition $\lambda < \omega_0$ [11] is

$$\delta V_m(t) = \frac{\delta P_0}{G_2\left[\left(\omega_0^2 - \omega^2\right)^2 + 4\lambda^2\omega^2\right]^{1/2}}\left[\sin(\omega t + \psi) - \frac{\omega}{\sqrt{\omega_0^2 - \lambda^2}} e^{-\lambda t} \sin\left(t\sqrt{\omega_0^2 - \lambda^2} + \theta\right)\right] \qquad (27)$$

where

$$\psi = \arctan\frac{2\lambda\omega}{\omega^2 - \omega_0^2}, \qquad (0 \geq \Psi \geq -\pi) \qquad (28)$$

$$\theta = \arctan\frac{2\lambda\sqrt{\omega_0^2 - \lambda^2}}{\omega^2 - \omega_0^2 + 2\lambda^2}, \qquad (0 > \theta \geq -\pi). \qquad (29)$$

δP_0 must be replaced by $\frac{B_{ef}}{V_A}\delta V_0$ when the volume oscillation $\delta V_{Drv}(t) = \delta V_0 \sin(\omega t)$ is applied at $\delta P_{Ext}(t) = 0$ (Fig. 3). In the beginning the measured signal is a superposition of two oscillations: one with the externally applied frequency ω and a second with the frequency $\sqrt{\omega_0^2 - \lambda^2}$. The second oscillation is damped with the characteristic time $1/\lambda$, i.e. the initial transitional regime of meniscus oscillation. At times much larger than $1/\lambda$ the meniscus oscillation is established, which is characterised by a constant phase angle ψ and constant amplitude, the factor before the square brackets in Eq. (27). The amplitude and the phase angle are frequency dependent. The amplitude has a maximum at the frequency $\omega = \left(\omega_0^2 - 2\lambda^2\right)^{1/2}$ which is the resonance frequency of the system.

When the meniscus oscillations are excited by external pressure oscillations ($\delta V_{Drv}(t) = 0$, Fig. 2) then according to Eqs. (21) and (22), the pressure oscillations in the cell are described by the same Eq. (27) with the additional factor B_{ef}/V_A on the right hand side. The pressure in the cell oscillates synchronously with the meniscus oscillation. When the meniscus oscillations are excited by volume oscillations in the cell ($\delta P_{Ext}(t) = 0$, Fig. 3) then the inverse transform of Eq. (22) for $\varepsilon(i\hat{\omega}) = 0$ and $\delta V_{Drv}(t) = \delta V_0\, sin(\omega t)$ gives

$$\delta P_A(t) = \delta P_{A0}\left[\sin(\omega t + \varphi) - \frac{\omega(\omega_0^2 - \omega_m^2)}{\sqrt{\omega_0^2 - \lambda^2}\left[(\omega_m^2 - \omega^2)^2 + 4\lambda^2\omega^2\right]^{1/2}} e^{-\lambda t}\sin\left(t\sqrt{\omega_0^2 - \lambda^2} + \theta\right)\right] \quad (30)$$

where

$$\delta P_{A0} = \frac{B_{ef}\left[(\omega_m^2 - \omega^2)^2 + 4\lambda^2\omega^2\right]^{1/2}}{V_A\left[(\omega_0^2 - \omega^2)^2 + 4\lambda^2\omega^2\right]^{1/2}}\delta V_0 \quad (31)$$

and

$$\varphi = arctan\frac{2\lambda\omega(\omega_0^2 - \omega_m^2)}{(\omega_0^2 - \omega^2)(\omega_m^2 - \omega^2) + 4\lambda^2\omega^2}, \qquad (0 \geq \varphi \geq -\pi) \quad (32)$$

are the amplitude and the phase angle of the established oscillation, respectively. In this case, the pressure in the cell results from the superposition of the meniscus volume variation and the applied volume variation, and therefore, it oscillates with a phase different from that of the meniscus.

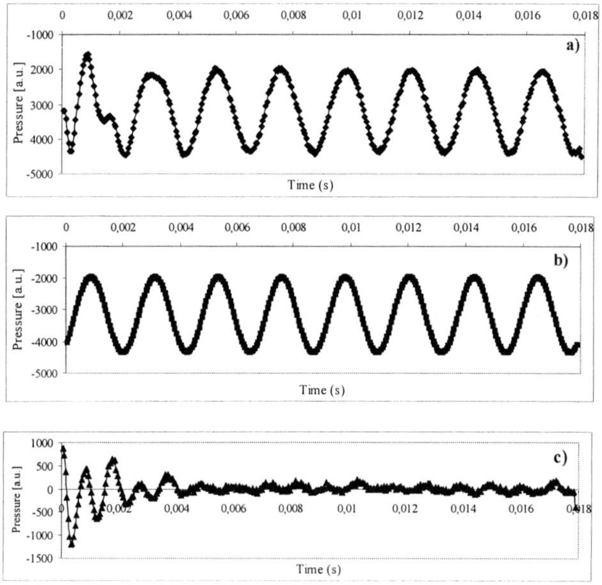

Fig. 6. The measured pressure oscillation in the closed cell (pure water/air at 450 Hz) (a) and its decomposition into established oscillation with the externally applied frequency (b) and damped oscillation with the bubble eigenfrequency (c).

The pressure oscillation (Eq. (30)) is also characterised by the initial transient regime and the established oscillation at $t \gg 1/\lambda$. These regimes are experimentally observed. Typical experimental data are shown in Fig. 6. The measured signal can be split into two oscillations: one with a constant amplitude and another damped with time.

5. STABILITY OF THE MENISCUS ATTACHED TO A CAPILLARY

The regimes of transient and established oscillations are observed for pure liquids as well as for surfactant solutions. For surfactant solutions the characteristic frequencies and the attenuation in the system depend on the relaxation processes in the adsorption layer and the system behaviour becomes more complicated. Many surfactants are characterised by a diffusion mechanism of the surface relaxation, and the complex dilatational modulus is given by [11]

$$\varepsilon(i\hat{\omega}) = \varepsilon_0 \frac{\xi\sqrt{i\hat{\omega}}}{1+\xi\sqrt{i\hat{\omega}}}. \tag{33}$$

Here $\varepsilon_0 = -d\gamma/d\ln\Gamma$ is the Gibbs elasticity modulus, $\xi = (d\Gamma/dc)/\sqrt{D}$, Γ is the adsorption, c is the bulk surfactant concentration, and D is the diffusion coefficient. Substituting $\varepsilon(i\hat{\omega})$ into Eq. (21) yields

$$\delta V_m(i\hat{\omega}) = \frac{\frac{B_{ef}}{V_A}\delta V_{Drv}(i\hat{\omega}) + \delta P_{Ext}(i\hat{\omega})}{G_2\left[\omega_0^2 - \hat{\omega}^2 + 2\lambda\cdot i\hat{\omega} + \kappa\varepsilon_0\frac{\xi\sqrt{i\hat{\omega}}}{1+\xi\sqrt{i\hat{\omega}}}\right]}. \tag{34}$$

For harmonic external pressure variations $\delta P_{Ext}(t) = \delta P_0 \sin(\omega t)$ at $\delta V_{Drv}(t) = 0$, and the inverse transform of Eq. (34) can be obtained [11]

$$\delta V_m(t) = A_0 \sum_{n=1}^{9} A_n\left[s_n e^{s_n^2 t}\left(1+\mathrm{erf}\left(s_n\sqrt{t}\right)\right) + \frac{1}{\sqrt{\pi t}}\right], \tag{35}$$

where $A_0 = \delta P_0/G_2$; erf(x) is the error function. The s_n are the roots of the equations

$$\xi s^5 + s^4 + 2\lambda\xi s^3 + 2\lambda s^2 + \left(\omega_0^2 + \kappa\varepsilon_0\right)\xi s + \omega_0^2 = 0 \tag{36}$$

$$s^4 + \omega^2 = 0 \tag{37}$$

and the coefficients A_n are given by the relations

$$A_n = \frac{\omega(1+\xi s_n)}{(s_n^4+\omega^2)(5\xi s_n^4+4s_n^3+6\lambda\xi s_n^2+4\lambda s_n+(\omega_0^2+\kappa\varepsilon_0)\xi)} \qquad (n = 1, \ldots, 5) \qquad (38)$$

$$A_n = \frac{\omega(1+\xi s_n)}{4s_n^3(\xi s_n^5+s_n^4+2\lambda\xi s_n^3+2\lambda s_n^2+(\omega_0^2+\kappa\varepsilon_0)\xi s_n+\omega_0^2)} \qquad (n = 6, \ldots, 9) \qquad (39)$$

When the volume oscillation $\delta V_{Drv}(t) = \delta V_0 \, sin(\omega t)$ is applied at $\delta P_{Ext}(t) = 0$ then $A_0 = \frac{B_{ef}\delta V_0}{V_A G_2}$. Eq. (35) describes both the transition regime and the regime of established oscillations for the diffusion controlled adsorption. The roots of the characteristic equation s_n and the coefficients A_n depend on the Gibbs elasticity modulus ε_0 and the parameter ξ which contains the parameters of the surfactant adsorption isotherm. The characteristic equation (36) has one real root and four complex roots. The analysis shows [11]: if ω_0^2 is positive the sum of the corresponding five terms in Eq. (35) always yields a decreasing (oscillating or aperiodic) contribution. The system is stable under these conditions. For pure liquids without surfactants this sum transforms into the second (decreasing) term in Eq. (27). If ω_0^2 is negative the real root of Eq. (36) becomes positive and the corresponding term in Eq. (35) increases with time to infinity. Hence, the system becomes unstable for $\omega_0^2 < 0$.

The sum of the terms corresponding to the four roots of Eq. (37) gives the non-damped contribution. At $t \to \infty$ this contribution transforms into [11]

$$\delta V_m^{est}(t) = \frac{A_0}{\sqrt{(\omega^2-\omega_0^2-\kappa\varepsilon_s(\omega))^2+\omega^2(2\lambda+\kappa\eta_s(\omega))^2}}\sin(\omega t+\Psi) \qquad (40)$$

$$tg\Psi = \frac{\omega(2\lambda+\kappa\eta_s(\omega))}{\omega^2-\omega_0^2-\kappa\varepsilon_s(\omega)}, \qquad (0 \ge \Psi \ge -\pi) \qquad (41)$$

where $\varepsilon_s(\omega)$ and $\eta_s(\omega)$ are the surface dilatational elasticity and viscosity, respectively, which correspond to the real and imaginary parts of the complex dilatational modulus $\varepsilon(i\omega) = \varepsilon_s(\omega) + i\omega\eta_s(\omega)$ [1, 2]. According to Eq.(33) we have

$$\varepsilon_s(\omega) = \varepsilon_0 \frac{1 + \dfrac{1}{\xi\sqrt{2\omega}}}{1 + \dfrac{\sqrt{2}}{\xi\sqrt{\omega}} + \dfrac{1}{\xi^2\omega}}, \qquad \eta_s(\omega) = \frac{\varepsilon_0}{\omega} \frac{\dfrac{1}{\xi\sqrt{2\omega}}}{1 + \dfrac{\sqrt{2}}{\xi\sqrt{\omega}} + \dfrac{1}{\xi^2\omega}} \tag{42}$$

For pure liquids, i.e. $\varepsilon_s(\omega) = 0$ and $\eta_s(\omega) = 0$, and Eq. (40) gives the first (non-damping) term of Eq. (27), and Eq. (41) transforms into Eq. (28). In the presence of a surfactant the elastic contribution of the surface is additive with other elastic contributions and therefore increases the resonance frequency. The viscous contribution of the surface increases the damping in the system.

The bubble or drop attached to a capillary inside a closed cell is stable when ω_0^2 is positive. This value depends on the meniscus shape at equilibrium conditions, the capillary geometry, the volume and effective elasticity modulus of the cell, and the properties of the contacting media, cf. Eq. (23). For capillaries narrow enough (i.e. with small Bond number $Bo = \Delta\rho g a_C^2 / \gamma_0$, where $\Delta\rho$ is the density difference, g the acceleration due to gravity) the influence of the gravity can be neglected and the meniscus shape can be approximated by a sphere. If the contact line of the meniscus is fixed at the edge of the capillary outlet then the meniscus volume and the radius of curvature are dependent on the meniscus height h_0,

$$V_m = \frac{\pi}{6} h_0 \left(h_0^2 + 3a_C^2\right); \quad a_0 = \frac{a_C}{2}\left(\frac{h_0}{a_C} + \frac{a_C}{h_0}\right). \tag{43}$$

Hence one obtains from Eq. (23)

$$\omega_0^2 = \frac{B_{ef} a_C^2}{V_A G_2'} \left[\frac{\pi}{2}\left(\tilde{h}^2 + 1\right) + A \frac{1 - \tilde{h}^2}{\left(1 + \tilde{h}^2\right)^2}\right] \tag{44}$$

where $A = \dfrac{4\gamma_0 V_A}{B_{ef} a_C^4}$ is a dimensionless parameter, $\tilde{h} = h_0 / a_C$ is the dimensionless meniscus height, and $G_2' = G_2 \dfrac{dV_m}{dh_0} = \dfrac{4\rho_B l}{3\pi a_C^2} \dfrac{dV_m}{dh_0} + \rho_A a_C \chi_2$. From Eq. (44) we see that the system behaviour is determined by the parameter A. If this parameter is smaller than the critical value $A_{cr} = 27\pi/2$ then ω_0^2 is always positive, and the meniscus is stable at any values of h_0. At

$A = A_{cr}$ the value ω_0^2 turns to zero for $h_0 = \sqrt{2}a_C$ and is positive for all other h_0. At $A > A_{cr}$ we have $\omega_0^2 > 0$ for $h_0 < h_l$ and $h_0 > h_2$, and $\omega_0^2 < 0$ for the interval $h_l < h_0 < h_2$. Here h_l and h_2 depend on A and satisfy the conditions $a_C < h_l < \sqrt{2}a_C$ and $h_2 > \sqrt{2}a_C$, i.e. the meniscus becomes unstable at $h_l < h_0 < h_2$. Stable oscillations are possible only for the meniscus positions $h_0 < h_l$ and $h_0 > h_2$. In open or infinitely large cells $A \to \infty$, and we have $h_l \to a_C$ and $h_2 \to \infty$. Hence, the meniscus is stable only for $h_0 < a_C$. For an open cell we have $\omega_0 = \omega_m$ (cf. Eq. (24)). Here, the parameter A has the same physical sense as the bubble stability number (BSN) introduced elsewhere in the theory of expanding drops [21-23].

The dependencies of the characteristic frequency ω_0 on the equilibrium meniscus height are shown in Fig. 7 for the conditions of an oscillating bubble. In this case the approximation $G_2' \approx \rho_A a_C \chi_2$ is valid (cf. section 7).

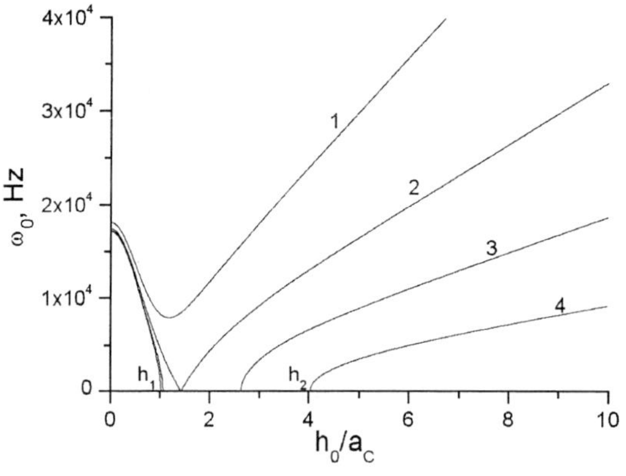

Fig. 7. Characteristic frequency of oscillating bubble in the closed cell ω_0 vs. equilibrium meniscus height h_0 at $V_A = 10$ cm^3 (curve 1), 30 cm^3 (curve 2), 100 cm^3 (curve 3), 400 cm^3 (curve 4); $B_{ef} = 2.2 \cdot 10^9$ Pa; $\gamma_0 = 73$ mN/m; $a_C = 0{,}01$ cm; $\rho_A = 1$ g/cm^3; $\chi_2 = 1$; gas inertia is neglected.

The characteristic frequency can be varied within many orders of magnitude depending on the equilibrium meniscus position. The meniscus position is adjusted by the initial setting of the bubble or drop, by cell volume ΔV_{Drv} or external pressure changes ΔP_{Ext}. For static conditions ($\omega = 0$) it can be obtained by integration of Eq. (21) with account for Eq. (44),

$$\Delta V_{Adj} = a_C^3\left[\frac{\pi}{6}\tilde{h}\left(\tilde{h}^2+3\right)+A\frac{\tilde{h}}{1+\tilde{h}^2}\right]. \tag{45}$$

ΔV_{Adj} is either ΔV_{Drv} or $\frac{V_A}{B_{ef}}\Delta P_{Ext}$; $\Delta V_{Drv} = V_{Cell}(h_0) - V_{Cell}(0)$ is the change of the volume of the measuring cell when the meniscus height grows from 0 to h_0 (usually produced by a separate syringe, cf. Fig. 5), and $\Delta P_{Ext} = P_{Ext}(h_0) - P_{Ext}(0)$ is the change of the external pressure at the same change of the meniscus height. We ignore here a small dispersion of B_{ef} caused by the difference between the isothermic and adiabatic liquid compressibility.

If $A < A_{cr}$ then Eq. (45) has one solution h_0 for any given ΔV_{Adj}. At $A > A_{cr}$ there is an interval of ΔV_{Adj} values where Eq. (45) has three solutions, i.e. the meniscus can take one of three positions at the same fixed external conditions. The analysis shows that one of the three positions is unstable and belongs to the interval $h_1 < h_{02} < h_2$ whereas the two others are stable and satisfy the conditions $h_{01} < h_1$ or $h_{03} > h_2$ [24]. This leads to a phenomenon observed experimentally: approaching position h_1 the meniscus suddenly jumps into position h_{03} with the volume increase ΔV_{Adj} or in turn into position h_{01} when it approaches the position h_2 with the volume decrease ΔV_{Adj}. This phenomenon is specific for a closed cell and explained by the finite compressibility of the liquid or a possible elastic cell deformation.

6. ESTABLISHED OSCILLATIONS

After the period of transitional oscillations an established oscillation remains. For established sinusoidal oscillations the amplitude- and phase-frequency characteristics can be obtained simply by substitution of the external frequency ω instead of the variable $\hat{\omega}$. It follows from Eq. (21) that the amplitude- and phase-frequency characteristics of the bubble or drop volume oscillations are described by Eqs. (40) and (41) where the constant A_0 is either $\delta P_0 / G_2$ or $\frac{B_{ef}\delta V_0}{V_A G_2}$ [24]. However, in the general case the functions $\varepsilon_s(\omega)$ and $\eta_s(\omega)$ are no longer given by Eq. (42) but should be determined from the surface relaxation mechanism of the specific surfactant system, which can deviate from the diffusion control. If one deals with the reverse problem the functions $\varepsilon_s(\omega)$ and $\eta_s(\omega)$ can be determined from the measured signal via Eqs. (40) and (41).

The amplitude of the established oscillation depends on the applied frequency with a maximum at the resonance frequency. The surface elasticity increases the effective elasticity of the system, and therefore, the presence of a surfactant at the surface increases the resonance frequency. The surface viscosity contributes to the total energy dissipation and increases the damping in the system. This leads to a decrease of the resonance maximum of the amplitude and to an increase of the phase shift Ψ. On the other hand, increase in the elasticity decreases the phase shift. Thus, the presence of a surfactant in the solution can lead to an increased as well as a decreased phase shift depending on the applied frequency. The phase shift Ψ varies from zero at small frequencies to $-\pi$ at large frequencies and passes through $-\pi/2$ at $\omega^2 - \omega_0^2 - \kappa\varepsilon_s(\omega) = 0$, which corresponds approximately to the resonance frequency.

It follows from Eq. (22) that the amplitude- and phase-frequency characteristics of the pressure oscillations inside the cell are also described by Eqs. (40) and (41) (with $A_0 = \frac{B_{ef}\delta P_0}{V_A G_2}$), if the excitation is produced by external pressure variations ($\delta V_{Drv} = 0$, Fig. 2) [24]. When the oscillations are excited by volume changes in the cell ($\delta P_{Ext} = 0$, Fig. 3) the amplitude- and phase-frequency characteristics of the pressure oscillations inside the cell are given by [29]

$$\delta P_{A0} = \delta V_0 \frac{B_{ef}}{V_A} \sqrt{\frac{\left(\omega_m^2 - \omega^2 + \kappa\varepsilon_s(\omega)\right)^2 + \omega^2\left(2\lambda + \kappa\eta_s(\omega)\right)^2}{\left(\omega_0^2 - \omega^2 + \kappa\varepsilon_s(\omega)\right)^2 + \omega^2\left(2\lambda + \kappa\eta_s(\omega)\right)^2}} \qquad (46)$$

$$\varphi = \arctan \frac{\omega\left(2\lambda + \kappa\eta_s(\omega)\right)\left(\omega_0^2 - \omega_m^2\right)}{\left(\omega_0^2 - \omega^2 + \kappa\varepsilon_s(\omega)\right)\left(\omega_m^2 - \omega^2 + \kappa\varepsilon_s(\omega)\right) + \omega^2\left(2\lambda + \kappa\eta_s(\omega)\right)^2}, \quad (0 \geq \varphi \geq -\pi) \qquad (47)$$

A comparison of Eqs. (40)-(41) and (46)-(47) shows that the law of the pressure variation inside the cell is very different when the oscillations are generated either by external pressure variations δP_{Ext} or by volume variations in the cell δV_{Drv}. In the first case, the law of the pressure variation inside the cell coincides with that of meniscus oscillations. In the second case, the pressure variation is a superposition of the applied volume variation δV_{Drv} and the meniscus volume variation δV_m which are not in phase. Therefore, the amplitude- and phase-frequency characteristics are different for pressure oscillations and meniscus oscillations.

When the volume of the liquid in the cell is sufficiently small and the cell walls are rigid then the meniscus oscillates with a small phase difference relative to the applied volume variation because of the small liquid compressibility. This case corresponds to a very large characteristic frequency ω_0. It can be supposed that in such cases the pressure in the cell varies with the same phase as for a meniscus oscillating in an open cell

$$\text{tg}\varphi \approx \frac{\omega(2\lambda + \kappa\eta_s(\omega))}{\omega_m^2 - \omega^2 + \kappa\varepsilon_s(\omega)} \tag{48}$$

However, this approximation is only valid under the condition

$$\frac{\omega^2}{\omega_0^2}(2\lambda + \kappa\eta_s(\omega))^2 << \omega_m^2 - \omega^2 + \kappa\varepsilon_s(\omega) \tag{49}$$

which is satisfied for large enough ω_m^2. In spite of large ω_0^2, the value of ω_m^2 can be very small when the meniscus shape is close to a hemisphere ($h_0 \cong a_C$), or even negative when the meniscus is larger than a hemisphere ($h_0 > a_C$). The meniscus radius of curvature has a minimum equal to the capillary radius for a hemisphere. Therefore, the derivative da_0/dV_m (cf. Eq. (24)) changes sign when the meniscus passes the hemispherical shape. In an open cell menisci larger than a hemisphere are unstable. In a closed cell, however, such menisci can be stable because its stability is determined by ω_0^2 and not by ω_m^2 (i.e. because of the bulk elasticity contribution). For small or negative ω_m^2 the approximation (48) is not valid. It is only valid for menisci much smaller than a hemisphere, when ω_m^2 is positive and large.

According to Eq. (47) the upper frequency limit of the phase angle of the pressure oscillations in the cell is always $-\pi$. That means, that at large frequencies the pressure in the cell always decreases when the cell volume increases. The lower frequency limit depends on ω_m^2. It is zero for $\omega_m^2 < 0$, $-\pi$ for $\omega_m^2 > 0$, and $-\pi/2$ for $\omega_m^2 = 0$. This result has a clear physical meaning. At small frequencies the meniscus volume changes simultaneously with the applied volume change. Increase of the cell volume and correspondingly of the meniscus volume leads to an increase in the radius of curvature for $\omega_m^2 < 0$, and to a decrease for $\omega_m^2 > 0$ (cf. Eq. (24)). In the first case, the capillary pressure decreases and the pressure in the cell increases (because the pressure outside the cell is constant). Thus, the pressure in the cell varies with the same phase as the volume of the cell. In the second case, the capillary pressure increases and the pressure in the cell decreases, i.e. it changes with a phase opposite to the volume of the cell. At frequencies where the denominator of Eq. (47) turns to zero, the phase angle φ is $-\pi/2$.

An example of the amplitude- and phase-frequency characteristics of pressure oscillations inside a closed cell with excitation in the cell are presented in Fig. 8. The characteristics are obtained for parameters of the Langmuir isotherm and the diffusion coefficient characteristic for the model surfactant dimethyl dodecyl phosphine oxide ($\Gamma_\infty = 4.39 \cdot 10^{-10}$ mol/cm^2, $b = 1.7 \cdot 10^8$ cm^3/mol, $D = 4 \cdot 10^{-6}$ cm^2/s, [30]). The parameter ω_m^2 is here assumed to be negative.

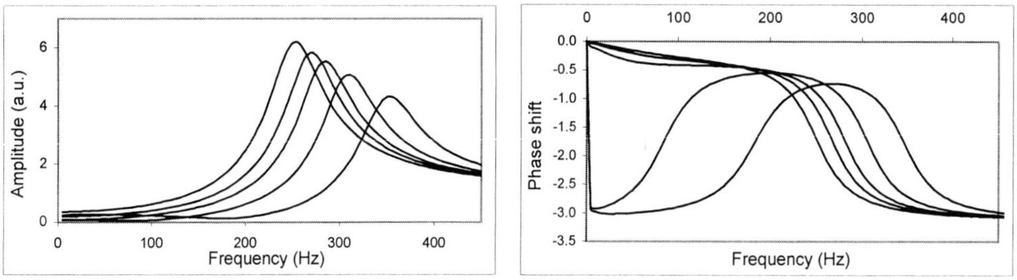

Fig.8. Amplitude- and phase-frequency characteristics of the pressure oscillations inside the closed cell according Eqs. (46)-(47); C_{12}DMPO concentrations: c = 0, 0.5, 1, 2 and 4 10^{-8} mol/cm^3; $\omega_0 = 1580$ s^{-1}; $\omega_m^2 = -9.03 \cdot 10^5$ s^{-2}; $\lambda = 176$ s^{-1}; $\kappa = 2.77 \cdot 10^7$ kg^{-1}.

The lower frequency limit of the phase shift is zero at all surfactant concentrations. The resonance frequency increases with surfactant concentration caused by the surface elasticity increase, whereas the resonance maximum of the amplitude decreases due to the surface viscosity increase. At small surfactant concentrations the denominator of Eq. (47) changes its sign approximately at the resonance frequency, which corresponds to the phase shift $-\pi/2$ (the factor $\omega_0^2 - \omega^2 + \kappa\varepsilon_s(\omega)$ changes its sign). At large concentrations the phase shift additionally passes two times the value of $-\pi/2$ at frequencies smaller than the resonance frequency.

Initially, the factor $\omega_m^2 - \omega^2 + \kappa\varepsilon_s(\omega)$ changes the sign with increasing frequency from minus to plus because of the surface elasticity increase, and then it changes the sign again because the term ω^2 increases whereas the surface elasticity approaches a maximum. This results in a strong increase of the phase shift at small frequencies and large surfactant concentrations. The discussed phenomenon is explained by the particular features of the capillary pressure change when the meniscus is larger than a hemisphere. In this case, an increase of the meniscus volume corresponds to an increase in the radius of curvature (cf. Eq. (24)), leading to a capillary pressure decrease. On the other hand, the surface expansion increases the surface tension which produces a capillary pressure increase. At small concentrations the first effect dominated whereas at large concentrations the second effect is more significant which means a change of the phase of pressure oscillations. This phenomenon is also specific for closed cells.

7. PARAMETERS OF AN OSCILLATING BUBBLE OR DROP SYSTEM

It follows from the analysis presented above that the behaviour of an oscillating bubble or drop strongly depends on the experimental conditions. The experimental parameters can be divided into several groups. The first one are geometrical characteristics of the capillary and the meniscus. The capillary radius and the equilibrium meniscus height influence all coefficients of the dynamic equations. The capillary length influences the damping coefficient and the inertial coefficient. The next are the cell properties: the volume of the cell and the walls rigidity influence the characteristic frequency and amplitude of the oscillations. The properties of the surfactants (adsorption isotherm parameters, diffusion coefficient, kinetic coefficients) influence the surface dilatational elasticity and viscosity. The bulk properties of gas and liquid (density, viscosity, compressibility) influence the damping coefficient, the inertial coefficient and the characteristic frequency.

The inertial coefficient G_2 is dependent on whether gas or liquid is inside the capillary (phase B). When the capillary radius is larger than 0.1 mm, the length is smaller than 10 mm, the meniscus height is of the order of the capillary radius and the phase B is a gas. Then the gas inertia can be neglected as compared to the inertia of the added mass of the liquid near the meniscus, as the liquid density is much larger than that of gas and the coefficient $\pi a_C^2 \chi_2 \frac{dh_0}{dV_m}$ is of the order of unity. For longer and thinner capillaries the gas inertia contributes to the inertial coefficient. The added mass coefficient χ_2 is not known exactly but can be approximated by $\chi_2 \approx \frac{1}{2\sqrt{2}}\left(1+h_0^2/a_C^2\right)^{1/2}$. When phase B is a liquid its contribution to the inertial coefficient is much larger than that of the added mass of liquid A near the meniscus because the capillary length is much larger than its radius. Thus, the inertial coefficient is much larger for a liquid in the capillary (oscillating drop) than for a gas (oscillating bubble). As a result the characteristic frequency for oscillating bubbles is much larger than for oscillating drops at the same capillary geometry.

The characteristic frequency has a wide limit of variability depending on the experimental conditions. Therefore, it is not always achieved in the experiments (the usual frequency interval is between 0.1 – 600 Hz). For a bubble in an open cell the characteristic frequency decreases with increasing capillary radius according to $a_c^{-3/2}$, and is almost independent of the capillary length [11]. For a drop in an open cell the characteristic frequency decreases with increasing capillary radius and capillary length according to $\left(a_c^2 l\right)^{-1/2}$. In a closed cell, the characteristic frequency is larger than in an open cell at the same capillary geometry and meniscus height, and it increases with decreasing cell volume (Fig. 7).

The characteristic frequency increases also with the effective cell elasticity modulus B_{ef} which depends on both the elastic properties of the wall material and the cell geometry (cf. Eq. (6)). The modulus B_{ef} influences also the parameter A which determines the conditions of the meniscus stability (cf. section 5). In practice, the cell geometry is rather complicated and it is difficult to calculate this modulus. It is much easier to determine this cell characteristic by a calibration experiment for the given cell with the studied solution. It can be obtained by direct measurement of the pressure in the cell at infinitely slow meniscus volume changes when the system remains in equilibrium. The pressure change produced by the meniscus displacement is maximum in the case of absolutely rigid walls and decreases with the walls elasticity [24]. The same is true for the dependency of pressure on the bulk elasticity of the liquid. Note, we have finite pressure changes in the liquid even in case of incompressible liquids ($B_A \rightarrow \infty$) although the volume change of the liquid tends to zero. In this case, the volume change produced by the meniscus displacement is equal to the volume change resulting from the displacement of the inner walls.

The difference in the viscosities of liquid and gas is usually not so large as the difference in densities. Therefore, the damping coefficient (cf. Eq. (25)) is determined practically always by the medium inside the capillary (if the liquid A is not extremely viscous, e.g., glycerol), since the capillary length is much larger than its radius. The coefficients χ'_{1A} and χ'_{1B} are not known exactly but the second term in Eq. (25) is usually small and can be neglected. The damping coefficient decreases with the capillary radius. For oscillating bubbles it increases with the capillary length. For oscillating drops it is independent of the capillary length because the inertial coefficient is increasing with the capillary length. The damping coefficient is usually much smaller than the characteristic frequency ω_0.

The exact calculation of some coefficients discussed above can be difficult, for example when the meniscus is not exactly part of a sphere or the contact line of the meniscus is not fixed at the edge of the capillary tip. In such cases, special calibration experiments can be used to determine the unknown coefficients [13, 15]. The fixation of the contact line of the meniscus can be improved by a sharp edge and special wetting properties at the tip of the capillary [4].

8. Experimental results of the STS-95 mission and ground experiments

The main conclusions of the theory presented here were proved in a series of experiments with the "oscillating bubble module", called FAST (Facility for Adsorption and Surface Tension), during the STS-95 space shuttle mission, and under ground conditions. The scheme of the experimental set up was analogous to that presented in Figs. 4 and 5. As a model surfactant, $C_{12}DMPO$ was studied at five solution concentrations and three different temperatures. A capillary of 0.243 mm radius and 10 mm length was used. The meniscus height was 0.792 mm. The oscillations were generated by volume variation in a closed cell by a piezo-driver (Fig. 3). A sinusoidal signal of a small amplitude was applied with frequencies ranging from 0.2 to 450 Hz. The pressure variations were measured in the cell by a pressure transducer. The experimental details are given in [29]. A typical pressure signal is presented in Fig. 6. It includes both the initial regime of transitional oscillations, and the regime of established oscillations at large times. For the largest frequencies the transitional time lasts over the first two oscillation periods only. For smaller frequencies transitional oscillations were practically not observed because of transitional times much smaller than the oscillation period.

From the data at large times the amplitude and the phase shift of the established oscillation are obtained at a given frequency. In Fig. 9 the amplitude- and phase-frequency dependencies are shown for pure water and four $C_{12}DMPO$ concentrations under ground conditions. The data are in good agreement with the conclusions of the theory discussed above.

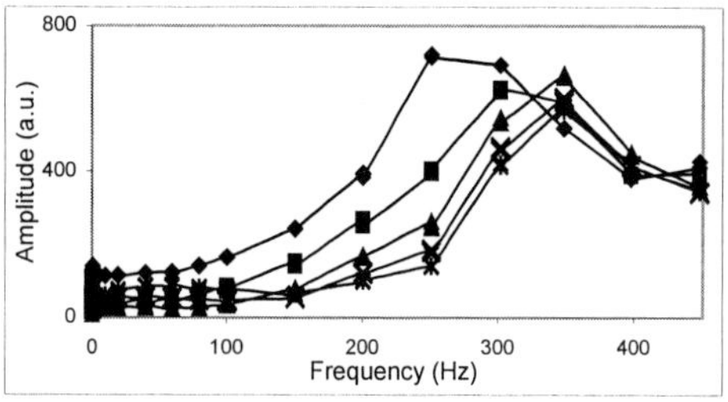

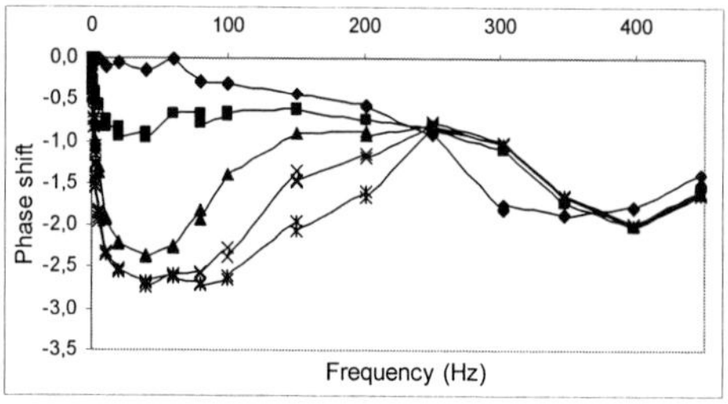

Fig.9. Amplitude and phase shift of the pressure variation in the cell vs. frequency in ground conditions. $C_{12}DMPO$ concentrations: 0 (♦), 0.5 (■), 1 (▲), 2 (×) and 4 (*) 10^{-8} mol/cm^3.

All data are obtained for the same meniscus size larger than a hemisphere. Accordingly, the value ω_m^2 is negative. Nevertheless the meniscus is stable because ω_0^2 is positive. The characteristic frequency $f_0 = \omega_0 / 2\pi$ is about 250 Hz for pure water and increases with concentration according to the increase in the surface elasticity. The amplitude has a maximum at approximately this frequency. This maximum decreases with concentration due to increasing surface viscosity. At small concentrations the phase shift increases with frequency and is $-\pi/2$ approximately at the resonance frequency. At large concentrations one can see the typical strong increase of the phase shift in the region of small frequencies (less than 200 Hz) because of the increased surface elasticity. The mechanism of this phenomenon was explained earlier (cf. section 6). Owing to this phenomenon, the phase shift of the pressure in the cell of an oscillating bubble set-up is a very sensitive characteristic to the presence of surfactants in the solution.

The amplitude- and phase frequency characteristics in Fig. 8 were calculated for the conditions of the experiment shown in Fig. 9. No fitting parameters were used, except the effective cell elasticity was chosen such that the characteristic frequency corresponds to the observed maximum of amplitude, because this parameter was not determined experimentally. A good qualitative agreement is obtained from the comparison of the experimental results with the theoretical calculations. Some quantitative differences may be explained by the uncertainty of the experimental parameters: the bubble was so large that significant deviations from the exact spherical shape are possible. Special calibration experiments should allow the improvement of the agreement. An obvious deviation from the theoretical predictions is observed only for large frequencies (above 400 Hz) because the phase shift does not tend to $-\pi$. Supposedly, this is connected with the frequency limitations of the oscillating bubble method. At large frequencies, the approximation of a quasi-stationary flow in the capillary and of incompressible gas are not valid (cf. section 3). For this case a more rigorous theory (see [26]) is required. Another limitation consists of the possibility of a non-synchronous deformation of the meniscus surface at large frequencies when different parts of the surface begin to oscillate with a large phase difference relative to each other. This effect can also give a large contribution to the measured phase shift of the pressure oscillations.

In Fig. 10 the results of the STS-95 space shuttle mission are presented obtained under the same experimental conditions (except of gravity).

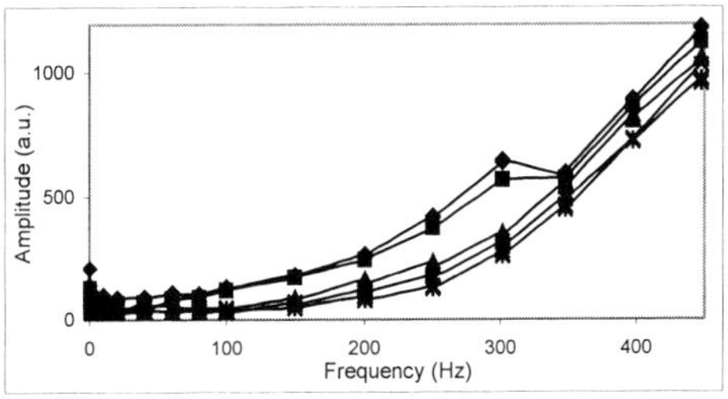

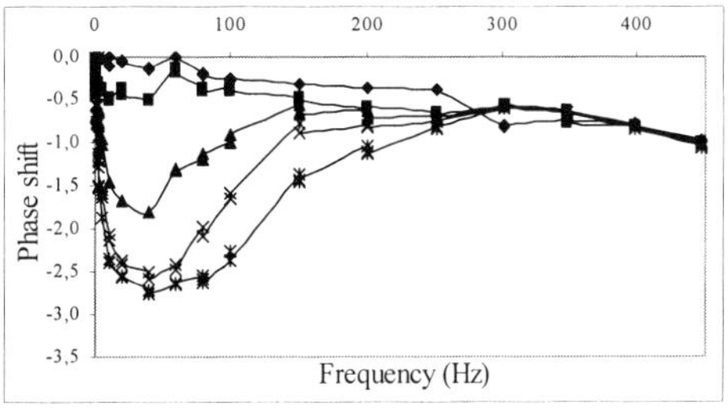

Fig.10. Amplitude and phase shift of the pressure variation in the cell vs. frequency under microgravity conditions. $C_{12}DMPO$ concentrations: 0 (♦), 0.5 (■), 1 (▲), 2 (×) and 4 (*) 10^{-8} mol/cm^3.

The main difference with the results in Fig. 9 is that the characteristic frequency is now much larger and appears out of the considered frequency range: 250 – 300 Hz under the ground conditions and 500 – 600 Hz under microgravity. This difference can be explained by a significant meniscus shape deformation on ground taking into account a strong dependency of the characteristic frequency on the meniscus shape. It is seen from Fig. 10 that the characteristic frequency obviously increases with addition of surfactant because of the surface elasticity increase. The same also concerns the phase shift which increases almost to $-\pi$ at large concentrations. Thus, the main regularities predicted by the theory are confirmed by space experiments.

9. CONCLUSIONS

The analysis of the dynamic behaviour of an oscillating bubble or drop provides the amplitude- and phase-frequency characteristics which take into account the dependency on the viscoelastic properties of the fluid-fluid interface. This dependency manifests itself through the frequency functions $\varepsilon_S(\omega)$ and $\eta_S(\omega)$, characterizing the relaxation processes in the interface. The corresponding experiment can be organized in different ways depending on the type of the measuring cell (open or closed), generation of bubble and drop oscillations (either by external pressure variations or volume variation in the cell), and the measured signal (the meniscus volume or the pressure in the cell). All experimental schemes can be considered in the framework of a common theory based on the application of Fourier transforms. The theory allows it to be solved as a direct problem, to describe the frequency dependence of the mechanical response to an external mechanical signal at given functions $\varepsilon_S(\omega)$ and $\eta_S(\omega)$, as well as the reverse problem, to find the unknown functions $\varepsilon_S(\omega)$ and $\eta_S(\omega)$ from the measured mechanical response, provided that all other system characteristics are known.

For a harmonic external signal the measured signal consists of two components, a damping with the eigenfrequency of the system and a non-damped oscillation with the applied frequency. Both of them contain information about the relaxation processes in the interface and can be used to study the dynamic response of the bubble and drop surfaces. In the transient regime, the damping time and the eigenfrequency can be measured, whereas in the regime of established oscillation the amplitude- and phase-frequency dependencies are obtained.

The method demonstrates a high sensitivity of the measured signal to change of the surface elasticity and viscosity, which depends on the presence of surfactants in the solution and on the mechanisms of surface relaxation. The resonance frequency increases with the surface elasticity and the maximum of the amplitude decreases with the surface viscosity. The phase angle changes significantly with both, surface elasticity and viscosity. For example, a strong increase of the phase shift of pressure oscillations in the cell is observed at frequencies where the dilatational surface elasticity and viscosity are increasing. The good qualitative agreement with experiments underlines that the theory is adequate to describe the main properties.

The performed analysis shows that the system behaviour varies significantly depending on the experimental conditions. The most important factors are the type of cell (open or closed), the cell volume and the walls rigidity, the capillary characteristics and the equilibrium meniscus height. By appropriate choice of the experimental conditions one can vary the main system parameters, such as characteristic frequency, meniscus frequency in the open cell or damping coefficient, and in this way control the system response. For a quantitative description of the dynamic behaviour of the system one should determine exactly all the system parameters. In practice, this problem can be solved via special calibration experiments.

ACKNOWLEDGEMENTS

The work was financially supported by the ESA MAP Project FASES.

10. REFERENCES

1. J. Lucassen and M. van den Tempel, Chem. Eng. Sci., 27 (1972) 1283.
2. J. Lucassen and M. van den Tempel, J. Colloid Interface Sci., 41 (1972) 491.
3. D.O. Johnson and K.J. Stebe, J. Colloid Interface Sci., 168 (1994) 21.
4. K.-D. Wantke and H. Fruhner, In “Studies in Interface Science”, Vol. 6, D. Möbius and R. Miller (Eds.), Elsevier Science, Amsterdam, 1998, p. 327.
5. R. Miller, E.V. Aksenenko and V.B. Fainerman, The elasticity of adsorption layers of reorientable surfactants J. Colloid Interface Sci., 236 (2001)
6. F. Ravera, M. Ferrari, R. Miller and L. Liggieri, J. Phys. Chem. B, 105 (2001) 195
7. B. A. Noskov, Colloid. Polym. Sci., 273 (1995) 263.
8. B. A. Noskov, D.A. Alexandrov and R. Miller, J. Colloid Interface Sci., 219 (1999) 250.
9. G. Loglio, U. Tesei and R. Cini, Colloid Polymer Sci., 264 (1986) 712.
10. R. Miller, G. Loglio, U. Tesei and K.-H. Schano, Adv. Colloid Interface Sci., 37 (1991) 73.

11. E.K. Zholkovskij, V.I. Kovalchuk, V.B. Fainerman, G. Loglio, J. Krägel, R. Miller, S.A. Zholob and S.S. Dukhin, J. Colloid Interface Sci., 224 (2000) 47.
12. G. Loglio, U. Tesei and R. Cini, J. Colloid Interface Sci., 71 (1979) 316.
13. K.-D. Wantke, K. Lunkenheimer and C. Hempt, J. Colloid Interface Sci., 159 (1993) 28.
14. C.-H. Chang and E.I. Franses, J. Colloid Interface Sci., 164 (1994) 107.
15. K. Lunkenheimer, C. Hartenstein, R. Miller and K.-D. Wantke, Colloid Surfaces, 8 (1984) 271.
16. D.O. Johnson and K.J. Stebe, J. Colloid Interface Sci., 182 (1996) 526.
17. D.O. Johnson and K.J. Stebe, Colloid Surfaces A, 114 (1996) 41.
18. C.-H. Chang and E.I. Franses, Chem. Eng. Sci., 49 (1994) 313.
19. C.-H. Chang, K.A. Coltharp, S.Y. Park and E.I. Franses, Colloid Surfaces A, 114(1996)185.
20. H. Fruhner and K.-D. Wantke, Colloid Surfaces A, 114 (1996) 53.
21. L. Liggieri, F. Ravera and A. Passerone, J. Colloid Interface Sci., 140 (1990) 436.
22. A. Passerone, L. Liggieri, N. Rando, F. Ravera and E. Ricci, J. Colloid Interface Sci., 146 (1991) 152.
23. L. Liggieri, F. Ravera, A. Passerone, J. Colloid Interface Sci., 169 (1995) 226.
24. V.I. Kovalchuk, E.K. Zholkovskij, J. Krägel, R. Miller, V.B. Fainerman, R. Wüstneck, G. Loglio and S.S. Dukhin, J. Colloid Interface Sci., 224 (2000) 245.
25. W. Zielke, J. Basic Eng., Trans. ASME, D, 90 (1968) 109.
26. V.I. Kovalchuk, J. Krägel, R. Miller, V.B. Fainerman, N.M. Kovalchuk, E.K. Zholkovskij, R. Wüstneck and S.S. Dukhin, J. Colloid Interface Sci., 232 (2000) 25.
27. H. Fruhner and K.-D. Wantke, Colloid Polym. Sci., 274 (1996) 576.
28. G.B. Thurston, Biorheology 13 (1976) 191.
29. J. Krägel, R. Miller, A.V. Makievski, V.I. Kovalchuk, L. Liggieri, F. Ravera, M. Ferrari, A. Passerone, G. Loglio, M. Cosi and Ch. Schmidt-Harms, Oscillating Bubble Experiments Performed During The STS-95 Mission in November 1998, Proceedings of the 1st International Symposium on Microgravity Research and Applications in Physical Sciences and Biotechnology, Sorrento 2000, in press
30. A.B. Makievski and D. Grigoriev, Colloid Surfaces A, 143 (1998) 233.

Subject Index